North America

North America

A Geography of
the United States and Canada

NINTH EDITION

J. H. PATERSON

New York   Oxford
OXFORD UNIVERSITY PRESS
1994

Oxford University Press

Oxford   New York   Toronto
Delhi   Bombay   Calcutta   Madras   Karachi
Kuala Lumpur   Singapore   Hong Kong   Tokyo
Nairobi   Dar es Salaam   Cape Town
Melbourne   Auckland   Madrid

and associated companies in
Berlin   Ibadan

Published by Oxford University Press, Inc.,
200 Madison Avenue, New York, New York 10016

Library of Congress Cataloging-in-Publication Data
Paterson, J. H. (John Harris)
North America : a geography of the United States
and Canada / J. H. Paterson. — 9th ed.
p. cm.   Includes bibliographical references and index.
ISBN 0-19-508058-0
1. United States—Geography.
2. Canada—Geography.
I. Title.   E161.3.P38   1994
917.3—dc20   93-13782

1 3 5 7 9 8 6 4 2

Printed in the United States of America
on acid-free paper

For Evangeline

Extract from the Author's Note
to the First Edition, 1960

The regional geographer is essentially a gatherer of Other Men's Flowers. It is not, usually, the transmission of new facts that justifies his authorship, so much as the rearrangement, into more significant patterns, of facts which are already known. This book represents a search for such rearrangements, as it traces new patterns emerging from the familiar facts of the geography of North America at mid-century.

In the course of such a search as this, the geographer inevitably finds that he has to revalue, or even to discard, some of the symbols of his shorthand which have served over past decades—such labels, for example, as the 'Manufacturing Belt' or the 'Cotton Belt.' I believe that, in some cases, their use has actually masked a more logical analysis of the geographical facts than they themselves provided, and I have tried to show how and why they have served their time. So much has changed in North America since they were first invented.

The later chapters of the book deal with the regions of North America, and all of these regions are familiar to geographers. I have not, however, devoted much attention to their exact definition on the ground, nor attempted to map them. Rather, I have allowed them to define themselves, in terms of the regional problems which confront them. Among a forward-looking people like the North Americans, the bonds of regional sympathy are tied less by topography or climate than by a shared concern for the future of the community.

In dealing as a geographer with these problems, however, I have tried to keep in sight the fact that they are human problems. Thus, for example, in the sections on manufacturing I have usually cited the statistics for employment rather than output. Although the employment figures admittedly tell only a part of the story, and tend to mask the effects of automation and increased efficiency, nevertheless they serve as a better expression of the human side of the ever-changing relationship between man and environment.

With regard to some of these regional problems, I have merely raised questions which I hope may provoke thought and discussion. With others, I have hinted at a judgment which is, obviously, personal and therefore liable to challenge. To an extent unknown in the longer-settled lands of Europe, however, the problems of resource use and geography are, in newer lands, subjects of public interest and debate, and millions of Americans have made their own personal judgments on these technical problems. In making mine, I have drawn upon as wide a range of literature and opinion as possible, besides incorporating, in two or three places, findings based on my own research.

Preface

The first edition of this book was prepared in the late 1950s and published in 1960. I had no thought, at the time, that thirty-three years later I should still be pre-occupied with it. But there is no denying the fact that, in the intervening years, tremendous changes have taken place in North America, changes great enough to justify a whole series of updates of what I originally wrote. Consequently, every four years or so my editor at Oxford University Press and I have produced a new edition.

The first effect of these changes has been to remove from everyday American life names and institutions which we used to believe were eternal. Where now are the Pennsylvania and New York Central railroads? Where is United States Steel, archetype of the great corporations that made the American economy a marvel of productivity? Pittsburgh is a steel city, yet today it makes very little steel. Milwaukee is a beer city, yet today it has only one brewery, owned not by any of its famous German families but by a tobacco company. And the largest coal producer in the United States is not West Virginia or Kentucky but that well-known home of the cowboy, Wyoming. In Canada, it is that equally well-known home of pine and fir, British Columbia. Truly, the world has stood upon its head.

The changes, moreover, do not concern merely the big, productive elements of North American life. There are plenty of small changes that impinge more directly on the average citizen, and some of them represent a net loss of the joys of life. The lunch counters in drugstores, once among our simple pleasures, have been largely eliminated by someone whose narrow soul calculated only in income per square foot of retail space. Gone, too, are most of the milk bars, where you could go in and order something called a Sad and Lonely and rely upon the waitress to pass the word to any unattached member of

the opposite sex that you were—sad and lonely. Meanwhile, whole generations of North Americans have grown up for whom a ride on a railway train would be a weirdly exotic experience, like a visit to Disney World.

I detect, also, changes in the American mind, and these make it more and more difficult to write about the continent. As a geographer and a foreigner, I can claim neither the right nor the expertise to analyze these changes, but I do find myself looking with much reduced enthusiasm at nations where the urge to achieve has been replaced by a determination to join the endless queue of claimants for victimhood, and where going to school is regarded as an exercise in group therapy rather than education. To write about landscapes and societies, admiration is not essential, but sympathy is. I find my own much weakened.

Not, of course, that there is any shortage of changes for the good, too. Almost overnight, North America has become a continent of garbage collectors. The word *garbage* is, as it happens, far too narrow for the dozen or more categories of used products that have to be disposed of; garbage is only one of them, and woe betide anyone who gets confused in his or her categories. Whatever America's performance at the international level or the Earth Summit, its cities have gone green. Also, many people are driving smaller cars, although quite a lot of the latter are Japanese. How times change!

But, as I commented in the preface to the eighth edition of this book, keeping track of the changes in the geography of North America is only one half of an author's task (a task, incidentally, made more difficult on this occasion by the slow publication of census results from the 1990 and 1991 counts). The other half is keeping up with changes in geography itself. The thirty-odd years over which I am looking back have seen a quite bewildering variety of changes, bewildering if only because each major change seems precisely to reverse the one that went before. What is more, the *pace* of change seems also to be increasing, so that movements that might once have taken half a century to rise, mature, and wither are now disposed of by instant criticism in a tenth of the time—and before their implications have been fully worked out. I am told now that even postmodernism is out of date, although that sounds more like a misuse of language than a serious intellectual comment. Under these circumstances, we have little time for discussion or sifting ideas—neither, we had better remind ourselves, had the Gadarene swine.

Let me refer to just one of these changes, one which closely affects the writing of a regional geography like mine. Geography, as a subject, was once the very essence of earthboundedness and objectivity; it was the description in detail of the stage on which the human drama was being played out. But now, for many geographers, it has become abstract. By this I do not mean that there is disagreement as to whether London is really on the Thames, or Chicago on Lake Michigan, but that these facts have ceased to interest geographers precisely because they cannot be disputed or penetrated by further study. Today's geographers, we may cynically add, may not even be able to pinpoint on a map such locations as these, certainly some of their pupils cannot.

And yet, paradoxically, London and Chicago and the rest remain of enormous interest to these geographers, not as places in themselves but as constructs, as territories, in the minds of individuals—in the first place, in the minds of the 7 million or so people who live in London or Chicago. To each person viewing the city, it becomes a patchwork of routes and networks, opportunities and threats, clear paths and no-go areas. It is the geography of these individual territories that interests many of our colleagues, the way in which they depict a city that is real to each observer but also different for each. *Quot homines, tot sententiae.* Loosely translated, there are as many geographies as people, all of them possessing the status of a personal myth or, in modern terms, a TV set through which the individual makes his or her way, scene by scene, throughout the day. How, then—so the argument runs—can we speak of "the" geography of a place or region when their number is infinite?

This argument, it must be admitted, is not one calculated to cheer the likes of geographers like myself,

with our single description of Chicago, or Calgary, or the Corn Belt. What use, now, are our efforts? To that question there are two possible answers. One is that a regional geography like this book can provide a baseline or, if you prefer, a measuring stick against which other people's perceptions can be checked. Here are the dull, old facts—what we used to think of as the realities—against which we can measure the deviations that are other people's perceptions. Otherwise we have no standard by which to judge those perceptions; all becomes relative and, at that, largely pointless. When we have laid bare the geographies constructed by 7 million Chicagoans, what are we to do with them? We shall merely be plotting deviations from the mean and, frankly, one wonders if it is worth all that trouble.

The second light in which we might regard the kind of geography contained in this book is as a type of central myth among millions of other myths, a sort of Master Myth propagated by the priests of geography but now subjected, for the first time, to proper dissection.

This may be so, despite the way in which the description keeps bumping (as the ancient myths did) into places and things which we are fairly certain do exist. But we propagate the Master Myth just because it is *not* the myth of a single individual's geography but a collective myth. North America is a land of societies, of nations. They have created their homeland, and there is a general consensus as to what, and why, and how that persists and has gained acceptance.

In the library where I carried out a good deal of the work for this revision of my book, I chanced across an entire shelf of books, some of them quite recent, about the life and exploits of Billy the Kid. From this I draw the conclusion that some myths have much greater staying power than others. It is my belief and, certainly, my hope that the myth that there is a real place called North America, with this and that in it, will far, far outlast this final edition of my book.

In presenting this ninth edition of *North America*, I want to acknowledge with gratitude the help I received from others. First and foremost among these individuals is Joyce Berry, of Oxford University Press, New York, who has seen not only this edition but several previous ones into print. We seldom meet, but this does not seem to matter; the book is published on time just the same, and the telephone serves to keep us in touch. Among my British colleagues, Dr. Michael Bradshaw and Dr. Richard Goodenough have been most helpful in bringing their specialized knowledge to bear on topics where I possessed none. In North America I want to thank all of the following for their help—in some cases given over more than forty years: Professors Richard Hartshorne, Clarence Olmstead, Robert Sack, Gordon Robertstad, Rodney Steiner, and Lewis Robinson. And, behind them all, thanks to my old teacher Andrew Hurst, without whose inspiration I might never have become a geographer at all.

Newcastle Upon Tyne, England   J.H.P.
July 1993

Contents

1. The Setting, 3

   The Physical Divisions, 5
   The Glaciation of North America, 16
   The Climate of North America, 22
   The Soils of North America, 30
   The Vegetation of North America, 33

2. The People, 41

   The Native Americans, 42
   The Immigrants, 43
   The Melting Pot—or Not? 45
   The Case of French Canada, 48
   The African Americans, 49
   The Hispanic Americans, 54
   Distribution of Population: Canada, 55
   Distribution of Population: The United
      States, 57

3. The Cities, 63

   City Building, 64
   Counting the City Population, 67
   In the City: Form, Function, and Quality, 69
   Urban Problems and Urban Futures, 78

4. Government, National and Local, 83

   The Nation-States, 84
   A Federal Constitution, 88
   The Lower Levels of Government, 91

5. Government, Land, and Water, 93

   Public Land and Private Land, 94
   Land Disposal and Land Law, 97
   The Control of Land Use, 102
   Government, Land, and Leisure, 104
   Government and Water, 106
   Conservation of Resources, 108

6. Government and Economic Activity, 113

   Government, Transport, and Freight
      Rates, 114

Government and Agriculture, 116
Government and Industry, 119
Government and Energy, 121

7. Agriculture, 127

   The Formative Factors, 128
   Agriculture and the Present, 135
   Agriculture and the Future, 146

8. Industry, 151

   The Distribution of Industry, 154
   Coal, Iron, and Steel, 156
   Petroleum and Natural Gas, 161
   Electric Power and Nuclear Energy, 167
   Change, Growth, and Decline in North
      American Industry, 170
   Canadian Industry, 173

9. Transport, 177

   Waterways and Ports, 180
   Railroads, 185
   Roads, 187
   Airlines, 190

10. Regions and Regionalism, 195

   Bases of Regional Diversity, 196
   Government and Region, 200
   A Regional Framework, 202

11. The Middle Atlantic Region, 211

   Agriculture, 213
   Industry, 219
   Region and Nation, 221
   The Seaboard Cities, 221
   The Problems of Megalopolis, 232

12. New England, 235

   The New England Economy, 238
   Rural New England, 239
   Industrial New England, 242

13. The Canadian Heartland, 251

The Eastern Section: The Lower St.
    Lawrence Valley, 252
The Western Section: Southern Ontario, 263
One Core or Two? 266

14. The Interior, 269

The Agricultural Interior, 272
The Industrial Interior, 284
The Great Lakes and Their Cities, 290
The Inland Cities of the Interior, 303

15. The South, 311

The Old South, 312
From the Old South to the New: Elements of
    Change, 316
Workers and Industry in the South, 326
The New South in the Nation, 330

16. Appalachia, 335

Negative: Appalachia, a Problem
    Region, 336
Positive: Two Approaches to Regional
    Development, 340
TVA or Appalachian Commission: A
    Comparison, 349

17. The Southern Coasts and Texas, 353

The Physical Setting and Early Land
    Use, 354
Specialized Production of Subtropical
    Crops, 356
Sunshine, Vacations, and Retirement
    Living, 358
Commerce and Industry, 360
Texas, 365

18. The Great Plains and Prairies, 371

The Physical Circumstances, 373
Land Use in the Great Plains, 378
Present Problems, 382
Present Regional and Urban Patterns, 388
The Prairies of Canada, 393

19. Mountain and Desert, 403

The Region and Its Character, 404
Land and Livelihood, 405

Transportation and Cities, 416
Development or Preservation? 418

20. The Spanish and Indian Southwest, 421

Regional Identity, 422
Regional Boundaries, 424
Land and Livelihood, 427
Regional Mixture, 431
A Most Remarkable Border, 435

21. The Northern Atlantic Coastlands, 437

The Region, 438
Settlement and Economic Development, 440
The Character and Role of the Atlantic
    Provinces Region, 447
Choices and Chances, 451

22. The Northlands, 453

Physical Conditions, 455
Future Prospects, 458
Alaska, 469

23. California, 475

The Settlement of California, 478
California's Agriculture: Organization, 480
California's Agriculture: Crops and
    Regions, 483
California's Industry, 489
Central California: Cities and Industries, 490
Southern California: Cities and
    Industries, 493
The Future of Southern California, 497

24. Hawaii, 501

25. The Pacific Northwest, 505

The Region and Its Resources, 506
Agriculture in the Pacific Northwest, 508
Forests and Industries, 513
Basic Problems of the Pacific Northwest, 517

Index, 523

North America

1

The Setting

Niagara Falls, looking from the United States side toward the
Canadian.

The continents of North and South America stretch from 72° N latitude to 55° S, linked to each other by that narrow isthmus across which cuts the Panama Canal. There could hardly be a clearer geographical division into two; yet, by a long-standing tradition, we all tend to see not a twofold but a threefold division of this continuous landmass. North America we take to refer to Canada and the United States. South America refers to Venezuela, Colombia, and the countries to the south of them. Mexico, together with the small states between it and Panama, as well as most of the islands of the West Indies, we classify as Central America. The United States–Mexican border delimits this third region on its north side.

There are weighty reasons for making this border a dividing line. Ever since we began to use the terms *Third World* and *First World,* this has been one of the clearest and most abrupt boundaries between them; not merely two nations, or two cultures, but two worlds confronting one another across a river, or a line on the map. One world has been mainly English-speaking and wealthy; the other has been Spanish-speaking and, by most known standards, poor. Crossing the United States–Canadian border has involved virtually no culture shock, whichever way the traveler was moving. Crossing the United States–Mexican border, by contrast, has thrown up not one but a whole series of shocks—language, standard of living, sights, smells and all. Small wonder, then, that writers and students have usually decided to cut short their explorations of "North" America at the Rio Grande.

Some day soon, however—and the sooner the better—we must rethink this arbitrary division between North and Central America. For one thing, it does little justice to the recent development of Mexico. For another, movements of workers and industries across the border are blurring the contrast. For a third, much of the southwestern borderlands origi-

nally lay in Mexico, and have always been Spanish-speaking. For a fourth, the United States has now negotiated a free-trade agreement with Mexico, as it already has with Canada, and this, for better or worse, will in due course create a free-trade area stretching from Baffin Island to Yucatan, and where will our "Central" American category be then?

For the present, however, we shall confine our attention to Canada and the United States. That the two of them together form an area that is distinctive and worthy of study can hardly be doubted. During the nineteenth century and, especially, the first half of the twentieth, these countries built up an awesome lead over all others as the most prosperous, the most productive places on earth. Nowhere was the average standard of living so high, and nowhere did such apparent miracles of technical achievement take place. When World War II ended in 1945, the world was divided, quite simply, into two: North America was rich, and everybody else was poor.

To the geographer, as to any other thinking person viewing the situation, the question that presented itself was this: What had brought about this astonishing contrast? For the geographer, the question took a more specific form: Was it the natural wealth of the continent that held the secret? Was it because North American soils were more fertile, North American minerals more plentiful, North American forests more productive than those of other lands? Or was it the way in which the continent had been settled, developed, or organized? If the causes were natural, there was little or nothing that other countries could do about it: nature had simply played favorites. But if they were organizational, then there was something that others might learn or imitate. Either way, North America at mid-century was a unique and fascinating area for study.

We now know that the cause was not nature alone: it was mainly organization, or education, or know-how. We know this because, since 1945, a number of other countries have caught up with and overtaken the North American giants. And, most striking, they are not countries with great natural endowments of iron, or coal, or timber. On the contrary, they are countries notorious for the lack of these things, cramped for lack of space and soil. But Switzerland today has a higher income per person than the United States, and Japanese goods flood the American market.

Needless to say, to have once held a long lead over all others and then to have lost it does not make North America less interesting to geographers but rather more so. We shall want to ask why these changes took place; we shall also want to speculate (although only time will tell) as to whether the lead can be regained. In certain respects, of course, it has never been lost. North America is the greatest supplier of food to the rest of the world. It has put on the moon the only men who have so far been there. The United States is, by a wide margin, the dominant power of the political world. How has all this happened? With such questions in mind we can begin our survey of North America.

The whole continent of North America has an area of some 8.3 million sq mi (21.5 million sq km). The area with which this book is chiefly concerned is, however, less than that, for it excludes Mexico and the countries bordering the Caribbean, and virtually ignores the Arctic Archipelago—the almost uninhabited islands that lie within the Canadian sector of the Arctic. If we deduct those two parts of the continent from the whole, we are left with an area of 6.7 million sq mi (17.35 million sq km), of which Canada occupies 3.1 million sq mi (8.03 million sq km) and the United States, including the huge outlying state of Alaska, accounts for the remainder.

Translated into distances, these dimensions mean that the part of the continent we are considering stretches across almost 50° of latitude, from the Florida Keys to the northern tip of mainland Canada, and across 115° of longitude, from the east coast of Newfoundland to the Bering Strait. The railway distance from Halifax in the east to Prince Rupert in the west of Canada is more than 3750 mi (6000 km); from New York to San Francisco some 3000 mi (4800 km); and from Brownsville, on the United States–Mexican border to the 49th parallel that marks the boundary between Canada and the United States, 1600 mi (2560 km). Thus, Prince Rupert is as far from Halifax as the mouth of the Congo is from London, and New York is as far from San Francisco as it is from Ireland (Fig. 1-1).

The Physical Divisions

For ease of reference the North American continent may be divided, north of the United States–Mexican

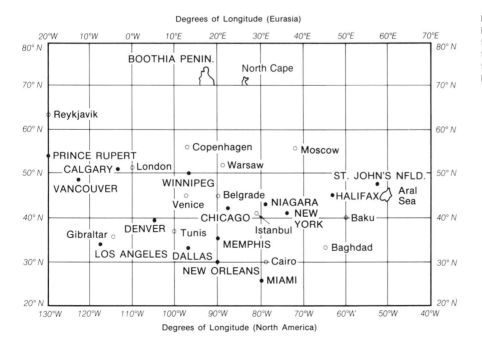

Fig. 1-1. North America and Eurasia: Comparative locations. North American locations are marked in capital letters; Eurasian locations, in lower case.

Degrees of Longitude (Eurasia)

Degrees of Longitude (North America)

border, into seven main physical provinces, as shown in Figure 1-2.[1]

shown in Figure 1-2.[1]

THE ATLANTIC AND GULF COAST PLAINS

From southern Texas to New York, the North American continent is bordered by coastal plains. In Texas, these plains are as much as 300 mi (480 km) wide; northeastward from the Florida peninsula they gradually narrow or, more accurately, an increasing proportion of them lies submerged beneath the sea. In the neighborhood of New York, the plains come to an end, reappearing north of the Hudson only in parts of Cape Cod and the islands off the coast of southern New England.

The coastal plains are formed by a series of beds of geologically young materials, dipping gently away from the older rocks of the interior toward the coast. The oldest beds forming the plains are Cretaceous, and lie at the inland margin. The successive layers form belts with low scarp edges facing inland, but the general slope of the plains' surface is so gentle that no true coastline can be formed. Swamps, lagoons, and bars, fronted by almost continuous sand reefs, characterize a transition zone between land and sea. Through it wind rivers that are highly liable to flood.

Not only is this southeastern margin of the continent especially subject to year-by-year changes in its present detail, but it also evidently has a long history of changes in level. These are indicated by the terraces of former shorelines, recognizable more than 250 ft (76 m) above the present sea level, and by the drowned valleys of Chesapeake Bay and the Virginia shore.

Midway along the coastal plains and separating the waters of the Atlantic from the Gulf of Mexico is the peninsula of Florida, projecting southward for some 350 mi (560 km). It is surrounded by shallow seas underlain by the continental shelf that have probably existed since Tertiary times; Florida itself, however, became part of the continental landmass only in recent geological times. Like the plains, Florida is flat (the highest point in the state is only 325 ft [99 m] above sea level), lake strewn, and swampy. In the south, an area of over 5000 sq mi (12,900 sq km),

known as the Everglades, is composed of marsh and swamp only a few feet above sea level. Much of the peninsula is underlain by limestone formations, and on its surface, solution hollows and sinkholes, often containing lakes, are common features. Sandbars and lagoons line the coast, and in the extreme south, coral reefs have formed in the warm waters of the Gulf.

Near their western end, the coastal plains spread inland to include the Mississippi Valley south of Cairo. The telltale scarps of the belted plains swing north in a great inverted V to embrace the valley, a structural depression that has been deepened by erosion. The valley bottom is wide, flat, and easily flooded, and through it meanders the Mississippi, carrying the vast silt load that year by year builds its delta further out into the shallow coastal waters of the Gulf of Mexico.

THE APPALACHIAN SYSTEM

In eastern North America, mountain-building processes went on at intervals throughout Ordovician, Devonian, and Permian times. The area affected by these processes forms a wide belt, with a marked northeast to southwest trend, from Newfoundland to central Alabama. This area, diverse in present character, but unified by its physical history, may be called the Appalachian System. It falls into three sections: a southern section, stretching from Alabama to the valley of the Hudson and including the Appalachian Mountains proper; a New England section, from the Hudson to central Maine; and a northeastern section, covering northern Maine and the Maritime Provinces of Canada.

The Southern Section

In the south, the system is formed by two parallel belts. On the east is that of the "old" Appalachians, made up of Precambrian igneous and metamorphic rocks. On the west is a "new" belt, formed by the upthrust edge of the great Paleozoic floor that underlies the central lowlands of North America. Each of these belts has an eastern edge different in form from the western, and this further subdivision gives the system four provinces (Fig. 1-3).

On the eastern, or seaward, edge the old rocks have been severely eroded to form a gently sloping, dissected plateau or irregular plain known as the Piedmont. From its junction with the coastal plains at the Fall Line (see p. 213), it rises gradually to 1200–1500 ft (365–455 m), where it merges with the

at the Fall Line (see p. 213)

[1]Some basic references are: W. D. Thornbury, *Regional Geomorphology of the United States* (New York, 1965); C. B. Hunt, *Physiography of the United States* (San Francisco, 1967), and *Natural Regions of the United States and Canada* (San Francisco, 1974); J. B. Bird, *The Natural Landscapes of Canada* (Toronto, 1980); and W. L. Graf, ed., *Geomorphic Systems of North America* (Boulder, Colo., 1987).

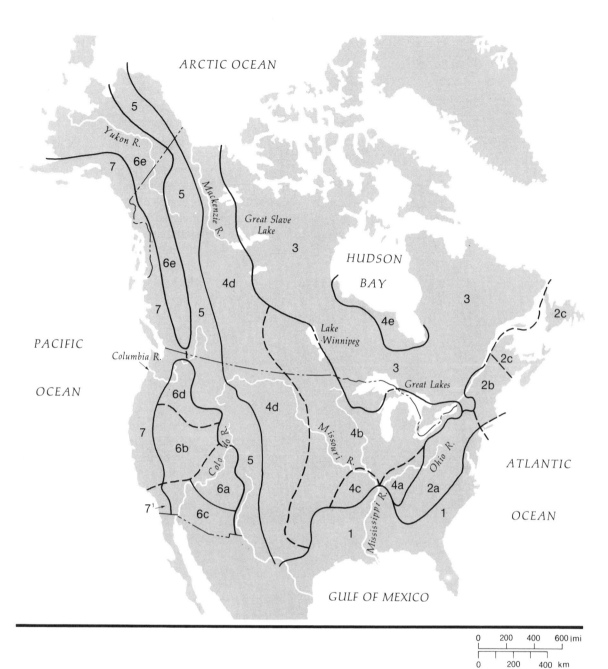

Fig. 1-2. Physical provinces of North America. The provinces are:

1. The Atlantic and Gulf Coast Plains

2. The Appalachian System
   (a) The South
   (b) New England section
   (c) Maritimes–Newfoundland extension

3. The Canadian or Laurentian Shield

4. The Central Lowland
   (a) The Eastern transition belt, or Interior Low Plateaus
   (b) The Mississippi–Great Lakes section
   (c) The Ozark–Ouachita section
   (d) The Great Plains
   (e) Hudson Bay Lowland

5. The Cordilleran Province

6. The Intermontane Region
   (a) The Colorado Plateaus
   (b) The Great Basin
   (c) The Desert Basin and Range
   (d) The Plateaus and Basins of the Columbia and Snake
   (e) The Plateaus of British Columbia and the Northwest

7. The Pacific Coastlands

7^1 The Los Angeles Appendix

Fig. 1-3. The Appalachian System: Physical divisions referred to in the accompanying text. They are: I. The Piedmont, II. The Blue Ridge, III. Ridge and Valley, and IV. The Appalachian Plateaus.

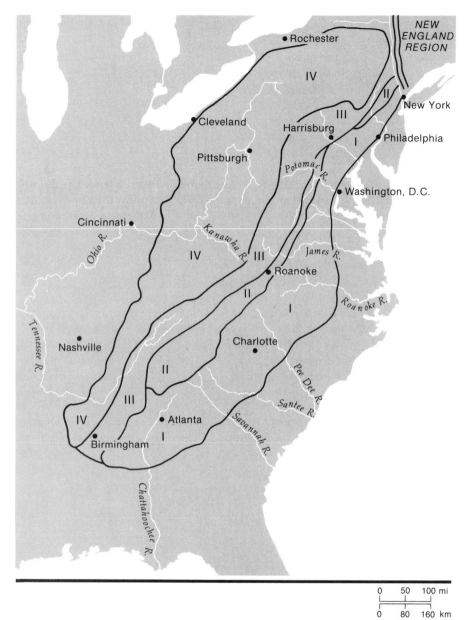

wooded mountains. These mountains represent the western half of the "old" Appalachians, an area less heavily eroded than the Piedmont, and so rising in places to 6000 ft (1825 m) or more. In the Great Smoky Mountains and the Blue Ridge of the southern states, this mountain barrier is virtually uninterrupted, but it subsides and narrows northward and, in northern Virginia and Pennsylvania, it becomes low and discontinuous.

When the "new" belt of the Appalachians in the west was upthrust, the Paleozoic strata, which formerly lay almost horizontal, were severely contorted along the line of junction with the older formations to the east. This junction zone now shows a remarkable series of folds running parallel with the trend of the system and known as the Ridge and Valley Province. This "corrugated" area is 25–80 mi (40–128 km) wide, and through it run the north-flowing Shenandoah and the south-flowing Tennessee with their tributaries, while east-flowing rivers like the Potomac are forced to cut through the ridges in a series of gaps.

West of the fold zone and above a high scarp face (known in Pennsylvania as the Allegheny Front) lies the fourth, or inland, province of the system. It is a plateau section, where the Paleozoic formations retain almost the same undisturbed bedding as they had before the upthrust. The Cumberland Plateau in Tennessee and Kentucky lies at about 2000 ft (600 m), and levels rise to twice that height in West Virginia. Northward the plateau stretches almost to the southern shores of the Great Lakes; westward it drops away to the lowlands. Among its sandstones and limestones lie the vast coal measures of the great Appalachian coalfield.

The New England Section

North of the Hudson, the Precambrian and Paleozoic belts of the Appalachian System continue into New England. The first of these is represented by the heavily eroded uplands of New England that correspond to the Piedmont. The Ridge and Valley Province of the Paleozoic belt, on the other hand, can be traced running north through the Hudson Valley, and so by way of the Champlain Lowland to where it merges with the down-faulted St. Lawrence Valley. On the west the Plateau Province terminates in the Catskill Mountains.

The uplands of New England form a plateau surface that drops gently eastward to the sea. They are surmounted by groups of peaks that represent harder masses of rock, which have resisted the attacks of three or perhaps four cycles of erosion. Such peaks have received the name of monadnocks, after Mt. Monadnock in New Hampshire. The uplands have two axes, separated by the lowlands that extend north through the Connecticut Valley. The western axis is marked by the line of the Green Mountains and is carried north by that of the Notre Dame and the Shickshock Mountains, which stretch into the Gaspé Peninsula of Québec. None of these is as much as 4500 ft (1370 m) high. The eastern axis runs through the White Mountains (where Mt. Washington reaches 6288 ft [1929 m]) and into Canada's Maritime Provinces. The whole surface of New England has been heavily glaciated, and much of the detail of the landscape is due to the action of the ice.

The Maritimes–Newfoundland Extension

Although the Appalachian mountain-building processes extended their influence to northern Maine and the Maritime Provinces, and traces of the system persist as far east as Newfoundland, the present structure of the area bears little relation to that further south. Huge igneous intrusions have occurred to form the batholiths that constitute the Central Highlands of New Brunswick and the hills of the "arm" of Nova Scotia and Cape Breton Island. Between these old formations in Nova Scotia and the equally old rocks of the Canadian Shield, a wide basin has been formed in which younger beds—Permian and Carboniferous—are found. These beds underlie the lowlands around the Bay of Fundy and Prince Edward Island. On the southern shore of the Bay of Fundy, the same feature that forms the Connecticut Valley—a band of young sandstones and shales of Triassic age—occurs again to form the Annapolis Valley of Nova Scotia.

THE CANADIAN OR LAURENTIAN SHIELD

Much of eastern Canada is covered by more than a million square miles of metamorphosed Precambrian rocks that form a vast block known as the Canadian, or Laurentian, Shield. Beneath the Paleozoic strata bordering it on the south, on the west, and around Hudson Bay, there can be traced the continuation of the Shield, which underlies much of central Canada and the Great Lakes region as well (Fig. 1-4).

Wide areas of this ancient rock mass have been severely compressed and contorted, and in the zones of disturbance a large variety of minerals is to be found. The present surface of the Shield, however, bears little relation to the previous disturbance. Through a long history of erosion culminating in severe glaciation, it has been converted into a rough plain that dips away from a level of 1700–2000 ft (500–600 m) along its southeast (or St. Lawrence) edge until it disappears below younger formations just south of Hudson Bay.

The Shield is nearly all Canadian, but two southward extensions into the United States share its main characteristics. One is in the Adirondack Mountains of northern New York State, and the other forms the Superior Upland in northern Minnesota, Wisconsin, and Upper Michigan. On the west, the Shield sinks beneath later materials along a line that runs through Lake of the Woods, Great Slave Lake, and Great Bear Lake to the Arctic seas. On the east, it includes Labrador, with its monadnock ranges, and it can also be traced to Newfoundland.

The tremendous forces of glaciation scoured the eroded surface in almost unobstructed action. Today the result of this action remains in a thin or nonexistent soil cover and in a chaotic drainage system. Gla-

Fig. 1-4. The Canadian or Laurentian Shield: The exposed
central core and the bordering
outer margin buried beneath
platform rocks. (Reproduced
by permission from J. B. Bird,
*The Natural Landscapes of
Canada* [Toronto, 1980], Fig.
12.1.)

Fig. 1-4. The Canadian or Laurentian Shield: The exposed central core and the bordering outer margin buried beneath platform rocks. (Reproduced by permission from J. B. Bird, *The Natural Landscapes of Canada* [Toronto, 1980], Fig. 12.1.)

▨ Exposed Shield

▨ Buried Shield

0   1000 km

cial detritus has dammed and diverted streams, creating a rocky landscape strewn with lakes and swamps. Sometimes, however, these lakes have disappeared, leaving behind them clay-filled beds and offering some prospect of fertility in this otherwise infertile world of the Shield.

THE CENTRAL LOWLAND

Between the Appalachian System in eastern North America and the Rocky Mountains in the West is a vast lowland area. As far as relief is concerned, the Gulf Coast Plains and much of the Laurentian Shield, which have already been described, also belong to this lowland. It is only on geological grounds that they are separated from it, for one is younger and the other older than the region we are now considering.

Even with the coastal plains and the Shield excluded, the remainder of the lowland area is of enormous extent. It is 1200 mi (1920 km) from the western edge of the Appalachian Plateau in Ohio to the foot of the Rockies at Denver. From north to south, the dimensions of the region are even more impressive—40° of latitude, or over 2500 mi (4000 km), from the edge of the coastal plain in Texas to the delta of the Mackenzie at 70° N.

Spread across the Central Lowland are Paleozoic beds of great thickness, evenly deposited on the floor of a former sea. In the northeast, these beds overlap the Shield, while on the south, they are themselves submerged below the later deposits of the coastal plain. In the east, as we have seen, they end in violent contortions against the wall of the "old" Appalachians. In the west, they extend, beneath a cover of Cretaceous, Tertiary, and recent materials, until they terminate against the wall of the Rockies, and on the north, they reach the Arctic Ocean.

The Central Lowland comprises several sections. We can describe these as the Eastern Transition Belt, the Mississippi–Great Lakes Section, the Ozark–Ouachita Province, and the Great Plains.

The Eastern Transition Belt

Between the Appalachian Plateau, where the Paleozoic formations have been upthrust, and the Mississippi Valley, where they lie even and almost undisturbed, there is a transitional area sometimes called the Interior Low Plateaus. As the effects of the Appalachian mountain-building processes spread further afield, they became less violent, and in the Transition Belt merely created some slight folds and a gentle westward dip away from the mountains. This area has, for the most part, a sandstone cover underlain by Carboniferous limestone. The latter gives rise to areas of karst topography and to such limestone features as the Mammoth Cave of Kentucky. Of more importance to the economic life of the region is the effect of this folding in creating two domes whose sandstone cover has been removed by erosion, leaving limestone basins—the Bluegrass Basin of Kentucky and the Nashville Basin of Tennessee. Both of these have long been famous for the excellence of the pastures on their lime-rich soils.

The Mississippi–Great Lakes Section

This section constitutes the eastern half of the Paleozoic lowlands. It is a true structural plain whose flatness is emphasized by a mantle of glacial drift as far south as the Ohio and Missouri rivers. Here, as on the Shield, glacial action has affected the drainage system, and the northern part of the area abounds with

lakes. Before the coming of the cultivator, who drained the land, swamps were also widespread. Slight folding accounts for the area's only significant physical features. These are synclines in Ohio and Illinois that contain coal measures, and a basin centered in southern Michigan whose rim is marked by cuestas that form an almost perfect semicircle around it. The best known of these is the Niagara Cuesta, a "rim" of magnesian limestone underlying Niagara Falls that runs northward across a string of islands in Lake Huron, then west and south through Upper Michigan into Wisconsin.

West of the Mississippi the strata dip very gently westward while, equally gently, the land surface rises. The result is a series of east-facing scarps which become progressively younger to the west. The most prominent of these scarps, the Missouri Coteau or Escarpment, is usually taken to delimit the section on the west. In the Southwest, the Balcones Escarpment provides a similar regional boundary between the Gulf Coast Plain and the Great Plains.

The Ozark–Ouachita Section

West of the Mississippi, and near where it merges with the Gulf Coast Plain, the Central Lowland is interrupted by a group of low mountains sometimes called the Interior Highlands. These highlands are formed by two separate features. In the north is a dome flanked by rocks of Carboniferous age and at whose crest Precambrian granites are exposed. This area forms the Ozark Mountains and comprises the Salem and Springfield plateaus and the Boston Mountains. In drainage pattern and landscape it corresponds to the Appalachian Plateau region. To the south, beyond the Arkansas River, lie the Ouachita Mountains, a folded belt that closely resembles in structure the Ridge and Valley of the Appalachians; indeed, they were created by the same mountain-building movements. The highest point in the Ouachitas is 2700 ft (820 m) and in the Ozarks, a little over 2000 ft (610 m).

The Great Plains

The eastern edge of the Great Plains, in the neighborhood of the Missouri Coteau, lies at about 2000 ft (610 m) above sea level. The western edge, where the Front Range of the Rockies abruptly rises, is at an elevation of 4000–5000 ft (1215–1520 m), giving an average slope of 8–10 ft per mi (1.5–1.9 m per km) across the area. Yet these plains are structurally part of the Central Lowland. They extend from Mexico northward until, narrowing between the Shield and the Rockies, they reach northern Alaska.

As in the western part of the Mississippi–Great Lakes Section, the strata underlying the Great Plains continue to dip westward toward a trough at the foot of the Rockies. The eastward slope of the land, however, is caused by the recent deposition of a mantle of often quite unconsolidated materials from the Rockies, transported and spread by the east-flowing rivers. This relatively smooth surface mantle is disturbed only by the effects of erosive agents upon it. Wind and water have cut deeply into the soft, loose materials washed down from the mountains. The plains west of the Missouri River are especially dissected. In Nebraska, a belt of sand hills was formed by the force of violent winds during the glacial epoch, whereas in the Dakotas—and to some extent in all the river valleys that cross the plains—water erosion has created the fantastic badland topography for which the plains are famous.

Only one major break interrupts the evenness of the plains—the Black Hills of South Dakota, which rise to over 7000 ft (2130 m). Here the old, crystalline continental bedrock breaks through the surface in a domelike swelling and forms a welcome change from the monotony of the plains.[2]

THE CORDILLERAN PROVINCE

Late in Cretaceous times tremendous mountain-building processes disturbed the western half of the North American continent, where a long series of sedimentary beds lay evenly spread over the ancient continental floor. This disturbance, which was accompanied by volcanic activity, is known as the Laramide Revolution. It resulted in a great uplift of the sedimentary beds, accompanied by folding and faulting, in the areas now known by the general name of the Western Cordillera.

The uplift led to a much-intensified attack by erosive agents in the Tertiary period that followed. From the highest parts of the West, thousands of feet of sedimentary cover were removed, and the Precambrian floor was exposed. Elsewhere, however, the cover has been preserved—usually by downfaulting—and it is the various strata of this sedimentary covering, with their characteristic horizontal bedding, that give to the landscape of the West many of its particular splendors.

On the eastern (Great Plains) margin of the Cordillera, the Laramide Revolution threw up two roughly parallel ranges, running generally from north to south. These ranges are mainly granitic, but from Montana northward they are formed of towering sedimentaries, many of which have retained horizontal bedding. It is these ranges along the eastern margin of the Cordillera that form the Rocky Mountains proper, although the name is often—and wrongly—applied to much of the Mountain West. The Rockies extend from about 35° N in the United States as far as the Liard River in Canada, where other ranges extend the line, via the Brooks Range, into Alaska. The peaks of these Front Ranges rise abruptly above the Great Plains to heights of 10,000–14,000 ft (3040–4250 m).

The Front Ranges are backed by an area of scattered mountains, interspersed with plateaus. The whole province has an east–west width of between 100 and 300 mi (160 and 480 km). Most of the mountain ranges follow the general north–south trend of the Rockies, a trend that is accentuated by faulting. Across the boundary between the United States and Canada, for example, runs the remarkable Rocky Mountain Trench, a fault valley 1100 mi (1800 km) long, occupied in turn by Flathead Lake and the Kootenay, Columbia, Fraser, Parsnip, and Finlay rivers. Elsewhere, however, volcanic activity has created great structureless batholiths, like the mountains of central Idaho, where granite forms a wilderness area that rises above 12,000 ft (3650 m) and remains almost impenetrable.

All through the Rocky Mountains, indeed, the effects of vulcanicity have been great. Lava flows cover much of northern New Mexico, and visitors flock to Yellowstone National Park to watch over 3000 geysers and hot springs that are still active on a 7000-ft (2130-m) plateau in the heart of the mountains.

---

[2]First to appreciate the beauty of the Black Hills after dusty days on the Great Plains appear to have been members of General Custer's expedition of 1874, the first organized white penetration of the Hills: "An Eden in the clouds—how shall I describe it?" wrote a reporter from the New York *Tribune* whom Custer had thoughtfully taken along with him. The general himself reported, "In no private or public park have I ever seen such a profuse display of flowers. Every step of our march that day was amid flowers of the most exquisite color and perfume."

It is in the granite of the Black Hills, exposed at Mt. Rushmore, that a gigantic monument has been carved, showing the heads of four American presidents.

The Canadian Rockies: The Rocky Mountain Trench. Following the general north–south trend of the mountains, this fault-line valley extends 800–900 mi (1287–1448 km), from the Liard River in Canada to Flathead Lake in Montana. *British Columbia Government*

In the Cordilleran barrier there are few breaches. One of them is in the far north—at the southern end of the Mackenzie Mountains, where the Liard River flows through the break—and is followed by the Alaska Highway. The other is a breach of the utmost importance for transcontinental communications and occurs between the southern and central Rockies where they fall back to form the Wyoming Basin. The old Oregon Trail (see p. 178) made use of this route, on which wagons could be hauled over the Continental Divide without difficulty—without, indeed, more than a distant glimpse of the mountains. Today, rail and road routes across Wyoming are equally gently graded; in fact, without the roadside signs it would be difficult to identify the Divide.

THE INTERMONTANE REGION

The Intermontane Region is a vast area, broad in the United States but narrowing to the north, which lies between the Rocky Mountains and the easternmost of the Pacific Coast mountain chains. The broad southern part may be divided into three sections.

Most spectacular in scenery are the Colorado Plateaus centered on the intersection of Colorado, Utah, Arizona, and New Mexico. They are composed of uplifted and deeply dissected layers of sedimentary rocks whose cliff steps range from 11,000–5000 ft (3340–1520 m) above sea level. Into these many levels the Colorado River and its tributaries have eroded spectacular canyons and left isolated mesas and buttes. In the Grand Canyon, the river has cut down through the whole sedimentary cover and into the Precambrian floor, exposing a giant cross section of the continent's physical history. Still greater diversity is provided by many volcanic outcrops that occur in parts of the Plateaus. The intermingling of Spanish and Anglo-American cultures with those of the Pueblo, Hopi, and Navajo peoples adds yet more interest to this unusual area.

To the west and south of the Colorado Plateaus is the Basin and Range Province. Its characteristic features are fault-block mountain ranges separating alluvium-floored basins. The province may be conveniently divided into two parts. The Great Basin, centered on Nevada, is aptly named in view of its basin-like position between the high Colorado Plateaus and Wasatch Range on the east, and the still higher Sierra Nevada Range to the west. The two mountain ranges mentioned are, like those within the Great Basin itself, great fault blocks, each with its steep faulted edge facing the Basin. The Great Basin is composed of many smaller basins, separated from each other by short ranges, rising to 2000–3000 ft (600–900 m) above the basin floors, and orientated for the most part north to south. The Basin and Range landforms extend southward into their second part in southeastern California, southern Arizona, and Mexico. Here, the physiography differs from that of the Great Basin in two respects. First, the fault-block ranges tend to be much smaller and less regularly orientated. Second, they rise from a much lower base level. Lowest of all is Death Valley, with a base level of 282 ft (85.7 m) below sea level. Since, in the West, precipitation generally diminishes both at lower elevations and in lower latitudes, the basins and ranges are even more desert-like than those in the generally arid Great Basin.

There is a fourth intermontane section in the United States, centered on the basins of the upper Snake (southern Idaho) and middle Columbia (eastern Washington) rivers. But it includes intervening and adjacent lava plateaus and even mountains. The great outflows of lava, especially from the volcanic Cascades to the west, have formed and been eroded into many levels so that parts of the region resemble the Colorado Plateaus. In fact, the canyon cut by the Snake River between Oregon and Idaho is as deep as the Grand Canyon of the Colorado, although not as spectacular in its coloration or sculpturing.

The Canadian–Alaskan section of the Intermontane Region is diverse. Parts of the valleys of the Fraser and Yukon rivers are fairly open and flat. But much of the area is composed of dissected plateaus, and the mountainous character of some areas is only a little less pronounced than that in the Rockies chain to the east and north or in the Pacific chains to the west and south.

THE PACIFIC COASTLANDS

The most westerly physiographic province of North America is filled with a remarkable variety of natural wonders, and it is, indeed, only a common history that links its varied landscapes. In or adjacent to this province are found the highest and lowest points of the continent, some of its deepest valleys, and its only remaining (recently) active volcanoes.

On the American shores of the Pacific, the trend of the land features is almost everywhere parallel to the shoreline. These features form three belts, and this triple-banded effect can be traced from Mexico to Alaska, with only a short break in the region around Los Angeles.

Landforms of the High Plateau Province: A landscape in northeastern Arizona, now the Hopi Indian Reservation. The picture emphasizes that the plateau surface owes its present forms to both desert erosion and volcanic activity—many of the striking features represent volcanic necks, of which there are hundreds. View of Hopi Buttes, looking toward the San Francisco Mountains. *John S. Shelton*

The eastern belt of the province is a mountain chain comparable in size and splendor with the Rockies. From southern California, this chain, the Sierra Nevada, stretches north to become the Cascade Range in Oregon, and the Coast Mountains when it reaches British Columbia. It culminates in the Alaska Range in North America's highest peak, Mt. McKinley, at 20,320 ft (6178 m). Throughout its great length, the chain is cut by only three low-level crossings: the valleys of the Columbia, Fraser, and Skeena rivers. It is formed mostly of old crystalline materials occurring in the shape of batholiths. The crestline of the Cascade Range is crowned by scores of volcanic cones. These cones, whose lavas lie thick over the ranges, form outstanding peaks like Mt. Shasta in California and Mt. Rainier in Washington (both over 14,000 ft [4260 m]). But height is not all in a volcanic region, where eruptions may blow the tops off the volcanoes and actually reduce their height or, as at Crater Lake in Oregon, the volcanic cone can collapse. The most awesome display of volcanic power in the modern period, however, occurred in May 1980, when Mt. St. Helens in Washington erupted, spreading destruction down its flanks and volcanic ash thickly for scores of miles eastward.

Much of the eastern side of the mountains is faulted and very steep. In distance, Mt. Whitney to Death Valley is only 70 mi (112 km), but the difference in elevation is nearly 15,000 ft (4560 m). The west side of the mountains, particularly the Sierras, is notable for its ice-carved valleys—Yosemite, with its waterfalls and its 3000-ft (900-m) cliffs, and King's Canyon, with its groves of sequoias.

West of the mountains is a line of depressions—

15

the Central Valley of California, the Willamette Valley, Puget Sound, and the coastal channels of British Columbia. Created in conjunction with the uplift of the mountains, these depressions have been filled at various times with ice, mud flows, and alluvia. Today they are submerged in the north but dry in the south, where up to 2000 ft (600 m) of materials have been deposited to form fertile farmlands, the agricultural heart of the West.

Bordering the coast itself is yet another mountain chain. In California it is composed of rolling hills 2000–4000 ft (600–1200 m) high called the Coast Ranges. These enclose the Central Valley on the west, except where the sea breaks through them to form the bay at San Francisco. The coastal chain continues into Oregon, but north of the Columbia, it becomes higher and more rugged until it reaches Puget Sound and the 8000-ft (2430-m) Olympic Mountains. North of the Strait of Juan de Fuca the ranges are discontinuous and are represented by Vancouver Island and by the long chain of islands that flanks the coast beyond as far as the Aleutians.

This triple belt of the Pacific coastlands is recognizable everywhere except in the extreme southwestern United States. Here it is interrupted by faulting in an east–west direction. The Tehachapi Mountains link the Sierra Nevada to the Coast Ranges, enclosing the southern end of the Central Valley of California. Further south, the east–west oriented San Gabriel and San Bernardino (transverse) ranges separate the small coastal lowland at Los Angeles from the Intermontane Region.

The Glaciation of North America

Of decisive importance in shaping the physical and cultural patterns of present-day North America was the epoch in the continent's history when some three-quarters of its surface was covered by ice. Glacial erosion smoothed its rugged relief and reformed its drainage patterns; glacial deposition covered 1 million sq mi (2.5 million sq km) with drift, which filled former valleys and provided materials for some of the continent's most fertile soils. On the farmlands of the Midwest, it is to the Ice Age as much as to any other single factor that modern farmers owe their prosperity.

Although the ice was once continuous from coast to coast, it is necessary to distinguish between two separate parts of this great cover. Over the northeast of the continent there spread the vast Laurentide Ice Sheet. In the Northwest and West, on the other hand, a separate system of glaciers—usually known as the Cordilleran System—formed in the mountains, from which it spread to cover the northern intermontane zone. Although remnants of this western system are still to be found in the northern Rockies and in the mountains of the northern Pacific coast, the Laurentide Sheet, although much greater in extent, has disappeared entirely from the North American mainland and today remains only on the mountains of Baffin and Ellesmere islands.

THE LAURENTIDE ICE SHEET

The ice probably originated in the Hudson Bay or Labrador Region. From there it spread across eastern North America almost to the foothills of the Rockies, and at its maximum extent covered some 4.8 million sq mi (12.4 million sq km). It is suggested that it originated in a series of valley glaciers that grew slowly until they became sufficiently widespread to affect the climate; thereafter they grew with increasing rapidity, expanding in the direction from which most precipitation was being received—in this case, the west. Over so large an area ice probably accumulated at varying rates, and so domes, or "centers," of ice built up in the areas of most rapid growth. There seem to have been two such centers—the Labrador and the Keewatin—one to the east and the other to the west of Hudson Bay, but these centers may have shifted their position with the passage of time. From them the ice flowed outward in all directions.

Southward, the ice spread over New England and the Adirondacks and as far as northern Pennsylvania. But on this all-important southern edge of the sheet, where the force of the ice movement was becoming weaker, the direction of the movement was influenced by the pre-existing relief features, such as the Superior Upland and the Niagara Cuesta. Diverted by these, the southern edge of the ice formed a series of southward-moving lobes. Under the influence of climatic fluctuations, these lobes advanced and retreated several times during the glacial epoch, so that the southern limit of the sheet varied in position from time to time. On the whole, however, its maximum southward extent is marked today by the line of the Missouri and Ohio rivers, which first developed as streams flowing along the ice front.

Beneath most of the vast area of the Laurentide Sheet, the principal action of the ice was erosive. Over most of the area north of the present Great Lakes there was little glacial deposition, but rather a

Baffin Island, Canada: The formation of a terminal moraine along the edge of the Barnes Ice Cap. *J. D. Ives*

widespread scouring. Relief was smoothed, soil cover was removed, and a new and sometimes inconsequent drainage pattern was imposed on the area. Today, as we have already seen, the features of this region are an infertile surface and a maze of swamps and lakes.

Along the southern margin of the sheet, however, erosion gave place to deposition, and scouring to infilling. Here the action of the ice produced a series of important landscape features. Foremost among them is the mantle of glacial drift that covers virtually the whole area. Varying in thickness from a foot or two to 150 ft (50 cm to 45 m), and in places much greater depths, the drift has completely re-formed the topography of central North America.

It is this drift cover—the ground moraine, or till, left by the ice sheet—which serves as the best guide to the advances and retreats of the ice under the influence of climatic fluctuations. Each new advance produced a fresh drift. From the evidence these provide, it is deduced that there were four main periods of southward advance, or "glacial stages," separated by interglacial periods. The earliest glacial stage, the Nebraskan (in each case the name is borrowed from one of the states most affected) was followed by the Kansan and the Illinoian and finally the Wisconsin. As the most recently deposited drift, the Wisconsin overlays the older mantles, which protrude at its edges. Providing as it does the most abundant materials for study, the Wisconsin stage has been further divided into four substages, which seem to correspond to minor fluctuations within the major advance.

The interglacial periods are also important. They

Alaska: The Columbia glacier, photographed in 1979. By 1986 the glacier front had retreated (i.e., melted back) over a mile. *Austin Post, U.S. Geological Survey, Tacoma, WA.*

were evidently long enough to permit both the development of a vegetation cover and the formation of wide belts of loess. The windblown loess accumulated on the ice-free surfaces after each retreat, and locally, as in the "loess state" of Iowa, it is of great value to agriculture.

Thus the glacial and interglacial deposits are often interleaved, and together form a thick cover, which may bear little resemblance, either in relief or in soil type, to the earlier surface that can be identified beneath it. Although the material of which the till is composed ranges from boulders to fine clays, on the whole it produces excellent soils.

The second main feature of this zone of deposition is the great number of moraines and hills of glacial origin. The edges of most of the drift belts are marked by end moraines. On the map these often appear—especially on the till plains south of the Great Lakes—as semicircular ridges that reveal how the ice front consisted of a series of lobes at the limit of its advance. Further east, in New England, these same end moraines are today represented by the islands

South Dakota: A landscape of morainic deposit, near Rosly. *John Shelton*

that fringe the region's southern coast—Nantucket, Martha's Vineyard, and Long Island. Elsewhere, as in Wisconsin, the moraines form long ridges running generally from north to south. The most numerous features of all are the drumlins, which are low, oval hills shaped like the hulls of overturned boats, their long axis representing the direction in which the ice that deposited them was moving. They are especially common in the Great Lakes shorelands of Wisconsin, Michigan, and western New York State. They generally appear in clusters of dozens together, orientated roughly north–south by the ice.

Lying as it does in the heart of the Great Lakes area, the peninsula of southern Ontario possesses these morainic features in very large numbers. The center of the peninsula is covered by glacial drift with drumlins, and this central area is almost surrounded by a horseshoe of morainic hills formed by lobes of ice pressing south into the lake basins on either side of it. Outside this horseshoe lie clay and sand plains; these are relics of the beds of the far larger lakes that formerly occupied the basins of Ontario, Erie, and Huron, surrounding an Ontario "island" much smaller than the present peninsula.

The third feature of note is the effect of glaciation on the drainage pattern. Before the glacial epoch, the drainage of much of the Midwest was probably eastward and then northward toward Hudson Bay. When the ice began to retreat northward, the residue of the sheet blocked this route to the sea and meltwater accumulated along the ice front in a series of huge lakes. These reached as far west as glacial lakes Agassiz and Regina on the Canadian prairies and together covered an area several times larger than that

The Atlantic Coast: Nantucket Island. Lying 25–30 mi (40–45 km) south of Cape Cod, Nantucket is one of the numerous islands formed along the Atlantic Coast from submerged morainic materials and sculptured into curious shapes by tide and ocean current. At the top of this aerial photograph Great Point projects northward. At the lower left are Nantucket Town and the entrance to its lagoon harbor. *EROS Data Center, U.S. Geological Survey*

now occupied by the five Great Lakes. The development of the Missouri–Ohio drainage system, although it provided an outlet southward for some of the meltwater, further dislocated the earlier drainage pattern, and widespread damming and river diversion occurred.

As the location of the ice front changed, so the form of these lakes varied. First one outlet and then another was uncovered by the ice, giving a fresh escape route for the meltwater. As each outlet was cleared and deepened by erosion, the lowering of the water level resulted in the abandonment of an earlier, higher route. At one period or another, this great expanse of water seems to have drained away, either southwest and southward to the Mississippi or southeastward into what are now the Susquehanna and the Hudson, or northeastward into the St. Lawrence, when that route was eventually cleared. Today the Great Lakes represent a remainder of this great body of water whose past presence is attested by the old lake beds that provide farmers of Minnesota, Manitoba, and Saskatchewan with some of the world's flattest and most fertile farmlands.

Photographs of these lake beds along the Red River or on the Regina Plains often create the impression that the whole of the northern plains and prairies is plowed in endless straight lines over a uniform surface. The reality, however, is rather different, for over most of the region the effect of ice and drift has been to dot the surface with innumerable kettle holes and patches of swamp. Water accumulates in these in wet weather to form sloughs, or "slews." Few prairie fields are without one or two, however small, and the farmer must take the plow around them. From the air, the landscape, for all the regularity of its cultural patterns, including roads that run to the points of the compass, has a pockmarked appearance.

In all this drift-covered expanse, one small part stands out as an exception—the Driftless Area of Wisconsin. By some trick of relief—probably the sheltering effect of the Superior Upland—this area seems to have been bypassed by the ice.[3] Walled in on two sides by the Wisconsin terminal moraine, its

[3]For a summary of evidence suggesting, however, that the Driftless Area may *not* have escaped all glacial action, see J. C. Frye et al. in H. E. Wright and D. G. Frey, eds., *The Quaternary of the United States* (Princeton, N.J., 1965), pp. 54–55. This book is a good general reference on the glaciation of the United States. For Canada, see Bird, *The Natural Landscapes of Canada* especially Chaps. 2–4.

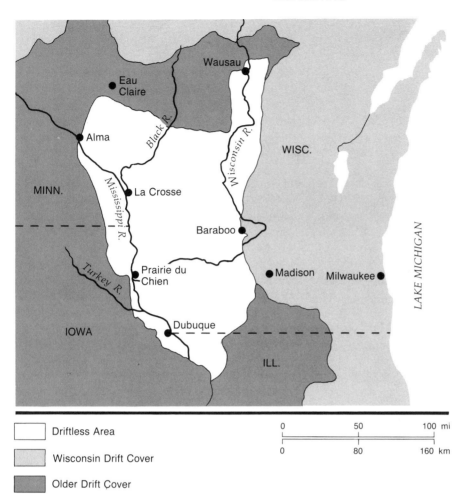

Fig. 1-5. The Driftless Area of Wisconsin, Iowa, and Minnesota. (Based on L. Martin, *The Physical Geography of Wisconsin* [Madison, Wisc., 1965], Fig. 26.)

☐ Driftless Area

▨ Wisconsin Drift Cover

▮ Older Drift Cover

```
0        50       100  mi
|----|----|----|----|
0        80       160  km
```

features are water- and not ice-formed. Its relief is more broken than that of the surrounding areas; its soils are the poorer for being drift free (Fig. 1-5).

THE CORDILLERAN GLACIATION

The Cordilleran System ultimately covered almost the whole of what is now western Canada and much of Alaska, and isolated ice caps spread over the higher areas as far south as the mountains of northern New Mexico. With its mountains and its heavy precipitation, the Northwest—particularly British Columbia—was a center of accumulation. From the heights of the two main mountain chains the ice tended to flow both east and west, so that eventually the westward-flowing ice from the Rockies and the eastward flow from the Cascades and Coast ranges coalesced to cover completely the intermontane zone.

Further south, where the ranges of mountains

were not so close to each other, this did not happen, but with the ending of the glacial epoch many of the intermontane basins filled with meltwater. As in the east, drainage was dislocated: outflow to east and west was barred by the mountains; northward the remains of the ice blocked all outlets. Only in the extreme south was there an open route to the sea. The Great Basin of Utah and Nevada (Fig. 1-6) became the bed of many lakes, the largest of which was glacial Lake Bonneville. It is of this great body of water, ponded between the southern Utah divide and the retreating ice front, that the present Great Salt Lake is merely a relic, its shoreline lying 1000 ft (300 m) below that of its predecessor. Ultimately, Lake Bonneville drained off northward into the swollen Columbia. Although today the Columbia is one of North America's largest rivers (see Chapter 24), it must have been, during the meltwater period, many times larger. The Dry Falls in the Grand Coulee of

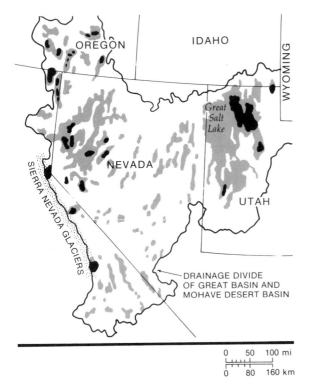

Fig. 1-6. The Great Basin of Utah–Nevada: Extent of pluvial lakes in post-Sangamon (late Pleistocene) time. Former lake beds are stippled; surviving lakes are marked in black. The boundary of the drainage basin shown does not precisely correspond with that of the Great Basin of today. (Based on a map in H. E. Wright and D. G. Frey, eds., *The Quaternary of the United States* [Princeton, N.J., 1965], p. 266.)

Washington, a water-cut valley representing only one of this earlier Columbia's many channels, have a brink 3.5 mi (5.6 km) in length.

Of this Cordilleran System, the ice fields of Alaska remain as substantial reminders. Further to the south, small glaciers still flank the peaks of British Columbia and the state of Washington as far south as Mt. Rainier, while a few remnants remain in the Colorado Rockies.

The Climate of North America

The Florida Keys, the southernmost point of the United States, lie just north of the Tropic of Cancer. The northern tip of the Canadian mainland lies at 72° N. With a latitudinal spread of nearly 50°, therefore, the area we are considering has a wide variety of climates. At one extreme are the frostless islands of the Gulf of Mexico; at the other the arctic conditions of the Canadian Northlands. The east–west di-

mensions of North America are equally significant. On the Pacific coast, there are stations that record over 200 in (5000 mm) of precipitation per annum, whereas a few score miles inland there are areas with less than 10 in (250 mm). There are points in the central plains more than 1000 mi (1600 km) from the sea, and they experience the climatic extremes of a continental interior. On the other hand, within this great land area there is room for a chain of inland seas— the Great Lakes—over 700 mi (1120 km) from east to west and for the 750-mi (1200-km) penetration of Hudson Bay southward from the Arctic Ocean, both of which have important effects upon the climatic regime of eastern North America.

AIR MASSES

For all its variety, however, the climate of North America is the product of relatively simple controls. Over the greater part of the continent east of the Rockies, it is the behavior and interaction of two air masses that are responsible for weather changes. These are the continental air mass centered on the Canadian Northlands and the tropical maritime air mass of the Atlantic–Gulf of Mexico area. West of the Rockies and on the Atlantic Seaboard, four other air masses play parts of varying importance in weather control: the polar maritime air of the northern Pacific and of the northern Atlantic; the tropical maritime air mass over the Pacific; and the tropical continental south of the Rio Grande, which affects the southwestern United States and Mexico.

Polar, or arctic, continental air from the first of these air masses spills out from its source area in northern Canada and is associated in summer with cool, clear weather. In winter, cold waves bring spells of bitter weather to the northern interior, and frost danger to the usually milder regions further south. Tropical maritime air from the Gulf, on the other hand, flows northward from its source region over a warm ocean. In summer it is associated with high temperatures and oppressively high humidity over much of eastern North America; in winter, Gulf air brings mild spells and rain or fog to the cold interior.

As the general position of these air masses shifts with the sun, so they tend to dominate different parts of the continent in summer and winter. There is at all seasons a zone over which they are in conflict; a zone in which the effects, sometimes of one and sometimes the other, are dominant. In summer, this zone of conflict runs south of the Great Lakes, so that the Upper Midwest experiences warm, humid weather

Florida: Hurricane damage near Coral Gables, November 1992. *U.S. Corps of Army Engineers, Jacksonville District*

of the type produced by Gulf air, alternating with cooler, less humid weather introduced from the north. In winter, on the other hand, when the whole system shifts south, the zone of conflict lies over the Gulf Coast. Then most of the interior is dominated by the cold air flow from the northern high-pressure area, with occasional breaks from the south, which may have the welcome effect of relieving winter cold, but may also cause fog and brief, dangerous thaws.

This means that in the Midwest there is a period in spring and again in autumn when the zone of conflict is in transit. At such times, weather watchers study the depression tracks closely, trying to determine whether they will pass by on northern or southern routes. In spring, once the tracks move north of, say, Chicago and its latitude, summer cannot be far behind.

This zone of conflict between North America's two most important air masses is an area where atmospheric disturbances are frequent; consequently it is crossed by an unusually large number of storms, even for a mid-latitude region. Cyclonic storms are generated all along the contact zone between the two air masses and move across the United States and southern Canada from west or southwest to east. The expanse of the Great Lakes seems to be a factor in their convergence on the line of the St. Lawrence Valley and the New England–Maritimes area.

It is, then, the interaction of these two air masses that produces the weather experienced by southern and central North America, and so by a high proportion of the continent's population. West of the Rockies, however, the situation is quite different. Penetrations of Gulf air are rare, and the main interaction of air masses is between polar continental air from the Northlands and polar maritime air from the Pacific.

In winter, the Pacific air brings wet and (for these latitudes) mild weather to the coastlands: the amount of precipitation falls off southward. Occasionally the Pacific air penetrates, though now cooler and drier, to the east of the Rockies. In summer, the northern Pacific is dominated by a high-pressure system, and the influence of polar maritime air on West Coast weather becomes stronger. At this season, however, it is associated with dry, rather cool weather.

The southwestern United States lies within the sphere of influence of two tropical air masses. Tropical maritime air from the Pacific is dominant along the coast of Baja California, where it is associated with the warm, dry conditions along the desert shore. But it is kept from extending its influence far to the north by the strong flow of polar maritime air throughout most of the year. Tropical continental air, in the same way, spreads northward from Mexico during the summer, but usually affects only a small area of the United States beyond the border.

On the Atlantic coast, in a similar way, the weather is produced by the interaction of three air masses—the polar continental and the Gulf air, which influence most of eastern North America, together with polar maritime air from a source area near Greenland. The southward penetration of polar maritime air affects the coastlands of the Northeast, bringing cold and drizzle in winter and lower temperatures with moderate precipitation in summer. Its sphere of influence, however, does not extend southward much beyond Chesapeake Bay.

The weather in North America, then, is subject to seasonal differences because of the shift of air masses northward and southward, with the apparent path of the sun. We have emphasized, also, the often abrupt, short-term changes in weather that occur with cyclonic disturbances. A critical element in the world atmospheric circulation, and a key control to the particular locations of the air masses and of the cyclonic storm tracks, is the high-altitude jet stream. This west-to-east stream of air along the poleward edge (polar front) of the mid-latitude westerlies attains speeds of 50–200 mi per hour (80–320 km per hour) at elevations of 35,000–50,000 ft (10,500–15,000 m). It encircles the globe, not in a straight east–west line, but rather with three or four great northward and southward waves, or curves. The waves tend to persist, but may change in number, amplitude, and position. They shift northward in summer and southward in winter along with the whole atmospheric circulation system. But shorter shifts may occur, especially in winter, when circulation, velocity, and wave lengths are strongest.

The west-to-east jet stream, with its northward and southward waves, tends to guide air masses and cyclonic storm tracks below, and consequently, to determine whether cold polar air or warm tropical air is drawn into the Midwest and the eastern United States and southeastern Canada, especially in winter. For example, during the very cold winters of 1977–78, 1978–79, and 1981–82, the jet stream curved far northward over the Pacific Coast, bringing mild temperatures to southern Alaska; it then curved far to the south, along the eastern margin of the Rocky Mountains, persistently guiding polar air into the Midwest and the East.

TEMPERATURE

The wide range of climatic conditions in North America is clearly marked in the temperature statistics. The Gulf Coast and Florida have a seasonal regime of warm winters and hot summers; the mean January temperature is more than 50°F (10°C) and the mean July figure is above 80°F (27°C). At the other extreme, northeastern Canada experiences the cold winters and cool summers (with a mean of less than 50°F [10°C] for all twelve months of the year) of a polar climate. Over most of the land mass, the chief feature of the temperature regime is its continentality, with a large annual range of temperature (up to 80°F, or 45°C, in northwestern Canada), a wide diurnal range in winter, and alternating cold winters and hot summers separated by only brief transitional seasons.

The majority of North Americans live, therefore, in areas where the mean January temperature is below 32°F (0°C). The January isotherm of 32°F runs from the Atlantic coast just south of New York, past Philadelphia, and then almost due west through southern Indiana and Illinois to southeastern Colorado, while almost all of the Mountain West experiences a January mean well below the freezing point. The January mean in Chicago is 25°F (−4°C); in Montréal it is 14°F (−10°C), and at Winnipeg, the coldest of North America's large cities in winter, the mean temperature for the month is −3°F (−19.5°C). The average figures, however, conceal the characteristics of this winter cold, which are, first, that during the day the temperature will frequently rise above the freezing point, only to fall to 10 to 0°F at night (−12 to −18°C), and second, that the means are depressed by extremely low temperatures occurring in periodic cold waves, rather than by consistently cold weather throughout the winter.

By contrast with the continental interior, the southern and western coastlands—from Virginia to Oregon—everywhere enjoy a January mean of more than 40°F (4.5°C), and the moderating influence of the ocean gives the smaller annual temperature range characteristic of maritime climates. The smallest temperature ranges in North America occur on the coast of central California (where in exceptional locations the annual range is less than 10°F or 6°C), at Key West, Florida (a range of 13°F or 7.5°C), and on the coast of Oregon and Washington (14–16°F or 8–9°C).

If most Americans spend at least one month of winter at a mean temperature of 32°F (0°C) or below, it is also true that, in almost every major city, they languish through one or more summer months with a mean of 70°F (21°C) or above. The 70°F July isotherm follows fairly closely the boundary of dense settlement across the central Great Lakes; Chicago experiences a July mean of 74°F (23.5°C), and Montréal one of 70°F. Further south, July means of 80°F (27°C) or above are general in Oklahoma, Arkansas, and the states lying to the south and southeast of them. Where these high summer temperatures are associated with high humidity—as is the case almost everywhere east of the Mississippi—the resultant summer weather is extremely oppressive, both by day and by night. To find the relief of lower humidity and wider daily temperature ranges, it is necessary to travel either north or to the Mountain West.

In the Mountain West, summer conditions vary with altitude, but days are generally warm and nights pleasantly cool. On the Pacific Coast, there is a steady decrease of summer temperature from south to north; the July mean at San Diego is 67°F (19.5°C) and at Victoria, B.C., 60°F (15.5°C). The moderating influence of the Pacific extends, however, only a short distance inland; the Central Valley of California experiences July means in excess of 80°F (27°C), while less than 200 mi (320 km) inland lies Death Valley, with its forbidding July average in excess of 100°F (38°C).

PRECIPITATION

The parts of North America that receive the most precipitation are, first, the mountains of the northern Pacific Coast and, second, the southern Appalachians and the southeastern states. Virtually all the land mass lying east of 98° W longitude and south of the 60th parallel has 20 in (500 mm) or more of rainfall per annum, while north and west of these lines is a dry belt that stretches to the Arctic and westward almost to the Pacific mountains.

Rain-bearing winds reach the continent from the North Pacific, and from the Atlantic and Gulf of Mexico, where these seas border the continent on the southeast. The winds of the North Pacific immediately encounter the abrupt rise of the great mountain ranges of the Pacific Coast; they quickly shed their moisture, leaving the greater part of the area between the coastal mountains and the Rockies in the rain shadow. The behavior of the rain-bearing winds of the Southeast is governed almost equally by factors of relief, but in this case, by the *absence* of any significant mountain barrier to their advance, so that their effect is felt over the whole continent east of the Rockies and south of the Canadian Shield.

The pressure systems on which these rain-bearing winds depend create seasonal variations in rainfall (Fig. 1-7). West of the Rockies, in the sphere of influence of North Pacific air, there is everywhere a marked winter maximum. Even along the British Columbia coast, where the annual rainfall is more than 100 in (2500 mm), July and August have little rain, while further south in California, where the total amount of precipitation is much smaller (20 in [500 mm] at San Francisco and 10 in [250 mm] at San Diego), there may be as many as four summer months that are completely rainless, after the familiar pattern of Mediterranean-type climates. Further inland, in the desert basins, although the amounts of rainfall received are even smaller than on the California coast, much the same regime holds good; in the West, summer rains, apart from thunderstorms, are experienced only in the higher mountains.

East of the Rockies the situation is reversed; here only a small area has a winter maximum. Everywhere else, if there is a significant maximum, it occurs in the warmer months. It may occur, however, in spring, as in the central Corn Belt, or in early summer, as in the Great Plains, or in late summer, as in northeastern Canada, and this distinction is of great importance to the farmer. In the Agricultural Interior and on the Great Plains, early rains favor the growth of crops. In areas of late summer rain, on the other hand, the maximum occurs too late to be of much agricultural value.

The only area of winter maximum east of the Rockies lies on the eastern coast of Labrador, Newfoundland, and the Maritime Provinces. Between the Great

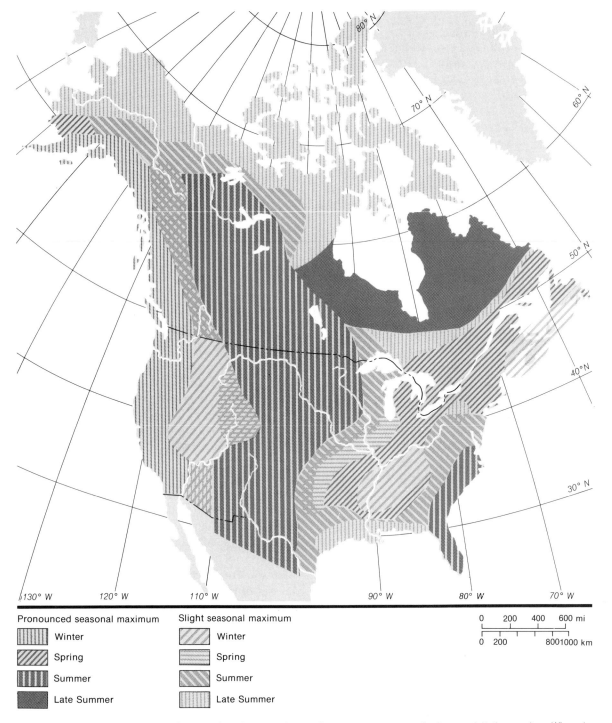

Pronounced seasonal maximum	Slight seasonal maximum
Winter	Winter
Spring	Spring
Summer	Summer
Late Summer	Late Summer

0 200 400 600 mi

0 200 8001000 km

Fig. 1-7. The climate of North America: Seasons of maximum precipitation. The map is based on data for some 250 stations, and shows the season or seasons having precipitation maxima. Where two maxima occur in the year this is indicated by cross-hatching.

Lakes and the Atlantic Ocean, however, rainfall is everywhere plentiful throughout the year because these water bodies tend to attract to the area depressions traveling on almost all the continent's eastbound storm tracks. Thus, New York has 3–4 in (75–100 mm) of rainfall in every month and Toronto, 2–3 in (50–75 mm). But where, in addition, the presence of polar maritime (Atlantic) air makes itself felt along the northeast coast, the winter's rainfall is supplemented from this source and exceeds that of the other seasons.

CLIMATIC REGIONS

Using these details we can now distinguish seven climatic regions (see Fig. 1-8):

1. *The northern fringe of the continent is an area of polar climate that finds its counterpart in northern Siberia.* Summers bring a brief surge in growth of tundra plants, but are too short and cool for tree growth; no month has an average temperature above 50°F (10°C). Precipitation is meager (generally below 10 in or 25 cm) but sufficient for the tundra growth in summer and for a shallow snow cover during the very long and cold winter.

2. *The Subarctic, the region of the great boreal forest, is dominated by polar continental air, especially in winter.* Winters are long, bitterly cold, and relatively dry—snow cover is not deep, but is of long duration. Summers are short but with long, often warm days. Precipitation is low to moderate, but sufficiently effective, given the cool temperatures, for tree growth.

3. *The Pacific coastlands are notable for an absence of cold temperatures.*

(a) North of San Francisco is a coastal belt with a cool temperate marine climate, which resembles that of northwestern Europe or New Zealand. Yet summers are drier, so that British expatriates often feel that in coastal British Columbia they have discovered an idealized version of the climate of their homeland. Migrants from eastern America, on the other hand, may delight in the cool, dry summers but find depressing the cloudy, rainy weather that accompanies the frequent cyclonic storms of winter.

(b) The Mediterranean-type climate of California is transitional between the climates of the temperate marine belt to the north and of the deserts of the south and the interior. In fact, it is simply a combination of temperate marine winters and desert summers.

Looking again at the coastal marine belt, as a whole, we see that the length of the rainless summer

period increases throughout the belt southward from Alaska. Inland, and throughout the belt in general, also, the characteristics we have emphasized are modified, especially by increasing elevation or exposure to the westerly winds.

4. *Local variation is the hallmark of the climatic region of the western mountains* (Fig. 1-9). This is the result of a decrease in temperature with altitude and an increase in precipitation with both elevation and exposure to moisture-bearing winds, especially from the west.

5. *The arid lands lack moisture* because they are seldom reached by either of the two air masses that carry most of the moisture to the continent north of Mexico. Tropical Atlantic air seldom penetrates so far to the west, and polar maritime Pacific air loses most of its moisture in the Pacific coastal belt and on the western slopes of the higher mountain chains.

(a) The Desert Southwest has dry, hot summers and mild, sunny winters. Amounts of sunshine are the highest on the continent. Because the air is dry and cloudless, very hot days may be followed by cool nights.

(b) The semidesert and steppe lands, with their greater latitude, altitude, and distance from the oceans, are somewhat less hot and dry in summer and considerably colder in winter than the Desert Southwest. The northern Great Plains are especially subject to winter invasions of very cold polar continental air from the north.

6. *In the Southeast, the climate is humid and subtropical,* dominated by tropical Atlantic–Gulf air, especially in the summer. Thunderstorms, with heavy convectional rain, are frequent in summer, often occurring daily. In winter, cyclonic storms may move eastward along the southern edge of invading masses of polar continental air, bringing rain and snow followed by freezing temperatures. This climate finds its counterpart in southern Brazil and the Plate River region of Argentina, or, allowing for a greater monsoonal effect there, in southeastern China.

7. *The humid continental areas are transitional between the subarctic North and the subtropical Southeast.* Because the weather takes its characteristics mainly from polar continental air surging southward or tropical Atlantic air moving northward, and from eastward-moving cyclonic storms born of their conflict, seasonal diversity and extremes are great. The two subregions differ according to which of the two major air masses tends to dominate.

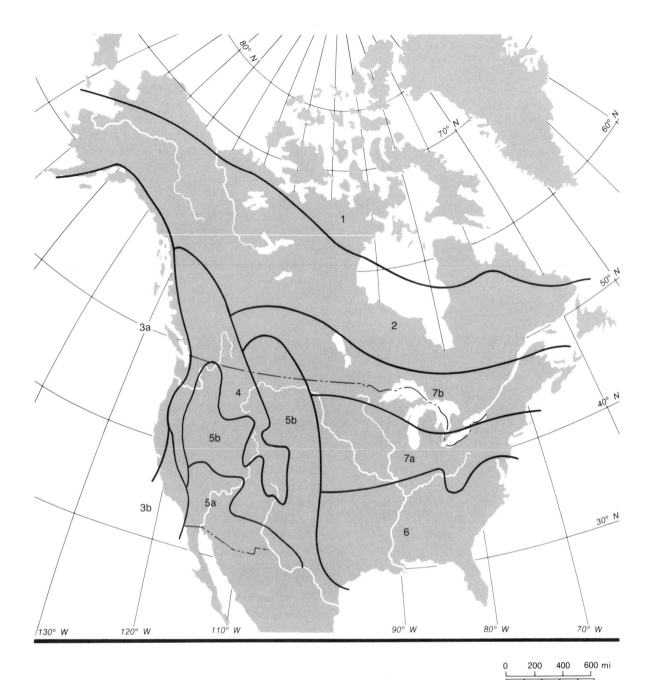

Fig. 1-8. Climatic regions of North America: 1. The polar fringe; 2. The Subarctic; 3. The Pacific coastlands (a) Temperate marine, (b) Mediterranean; 4. The western mountains; 5. The arid lands (a) The Desert Southwest, (b) Semidesert and steppelands; 6. The Southeast; 7. The humid continental areas (a) Hot summer subregion, (b) Cool summer subregion. The regions are described in the accompanying text.

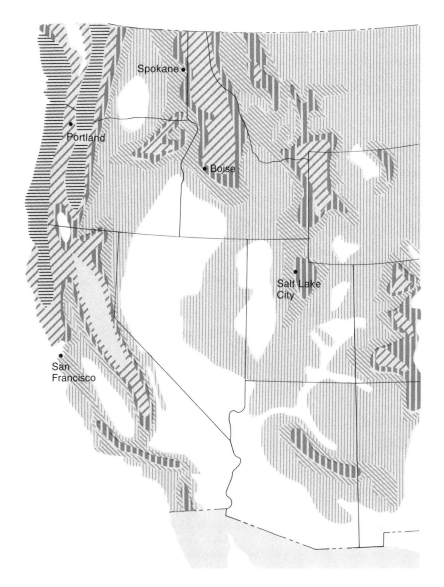

Fig. 1-9. The Western United States: The climate west of the Cordillera, according to the classification of C. W. Thornthwaite, and showing the wide variety of climatic conditions induced by the relief. (Based on a map in *Geographical Review,* vol. 38 [1948], p. 94, by permission.)

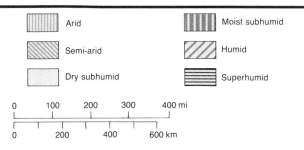

Arid

Semi-arid

Dry subhumid

Moist subhumid

Humid

Superhumid

0	100	200	300	400 mi

0	200	400	600 km

(a) The warm summer subregion falls generally under the dominance of tropical Atlantic air, especially in summer. High temperatures and humidity and frequent thunderstorms are common, as they are in the Southeast. In winter, the eastward passage of a cyclonic storm, often along the lower Missouri and Ohio river valleys, may bring rain, followed by snow, rapidly falling temperatures, and several days of cold, clear continental air.

(b) The cool summer subregion is dominated by polar continental air, especially in winter. In effect, it then becomes an extension of the subarctic region, with clear, pleasant days and very cold nights. In summer, the passage of cyclonic storms along the Great Lakes and St. Lawrence Valley may be preceded by warm, humid air and thunderstorms, and followed by weather that is cool, dry, and clear.

The two air masses described are not the only ones to affect the humid continental areas. When a cyclonic storm moves along the coast of New England or Atlantic Canada in winter, it may bring in its wake a northeasterly flow of polar Atlantic air whose cold mist or rain leaves a coat of ice on everything exposed, be it a windshield or a fisherman's nose. In contrast, an eastward-moving mass of polar Pacific air, which may have deposited heavy rains and snows on the Pacific coastal and mountain belts, may be relatively warm and dry as it descends from the Rockies and moves toward the Great Lakes. Such air may, in the autumn season, bring the period of delightfully warm, sunny days and cool, clear nights known in both humid continental regions as Indian summer.

The Soils of North America

Insofar as climate is responsible for soil characteristics, the climatic factors we have just been considering assure the North American continent of a wide variety of soil types. But since a number of other factors are involved in soil formation, the variety is much greater than it would be if determined by climate alone: soil is also a product of different parent materials and of processes such as glaciation. So wide is the variety that its classification becomes a major problem.

Some attempts at classification made in the past are American in origin and deserve our attention in passing, if only because references to them are still frequently found in textbooks. One such useful generalization, native to North America and bearing an obvious relationship to local conditions, was that made by C. F. Marbut, who proposed a broad classification of soils into *pedocals* and *pedalfers*. Pedocals (the syllable -*cal* is an abbreviation for "calcis" or "calcium") are soils in which carbonate of lime accumulates. Pedalfers (the -*al* indicates aluminum and the -*fer*, iron) are soils in which compounds of aluminum and iron accumulate. The factors that decide which of these two processes takes place are mainly climatic. Lime accumulation is a feature of the soil profile of semiarid and arid regions; the depth at which the accumulation takes place varies with the rainfall, so that in the driest areas accumulation takes place near or even on the surface, while with increasing rainfall it occurs at greater depths. In humid regions, however, the process of leaching tends to remove the calcium carbonate from the soil and to lead instead to the formation of a layer of compounds of aluminum and iron. Thus Marbut used the terms pedocal and pedalfer to distinguish the soils of humid regions from those of semiarid and arid regions; in broad terms, of course, this meant in North America a division of the soils into those of east and west, with most of the soils of northern regions falling into the eastern, or pedalfer, group.

Another type of framework is offered by the division of soil types into three categories: *zonal, intrazonal,* and *azonal.* This classification presupposes that the basic control of soil is climate, and that soil types characteristic of each climate zone will develop in that zone unless other circumstances intervene. Particular conditions of relief or parent material may modify the *zonal* soil, giving it a different character from that which might be anticipated within its zone. Such soils are known as *intrazonal.* An example of intrazonal soils is provided by the Rendzinas of central Texas and central Oklahoma and the High Lime soils (or Gray Rendzinas) of eastern Manitoba. These are groups of soils developed on a soft limestone or chalk base which have, in consequence, a high lime content in spite of the effects of leaching; from their chemical composition they appear to have been formed in a drier climate than is actually the case. A second example is provided by soils developed on the loess and drift plains of Illinois, Iowa, and Missouri where, in the postglacial period, deposition has created an exceptionally flat surface and affected drainage; poor soil drainage and absence of surface erosion have created a soil in which there is a gley (sticky clay) horizon, and the subsoil includes a layer of clay hardpan.

In other areas, soils are immature; that is, they are composed of materials so recently accumulated or so lightly bound that no true soil profile has developed. Such soils are classed as *azonal*. Materials in this class cover a significant fraction of North America and include both lands of agricultural value, such as the alluvial plains of the Mississippi Valley and the Central Valley of California, and unimproved lands like the Sand Hills section of Nebraska. In the Mountain West azonal materials include the thin, stony cover of the mountains, the alluvia collected in the desert basins, and the most recent of the lava flows on the lava plateaus of the Pacific Northwest.

Beyond these few generalizations, however, most of the earlier systems of classification descend, at their lower levels, into mere description. In any case, they tell little about the structure or potential of the soil: they are too heavily weighted on the genetic side. In the United States, therefore, soil scientists worked for nearly thirty years on a fresh classification. They approached the problem by way of a series of *approximations* to a complete solution; the 3rd Approximation, for example, appeared in 1954 and the 4th in 1955. By 1960, they had reached the 7th Approximation and it, with various amendments, forms the present basis of the classification used in the United States.[4] It introduced an entirely new system based, so far as is possible, on measurable properties of the soil.

On this basis, there are recognized 10 soil orders, each with a name ending in -sol. There are then 47 suborders, 185 great groups, 970 subgroups, about 4500 soil families, and some 10,000 soil series (or what were formerly known as soil types). The naming of these various individuals in the classification is standardized; each suborder, for example, has a name of two syllables, the second of which identifies its parent order. If we can learn the language, therefore,[5] we now have access to a system that is de-

signed to depend, at last, entirely upon what the soil is like according to measurable scales of values, and in which the nomenclature is consistently and logically built up. The ten soil orders are given below, together with some familiar names from earlier classifications (Fig. 1-10).

U.S. Soil Survey Order	Includes soils previously named
1. Entisols	Most azonals
2. Vertisols	Grumusols and some alluvia
3. Inceptisols	Brown forest; *sol brun acide;* humic gleys
4. Aridisols	Desert and desert-saline
5. Mollisols	Chestnuts; chernozems; prairie; rendzinas
6. Spodosols	Podsols; brown podsolic; groundwater podsolic
7. Alfisols	Gray-brown podsols; grey-wooded; some planosols
8. Ultisols	Red-yellow podsols; humic gleys (planosols)
9. Oxisols	Lateritics; latosols
10. Histosols	Bog soils

While the new U.S. classification certainly represents an advance upon the old, haphazard systems of nomenclature, it does have its drawbacks. One is that the ten soil orders are so wide as to conceal more than they reveal. To see that this is so one has only to notice, for example, that the Mollisols include what used to be called chestnut brown soils and chernozems *and* prairie *and* rendzinas. It is necessary, in fact, to go at least as far as the suborders, if not the great groups, before we learn anything very specific about a given soil. A more serious drawback is that while the U.S. Soil Survey has been developing its taxonomy, the United Nations through the Food and Agriculture Organization (F.A.O.) has been busily producing an entirely different, international classification. It has not ten orders but twenty-six; its nomenclature is a little less ruthlessly logical (but also less tongue twisting) than that of the U.S. classification, and only here and there do the American and international schemes coincide. Since the F.A.O. system has already been used to map not only the whole of North America but the rest of the world as well, in the United States the two schemes are in direct competition with each other. It seems unlikely, in fact, that there will be any early ending to the long-standing problem of getting general agreement for any single system of soil classification.

[4]The relevant documents are: Soil Survey Staff, U.S. Dept. of Agriculture, *Soil Classification, A Comprehensive System, 7th Approximation,* and *Soil Taxonomy: A Basic System of Soil Classification for Making and Interpreting Soil Surveys* (Washington, D.C., 1960 and 1975, respectively).

[5]In itself no light task, although a knowledge of Latin will help. At the third level of subdivision we have, for example, a great group of the order Inceptisol called *Anthrumbrepts* and a great group of Alfisols called *Fraglossudolfs,* after which the student goes on to *Quartzipsamments,* and *Haploxerults*—hardly conversational stuff, but with each syllable telling its own story.

Fig. 1-10. The soils of North America: A generalized reference map of
the soils according to the 7th Approximation.

1 Entisols
2 Vertisols
3 Inceptisols (Aquepts)
4 Aridisols (Argids)
5 Mollisols
6 Spodosols
7 Alfisols
8 Ultisols (Aqualts and Udults)
9 Histosols
10 Mountain and icecap areas; soils often absent

On the basis of the U.S. classification, however, it is possible to make a simple subdivision of the North American continent into five major soil regions:

1. *A broad belt of Inceptisols* north of 60° N latitude. Inceptisols are wet or frozen according to season and have little or no horizon of accumulation.

2. *A northern belt of Spodosols, Histosols, and Entisols,* occupying most of eastern Canada south of 60° N and covering much of New England and the Upper Great Lakes area of Wisconsin and Michigan. The Spodosols are the soils of the northern forests (which were formerly broadly classed as podsols); they are generally heavily leached and acid. The Histosols are boggy or peaty, with very slow rates of decay of organic matter; they occupy a broad area around Hudson Bay. The Entisols form the thin and patchy soil cover of Labrador, where glaciation has removed whatever original soils the region may have possessed.

3. *The Central Interior* (which is also the Agricultural Interior). This region is largely surfaced with soils of two orders:

(a) Alfisols, grading from what were formerly classed as gray-brown podsols and gray-wooded soils to less acid gray-brown types. Alfisols are usually moist and contain lower horizons of clay accumulation. They extend across the southern Great Lakes region, south of the Spodosols, and down the flanks of the Mississippi Valley as far south as Texas.

(b) Mollisols, which underlie the fertile farmlands of the western Interior and the grasslands of the Great Plains, from the Mexican border to the southern Prairies. Rich in organic matter and bases, the Mollisols account for most of the soils formerly classed by their color as black, chestnut-brown, or brown.

4. *A southeastern region of Ultisols,* stretching from Maryland in the north, west into Kentucky and southern Missouri, and southwest into eastern Texas. Low-latitude temperatures and high rainfall produce a range of soils which are subject to heavy leaching and clay accumulation so that, although they are widely cultivated in the southeast, they are not naturally very fertile; they are basically forest soils. They are often yellow or red in color, and some of them have weathered far enough to resemble the tropical laterites or latosols of the Oxisol order.

5. *A western belt of Aridisols.* As their name suggests, Aridisols are the soils of the deserts, stretching over the Southwest and the intermontane plateaus.

The Vegetation of North America

The varieties of rock type, relief, soil, and climate we have so far considered influence, in their turn, the vegetation of the continent.[6] To the early European settlers, North America was a land of forests. As the French along the St. Lawrence and the English on the Atlantic coast moved inland, they found little break in the monotony of the trees, and their economy became a forest-based economy, like that of the Indians of the area. Only after settlements moved far inland did the pioneers emerge from the forest into the "oak openings" and small prairies of the area south of the Great Lakes. Since, by this time, they had come largely to depend on the forest and its animal life for their livelihood, they regarded the treeless horizons with considerable suspicion, whereas the true grasslands further west found them, as we shall later see, almost wholly unprepared.

Only in the southwestern part of what was later to become the United States did the Spaniards, pushing north from Mexico, encounter open country. In southern Texas they entered, at its southwestern corner, a great triangle of grassland whose apex was in the oak openings of the Midwest and whose base lay along the foothills of the Rocky Mountains northward roughly as far as the 52nd parallel. To what extent, or in what sense, this area, especially its eastern margins, could be described as "natural" grassland is uncertain, for it seems clear that occupance by the Indians had altered its character; in particular, that burning had occurred. The coming of European settlers produced further changes—the clearance for agricultural use of much of the transitional forest zone bordering the grassland but, to balance this, an end to the burning practiced by the Indian hunters. Through all these changes, however, the basic climatic control of vegetation patterns can still be recognized.

THE FORESTS

Extent and Character

Before the coming of the Europeans it seems that forests covered slightly less than half of the area of what is now mainland Canada and the United States; that

[6]See J. L. Vankat, *The Natural Vegetation of North America* (New York, 1979), and a number of the yearbooks issued by the U.S. Dept. of Agriculture, such as *Trees* and *Grass.*

is, some 3 million sq mi out of 6.1 million (7.7 million sq km out of 15.8 million). In addition, rather more than a third of Alaska was forest-covered. Today this area has been reduced by exploitation and destruction to about 2.5 million sq mi (6.47 million sq km).[7]

This original forest cover was diverse in character, but almost continuous in extent over the eastern part and much of the north of the continent. Indeed, it is simplest to state that the forest extended north and west from the middle Atlantic Seaboard until it encountered one or another of three limiting circumstances: cold, dryness, or high altitude.

On the north, the forest thins out and gives place to the heaths and mosses of tundra vegetation, roughly along the boundary of the region of polar climate (that is, where no month has a mean temperature of more than 50°F, or 10°C). Here, the governing factors are the lack of summer heat, the dryness of the western Arctic, and the presence of the permafrost layer, which would prevent the roots of trees from penetrating to a depth sufficient to support growth. The northern forest boundary runs from the Mackenzie Delta to the middle of Hudson Bay and, from there, into northern Labrador (Fig. 1-11).

South of this line, a belt of forest spreads across the whole breadth of the continent, except in the West, where the cover is broken by the rise of North America's loftiest mountains—those of Alaska and the northern Pacific Coast. In Alaska and northern British Columbia, in consequence, the forests are largely confined to the interior valleys and the coastlands. This continent-wide forest cover extends south, with no other serious interruption, until it reaches the tree frontier imposed by increasing aridity in the Southwest.

This dry and largely treeless area lies south and west of a line running from Alberta, near Edmonton, to central Illinois. Along this boundary increasing aridity, especially seasonal drought, discourages tree growth. Moreover, where grassland has become established, young shoots of many tree species compete unsuccessfully with quicker-growing grasses.

From the apex of the triangle in Illinois, a southern boundary of forest–grassland transition runs west and south to central Texas. Along this boundary, as on the northern side, the luxuriant tree growth of the eastern Gulf Coast and the southeastern states is thinned by diminishing precipitation, increasing evaporation rates, and markedly seasonal rainfall, which favors the growth of grasses against that of trees. Until comparatively recent times, thinning of forests in the Southeast was also brought about deliberately by burning.

The remaining forest areas of North America can best be described as southward penetrations of the forests from north of the treeless area. They correspond to the areas of higher rainfall and reduced summer temperatures along the Rocky and Pacific mountain systems, and together they form an important part of the commercial forest resources of the continent.

The Rockies are, in general, tree covered on their middle slopes. The lower limit of the forest—represented by scattered and bushlike pinyon and juniper—is encountered between 4000 and 6000 ft (1200 and 1800 m), and the treeline is, according to slope and exposure, between 9000 and 11,000 ft (2750 and 3350 m), with alpine meadow and bare rock above it. Various forms of open forest are also found on the higher sections of the intermontane plateaus wherever rainfall is sufficient.

In the great ranges of the West are found the continent's largest and most valuable stands of timber. In the Pacific Northwest, the forests are continuous from sea level up to the timberline at 6000–8000 ft (1800–2400 m). In California, where there is less precipitation, an intermediate vegetation belt, known as chaparral and composed of bushes and small trees, intervenes before the true forest begins, about 2000 ft (600 m) above sea level.[8]

The coastal hills themselves are tree-clad roughly as far south as San Francisco Bay and intermittently beyond the Bay. In the Pacific Northwest, the coastal forests, like those of the Cascades inland, are composed largely of Douglas fir. But the distinctive feature of the California coast is the redwood belt, the

[7]See T. R. Cox and others, *This Well-Wooded Land: Americans and Their Forests from Colonial Times to the Present* (Lincoln, Nebr., 1985).

[8]M. G. Barbour and J. Major, eds., write, in *Terrestrial Vegetation of California* (New York, 1977), p. 420, "The Mediterranean type climate is the overriding environmental factor in the ecology of California chaparral." It covers 8.6 million acres (3.5 million ha), or one-twentieth of the state.

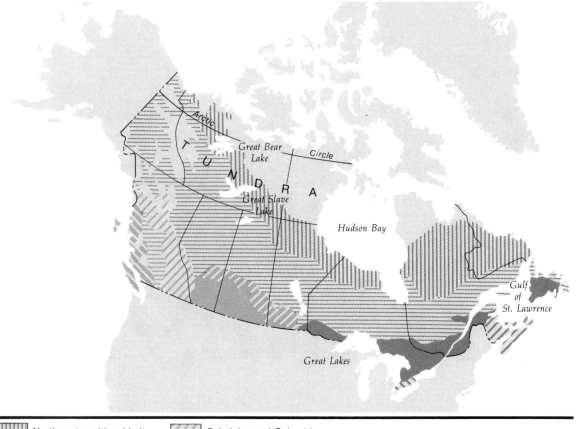

Fig. 1-11. Canada: Forest regions. The principal species represented in each forest division are: *Acadian*—spruce, balsam, yellow birch, maple; *Great Lakes–St. Lawrence*—pine, spruce, yellow birch, maple; *Deciduous*—assorted hardwoods; *Boreal*—spruce, balsam, white birch, poplar; *Sub-alpine and Columbian*—Engel- mann spruce, lodgepole pine, Douglas fir, cedar; *Montane*—ponderosa and lodgepole pine, spruce, Douglas fir; *Coastal*—Douglas fir, cedar, hemlock, spruce. (Based on the official map published periodically in the *Canada Yearbook*.)

habitat of the continent's largest tree species.[9] In the early days of California's settlement, these huge trees were obvious targets for lumbermen, and one of the

[9]There are actually two belts of redwoods, one along the coast, where the predominant species is *Sequoia sempervirens*, and the other midway up the western slopes of the Sierra Nevada, where *Sequoia dendron giganteum* is found. The former species is taller and exists in much larger, denser stands; coveted for lumber, they are being rapidly diminished. The smaller stands of the latter species are mainly preserved in national parks. The individual trees are shorter, but of much greater girth and age.

continent's first conservation movements sprang up to protect them. Today, the finest specimens are preserved, but in the coastal belt, which extends about as far south as Santa Cruz, and which seems to owe its presence to the humid conditions created by coastal fogs, lumbering is still steadily reducing this unique forest resource.

The northern forests are almost entirely coniferous in character. Those of the Great Lakes region and southern New England are mixed; like those of the Appalachians, they mark a transition to the deciduous forests of oak, beech, and hickory found in the

Vegetation types of North America: (1) The California chaparral. *U.S. Forest Service*

southern interior of the continent. In the Mountain West, the distribution of species—most of them coniferous, such as spruce, fir, pine, and larch—is governed largely by altitude. Yet no inventory, however brief, can overlook the magnificent stands of Douglas fir that spread densely over the whole coastal belt of the Pacific Northwest and that represent the most valuable single forest resource of North America.

The Forests as Resources

Of the 1 million square miles (2.5 million sq km) of forests remaining in the United States, some two-thirds are classed as "commercial," whereas in Canada, about 75 percent of 1.6 million sq mi (4.1 million sq km) of forest are said to be potentially productive. These forest resources have been tapped to supply the needs of a continent that consumes more than 50 percent of the pulp and paper production and over 40 percent of the lumber production of the world. Both nations began by possessing roughly equal areas of forest, but since both population and demand are ten times greater in the United States than in Canada, it is not surprising to find, first, that after three centuries of exploitation the United States has

only 70 percent of its cover and 10 percent of its virgin sawtimber left, and second, that it has come to rely heavily on the production of Canadian forests to meet its own gigantic demand for wood products. In fact, the annual drain on the sawtimber resources of the United States is 5 percent greater than the annual growth, while in the most valuable part of the nation's forests—the great softwood stands of Oregon and Washington—the drain has been running close to twice the annual growth.

In Canada, by contrast, the annual cut is only about one-half to two-thirds the annual growth, if only because half of the potentially productive forests remain to be exploited. The factor that governs their future use is not demand—that may be taken for granted—but accessibility. Exploitation naturally began in the areas of easiest access, along shorelines and the southern edge of the great boreal forest, whereas tomorrow's reserves lie, to a large extent, either in the western mountains or north of the St. Lawrence–Hudson Bay watershed, from where the rivers on which the logs would be floated run north into the wilderness.

The exploitation of the continent's forests has created industries of great size. Although the actual logging operations employ a good deal of labor, much of this employment, especially in eastern North America, is only seasonal. It is the processing of the forest products that has become so important. In Canada, wood products—lumber, wood pulp, and paper—form the basis of the country's largest group of industries, a group employing more than one-quarter of a million industrial workers, and contributing more than one-eighth of the value of the country's exports. In the United States, the forest-products industries account for 6–7 percent of the nation's industrial employees. Locally, of course, these industries play a far more important role than even these figures suggest. In Oregon, for example, they employed in 1982 nearly 38 percent of the state's industrial workers.

THE UNFORESTED WEST

In the semiarid and arid West, the natural vegetation is grass and shrubs.[10] Only on the mountain slopes, where precipitation is higher, or along rivers, where moister conditions are found, is tree growth encountered. For the rest it is the quantity of precipitation and its seasonal distribution that govern the character of the vegetation. This means that there is a sequence of vegetation types that holds good, either between the desert heart of the area and its more humid fringes, or between the lowest levels of the western basins and the lower tree limit on the mountains. Within these sequences it is possible to distin-

[10]See H. B. Sprague, ed., *Grasslands of the United States* (Ames, Iowa, 1974).

Fig. 1-12. Vegetation of the United States.

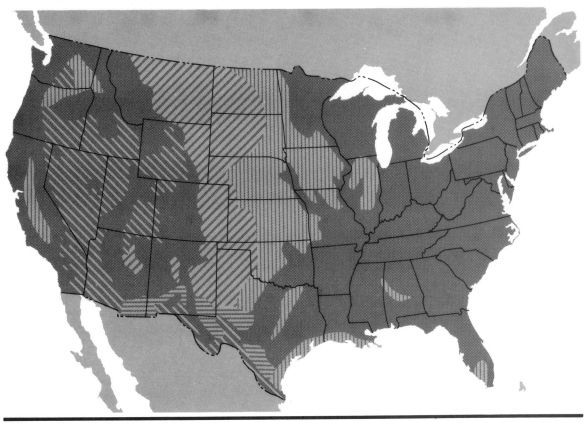

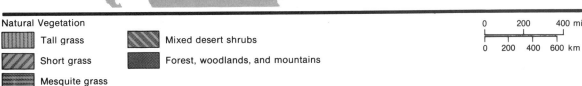

Natural Vegetation

- Tall grass
- Short grass
- Mesquite grass
- Mixed desert shrubs
- Forest, woodlands, and mountains

0   200   400 mi
0   200   400   600 km

Vegetation types of North America: (2) The saguaro or seguaro cactus, Tonto National Forest, Arizona. *U.S. Forest Service*

guish four main types of natural vegetation: tall grass, short grass, sagebrush with grass, and desert shrub (Fig. 1-12).

The tall-grass prairies, most of which have now been brought under the plow, occupied the eastern apex of the treeless triangle, where annual rainfall is between 20 and 35 in (500 and 875 mm). The vegetation of this region consisted of both tall and short grasses growing together, with such varieties as the bluestems *(Andropogon)* reaching 6 ft (180 cm) or more in height. The dense root network developed by these abundant grasses has produced the deep humus layer that makes the black soils of the tall-grass prairies famous.

Westward across the Great Plains, the natural tall-grass prairies gave way to grasses of medium and shorter height. Among a wide variety of species, some of the most common were wheatgrass *(Agropyron)*, grama *(Bouteloua)*, and buffalo grass *(Buchloë*

dactyloides). These native grasses can grow from a few inches to 3 ft (15–90 cm) high; their height is restricted by the shallower percolation of water in the 12–20-in (300–500-mm) rainfall zone and the consequent limitation of root development. As the grasses generally decrease westward in height and density, the soils that develop beneath them tend to decrease in humus content and, consequently, darkness of color and natural fertility. The flatter, more fertile areas are cultivated for wheat, often by dry-farming techniques: the vast, uncultivated, rougher or drier areas provide grazing for beef cattle.

Although the main area of short-grass vegetation is on the Great Plains, this type of vegetation is to be found in modified form in two other areas of significant extent in the West. One of these is in the southwestern United States, between western Texas and Arizona, where the vegetation is classified as semi-desert grass. It differs from the true short-grass prai-

Vegetation types of North America: (3) The tundra—a scene on Baffin Island. *National Film Board of Canada*

rie in its composition—dominant species are grama, dropseed *(Sporobolus)*, and curly mesquite *(Hilaria belangeri)*, growing, as on the Great Plains, from 1–3 ft (0.3–0.9 m) tall—and in that a scattering of shrubs or even small trees, particularly mesquite and creosote, generally accompanies the grass. Some 50 million acres (20.2 million ha) of semi-desert grass, unlike the areas of true desert adjoining them, receive sufficient summer rainfall (from the Gulf of Mexico) to support grasses; the combination of grass and shrubs thus marks a transition between the true grasslands and the scrub of the deserts.

Short-grass prairies are also found in the Pacific coastlands, especially the foothills surrounding the Central Valley of California (where they lie below the chaparral in altitudinal sequence) and in the lower Columbia–Snake River plains and basins of eastern Washington and Oregon. These are areas of Pacific bunchgrass, where grasses similar to those found on the Great Plains originally covered some 60 million acres (24.3 million ha). Most of these grasslands, however, have by now been plowed up or have been overgrazed by stock and invaded, in consequence, by sagebrush and shrubs.

The combination of sagebrush and grass is as much a feature of the Intermontane Region as short grass is of the Great Plains. It is a combination estimated to cover 250 million acres (100 million ha) of the West. The Intermontane Region, as we saw in an earlier section, generally has a winter–spring rainfall. Conditions that favor the growth of woody shrubs rather than grasses occur where the annual precipitation rate is low and the temperature and evaporation rate are high. The typical cover is, therefore, a combination of sage and grass, in which the proportions of the two vary according to rainfall and temperature, the grasses disappearing altogether on the desert margins.

Where the grasses maintain so precarious a hold, however, their disappearance may equally well be the result of overgrazing by livestock. The sagebrush, of which there is a great variety of types, each with wide climatic tolerance, offers little forage to stock (although game animals feed on it); it flourishes while the grasses decline. Western ranchers spray and beat the brush to try to keep the scrub down, but only careful management (discussed more fully in a later chapter) will allow the grass cover to flourish.

In the lowest and driest parts of the West, if and where a vegetation cover exists, it is composed of scattered desert shrubs, mostly woody in character. Over most of the dry area, creosote bush is common, while on the highly alkaline surfaces of salt deserts and old lake beds, varieties of sagebrush and other alkali-tolerant shrubs are found. In the deserts of southern California and southern Arizona are also to be found a variety of large cacti, such as saguaro and cholla, whose bizarre shapes make them one of the tourist attractions of the Southwest.

2
The People

Immigration to the United States: Candidates for citizenship recite
the Pledge of Allegiance during the naturalization ceremony, New
York, 1991. *U.S. Immigration and Naturalization Service*

The Native Americans

The earliest European immigrants to the New World found a native population already in possession. It is not clear how large this population was, but it soon became apparent that, as far as North America was concerned, most of it was concentrated on the plateau of Mexico. In the rest of the continent there were probably about 1 million inhabitants, of whom some 200,000 lived in what is now Canada. These latter included a few thousand Eskimos.[1]

Among these original Americans, however, there were wide differences of culture and language. Living close to nature, they had adapted their economies to their environment. In the North, the Inuit lived by hunting and fishing, whereas in the East, most of the Indian tribes combined hunting with a primitive agriculture that produced corn and squash. In the Southwest, by 1500, a remarkable urban culture—that of the pueblo-dwelling Indians—had reached and passed its climax, whereas elsewhere roamed tent-dwelling pastoralists or food-gatherers, such as those of the Great Basin of Utah.

These cultural variations, moreover, were not permanent. Neither were the locations of the tribes. The pressures of war, famine, or disease would force a tribe to move its hunting grounds, abandon its fields, or adopt a new form of economy. One of the most striking of these changes must have occurred about the year 1300, when a number of the largest pueblos in Utah and Colorado were apparently abandoned, swiftly, and for reasons—war or famine—that can only now be guessed at. But the greatest changes of all, both in location and in livelihood, were brought about by the use of the horse.

The Great Plains formerly had been the home of a

few pastoralist tribes. Across these plains roamed herds of millions of bison. So numerous were they as to be unaffected by the attacks of hunters moving on foot. But about the year 1600, horses—which had been introduced into the New World by the Spaniards—became available to the Indians, making possible for them a new way of life. Traveling on horseback and hunting the bison in groups, the plains Indians became the "new-rich" of seventeenth-century America. So strong was the attraction of this new way of life that other tribes moved into the area. From the eastern woodlands the Blackfoot and Cheyenne tribes trekked west, followed later by the Sioux and others; from the Rocky Mountain foothills came the Comanche; from the Southeast came tribes that abandoned agriculture for the new life of the plains. And hard on the heels of the last arrivals and, indeed, often responsible for the movement of the tribes, came the Europeans, to inaugurate a new era in the West.

From the eastern seaboard the tide of settlement flowed west. Whereas within the Spanish and French spheres of influence contact with the tribes had revealed a variety of motives—religious and strategic, as well as commerical—the English thrusts into the interior became increasingly land-orientated as time revealed the riches of the continent. As was so often the case in the era of European settlement overseas, it was the inability of European and Native American to understand the distinction between each other's views of *occupance* and *ownership* that underlay the culture contact and its often violent results. To the Europeans, the whole point of colonization was to convert land into private property. To the Americans, land was not "property" in any sense; it was simply an area within which a tribe might have a right of use, sometimes only for a short season before it moved on. As William Cronon has written of the earliest cultural contacts in New England, "English

---

[1]Most Eskimos call themselves "Inuit," which simply means "People." This name is coming into official and common usage as a preferred term.

fixity sought to replace Indian mobility; here was the central conflict in the ways Indians and colonists interacted with their environments. . . . More than anything else, it was the treatment of land and property as commodities traded at market that distinguished conceptions of ownership from Indian ones."[2]

The fate of the tribes whose land lay in the path of the white advance varied. In Canada, contact was on the whole peaceful. There were few "Indian wars," although it is possible to argue that this was not the result of any idealism on the part of the Canadian government, but rather that it simply reflected the fact that destruction of the bison had already reduced the Indians to starvation and docility before the prairies were settled. In the United States, conditions were less satisfactory. The westward thrust of settlers forced tribe after tribe off its hunting grounds, and although the policy of the government was, at most times, reasonable, its good intentions were constantly overtaken by the swiftness of the settlers' advance; another war followed and another tribe withdrew, broken, to the west. Lands reserved for the Indians were subsequently "needed" for settlers. Ultimately, the tribes were assigned reservations in the least desirable—which generally meant arid—sections, and the hunter–nomads were encouraged to become farmers, a change that nothing in their previous experience had prepared them for. Inevitably, the extensive system of Indian land use had to give way before the demands of the nineteenth century. But the manner of the change leaves abundant cause for regret.

The Native American population reached a low point at the end of the nineteenth century. Since then, it has expanded again to something over 1.7 million, although intermarriage has made definition difficult and, in Canada, has produced a distinctive population of *métis*, representing the mixture of Indian and European stocks. About 52 million acres (21 million ha) in the United States and 6 million acres (2.4 million ha) in Canada are now in Indian reservations; that is, lands secured to the Indians by treaty, but overseen by federal governments. On these reservations, the preservation of the Indian culture and life-style is both permitted and encouraged; yet that simple statement itself conceals not one, but

a whole series of dilemmas for the Indian. For one thing, "Indian life-style" is often a synonym for poverty—the reservation and its resources are inadequate to support a population at a standard of living in any way approaching Anglo-American norms. The dilemmas then unfold: To leave the reservation or to remain? To preserve the old way of life or to adapt to that of the white majority? To preserve the old ways for their own sake or as a tourist attraction? To maintain individuality, but at the cost of becoming living museum pieces? To pretend that there are simple answers to these questions would be naive in the extreme.

But there is a further complication. On some of the reservations, minerals have been discovered—oil, coal, or uranium—in significant amounts. There is a fine irony in the fact that lands the Indians received because no one else wanted them should now prove to be enormously valuable. It is an irony made more poignant by the impact of this wealth on an already weakened tribal structure and culture.

Nor is this all. The treaties made with the Indians by newcomers during three centuries can today be seen to have favored the latter; some lands were grossly undervalued, some rights were simply disregarded, and some treaties were imposed upon the Indians with no compensation. In the United States, there has existed, at least since World War II, an official means (the Indian Claims Commission) by which the Indians can claim compensation for past losses. Huge sums of money have been called for, and some much smaller sums have been awarded. Indians have also set in motion legal processes to establish historic title to the ownership of tribal lands, an action that, pending settlement in the courts, calls into question other land titles within the disputed areas, all the way from Maine to Alaska.[3]

The Immigrants

In the year 1800, there were about 5.5 million people in the territories now covered by the United States and some half million in Canada. Immigration from Europe had been proceeding for 200 years, to the French areas along the St. Lawrence, to the Spanish lands in the South and Southwest, or to the Atlantic Seaboard, where English culture and institutions

---

[2]W. Cronon, *Changes in the Land: Indians, Colonists, and the Ecology of New England* (New York, 1983). Quotations from pp. 53 and 75. See also Imre Sutton, ed., *Irredeemable America: The Indians' Estate and Land Claims* (Albuquerque, 1985).

[3]For general references on the North American Indian, the reader may be referred to the two works by C. Waldman, *Atlas of the North American Indian* and *Encyclopedia of Native American Tribes* (New York, 1985 and 1988, respectively).

dominated a cosmopolitan society that included Germans, Dutch, and Scandinavians.

For the first few years of the nineteenth century, the tempo of immigration remained slow. But for the United States, the hundred-year period between the end of the Napoleonic Wars and World War I was the century of immigrants—over 30 million of them arrived. In Canada, the main immigration came later. The first decade of the twentieth century saw 1.8 million arrivals and, since that time, only the two world wars and the depression of the 1930s have halted the flow.

M. L. Hansen, chronicler of the "Atlantic Migration," has pointed out that there were three main periods in the "immigrants' century."[4] From 1830 to 1860, the movement was largely one from the Celtic fringes of the British Isles and from the middle Rhine, especially Hesse. Scottish crofters, dispossessed by the advent of sheep farming, joined with Irish peasants, dispossessed by their landlords. This movement reached its peak after the terrible famine years of the 1840s. Then, from 1860 to 1890, Englishmen mingled with German and Scandinavians in the second great wave. Finally, between 1900 and 1914 (and in Canada between 1900 and the present day), the majority of the immigrants came from the Slavic countries of Eastern Europe and from the Mediterranean lands. In the United States a peak was reached in 1907, when 1,285,000 immigrants were admitted.

World War I created upheavals in Europe on a gigantic scale. Immigration to the United States, which had fallen off during the war years, was, by 1921, up again to 805,000, and, had emigration been unrestricted, there is no doubt that millions would have left Europe for America. In view of this prospect, the United States felt it necessary to limit, by quota, the number of immigrants. The quota, as fixed in 1927, permitted the entry of only 150,000 immigrants each year. Only for persons born in Canada, Mexico, and the Latin American countries was entrance to the United States unrestricted by quota.

Since 1927, a quota system in various forms has continued to operate, but special provision has been made for the admission of political refugees and relatives of those already admitted. As a result, in 1989 total immigration to the United States stood at 1,091,000.

But there is much more to this story than the round

[4]M. L. Hansen, *The Immigrant in American History,* and *The Atlantic Migration, 1607–1860* (Cambridge, Mass., 1948 and 1951, respectively).

figure of a million-plus. We have described the years from 1815 to 1914 as the immigrants' century. One day, historians will have to decide what to call the years from, say, 1970 onward. It is unlikely to be a second "century" for, long before a hundred years have passed by, the U.S. government will probably have been obliged to shut the doors. What is certain is that this new period of immigration is, and will be, quite different from the earlier one. The great movements of the nineteenth century were almost wholly made up of Europeans. In 1989, however, the latter made up less than 8 percent of all immigrants. The new tides are flowing from two other continents: Asia (312,000 arrivals) and America north of Panama (607,000 in 1989, of whom 405,000 were from Mexico). And whereas the Europeans arrived mainly at the ports of the Atlantic coast, spreading inland from there, the new tides affect, above all, entry points on the Pacific coast and the Gulf of Mexico. Nowadays, Los Angeles admits two to three times as many immigrants annually as does New York. To these facts we shall have later to return.

There have been Asian arrivals on the Pacific shore for more than a century. Chinese labor virtually built California between 1850 and 1880, and both San Francisco and Vancouver have large, long-established Chinese communities. But the scale of the new movement is entirely different, and its ethnic makeup much more diverse. Not for nothing has Los Angeles been referred to as the Ellis Island of the 1980s, the point of arrival in the New World of today's immigrants, just as New York was for those of the past (see Chapters 11 and 23).

In Canada, a policy designed to foster the growth of population brought in two main waves of immigrants. The first of these was between 1900 and 1914, when the prairie wheatlands were being opened up, and it reached a peak of 400,870 arrivals in 1913. It was largely British, but with many eastern Europeans, and by the time it subsided, the most common languages on the prairies, after English, were German and Ukrainian. The second wave of arrivals came in the years after 1945: up to the end of 1969, this wave had brought into the country nearly 3.3 million immigrants, with a peak of 282,000 in 1957. About one-third of this total had come from the British Isles, but the British element was becoming smaller in proportion to the whole. In the past decade, in any case, immigration from Europe has declined, whereas that from Asia has increased, as it has in the United States. Canada's Immigration Act of 1978 dispensed with all ideas of national quotas or countries of origin: kinship ties and the possession of marketable skills are the criteria now used to admit between 100,000 and 250,000 each year. In 1989, the figure was 190,000 of whom roughly 40,000 were from Europe and 80,000 from Asia. In every decade, however, a considerable number of Canada's immigrants have subsequently moved southward and settled in the United States.

Where, in fact, in the New World were these immigrants to settle? Although the ports of entry had a cosmopolitan population in transit, away from the coast there was a tendency for immigrants to settle in "national" areas. West of Lake Michigan, for example, people of Scandinavian stock are in a majority over much of Wisconsin, Minnesota, and Iowa, whereas other areas are markedly German or Finnish (Fig. 2-1). In some cases, this is simply because the earliest arrivals encouraged their friends at home to join them. In others, it is a product of organized, or group, emigration like that of the Pennsylvania Dutch or the Mennonites. In the emigration zones of Europe, there was a system of recruiting that provided everything needed for door-to-door resettlement. In yet other cases, it reflects the *time* of arrival in North America. As settlement extended across the continent, so each decade encountered its own economic frontiers. For example, by 1850, the westward movement in Canada was temporarily halted against the margins of the inhospitable Laurentian Shield north of the St. Lawrence Valley and Ontario Peninsula. During the following years, as a consequence, agricultural settlers tended to be diverted southward into the U.S. Midwest, until a truly Canadian farm frontier was reopened on the Prairies in the 1880s.

The 1860s saw the beginning of the Scandinavian immigration. Those years saw also the exploitation of the timber resources of the Great Lakes area, and the Swedes and Norwegians went there as lumberjacks and frontier farmers. The Ukrainians, who arrived in the 1870s, were recruited to work in the rapidly expanding coal-mining areas of Pennsylvania. And after 1900, when agricultural lands in the United States had been largely taken, settlers poured into the newly opened Canadian Park Belt and prairies,[5] not only from eastern Canada and eastern Europe, but also from the United States (see Chapter 18).

Wherever possible, the immigrants sought work and conditions comparable with those they had left in the Old World. But the great immigrant problem was that so many of the newcomers were European peasants possessed of a single skill, and that skill was in limited demand in a society that was both mechanizing its agriculture and rapidly becoming industrialized. All that many could offer was the strength of their arm, and so they tended to drift into poorly paid labor gangs and were often separated from their families. For many, the long process of settling in the paradise of the West was hard.

The one area of the United States where they might have practiced their peasant farming was the South. But in that region, the climate and the system of agriculture were uninviting to northern Europeans, and cheap farm labor was provided by the blacks. For these and other reasons, very few nineteenth-century immigrants settled in the southern states. The effect of this virtual exclusion of immigrants from the southeastern states is graphically illustrated by Figure 2-1, which is taken from the atlas accompanying the census of 1870.

The Melting Pot—or Not?

The well-worn phrase "the melting pot" describes the way in which elements of the population of four continents have been set down in the United States and have there merged into one distinctive American society. The phrase rings less true for Canada, for there the melting pot, although real enough, has been devoted to preserving two particular elements—the British and French connections. In Canada, people of British and French stock make up 70 percent of the population. In the United States, where in 1790 people of British origin represented 80 percent of the population, the 1980 figures[6] were:

English and Scots	26 percent
Germans	21 percent
Irish	17 percent
African-Americans	9 percent

[5]Which they chose—Park Belt or prairie grass—depended to some extent on their background. Most chose the Park Belt, with its open spaces interspersed with woodland; that is, with timber for fuel and building. The treeless prairie was a more daunting proposition. Some of the earliest undaunted settlers, however, were Russian Mennonites who settled in the late 1870s and succeeded (1) because of their Russian experience of breaking grasslands, and (2) because they settled in groups and tackled their tasks communally.

[6]If to these figures we add those for the *foreign-born* population of the United States at the 1980 census, then we find that there were 14.1 million foreign-born residents, the largest numbers from (1) Mexico, 3.0 million; (2) Germany, 0.85 million; (3) Canada, 0.84 million; (4) Italy, 0.83 million; (5) United Kingdom, 0.67 million; and (6) Cuba, 0.61 million.

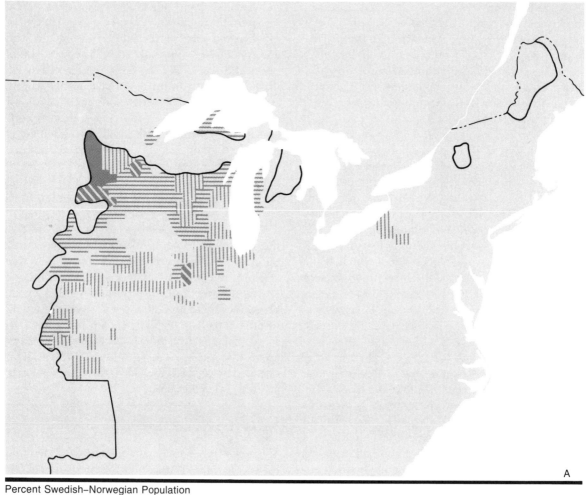

Percent Swedish–Norwegian Population

	1–4%		Over 25%
	4–15%	——	Limit of Census Area
	15–25%		

A

There has always been a certain opposition in the United States to the admission of new immigrants. On the whole, however, this reaction has been surprisingly slight, considering how unprepossessing was the appearance of many of the new arrivals, bewildered peasants disembarking after a nightmare Atlantic crossing. At the times of greatest influx it must, indeed, have appeared that American culture and institutions were in danger of being swamped by aliens. There have, in fact, been efforts to halt immigration—by old colonial elements fearing the introduction of so much Mediterranean and Slavic blood into Anglo-Saxon America and, more reasonably perhaps, by those who feared that the introduc-

tion of so much labor into the economy would undermine the position of the American worker. That this fear has never become reality is probably because the growth of the United States has, in the past, been so rapid that labor has generally been in short supply.

Earlier immigrants readily adapted themselves to American political and social ideals and were anxious to prove themselves good citizens—some hundreds of thousands did so in the Union armies in 1861–65. At the same time, wherever a group of fellow-countrymen congregated, they kept alive some remnants of the culture of their homeland—a church, a newspaper in their own language, or a group for folk

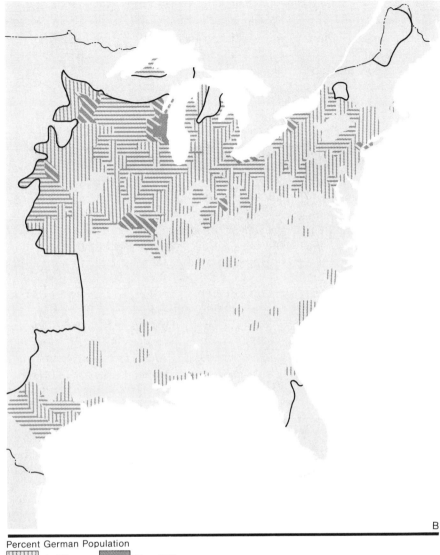

Fig. 2-1. Foreign-born population of the United States, census of 1870: (**A**) Swedish–Norwegian-born and (**B**) German-born. These maps, from the *Statistical Atlas of The United States*, published in 1874 to embody the results of the Ninth Census, show two features of the population distribution: (1) the tendency of European immigrants to congregate in particular areas, as a result of the work opportunities available then or the influence of earlier arrivals, and (2) the tendency of European immigrants to avoid the South. Even allowing for the fact that the Civil War had just ended, the almost complete absence of a foreign-born population in the region is striking.

B

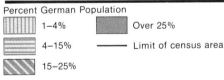

Percent German Population

‖‖‖ 1–4%	▨ Over 25%	
▤ 4–15%	—— Limit of census area	
▥ 15–25%		

dancing or singing. In general, the larger the alien community, the slower the process of assimilation and the stronger the nostalgic nationalism. But for the immigrants' children, the situation was very different. With no language barrier to overcome, they made wider social contacts than their parents. For them the cultural relics of the homeland, like their parents' accents, were largely curiosities. The influence of the school quickly overcame the alien influ-

ence of the home and produced, as it was intended to produce, a new generation of Americans.

In the 1970s and 1980s, however, this situation altered dramatically. There have been two main reasons for the change. The first is that the immigrant is much easier to recognize, now that he or she comes from Vietnam or Korea, rather than from the countries of white Europe. After a few weeks in North America, the nineteenth-century immigrant had every possibility of passing as American: nothing in his or her appearance prevented this. But the new wave is highly visible—and in consequence, more easily resented. Part of the resentment arises, of course, because a large number of these immigrants, from both Asia and Central America, have been admitted as political refugees, helped in their arrival and settlement by the government. Meanwhile, many native-born Americans struggle, unsubsidized, for work and survival.

The second, and more widespread, factor of change is the growth of what is usually called a "new ethnicity." The North American peoples are seeing this matter of assimilation to a single, "national" character in a new light. Young countries foster nationalism as a defensive measure, but once they have consolidated themselves and proved their power and cohesion in the political world, it becomes clear that they can support a greater weight of cultural diversity than was at first thought possible.

Meanwhile, the nation's various ethnic minorities, becoming more self-aware but recognizing, for their part, that they will never individually become a majority, have chosen to stress their distinctiveness; that is, deliberately to distance themselves from the national model. They demand that their separateness be recognized by others; that their own history and culture shall form distinct features of the program in classroom and college; that they shall be represented on national bodies; and that only they can and shall speak for their people.

The number of such minority groups seems to increase year by year, making the "melting pot" image of North America more and more out of date. The last thing that these groups want is to be melted down to a common metal. And so there has occurred a process by which the new ethnicity has been variously described, in its national impact, as fragmentation, tribalization, self-ghettoization, or Balkanization—the disuniting of the once-united. The major groups involved are the French Canadians, the African Americans, and the Hispanics, but the minor players are numerous: in fact, this is a game any number can play.

One of the ironies of this situation is that while all these minorities are busy stressing their separateness, *together* they form a majority in many places. In 1990 in the United States, there were nearly 200 counties where the non-Hispanic white population was outnumbered by the *combined* minorities. Some forecasts predict that by the year 2060 the entire United States will have a "minority majority." But whether the minorities will ever be prepared to make common cause on the political level is extremely doubtful. The competition among them for jobs, housing, and status is too intense.

The next three sections of this chapter will consider the continent's three largest minority groups in more detail.

The Case of French Canada

To every mention of the melting pot and every suggestion of assimilation to a common cultural denominator, one group has remained doggedly opposed—the French Canadians. United culturally by language and religion and represented politically by Québec, the second most populous province of Canada, they have succeeded, as has no other national group, in maintaining their distinctiveness through every vicissitude of the continent's political history.[7]

That the French Canadians have achieved this privileged position is due, politically, to their bargaining power—their numerical strength and their concentration in one key province. This power they exerted in 1867 (when about one-third of all Canadians lived in Québec Province) to ensure that the new system for the government of all Canada should be federal and that the new Province of Québec should preserve intact its distinctive institutions. French Canada, then, is a product of a federal constitution and of the determination of a powerful minority, a minority to whom, as G. S. Graham wrote, "national survival had become the dominating passion."

The heart of French Canada, today as in the seventeenth century, is along the Lower St. Lawrence. The area reaches upstream through Montréal, which is a bilingual metropolis, to a short distance west of the Ottawa River. South of the St. Lawrence, most of the settlement was originally British, but this area has now been largely taken over by French Canadians, who have also spread into New Brunswick and over the border into New England. By contrast, the originally French settlements around the Bay of Fundy

[7]There is a large literature on the French Canadians, though most of it is, naturally, in French. On the cultural side, which is dealt with in this chapter, see F. Mason Wade, *The French Canadians* (rev. ed., Toronto, 1968); R. Breton and others, *Cultural Boundaries and the Cohesion of Canada* (Montréal, 1980); R. Lachapelle et al., *The Demolinguistic Situation in Canada: Past Trends and Future Prospects* (Montréal, 1982). For other references, see Chaps. 4 and 13.

(Acadia) are now nearly empty of French population. Elsewhere, there are clusters and pockets of French Canadians. St. Boniface, for example, across the Red River to the east of English-speaking Winnipeg, retains much of its French-Canadian culture and inheritance, including the long lots running down to the riverbank that reproduce the landholding pattern of Québec. In northern Alberta, similarly, there are French communities, the product of a series of migrations westward from Québec in the period 1900–50, many of which were organized and led by the priest of the home parish in the east. Thirty percent of all Canadians have French as their mother tongue. In Québec Province the figure is 83 percent, in New Brunswick 33 percent.

This cleavage at the heart of the Canadian federation is vitally important as a cultural reality. How important it is as a *political* reality rather depends on the date, for its strength has tended to wax and wane over the years. We shall return to this subject in later chapters. For the moment, we need only note that it grew out of the conflicts of colonial days, and was given its dimensions not only by language but also by religious differences and by two separate school systems. Originally, it was a division between a French Catholic rural population and a British non-Catholic population that dominated much of the urban and commercial life of the region. Yet, although the church and especially the countryside no longer exercise so strong an influence on the French-speaking population, the cleavage continues to be real and, sometimes, bitter.

There are several interesting aspects of this special position of French Canada. One of them is that among the French Catholic families the birthrate has traditionally been high—a good deal higher than for Canadians as a whole. (During the period 1910 to 1930, the differential between Québec and the nation was about eight per thousand.) There seemed a possibility that the French population might ultimately grow to outnumber the remainder, and the statistics of population increase were watched with close attention, especially in the years following World War II, when many of the immigrants from Europe were Roman Catholic. However, during the past two decades, the Québec birthrate has been falling, and since 1961 has been consistently below the national average in every year.

Militating against this numerical increase is the second factor of change in the situation. In the past, the strength of French-Canadian culture lay in rural settlements, with their traditional way of life. In the villages of the French *habitants*, the two most powerful influences were the church and the family, and there was little to challenge them. But the increase in population has forced sons and daughters to seek employment elsewhere, while the cities offer to them—as to the younger farm population everywhere—a wider variety of interests and prospects than is to be found at home. Today Québec's population is more than 75 percent urban.

So there has been a move to town, which has inevitably weakened the ties with the traditional culture. For one thing, the old controls are relaxed; for another, much of the industry of the cities is owned and managed by non–French Canadians.

There is a third factor—the choice of language in everyday affairs. In French-Canadian homes, and in the heartland communities of Québec and northern New Brunswick, there is no threat to the use of the French language; no difficulty in maintaining it. But in the world of commerce or industry it becomes a problem: the French Canadian is doing business in a continent almost wholly English-speaking. Canada is officially bilingual: North American business is not. When that business is done in Québec, French Canadians can, if they wish, insist on its being transacted in French. But if they, in their turn, wish to buy or sell, advertise or export, in the rest of a continent where they form only 2 percent of the population, they must switch to English to do so.

Even more difficult is the issue of education. Where French speakers predominate, it is entirely proper that French should be the language of the school. But what happens where there is a rough balance between French- and English-speaking children? And what is to happen to an immigrant family from, say, eastern Europe that settles in Québec? The children must learn French as their first "American" language, but it is obvious that if, as so many do, these immigrants move on further into the continent, English will become the language of opportunity for them. The cost of cultural separatism is high.

The African Americans

In 1990, there were almost 30 million African Americans or blacks in the United States, and they formed 12.1 percent of the population, compared with 10.5 percent in 1960. They were the descendants of the slaves brought into North America by the slave traders of Spain, Portugal, and England from 1600 on-

Immigration to the United States: Ellis Island, in New York Harbor, where millions of immigrants sought entry into the United States during the period of maximum immigration. The picture was taken in 1905 at the height of the influx. Today the main building has been turned into a museum. *Library of Congress*

ward, until the trade was outlawed in the nineteenth century.

In the early days of the colonies, there was a pressing need for labor. We have already seen that, in North America, as opposed to mainland Middle America, the native population was sparse; moreover, it was generally hostile toward being enticed or forced into unfamiliar, regimented labor. The Spaniards early resorted to importing African slaves into their Caribbean colonies, but in the English colonies, the process was more gradual. Originally, the labor shortage was met by the *indenture* system. An individual would contract to work for a colonist for a limited period in exchange for his passage and subsistence, and at the end of his term, he would become a free settler. It was only gradually that this temporary servitude of Europeans came to be distinguished from that of the blacks, for whom there was no terminal date. The first Africans were brought to Virginia in 1619, but slavery was not made legally hereditary until 1662. For Massachusetts, the comparable dates were 1636 and 1641, and in 1705 a law was passed classifying slaves as a form of real estate.

By 1790 blacks made up almost 20 percent of the U.S. population.[8]

It was in the plantation states of the Southeast that slavery flourished. There, in the hot, humid climate, the plantation owners needed abundant and acclimatized hand labor for the cultivation of tobacco, rice, and cotton. Any questions that existed in the Southerners' minds about the moral legitimacy of slavery were answered by the demand for labor, as cotton growing spread across the southeastern states in the early nineteenth century. Perhaps the greatest tragedy of all for the blacks was that the South committed itself so completely to slave-based cotton cultivation that, in the end, in spite of many a voice raised in protest against the system, it was widely believed that it could not discard slavery and survive.

The Civil War and emancipation left the black

[8]Probably the two best studies of the black experience in the United States for our purposes remain as J. Hope Franklin, *From Slavery to Freedom* (New York, 1952); and C. Vann Woodward, *The Strange Career of Jim Crow* (3d ed., New York, 1974). See also R. T. Ernst and L. Hugg, eds., *Black America: Geographic Perspectives* (New York, 1976).

population solidly concentrated in the Southeast; nominally free, but tied economically to the same cotton lands as before the war. Today, over a century later, there is still a concentration of blacks in the Southeast, but it is much less marked. One of the most notable population movements of post-1914 North America was the migration of blacks from southern farms to northern cities. The beginnings of this movement belong to the period before the Civil War, but the event that may be said to have totally altered its scale was the drying-up of immigration during World War I, which created a shortage of labor in the expanding war industries of the North. Large-scale urbanization of the blacks dates from this time: in 1920, there were almost a million more blacks classified as urban dwellers than there had been twenty years earlier.

World War II gave a fresh impetus to the movement, which continued for the next three decades. It produced large black populations in all the major northern cities (see Table 2-1), and whereas in 1910 only 27 percent of the black population was classified as urban, the figure for 1980 was 82 percent, which was above the average for the United States as a whole. And whereas in 1910, 89 percent of the black population lived in the South, in the old slave states, by 1980, only about one-half did so (Fig. 2-2).

Now times have changed again. The 1980 and 1990 U.S. censuses made it clear that the black migration northward has slackened, halted, and reversed. This is not to say that no blacks are moving north, but that they are outnumbered by those returning south. To general astonishment, the 1990 census revealed that Chicago, of all places—Chicago, with its 1.3 million African Americans—had lost black population during the decade.[9]

[9]For fuller commentary on this, see J. Cromartie and C. B. Stack, "Reinterpretation of Black Return and Nonreturn Migration to the South, 1975–80," *Geographical Review*, vol. 79 (1989), 297–310; and K. E. McHugh, "Black Migration Reversal in the United States," *Geographical Review*, vol. 77 (1987), 171–82.

The Canadian ethnic mosaic: Ukrainian church in eastern Alberta. More than half a million Canadians are of Ukrainian origin. *J. H. Paterson*

Table 2-1. United States Metropolitan Areas* with the Largest Minority Populations, 1990 (numbers in millions)

Area	African-American Population	Area	Hispanic Population	Area	Asian Population
1. New York PMSA	2.25	1. Los Angeles–Long Beach PMSA	3.35	1. Los Angeles–Long Beach PMSA	0.95
2. Chicago PMSA	1.33	2. New York PMSA	1.89	2. New York PMSA	0.56
3. Washington, DC MSA	1.04	3. Miami-Hialeah PMSA	0.95	3. Honolulu MSA	0.53
4. Los Angeles–Long Beach PMSA	0.99	4. Chicago PMSA	0.73	4. San Francisco PMSA	0.33
5. Detroit PMSA	0.94	5. Houston PMSA	0.71	5. Oakland PMSA	0.27
6. Philadelphia PMSA	0.93	6. Riverside-San Bernadino PMSA	0.69	6. San Jose PMSA	0.26
7. Atlanta MSA	0.74	7. San Antonio MSA	0.62	7. Anaheim–Santa Ana PMSA	0.25
8. Baltimore MSA	0.62	8. Anaheim–Santa Ana PMSA	0.56	8. Chicago PMSA	0.23
9. Houston PMSA	0.61	9. San Diego MSA	0.51	9. Washington, DC MSA	0.20
10. New Orleans MSA	0.43	10. El Paso MSA	0.41	10. San Diego MSA	0.20
11. St Louis MSA	0.42	11. Dallas PMSA	0.37	11. Seattle PMSA	0.14
12. Newark PMSA	0.42	12. Phoenix MSA	0.35	12. Houston PMSA	0.13

*For definitions of those areas, see Chapter 3.

We may explain in a variety of ways the halting of this, one of recent history's great population movements (see Table 2-2.) It is evident that the black population of the United States has its own structure and dynamics. The original move north could readily be explained in terms not only of job opportunities, but also of word-of-mouth encouragement to individuals by such stimuli as the "On to Harlem" movement (see p. 227). But such a move simply spread the problems of race relations to new areas and introduced to the northern states (where, in the past, most people had been free to theorize about race) the questions that had long dominated the South: whether blacks can compete for jobs on equal terms with white workers, and whether blacks should be free to live on any street where they can afford to buy a home.

Confronted with these questions, and the blacks whose presence posed them, the North reacted defensively. At the same time, the supply of jobs began to dry up: as we shall later see, urban-industrial job opportunities were no longer in the North, but rather in the South and the West. The great migration north lost its economic point with the regional shift in employment, and it lost its social point when, because of a curious combination of Civil Rights legislation and northern defensiveness, the conditions of life in northern cities proved to be no better, and often worse, than in southern towns. It looks, in other words, as if the 1970s and 1980s will prove, in retrospect, to be the decades of "Back from Harlem."

Not that these blacks will be returning to their old homes, or old occupations, in the South. As we shall see in Chapter 15, there have been great changes in the old slave states, and one of them, perhaps the greatest, has been the disappearance of the small farms from which the blacks in past decades left for the North. The returning flow is to the cities and towns where, if anywhere, there are jobs and homes.

The blacks of the United States form, in fact, a population within a population. This shows, firstly, in their demographic rhythms. For the past twenty years, they have experienced an annual rate of increase 0.5–0.6 percent higher than for the population as a whole. In 1985, their birthrate was 21.1 per thousand, compared with 14.7 for whites. Their overall death rate was lower than for whites (8.3 against

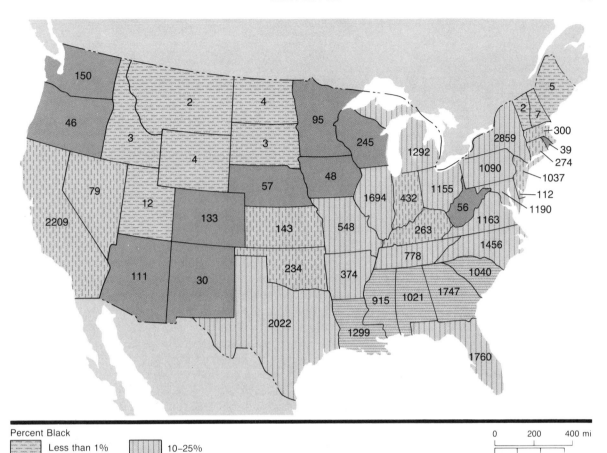

Percent Black

Less than 1%   10–25%

1–5%   More than 25%

5–10%

0   200   400 mi

0   200   400   600 km

Fig. 2-2. The United States: Distribution of African-American population, census of 1990. The cross-hatching indicates the percentage of each state's population that was African American in 1990: the number in each state gives the actual black population, in thousands.

9.0), but only because they are a younger population: the infant mortality rate among blacks is 1.9 times that among whites, and their life expectancy is about six years less than for whites of the same sex. They resemble whites, Hispanics, and Indians in that their rate of natural increase is linked to their educational level, and especially to that of their women: the more education, the fewer children (Table 2-3).[10]

[10]Nearly all these statistics are found in, or can be calculated from, the census, but it would be ungrateful of the present author not to recognize that much of the hard work of assembling them has been done by others, e.g., the U.S. Bureau of the Census, *The Social and Economic Status of the Black Population of the United States . . . 1790–1978* (Washington, D.C., n.d.), and J. Reid, *Black America in the 1980s* (Washington, D.C., 1982), from which Tables 2-1 and 2-2 are taken.

Table 2-2. Net Migration of Blacks, by Region, 1910–20 to 1975–80 (numbers in thousands)

Period	South	Northeast	North Central	West
1910–20	− 454	+182	+244	+ 28
1920–30	− 749	+349	+364	+ 36
1930–40	− 347	+171	+128	+ 49
1940–50	−1,599	+463	+618	+339
1950–60	−1,473	+496	+541	+293
1960–70	−1,380	+612	+382	+301
1970–75	+ 14	− 64	− 52	+102
1975–80	+ 195	−175	− 51	+ 30

Table 2-3. Fertility of Black and White Women, by Education, June 1981

Years of School Completed	Black*	White*
Elementary		
0–7 years	3879	2512
8 years	4605	3209
High school		
1–3 years	3919	2988
4 years	3031	2536
College		
1–3 years	2510	2335
4 years or more	1730	1857

*Children ever born per 1000 women aged 35–44.

Economically, they form just as distinctive a group. For decades now, their unemployment rate has been double that for whites, and their average family income has never risen above 62 percent (in 1975) of the white level. A black person is about three times as likely as a white to be living below the poverty level. In 1985, the figures were: whites, 11.4 percent, and blacks, 31.3 percent (with Hispanics at 29 percent). Their involvement in the professions and higher grades of management is very small. Despite considerable advances in education and the removal of occupational barriers, there is little sign of change as yet in the critical economic indicators.

Socially, black distinctiveness is most prominently marked by the location of black homes; specifically, by the development of the *ghetto*. We shall consider this topic in the next chapter, when we deal with the American city.

In recent years, two counteracting influences have been at work on the black population of the United States, as a result of which their separate identity is at least as clearly marked as it was half a century ago, if not more so. One has been the passage of civil rights legislation, which has removed legal or formal barriers that disabled blacks, but left the informal, social barriers about which it is difficult, if not impossible, to legislate. The other has been the wish of the younger black population precisely to assert its distinctiveness; that is, to make a point of stressing the ways in which it differs from the white majority. In these days of a "new ethnicity," blacks, like a number of other groups far less easily recognizable within the American community, have rejected the idea of a mere surrender to the majority culture and demanded the right to be themselves.

The Hispanic Americans

The growth of a large population in the United States which is Spanish-speaking and born south of the Rio Grande has been one of the most striking demographic features of recent American geography (Fig. 2-3). Between 1980 and 1990, their census numbers increased from 17 to 22.4 million, which means that they form a minority not much smaller than the African American, particularly when we consider that the real total is a good deal more than 22 million. (Many of the Hispanics have good reasons, as we shall see in Chapter 20, for not wishing to catch the census taker's eye.) In the Los Angeles metropolitan area, which easily leads the nation's cities with a population of 3.3 million Hispanics (see Table 2-1), the single decade 1980–90 saw an increase of 1.3 million. California, Texas, and Florida are the states receiving the bulk of the newcomers: New York, Chicago, and Washington, D.C., are other major goals.

The pressure of Spanish speakers along the United States' southern borders is, of course, nothing new: much of the territory there formerly belonged to the Spanish Empire, anyway. What is remarkable is the speed of the recent buildup. The Hispanic newcomers divide into several groups—Mexicans who cross the border to move from poverty to imagined riches; immigrants, some of them refugees, from troubled Central American countries like Nicaragua; and Cubans who either escape or are encouraged to leave. It is the Central Americans who, on the whole, make for the cities of the Northeast, where they are increasingly in evidence serving in fast-food outlets with short, simple menus (for there is a language problem), or in hotels. The Cubans have settled in Florida, almost a million of them in the Miami area: five of the seven counties where Hispanics showed the largest percentage increases in the 1980s were in Florida.

It is probably because the source areas of this Hispanic invasion—and invasion seems to be the right word—are so scattered that there has not so far been a Hispanic group consciousness comparable with that of the blacks at the political level. The Californian, Texan, Floridan, and New York Hispanics share a common language and culture, and they have the strength of growing numbers, but they are clustered in separate groups across the whole breadth of the United States and, economically, they are preoccupied, competing with other minorities for

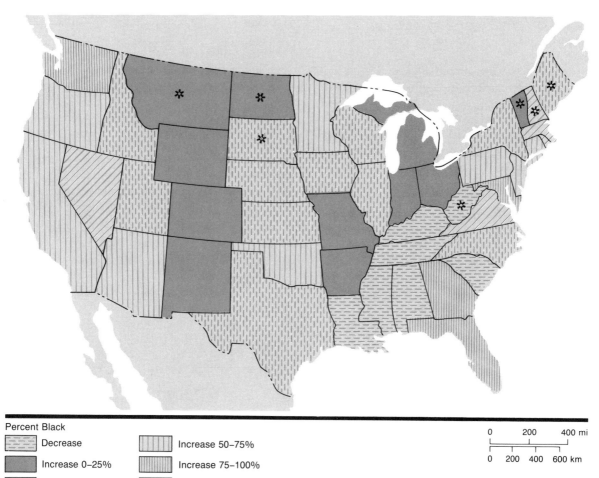

Percent Black

Decrease		Increase 50–75%	
Increase 0–25%		Increase 75–100%	
Increase 25–50%		Increase 100–135%	

0   200   400 mi

0   200   400   600 km

Fig. 2-3. The United States: Change in percentage of the population which is of Hispanic origin, 1980–1990. For the nation as a whole, the percentage increase was 53.0. For Alaska and Hawaii, the changes were 87.3 and 14.2, respectively. Some states, however, had in 1990 a negligible Hispanic population: these are marked with an asterisk on the map.

the more low-paid jobs. Furthermore, the status of thousands of them as illegal immigrants ensures that they avoid, rather than seek out, publicity or political action. This is not to say that, in the future, we may not witness the emergence of a united Hispanic front, merely that it has not yet happened. We shall discuss this question further in Chapter 20.

Distribution of Population: Canada

The population of Canada at the census of 1991 was 27.3 million, or 3 million more than at that of 1981. For 3.8 million sq mi (9.8 million sq km) of territory, however, this represents a very low average density of population. And for vast areas of the country,

even this figure is misleading, for there are population clusters along the southern Canadian border, with huge northern expanses virtually uninhabited.

From either a strategic or an economic point of view, this distribution of population is, in certain ways, highly unsatisfactory. The Maritime Provinces are linked to the rest of populated Canada only by a narrow corridor of Canadian territory between the St. Lawrence and the border of Maine. The capital of Canada's most eastern province, Newfoundland, lies over 1000 mi (1600 km) by sea from Québec and more than 600 mi (960 km) from Halifax. The broad expanse of the Laurentian Shield along the northern shores of Lake Superior and Lake Huron virtually cuts Canada in two. Beyond the Prairie Provinces rise the Rockies and the Coast Mountains of British Columbia as further obstacles to movement and dense settlement.

The Prairie Provinces account for some 17 percent of Canada's population, and the Atlantic Provinces for 8.5 percent. Over 60 percent of the total, however, are to be found in Ontario and Québec; to be precise, in the southern fringes of these two provinces. Along the St. Lawrence Valley and in the peninsula of southwestern Ontario is more than one-half of Canada's population (Fig. 2-3).

That these widely separated groups of settlers ever formed the Canadian Federation may be taken as a tribute to skillful diplomacy and as a measure of the fear inspired by the military strength of the United States after 1865. It has often been remarked that it is easier for each of these Canadian population clusters to communicate southward with the United States than with its neighbor on either side; it is certain that, up to 1962, the normal highway route from east to west, for Canadians, ran south of the border. Moreover, even when the nation has solved all the problems of communication between its existing parts, there still remains the task of integrating with the remainder of the country the great empty area of the North.

In the East, then, the population has been increasing relatively slowly. Settled early, these areas have been the scene of no striking recent development, and have attracted only a small number of immigrants. Apart from Halifax, there is no large city, and today's immigrant almost always makes for the towns, with their industrial and commercial opportunities, rather than for the remote farmsteads of the region.

In Québec, a high rate of natural increase in an essentially rural population led, in the period up to World War II, to an expansion first into the Eastern Townships, east of Montréal, and then to the northern frontier, where pioneer farmers and lumbermen penetrated the Shield, especially in the Lake St. John and Abitibi areas. But in recent decades, the cities have been the growth areas: the Québec population has been urbanizing rapidly. In Ontario, similar trends are evident. Among recent immigrants, one out of every two has made for this province, contributing to the populations of Toronto, Hamilton, and London and spreading from there to the smaller centers. Along the northern edge of the Paleozoic formations, where they adjoin the Shield, there is again a frontier zone, a frontier with outliers in the various

clay belts on the Shield further north. But a geological map remains the best key to the distribution of Ontario's population, which is almost wholly concentrated in the southwestern peninsula.

In the Prairie Provinces, a steady trickle of workers has been leaving the farms. This migration, as we shall see, has occurred in many other farm areas, especially in the comparable Great Plains region of the United States. Although some of the migrants move to the towns, many have left the area to resettle in British Columbia. During the 1970s, the number of farm operators on the prairies fell by about 2000 each year. However, the loss from the farms has been more than offset by the growth of the prairie cities, a growth due to industrial development and, between 1950 and 1980, stimulated by the spectacular rise in oil and gas production.

In the West, the population of British Columbia, like that of other Pacific Coast areas, has increased very rapidly, but outside the metropolitan area of Vancouver, it is a population largely in clusters scattered over a huge area, in remote valleys and on islands. The Vancouver district has been the goal of many immigrants, but the province also includes a pioneer fringe along the Peace River, where the settlements have contributed to the increase in population.

Finally, the Northwest Territories and Yukon which between them constitute nearly 40 percent of Canada's area, possess only 0.2 percent of the population—some 75,000. Half of these people are Indians and Inuit, the Indians generally to be found in the forested areas of the North and the 16,000 Inuits along the seacoast. The other half of this sparse population is to be found in mining communities and trading posts. The white population of the Yukon, although it has increased since 1941, is far below the figure for the halcyon days of the gold rush, ninety or more years ago.

Canada's provincial populations change with two external circumstances: immigration from abroad and interprovincial migration (Table 2-4). As to the first of these, Ontario is, by a long way, the favored destination of the newcomers: in recent years, it has been receiving three times as many immigrants as either Québec or British Columbia. As to the second, Ontario is equally the prime target for interprovincial migrants. The 1991 census revealed that, since 1986, Ontario had received the following proportions of all out-migrants from other provinces: Québec Province, 69 percent; Newfoundland, 58 percent; Nova

Table 2-4. Canada: Factors of Population Change, 1951–81 (numbers in thousands)

	Total Increase	Natural Increase	Net Migration	Ratio of Migration to Total Increase (%)
1951–1956	2071	1473	598	28.9
1956–1961	2157	1675	482	22.3
1961–1966	1777	1518	259	14.6
1966–1971	1553	1090	463	29.8
1971–1976	1424	934	489	34.4
1976–1981	1288	978	310	24.1

Scotia, 44 percent; New Brunswick, 39 percent; Manitoba and British Columbia, 32 percent each (Fig. 2-4). In its turn, British Columbia was an important destination for migrants from the Prairie Provinces (e.g., Alberta, 46 percent of out-migrants; Manitoba, 26 percent). In the precarious political balance upon which Canada's federal government depends, all these movements play an important—and possibly an unbalancing—part (Table 2-5).

Distribution of Population: The United States

At the census of 1990, the United States had a population of 248.7 million, up from 226.5 million in 1980—an increase for the decade of 9.8 percent. Over the period since World War II, the increase in the nation's population was generally much more rapid than demographers of the 1940s and 1950s

Table 2-5. Canada: Population at the Censuses of 1986 and 1991, by Provinces and Territories

Province/Territory	1986 Population	1991 Population
Canada	25,309,331	27,296,859
Newfoundland	568,349	568,474
Prince Edward Island	126,646	129,765
Nova Scotia	873,176	899,942
New Brunswick	709,442	723,900
Québec	6,532,461	6,895,963
Ontario	9,101,694	10,084,885
Manitoba	1,063,016	1,091,942
Saskatchewan	1,009,613	988,928
Alberta	2,365,825	2,545,553
British Columbia	2,883,367	3,282,061
Yukon Territory	23,504	27,797
Northwest Territories	52,238	57,649

had predicted. For twenty years after the war ended in 1945, the birthrate ran high and, when it began to fall off, immigration replaced it as a source of growth. The annual increase due to immigration rose from one-fifth of the total in the 1960s to more than one-third today.

Between 1980 and 1990, four states and the District of Columbia lost population: Iowa, −4.7 percent; West Virginia, −8.0 percent; North Dakota, −2.1 percent; and Wyoming, −3.4 percent (see Fig. 2-5). The remainder gained population, although the rates of gain varied widely: the leaders, with their percentage gains, were Nevada, 50.1; Alaska, 36.9; Arizona, 34.8; Florida, 32.7; and California 25.7. In general, the highest rates were in the South and West, the lowest in the North.

The American population is notorious for its mobility; up to 20 percent of its moves house each year. To a degree unknown in older lands, Americans are geographically unattached, and are prepared to move long distances in search of better pay or more pleasant living conditions. Out of this mobility have come two kinds of movement in recent decades.

The first of these has been movement between country and city. Not only in North America, but in virtually all parts of the world, people have been leaving the countryside for the town as soon as there were towns to move to. This movement seems to have reached a peak of sorts, in the United States at least, between 1960 and 1970, when there was an exodus from the nation's farms equivalent to one-half of the 1960 farm population, and little return flow. Mechanized farms required less labor, and the towns and cities offered jobs and a more varied life-style. At the same time, the cities spread outward to engulf rural areas; country people became urban, for statistical purposes at least, without moving house. In both the United States and Canada, the census says that three-quarters of the present population are urban.

It comes as something of a surprise, therefore, to discover that there is a counter-tendency of the 1980s, and one strong enough in places to be called *counter-urbanization*. We have become accustomed to the fact that people will move from the city's center to its suburbs as and when their means allow: what is new is that they will move out of the suburbs and back into the country. Insofar as we can generalize about this movement, some of its causes seem to be: (1) a disenchantment with traditional suburban life-styles and a back-to-nature movement; (2) the

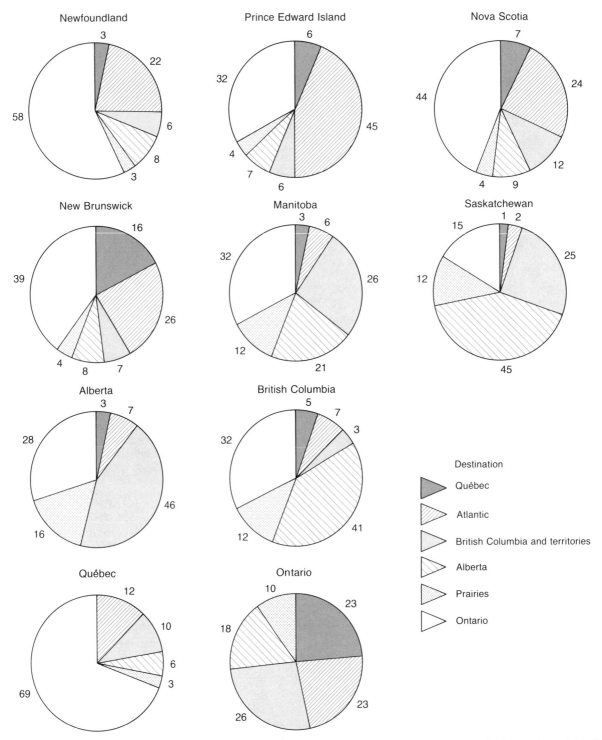

Fig. 2-4. Canada: Interprovincial population movements between the censuses of 1986 and 1991. The diagram shows the destinations of out-migrants leaving each province. The figures show percent- ages of totals. Note the importance as a destination of Ontario for all the eastern provinces, and of British Columbia for the Prairie Prov- inces. (Source: *Statistics Canada*.)

availability of rural housing at moderate prices, resulting from a dwindling farm population; (3) reduced commuting times from country areas, resulting from the spread of an express highway network; and (4) a general decentralization of business and services. As the years go by, counter-urbanization may prove to be as important a phenomenon of the late twentieth century as the "rush to town" was of the years of industrial revolution.

The second type of population movement is interregional. Ever since the first Europeans arrived, there has been a fairly regular shift in the regional balance; in particular, a movement from east to west. For decade after decade, the main goal of this movement has been California until, midway through the 1960s, it became the most populous state in the union. Draining out of the longer-settled states further east, lesser streams flowed into the Pacific Northwest and Texas. The movement became a tradition, a part of everybody's experience as an American.[11]

The latest figures show, however, that there have

[11]The U.S. Bureau of the Census calculates every ten years the "center of population" of the United States. Decade by decade, this has steadily moved westward across the northern states. In 1800, it was near Baltimore. In 1900, it had reached southern Indiana. In 1980, pulled a little southward by population growth in the Gulf states, it was west of the Mississippi for the first time, southwest of St. Louis.

Fig. 2-5. The United States: State populations (in millions) at the census of 1990, and percentage increase in population, 1980–1990, as shown by shading. The national increase during the decade was 9.8 percent, and the highest rate of increase was shown by Nevada (50.1 percent). For the states now shown, Alaska's population increased by 36.9 percent to 550,000; that of Hawaii by 14.9 percent to 1.1 million.

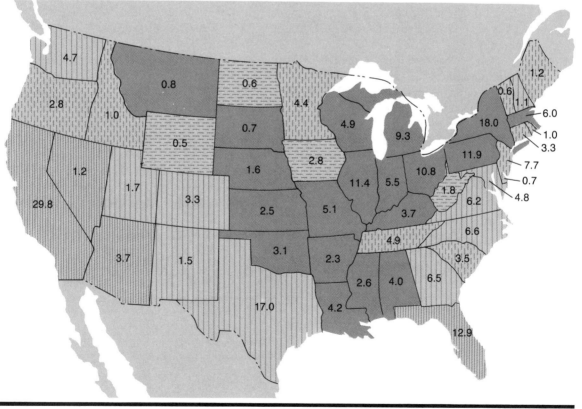

Decrease   Increase 12–25%

Increase 0–5%   Increase 25–50%

Increase 5–12%

0   200   400 mi

0   200   400   600 km

been changes. It is no longer a question of a simple westward movement, terminating in California because there was nowhere else to go. California, with 29.76 million people in 1990, may well be the most populous state in the Union, but it is certainly no longer the migrants' inevitable goal. What has raised its population by 2.5 percent a year in the last decade is not so much the historic trek west from the eastern states as the convergence upon it of the two main immigrant streams, the Hispanic and the Asian, described earlier. Majority Americans have in fact been *leaving* California: the state is getting overcrowded, at least in its urban areas; the costs of housing and living are high; and jobs have been disappearing in large numbers. Evidently, a fresh analysis of inter-regional movements is called for.

The traditional American incentives to move have been provided by the search for space, independence, and wealth. For many decades, all three of these were to be found in the same region—the West. Space and independence were provided by sparse population and cheap land; wealth by minerals and the thousand opportunities for speculation a frontier provided. But eventually, the frontier was stabilized and the space-seekers began to crowd each other out—which in California (which now has nearly 30 million of them) is what has happened. So the original movement has been modified in two ways:

1. *As the original target areas of the movement filled up, they have been replaced by others.* Since the most inviting areas were generally filled first, the new areas may lack one or more of the qualities of the original, but still compensate for that lack by their emptiness. In this way there has occurred a switch, for example, from southern California to Arizona—a kind of bounce back from the Pacific Coast. The same kind of rebound is now happening for Nevada, or for Washington State, which is starting to affect the Pacific Northwest as a whole. There are fads in migration goals as in everything else.

2. *To the old incentives to move has been added a new one: amenity.* In earlier centuries, when most people were producers of goods, they had necessarily to live where they produced—on their farm or close by their workshop. Their choice of location was limited by their employment, and some of them lived under appalling conditions simply to be near their work. But in North America today, most people are not tied in this way. Seven out of ten American workers are not producers in the old sense at all: they are employed in service occupations, most of which can be carried on wherever there are customers. And they are likely to choose a workplace for reasons quite unconnected with the occupation itself; in short, for reasons of amenity. There is nothing surprising about this, and no doubt Americans in the past would have greatly preferred, given the option, to stitch garments in pleasant small towns rather than in sweatshops in the slums of New York. The point is that they were not given the option, while today's worker generally is. He or she has personal freedom of movement, instant information about places to live, a job that can be done anywhere, and sufficient leisure to make recreation a major factor in planning. Under these conditions, as we might expect, millions of Americans have been exercising their options.

There are, of course, economic factors, too. With the steady drift out of farming that North America has experienced, agricultural regions have lost population. Even more widespread is another economic consideration: that it costs less to live in a warm climate than a cold one—less to heat a house or an office, less to buy warm clothes. If leisure and recreation invite Americans to move to a climate where they can carry on outdoor activities all year around, economic sense reinforces the attraction of such areas.

What America has experienced, therefore, is the development of a Sun Belt. It stretches, in most people's perception,[12] from the Carolinas to California, and it contained, at the last census, more than half of the population of the United States. For all the states contained within it, the figures of population growth have been striking.

To summarize the factors involved in this regional shift: there is freedom to move and, for most people, a choice of workplaces. There is the economic advantage of avoiding a cold climate (although this is offset to some extent by the probable need for air-conditioning). There is the attraction of an outdoors accessible and attractive during virtually twelve months of the year. And undergirding all this is the fact of an

---

[12]There is no "official" definition of the Sun Belt, although one could doubtless be agreed on the basis of such sunshine maps as our Fig. 20-4. Not only is the term rather too emotive for scholarly use, but one is left, in using it, with the problem of what to do about the rest of the country. There is a tendency carelessly to label the whole of this rest the "Frost Belt" or the "Snow Belt." But this grossly oversimplifies the situation of a whole range of "middle" states like Tennessee or Oklahoma, not to speak of states like Colorado, which have sustained rapid growth precisely because, among other things, they offer both snow and sunshine in an attractive combination.

economy with a sufficient margin of productivity not to require every business to be carried on in the absolutely least-cost location, an economy in which the contentment of the work force is a factor which can be, and is, taken into account.[13]

Most of these considerations apply very particularly to one group of Americans—those who have retired. Welfare schemes and Social Security are institutions whose coverage has grown greatly in recent years. A worker who retires on a pension can collect payments in Florida as easily as in Chicago; he or she is certainly interested in living cheaply, and has all the leisure time in the world. The drift to the Sun Belt therefore involves great numbers of the retired, and as we shall see again in Chapters 17 and 20, there are communities in the South and West that cater particularly to them.[14]

But there is more to it than this: not all the retired folk go south permanently. McHugh and Mings have called attention to the scale of a *seasonal* migration to the south, creating what they call "seasonal retirement communities." The key to this movement is the recreational vehicle (RV), in which seasonal visitors drive south from as far away as Canada to spend an average 4.4 months of the winter in the South without incurring the cost of a second home. Already, it appears, in 1989 Phoenix, Arizona, had standings for over 41,000 RVs. "Older Americans and Canadians," it is reported, "are defining new lifestyles and forms of retirement living based on seasonal movements in RVs."[15]

There is still, therefore, an inter-regional migration from East to West. The new feature is a movement from North to South. For a century past, the South has been a net loser, and in the decade of the 1960s, five southern states were among the eight biggest losers in the nation. In the 1970s, this began to change, and has since changed with a vengeance. By the end of the century, we may well see a new population geography of the United States, with new gainers and new losers.

---

[13]To complete this reference to the Sun Belt, we should add that, in an era of serious unemployment, the Belt has been the subject of two obviously conflicting perceptions: (1) that there are jobs in the Sun Belt where there are none in the North, and (2) that even if there are not, it is still worth moving south because it is better to be unemployed and warm than unemployed and cold.

[14]It is estimated that more than 30 percent of today's U.S. population is supported by "non-earnings" income—pensions, welfare payments, and dividends. As to where such people settle, the reader may wish to consult T. O. Graff and R. F. Wiseman, "Changing Patterns of Retirement Counties Since 1965," *Geographical Review*, vol. 80 (1990), 239–51.

[15]K. E. McHugh and R. C. Mings, "On the Road Again: Seasonal Migration to a Sunbelt Metropolis," *Urban Geography*, vol. 12 (1991), 1–18.

3
The Cities

Hartford, the capital city of Connecticut, and one of North America's
leading insurance centers. *McConnell-McNamara and Co.*

City Building

Three-quarters of the population of the United States and Canada live in cities and towns. The one-quarter who do not nevertheless transact much of their business there, obtain their news and television programs from the cities, and receive from urban areas most of their cultural and educational influences. It is therefore only logical to follow our general survey of North America's people in the last chapter by a closer look at the cities in which so many of them live.

Cities and towns began to grow up when people first found it necessary to gather regularly in larger-than-family groups. The earliest and most usual reasons for these gatherings were defense, religion, and the exchange or transshipment of goods. In the Old World, this trinity of causes is represented, down to the present day, by the grouping together in a thousand towns of castle, church, and marketplace. In the New World, although in a rather different manner, the same trinity was recognized by the Spanish Empire with its *praesidio,* or military base, *mission,* or religious gathering point, and *pueblo* for the exchange of goods.

Later on, there arose other reasons for bringing large numbers of people together. One of these was the processing of goods and their storage for processing; that is, industry and warehousing. Another was administration of law and government. At the same time, some of the original causes of town growth lost importance: places once chosen as defensive sites no longer needed defending, and the religious causes that built up Old World pilgrimage centers or New England settlements today contribute to only a handful of cities, of which the most obvious North American example is found in Utah, beside the Great Salt Lake.

Because urban North America was the creation of immigrants from Europe,[1] the earliest cities were ports on the Atlantic coast. And because movement into the interior was easiest by water, the second-generation cities were river and lake ports. Cities that combined these two functions of oceanic and inland transport—New York, Montréal, New Orleans—emerged as the earliest leaders. The river ports of Pittsburgh, Cincinnati, and St. Louis came to dominate the interior. Later, cities grew up in which one or other of the town-forming functions was dominant—industrial towns at water power sites or on coalfields, and the administrative centers that were the state capitals, headed by Washington, D.C., a city built for one purpose alone: to house the federal government.

Overwhelmingly, however, the cities of North America have grown as *commercial* centers. With a mobile population able to commute to work instead of having to live under the shadow of the factory walls, industry has lost a good deal of its city-creating strength: much of it today has moved from old urban centers to new peripheral, greenfield sites, and it is good that it has done so. Administration of law and government does not require many cities; certainly only a fraction of the number that exist in North America. The growth areas in city-building functions in the twentieth century have been in buying, selling, and servicing. To use a phrase now commonly employed, this is the age, and this is the continent, of the *transactional city.*[2]

We can express these changes in the role of the American city in other terms. In the first phases of

[1] The important word here is *North.* Readers will not need to be reminded that, at the date of the European arrival in the Americas, remarkable urban civilizations already existed in Central and South America.

[2] See Jean Gottmann, *The Coming of the Transactional City* (College Park, Md., 1983).

European settlement in North America, up through the first quarter of the nineteenth century, the chief function of the cities was to serve as supply bases. Much of the equipment needed to settle the continent came from Europe: the ports of entry formed the primary bases and, as the frontier moved west and distance from the coast increased, other bases further inland took over the supply and storage roles. This phase is often described as the mercantile period; that is, the cities acted mainly as trading posts or shopkeepers for a still largely rural population of westward-moving settlers.

Some time about 1830 or 1840, a new phase began, with the growth of native North American industry. The cities underwent a surge of growth as they now not only provided imported supplies, or acted as marketplaces, but themselves manufactured those supplies and processed the output of the settlers' farms. To this period belongs the growth of a new generation of cities—the mill towns of New England; Cleveland, Chicago, Detroit—as well as new development of older supply centers like Pittsburgh and Cincinnati.

The second half of the nineteenth century saw the heyday of industry-induced urban growth, in new areas of the West and South as well as the old core areas, and two world wars each added a renewed stimulus. But the twentieth century as a whole has seen the emergence of what is now sometimes called the *post-industrial city*, which is, in simple terms, a city developing through the volume of the business transacted in it, without any necessary contribution from industry at all.

This being the case, a great continent like North America is likely to develop a large number of commercial cities, growing more and more to resemble one another, and providing a range of services to their surroundings or hinterlands. If we ask *how many* such cities there will be, the answer clearly depends on two things:

1. Distance from center to center, judged in terms of people's ability to cover that distance. A population limited to movement on foot, or by horse and cart, requires a much denser network of service centers than one using motor vehicles. In fact, as the mobility of the population increases, some of the older centers will find themselves to be redundant, and will fade away.
2. The volume of demand arising from the surrounding area. A sparse population, producing

little except its own necessities, will require the services of only a few centers: there will simply not be enough business to support many cities.

If we then go on to ask *what selection of services* the city is to provide, the answer is that some services and goods are needed frequently, and should be locally available, but others are rarely used and so can only be supported by a large population or a wide area. When they are needed, however, they are worth traveling a long way to obtain. Some, too, are simple to provide, while others depend on sophisticated skills or on equipment only found in a few centers.

From this we go on to develop the idea of a hierarchy of centers, graded according to the number of services they provide and the area or *range* over which each particular service is offered. Given some basic information about the character of a population and its economy, it is the function of *central place theory* to calculate or forecast the number, spacing, viability, and ranking of a region's cities.

A city's position in the hierarchy of central places, then, will depend on the volume and nature of its transactions—commercial, financial, legal, medical, administrative. On this basis, the major cities of North America can be classified in four or five categories. Such a classification of the larger cities, based on banking linkages, is offered by Maurice Yeates (Table 3-1).[3] Other criteria would no doubt produce different rankings in some cases, but over the membership of the first and second orders, at least, there can be little dispute. In the same way, the smaller centers can be subdivided into categories, down to the smallest "central place" of all, the crossroads, which offers a single, frequently used service, like a filling station, or a milk delivery point.

[3]M. Yeates, *North American Urban Patterns* (Silver Spring, Md., and London, 1980), p. 12. A selection of other titles from what has become a vast literature of North American urban geography would be: Risa Palm, *The Geography of North American Cities* (New York, 1981); R. J. Johnston, *The American Urban System: A Geographical Perspective* (New York, 1982); the Association of American Geographers' *A Comparative Atlas of America's Great Cities* (Minneapolis, 1976); and, for Canada, G. A. Nadar, *Cities of Canada*, 2 vols. (Toronto, 1975 and 1976); and M. Yeates, *Main Street, Windsor to Quebec City* (Toronto, 1975).

Two other significant works on Canada are M. A. Goldberg and J. Mercer, *The Myth of the North American City: Continentalism Challenged* (Vancouver, 1986), which argues forcefully that Canadian cities are different from those of the United States; and G. A. Stelter and A. F. J. Artibise, *Power and Place: Canadian Urban Development in the North American Context* (Vancouver, 1986).

Table 3-1. Hierarchial Structure of the Larger Metropolises in the North American Urban System (after M. Yeates)

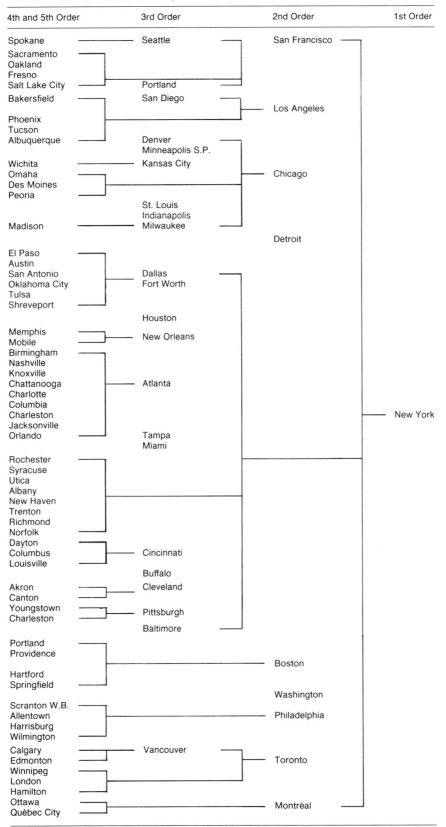

Notes: (1) Minneapolis S.P. = Minneapolis–St. Paul; (2) Scranton W.B. = Scranton–Wilkes–Barre

About the higher-order cities listed in Table 3-1 there are two things to be said. One is that, owing to the size of North America, most of the second-order cities are in a position to serve as regional metropolises, each one dominating its own full hierarchy of central places; in other words, there is plenty of space for a dozen regional centers to develop without their "shadows" necessarily falling across each other. Some of these regional metropolises have huge hinterlands, larger than whole countries in Europe or Central America. In Table 3-1 we see Chicago's sphere of influence stretching from Denver to Indianapolis, a distance of 1100 mi (1760 km), and Albuquerque in some sense tributary to Los Angeles, although it is 850 mi (1360 km) away.

The second comment is, however, the reverse of the first. In some parts of North America, cities are very close together—much closer than any central place theory would lead us to expect. This is especially true on the northern Atlantic coast and in the eastern Great Lakes area. For reasons connected with the density of population or the history of early town founding, with resource patterns or transport routes, cities have grown and survived in clusters rather than in isolation. These clusters we may recognize as *urban regions.* To quote Maurice Yeates once again,[4] there are nine of them in contemporary North America (Table 3-1), although some of the nine are less obvious candidates than others and are, certainly, different in character.

What is remarkable in these regions is the way in which the processes of urbanization have taken over, and individual cities have flourished, even under the shadow of second-order metropolises. In Florida, indeed, or on the Gulf Coast it is almost as if there has been some conspiracy to *prevent* the emergence of a single, dominant center. The rivalries are too sharp, the growths too recent, and the city-forming factors too general for one city to have established an undisputed lead over the rest.

To a small extent, of course, it is possible to engineer the shape of a hierarchy. A number of states and provinces did so in their early days, by sharing out the major functions of government and administration—few though they then were—among claimant cities. At lower levels in the hierarchy, the same process goes on today, with the much wider range of services in which governments are now involved: a research station here, or a hospital there, may make the

[4]Yeates, *North American Urban Patterns*, Fig. 1.9, p. 21.

Table 3-2. United States: Percent Population Urban, by Region, 1910, 1950, and 1980

Region	1910	1950	1980
New England	73.4	74.8	75.1
Middle Atlantic	70.2	75.6	80.6
East North Central	52.7	66.3	73.1
West North Central	33.1	49.9	63.9
South Atlantic	23.1	41.6	67.1
East South Central	18.6	35.5	55.6
West South Central	22.2	53.1	73.4
Mountain	35.8	49.1	76.4
Pacific	56.8	63.6	86.6

difference between survival and extinction for small cities in areas of declining population.

As we might expect in a continent settled by three major European groups—Spanish, French, and British—and possessing a wide variety of climates and resources, the rates of urban growth and size of cities have differed from region to region. The French, with their fur trade under royal monopoly and their limited agricultural base, had little reason to develop urban centers: a small number of forts and ports met their needs. The contrasting roles played by agriculture, commerce, and industry in New England and the Old South explain much of the difference in rates of urbanization and sizes of cities in those two regions (Table 3-2). The Spaniards, on the whole, were methodical town planters, just as were the western railway builders in a later era. Elsewhere, in the mainly Anglo-Saxon interior, town founding was a hit-and-miss affair of rivalries, skulduggery, and pure chance. And each of these groups had its favorite patterns of urban layout which, consequently, tended to appear again and again, and give to the cities they founded their distinctive appearance.

Counting the City Population

North America's first town was St. Augustine, Florida, founded by the Spanish in 1565. In the four centuries since then, which have become the continent's largest cities?

To answer this apparently simple question is far from straightforward, whether we are thinking of population or extent. The city of Chicago, or Los Angeles, or St. Louis, is an area whose size and population are both known, but each is surrounded by other built-up areas which are continuous with it, and whose populations can with every justification be counted as parts of a whole supercity. There is, in other words, a problem of definition.

To overcome this problem, the census takers in both Canada and the United States have created statistical units that cover the whole built-up area around a central city (and sometimes include remainders of counties that are desert, or forest, and are not built-up at all). In Canada, these are known as Census Metropolitan Areas (CMAs). The last census distinguished twenty-five of them, and together they contained some 60 percent of the population of Canada. Toronto CMA, with 3.7 million inhabitants, was the largest, having pulled ahead of Montréal (3.0 million) since the 1976 census. These two giants were followed by Vancouver (1.5 million), Ottawa, Edmonton, and Calgary.

In the United States, the situation is more complex. There are no less than four levels or units of urban population count: the city, the Metropolitan Statistical Area (MSA), and the Primary and Consolidated Statistical Areas (PMSAs and CMSAs). The MSA typically comprises a central city of over 50,000 population and one or more adjoining counties of suburban overspill. The MSA of Atlanta, Georgia, actually contains eighteen counties, with a 1990 population of 2.83 million, although the city of Atlanta itself had only 450,000 inhabitants within the city limits. The PMSAs, seventy-one of them, are urban areas that form parts of the largest units of all, the twenty CMSAs.

This means that, for the major cities, a series of population figures can be given. Here, for example, are the results of the 1990 census for the three largest urban areas (all figures in thousands):

	New York	Los Angeles	Chicago
City	7323	3485	2784
PMSA	8547	8863	6070
CMSA	18,087	14,532	8066

Among the CMSAs, each of which represents a whole conurbation, often stretching across state lines, New York and Los Angeles are first and second largest, with Chicago third and the San Francisco Bay Area fourth. But it is this technique of grouping together the populations of several adjacent cities that makes it difficult to produce a definitive listing of U.S. cities by size. The only certainty is that, up till now, whichever statistic we adopt, we find that New York City and its surroundings comes first. But with a considerable drain of population there in the 1970s out of the central PMSA (a loss of 802,000 between 1970

and 1980), even this certainty may not last much longer.

That counting population on the basis of metropolitan areas is a rather rough-and-ready procedure can easily be seen if we refer to Figure 3-1, a map of the metropolitan areas of southern California. Around Los Angeles the MSAs and PMSAs cluster tightly but, because all these areas are based on county boundaries, we find that one of the more recent MSAs, that of Riverside–San Bernardino–Ontario, is enormous in size, for no better reason than that San Bernardino County is enormous. In fact, it abuts on the Las Vegas MSA in Nevada, two metropolitan areas between them dividing the 230 miles (370 km) of bleak desert between the two cities. Riverside–San Bernardino actually includes the southern end of that desert to end all deserts, Death Valley, in its "metropolitan" area.

Of more interest to the geographer are two questions: which of these cities or built-up areas are growing and which are losing population? Both in Canada and in the United States, the fastest growth rates are at present being registered in the West or South; that is, away from the old urban core areas of the East Coast. In Canada, the runaway leaders in recent years have been Calgary and Edmonton in Alberta; they have eclipsed in their rate of growth even Vancouver, for so long a goal for westward-moving migrants. Toronto has grown substantially, if only because of the flow of immigrants who make for the city. Montréal, for so long Canada's leading city, has grown more slowly, and has lost its pre-eminence to Toronto.

In the United States, where a city and its surroundings are designated an MSA as soon as its population reaches the Census Bureau's threshold, new MSAs have been emerging in large numbers in Florida, Texas, and California. These states also contain the fastest-growing MSAs of the 1990s; growth rates of 3 percent per year are common, 5 percent the highest. There are now more than 260 metropolitan areas in the United States.

The list of losers in the decade 1980–90 focuses equally clearly on the Northeast. The metropolitan areas with the largest losses (percentages in parentheses) were: Davenport–Rock Island–Moline (8.8), Pittsburgh (7.4), Youngstown (7.3), Huntington-Ashland (7.1), Peoria (7.3), Charleston, W. Va., (7.1), Gary-Hammond (5.9), Buffalo–Niagara Falls (4.3), and Flint (4.4).

But gainers or losers, one tendency was virtually

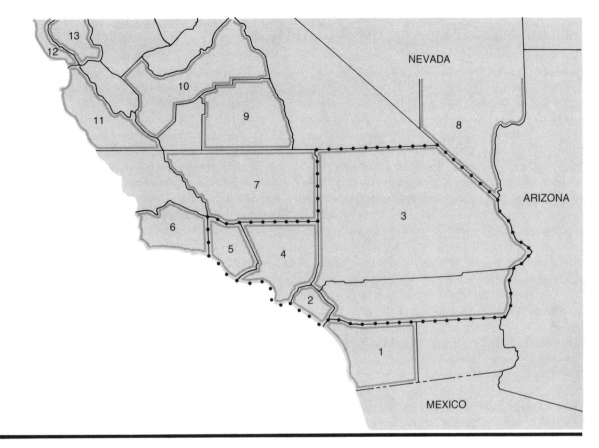

Fig. 3-1. Southern California: Metropolitan Areas. The map shows southern California by counties, its metropolitan statistical areas, and the consolidated metropolitan statistical area of Los Angeles–Anaheim–Riverside. (See accompanying text for further explanation of these terms.) Note that virtually every county on the map is included in a "metropolitan" area, despite the fact that deserts and mountains occupy much of the space: in fact, Area 3 contains the southern end of Death Valley.

Key to areas: (1) San Diego, (2) Anaheim–Santa Ana, (3) Riverside–San Bernardino, (4) Los Angeles–Long Beach, (5) Oxnard–Ventura, (6) Santa Barbara, (7) Bakersfield, (8) Las Vegas, Nevada, (9) Visalia–Tulare–Porterville, (10) Fresno, (11) Salinas–Seaside–Monterey, (12) Santa Cruz, (13) San Jose.

universal: the central city lost population and the suburbs within the metropolitan area gained. We shall return to this subject in the next sections; for the moment, we need only notice that some of these central city losses were quite spectacular. Between 1970 and 1984, the city of St. Louis lost one out of every three inhabitants, and so did Detroit. Buffalo and Cleveland lost one in four. What such losses mean to the city's tax base, or to its ability to attract federal support funds—which for certain programs are allocated to cities on a per capita basis—it is easy to visualize.

In the City: Form, Function, and Quality

Geographers study cities, whether they are in North America or anywhere else, by reference to three main variables: form, function, and quality. In this section, we shall consider each of these in turn, but before we do so, it will be useful to pause and ask whether there are any features that single out North American cities as a group; which are distinctively American as opposed, say, to European or Australian.

There are at least three such features.

1. *Rapid rise and rapid change.* The first of these features of urbanization is the speed with which it has

taken place. Not only have North American cities in the past century grown spectacularly fast, but the *rate of transformation* of the urban landscape has also been very rapid.

In Third World cities, growth may be equally fast, but it all too often produces sprawling shanty towns of homemade shacks, because the resources do not exist to house the incoming masses. Only in the crudest locational sense do these incomers live in the city. They share none of its services like water or sanitation; their sole advantage is proximity to what they see as a source of jobs, and their disadvantages are legion. In North America, in general, construction has kept pace, even in the fastest-growing cities, with the increase in population and, where it has not, mobile homes have filled the gaps in permanent housing.

The speed with which old features of cities are torn down and new ones are constructed is equally North American. It is a function not merely of the technical ability to raze and build, but of *wealth*. The truth is that *cities cost money*, and only a very wealthy society can afford to pull down what it built yesterday, and build again for tomorrow. Older and poorer societies have to adapt and make do.

2. *Personal mobility.* A second feature in the continent that first saw the mass use of the automobile is the great range of urban distance over which the individual North American enjoys freedom of movement. Where, in a pre-twentieth century city, all urban functions were necessarily grouped close to one another, the car makes it possible to locate them miles apart without seriously inconveniencing their users. Central Boston is, perhaps, the last big American city in which doing business is still possible, and even pleasurable, without using a car. Los Angeles was the first in which business became impossible without one.

3. *A wide variety of ethnic groups.* As a result of centuries of immigration and hospitality to refugees, North American cities house what is probably a wider range of ethnic groups than any others in the world, for a majority of these newcomers have at all times headed for the city and urban employment. In the nineteenth century, it was New York and Boston; after World War II it was Toronto and Vancouver; in the latest phase of arrivals it has been Los Angeles, Miami, and Houston. In 1980, when the population of the United States was classified as 73.7 percent urban, the proportion of urban *Asian*-Americans was

92 percent, and that of *Hispanic*-Americans 88 percent.

Sometimes by choice, and sometimes under pressure from outside, these groups tend to form a series of separate communities within the city, with their own commercial, retail, and residential areas. Whatever problems of space allocation the city as a whole may have—and there are very few cities that have none—their total has to be multiplied by the number of separate groups being accommodated.

Everybody, it would seem, knows about the problems of American cities. They have received a depressing amount of publicity, and we shall return to consider some of them in the following sections. But before we do so, it is only fair to place them in their context. These problems are arising in cities that are expanding, and changing their landmark features, so fast that their longtime inhabitants lose their way, let alone the thousands of new arrivals who cannot even speak enough English to ask for directions. These problems are arising in communities divided a dozen ways into ethnic groups, with all the mutual hostilities and rivalries those divisions may produce—hostilities, at that, brought from homelands far away but with which American cities must now learn to live. These problems arise in cities where the old concept of a community of neighbors, with a common loyalty to a street or area, has been undermined by a mobility that may take those neighbors off to work or to shop many miles away each day, and the word *neighbor* loses all meaning. The wonder is not that North American cities share with so many others a wide range of problems, but that they cope with them as well as they do.

URBAN FORM

The earliest North American cities, like the European counterparts on which they were usually modeled, grew without formal plan, in conformity to local topography or harbor frontages. Late in the seventeenth century, however, there first appeared what was to become the stylized North American city model, with straight streets crossing at right angles. It was a model borrowed from Europe's planners of the same period, but it dated back to classical times, and its early appearances in Charleston, South Carolina, and in William Penn's Philadelphia in 1683 were only the first of many others.

City followed city, and the habit of laying out squares or rectangles was strongly reinforced when, at the end of the eighteenth century, the new U.S. federal government chose a rectangular survey for all the public lands west of Pennsylvania (see Chapter 5). From then on, there was always a strong prior argument for laying out towns on the survey's north–south and east–west gridlines, an argument that, in due course, applied to western Canada as well. In older cities of the East—New York, Boston, and Montréal are examples—the random, or European-style, ground plan of the old core was replaced in later extensions by regular gridirons of streets. Sometimes, as in San Francisco, this gridiron pattern was imposed upon unsuitable topography to the point of absurdity. At least it kept the street plan simple, even if it did make diagonal movement tedious.

In the settlement of the great West as a whole, it became necessary between 1850 and 1900 to create a large number of towns almost overnight. It is small wonder, therefore, that the pioneers—in many cases the railroad builders—stuck to a regular ground plan, of which the railroad station was often the focal point. Some railroad companies laid out virtually the same town all along their lines: it was both logical and labor-saving to do so. And in that same West one may still come across the remains of the same pattern of streets, on which no building ever rose—towns that were platted more in hope than in wisdom, and that today cannot claim even the status of ghost towns, for there is nowhere for the ghosts to haunt.

The gridiron suited the relatively simple needs of

Indianapolis: The city center. Laid out according to the four points of the compass, Indianapolis, a "four-square" city, is the capital of Indiana. The photo shows the typical North American cityscape of a central business district, with tall modern buildings tailing off—sometimes quite abruptly—into an older landscape of low-rise homes. *Thomas P. Wolfe, Public Service of Indiana*

Harlem, New York City. Most views of Harlem focus on derelict or grossly substandard housing. While this is all too widespread, it should be remembered that within Harlem there is all the variety nor-mally found in a city. These houses, adjacent to the historic Jumel Mansion at 159th–160th streets, are also historic and beautifully pre-served by their occupants. *J. H. Paterson*

an urban population on foot: it is far less obviously suited to urban life today. One necessary modification as cities grew larger was the introduction of diagonal streets: their absence in a place the size of Chicago was adding miles to journeys across town. Another problem arose where the original city blocks were platted with short frontages: not only does this hold up modern traffic flow every few yards, but it means that the blocks are too small for the size of structures which it is now economical to build on them.

If the form of the city's central areas is in this way predictable, so, too, is that of its suburbs. Less regular—indeed, carefully irregular—in plan than the older, inner city, they sprawl out along the principal traffic arteries, giving to small North American towns a length or diameter that would often enclose a good-sized European city. The search for space which, a century ago, was creating the "streetcar suburbs" of older American cities has, in the automobile era, caused an explosion of urban spread, in which space

has become the prime symbol of conspicuous consumption, not only for individual homes, but for business and industry also.

What this suburban sprawl means in terms of land-use changes we shall later need to consider. There is at present little sign of its being halted, at least until the advancing suburbs of one city meet those of its neighbor. But it brings into prominence the belt or ring roads which increasingly link these far-flung parts of the city—and which become, in their turn, sought-after locations for new businesses, as in the case of Boston's famous Highway 128 or Toronto's Highway 401 (see Chapters 12 and 13).

URBAN FUNCTION

Cities are functional objects; that is, they do not exist in nature but have been created, as we have already seen, to perform particular social and economic functions. To the geographer, the questions of interest that arise from this fact are twofold: (1) How are these functions arranged on the ground, within the urban space, and (2) How does this spatial arrangement assist, or hamper, the efficient carrying out of those functions?

Already, in the earliest cities, functions tended to become segregated into different areas: craftsmen settled together in streets or quarters given over to their trade, and particular activities were banished outside the central area because of the nuisance they caused (which finds its counterpart today in the practice of zoning). There were, in other words, *functional clustering* and *functional exclusions.*

Functional clustering nowadays may occur for several different reasons. One is that firms or manufacturers benefit from linkages between them and choose, therefore, to locate together. Finance houses do this around New York's stock exchange, and high-tech industries follow the same pattern in California's Silicon Valley. The linking factors in other clusters are power or transport: they first made themselves felt in the textile mill towns along the Merrimack River in New England, and subsequently led to the clustering of industries at lineside along railroad routes across the continent.

But within the city itself the chief cluster-creating factor is simply the number of customers converging on its center. There is in every city, in theory at least, a point of maximum circulation; the place where most people pass by, most days. For any business involved with the public, that is the place to be. In practice, this point of maximum circulation is replaced by an area throughout which business levels are high: the central business district (CBD). Since location within the CBD is desirable, the cost of space there is high; so high that only a limited range of space-users can afford to occupy it.

Away from this central area, the density of circu-

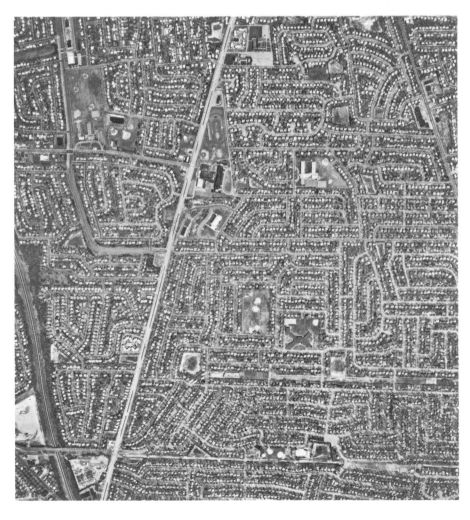

The North American city: Suburban sprawl and the single-family dwelling. An aerial view of the ever-extending suburbs of New York City, taken at Hicksville, Long Island. *Aero Graphics Corp., Bohemia, New York*

lation and the cost of space fall off, allowing the entrance of other urban functions. The way in which these other functions—industry, warehousing, transport terminals, recreational facilities, and administrative complexes, all of them large space-users—come to occupy the area surrounding the CBD has been intensively studied, and a number of models have evolved, with land use sectors or concentric rings, high-quality and low-quality housing. None of these models can be applied to all North American cities, nor should we expect this given the differences in age, site, and social composition between these cities. Functional exclusions increasingly come into play to zone industry out of residential areas and into industrial parks. Political boundaries limit growth in particular directions. Residential areas do not sort themselves neatly into high- and low-quality and, in the course of time, "high" may become "low," or vice versa.

This point leads us toward an answer to our earlier question: how *efficiently* is the city organized to carry out its functions? Are the functional clusterings in the right places? But this is a question that can only be answered for a particular moment in time. Over a period, the working of market forces in an urban area is likely to produce an optimum distribution of the commerical functions. But the single most profound force at work in cities is the force of change; change, that is, in the functions to be accommodated and in their relative importance.

If we return in thought for a moment to the city of the late nineteenth century in North America, it was the creation of two dominant functions: manufacturing industry and rail transport. To these two functions the city itself was, in many cases, little more than a backdrop: it existed because manufacturers and railroadmen had ordained it. Industry and railroad installations occupied the prime sites.

And now consider the changes that have taken place—first and foremost in the switch to the use of road and air transport, making many of the urban rail beds unnecessary, or so much waste space. With industry freed from its rail links, it has been able to relinquish city-center sites, taking its warehouses with it. In its place have come new functions, transactional and administrative; banks and conference centers replacing old mills as parking lots replace railroad yards. The CBD, with its high ground rents, has attracted commerce, whereas public services have had

to content themselves with peripheral sites, where the cost of space is lower.

So, our discussion becomes not so much about how efficiently the city is organized as to what extent its organization has kept pace with the changing demands upon its functional areas. Is the area of the CBD large enough for the volume of business now transacted there? For that matter, is it bigger than it need be, given that much of its former business may have migrated out to the suburbs? Has space made available by the closure of railroad stations and the departure of industry been *recycled* as quickly and usefully as possible?

In probably no city in North America can this last question be answered with a clear Yes. There is always a time lag in recycling urban space. And yet that space is the fundamental urban resource. To the extent that it is underused, or is used for purposes not essential to the life of a city in the 1990s, it is a resource being wasted.

Recycling may not, of course, be easy. With the drift out to the suburbs not only of upper- and middle-class residents but of the businesses that either employ or serve them, there may well arise the question of what the old CBD is actually needed for. Meeting point of routes it may be, but the activity that once gave purpose to it may well have migrated to more spacious sites, nearer the customers, where freeways give access and parking is no problem. A "dead eye" to the city, in a prestigious area which was once vibrant with life and business, poses the sternest of challenges to politicians and planners. The automobile, which has been the creator of the modern suburb, may, unless steps are taken in time, prove to be the destroyer of the city center.

URBAN QUALITY

It is greatly to the credit of social scientists—including geographers—that in recent years they have steadily been developing an interest in a third urban variable, apart from form and function. This we may call a concern for the quality of urban life. Instead of merely asking Who lives in the city? and Where do they live? they have shown increasing interest in answers to the questions In what conditions do they live? and How is the quality of life affected by *where* in the city they live?

With three out of every four North Americans living in urban areas, there are inevitably great contrasts in living standards among city populations.

Problems of urban space: The intersection of the Eisenhower and Kennedy expressways, just west of the central business district in Chicago. It is possible here to count the number of city blocks which have been sacrificed to make this one crossroads. *Illinois Department of Transportation*

Ever since the tide of immigration began to rise and the majority of the population no longer worked the land for a livelihood, there have been poverty, unemployment, and slums in New World cities no different in origin from those in Europe or Asia. The wealthy—or the earlier arrivals—controlled the supply of work and housing, and the newcomers had to take what they could get. This was usually either hand-me-down accommodation, which the better-off had abandoned for the suburbs, or "purpose-built" workers' housing with a minimum of services and an emphasis on cheap construction. It happened everywhere in the era of industrialization and, in North America, it went on happening in spite of public outcry and efforts to improve things because there was a free market in housing and a huge continuing influx of immigrants desperate for shelter of any sort. To accommodate, as the United States was expected to, almost 9 million extra people in the first decade of the twentieth century *by immigration alone* demanded a colossal construction program, and the last people in the world who could supply the capital for such a program were the immigrants themselves.

Workplaces in the cities were often as squalid as homes—were, in fact, often the same overcrowded, unhygienic urban spaces, with whole families engaged in piecework in the room where they lived. Labor laws and factory inspectorates gradually removed most of these abuses, but the residential problem remained. Its solution depended, theoretically at least, on one of two measures: some control of the housing market, or the construction of large amounts of public housing by the cities themselves. Both these ideas were alien to Americans at the turn of the century and, although the second solution was in due course gradually adopted, the first is today little more popular than it was then. Only a complete transformation of both thought and society could overturn a housing market that most North Americans have come to regard as part of the natural order of the universe.[5]

So we have, in North American cities, the familiar distinction between the upper and lower ends of the housing market, the upper end marked by a search for space, privacy, and status and the lower end by congestion, decay, and "misfit" housing, where wealthier homeowners have moved out of large properties, and these have been taken over for multiple occupance by lower-income groups. But now we must introduce another element into the situation: that of the ethnic variety found in so many American cities.

That a city contains rich and poor we may accept; Americans, in particular, may accept it because so many of their poor have become rich, to their own credit. But if it then emerges that all the poor belong to a particular ethnic group or groups, the situation becomes more complex. And if it further emerges that the disadvantaged are easily recognizable, on account of skin color or culture, then mere complexity is likely to grow into bitterness. What we then have is a ghetto.

It is important to remember that, ever since large-scale immigration began, North American cities have had ghetto areas of a sort; that is, quarters of the city in which particular ethnic groups were concentrated. For non-English-speaking immigrants, especially, these quarters served a most useful purpose. They were places where new arrivals could become acclimatized; transit camps where they could learn a new language and new customs before moving out into the wider society of North America, as millions of others had done before them, to become indistinguishable from their predecessors.

But this was only possible if the ghetto was "open-ended," or if the American immigrant truly became indistinguishable from those who had gone before.

And for some ethnic groups neither of these conditions was fulfilled. Social pressures could close off the ghetto exit, and racial differences could never be concealed. This was the case with the Chinese, who settled in their well-known Chinatowns in San Francisco, Vancouver, New York, and elsewhere—areas which, in due course, became tourist attractions. It has been the case, in recent years, with Puerto Ricans in New York and Cubans in Florida, although here the distinctive ethnic features, language apart, are less marked. But it has applied most of all to the American blacks.

A *ghetto*, as the term is understood in North America today,[6] is therefore neither a wholly economic nor a wholly social phenomenon. It is economic to the extent that it is usually an area of low-quality housing and since, as we saw in the last chapter, black Americans (like the Hispanic and the Amerindian) fall below the national averages of prosperity, there is likely to be a concentration of such groups in ghetto areas. It is social to the extent that ordinary market forces would permit wealthier members of these groups—of whom there are many—free exit from the ghetto, but that exit is blocked by hostility on the outside and fear within. And then it is economic again because, once a ghetto area can be identified, its inhabitants find themselves denied access to financial services enjoyed by the rest of the community; for example, in obtaining loans to buy a house or start a business (Fig. 3-2).

Canadian cities have largely escaped this particular problem, although many of them have their built-in national problem of French-or-English. In the cities of the United States, the black ghetto has

---

[5]On this point readers may wish to consult D. Harvey's well-known work, *Social Justice and the City* (Baltimore, 1973), and his more recent writings.

[6]Maurice Yeates and Barry Garner give the following definition of the word *ghetto* in *The North American City* (New York, 1971), pp. 303–4: "The term *ghetto* originally referred to the Jewish ghettos of eastern and southern Europe. . . . As applied to North America, a *ghetto* is a spatially contiguous area of the urban landscape in which the inhabitants have particular social, economic, ethnic, or cultural attributes that distinguish them from the majority of the inhabitants of the country in which they reside. Because of these differentiating characteristics, the inhabitants are, by and large, not permitted by the majority to reside beyond this well-defined area even if they wished, unless these differentiating attributes change sufficiently for them to be accepted. . . . The ghetto is thus a result of external pressure rather than internal coherence."

A different definition is offered in J. T. Darden, ed., *The Ghetto* (Port Washington, N.Y., 1981), p. 14: "A ghetto is an area or contiguous areas of a city where more than 50 percent of a racial, ethnic, or religious group live due to past and/or present discrimination in housing."

developed as a virtually constant element of the urban scene, one or more areas of solid black population, with little mixing except around the edges, where conflict between pressure of population inside and social resistance to expansion outside creates a narrow zone of transition. Within the ghetto there develops a city within a city: it has its own socioeco-nomic differentials (including some high-priced housing) and its own power structure.

The ghetto, in fact, fulfills a function today as it did in the past; in this case, a political function. It is the basis of such power as the black American is able to exert in a society where, overall, he will always be in a minority. Its disappearance, therefore, might not

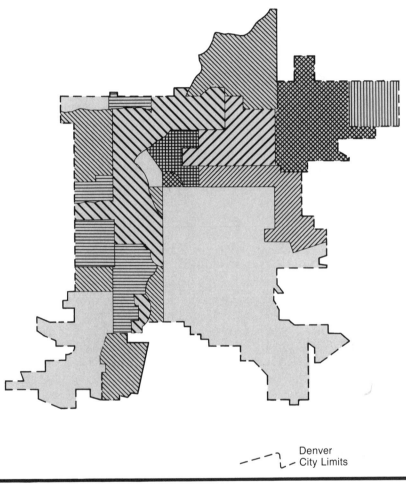

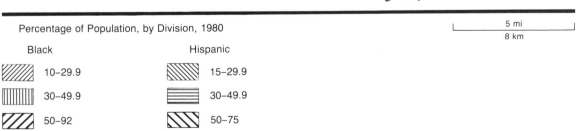

--- ⌐ Denver
   ⌐ ⌐ City Limits

Percentage of Population, by Division, 1980

Black		Hispanic	
▨	10–29.9	▨	15–29.9
▥	30–49.9	▤	30–49.9
▨	50–92	▨	50–75

5 mi
8 km

Fig. 3-2. Ethnic concentrations in Denver, Colorado. Denver PMSA possessed, in 1980, a population which was 5.3 percent black and 11.4 percent of Spanish origin. Both groups showed a high degree of concentration in particular areas of the central city and little extension into the suburbs although, as the map shows, the black population had a higher maximum of concentration (up to 92 percent of the divisional population) than did the Hispanic. The point of interest is the lack of overlap between the core areas of the two ethnic groups: Denver is, in effect, a two-ghetto city. The research underlying this map was carried out to establish the role real estate agents played in creating this ethnic situation. That role was shown to be substantial. (Based on Figs. 2 and 3 in Risa Palm, "Ethnic Segregation of Real Estate Agent Practice in the Urban Housing Market," *Annals*, Assoc. Amer. Geog., vol. 75 [1985], 58–68, by permission.)

necessarily work to the black advantage. But this is a very different thing from denying those who wish to venture out from the ghetto the right to do so.

This is not a problem that will disappear overnight. To disperse a ghetto of a million souls into the wider American society would take a long time, even if we were to make the assumption—the colossal assumption—that such a movement would be acceptable to the receiving majority. In most cities, the more modest goal is to rebuild the worst areas—to tear down the old structures and replace them with new housing. Yet this is a solution to only one part of the problem, and it leaves the racial imbalance within the city unchanged.

Urban Problems and Urban Futures

In the previous sections of this chapter, we have identified some of the features of North American urbanization, and hinted at the problems confronted by the continent's cities. In the present section we shall list these problems more specifically, and conclude by trying to predict future developments.

The urban problems can be divided into three groups: social, physical, and economic or administrative.

1. *Social problems*. The social problems emerge out of the urban life that we have just been considering. In North America, as we have seen, the ordinary social stresses of an urban environment are multiplied by such factors as rapid urban change—which would create social flux in the most stable of societies—and a great variety of ethnic groups. Paradoxically, too, civil rights legislation, although it has been designed precisely to help disadvantaged groups, has contributed in its own way to social problems. It has operated with some success in the economic and administrative spheres but, since social relationships are things about which it is impossible to legislate, these relationships have become a repository for all the hostilities and prejudices formerly spread throughout the community's life.

What appear to be the psychological effects of these urban conditions upon the inhabitants have now become clear. Where rapid change is creating so ephemeral a townscape, where landmarks disappear overnight and are replaced by structures that often have a very limited life span, the instability of the built environment seems to communicate itself to the population. People lose their sense of association with, or pride in, their city and district and they move

frequently, often for no other reason than to acquire a prestige address. Whites become insecure and move because the frontier of the ghetto has advanced too close for their liking, whereas the black or brown inhabitants across the "frontier" develop a ghetto psychology, which inhibits them from moving outside the ghetto to work, even when jobs are available elsewhere. The old concept of a home place quickly evaporates in a society where most people are either transients or siege victims.

2. *Physical problems*. The basic physical problems are twofold: how to fit into the available space the necessary functions and how to provide for the necessary volume of movement into and out of the city. Under "movement" we naturally include the inflow and outflow of people, but we must not overlook as equally vital to the survival of the city the inflow of water supplies and the outward flow of sewage and urban waste. New York's water supply has been precarious for decades, and both Los Angeles and San Francisco bring in their water from hundreds of miles away. And in highly urbanized areas, there is often nowhere to dump waste except in a neighboring community, which can hardly be considered a solution to the problem at all.[7]

Everybody knows that the standard modern solution for a physical lack of space in the city center is to build upward. Although this certainly is a solution to the problem of providing more office space in the central city without increasing its perimeter, it is a solution that generates more problems. For the movements to and from these high-rise buildings are still confined to earth. The workers—or the supplies—moving into these buildings must do so at ground level. In the so-called vertical city, the critical point is where the vertical and horizontal planes meet. Sooner or later, building upward will lead to impossible congestion on the ground.

It is rather curious that the North American concern for economy in using space is generally confined to the centers of cities, whereas the suburbs consist of detached homes, only one or two stories high, and

---

[7]These words were originally written for an earlier edition of this book: in them, the author was offering what he considered a light-hearted glimpse of the obvious. The words took on an altogether new meaning in 1987, however, when the barge *Mobro*, with 3100 tons of a New York suburb's relatively nontoxic rubbish on board, made an epic voyage of over 6000 mi (9600 km) in search of a dumping ground for the cargo, only to be refused dumping privileges by three countries and seven states. It then returned to New York, where the rubbish was incinerated.

stretch for miles. Here the space problem is the opposite of that in the city center: how to provide urban services for a very diffuse population (which, nevertheless, expects services of a high quality); how to provide, for example, transport when the density of population is so low as to make bus lines uneconomical to run, or how to collect garbage when the city extends for many miles in all directions.

Increasingly, however, there is another aspect of the physical problems of the city forcing itself on the inhabitants' attention. This is the sheer physical decay of the urban infrastructure. We have to remember that many of these cities, and their service systems of bridges, drains, and sewers, were built in bursts of rapid growth, and that the systems are now hopelessly overloaded or life-expired or both. Systems that were built speculatively for small cities are now required to serve large, stable, or growing communities. This is true not only of "old" cities like New York, but also of Los Angeles: runaway growth, sustained over decades, has placed an intolerable strain on the city's infrastructure. To rebuild or modernize these systems involves very large sums of money indeed.

3. *Economic and administrative problems.* As soon as we speak of "competition" for space within the city, we imply an economic factor, for this kind of competition is normally decided by bids made on the open market; that is, each urban space-user, whether as an individual or as an organization, decides what space is worth to him or her and offers accordingly for it. Space has become a very valuable commodity—so valuable, in fact, that usable space is being sought out and even the air space above a railway line or an old dock can be bought and sold.

But in most cities, this competition applies to only *some* of the area and *some* of the urban functions. Alongside the economic life of the city, which produces the kind of competition so far described, there is the administrative life—the necessary services of the community, which may not be revenue producing. They cannot compete for land, but the city cannot afford to be without them. Often, the only way to secure space for them is to exercise in some way a government's power of eminent domain—to override the workings of the open property market. This has, of course, become common practice today; cities have zoning ordinances and issue building permits. But it is often a complaint in city hall that the city cannot acquire, at reasonable prices, the land needed for new hospitals or schools. Compared with the private

and corporate individuals who make up its population, the city itself has become a pauper.

The poverty of the cities today stems from quite different sources: one is the tendency of the people who work in them to live, spend, and pay taxes outside the city. The flight to the suburbs steadily reduces the city's tax revenues; yet its expenditures cannot be proportionately reduced; the streets must still be lit, policed, and cleaned, and fires must be fought. What is particularly debilitating for the city is that higher-income taxpayers usually move outside the city and are lost to it, while inside the city there is a growing concentration of lower-income groups and welfare recipients. By this process, the tax base of the city is steadily eroded, while expenditure rises.

One obvious way to overcome this problem is to allow the city to increase its area so as to include its suburbs. Cities all over the world have been doing this for decades. Equally consistently, the suburban communities fight against annexation: they have no wish to give up local control, or to pay higher local taxes to help subsidize the administration of central slum areas. This is not an American refrain only; it is at least as characteristically European. But the only practical alternative to annexation, if the city is to find space enough and revenue enough to survive, will be the takeover, for public purposes, of central areas, where some commuters have their offices, a process that one can only feel would be even less popular to the suburbs than that of being swallowed up by the city. Urbanization today, on the North American scale, calls for new structures of local government and administration, and some metropolitan areas, such as Toronto, have already experimented with these.

Confronted as they are with so wide a range of problems, what does the future hold in store for North American cities? We can identify one or two probabilities but, after that, we can do no more than pose questions, for it is impossible to be certain about the direction that urban development will take.

What seems fairly certain is that the city of the future will:

1. *Pay increasing attention to amenity considerations,* whether we express these in terms of the range of facilities and services the city provides, or the amount of urban space per inhabitant. There is an interesting history to be written of changing attitudes toward amenity, from the time when only the rich possessed access to any amenities at all (and those private to themselves and jealously guarded—woe betide the

Edmonton, Alberta: The West Edmonton Mall. With 5.2 million sq feet of floor space and 15,000 employees, this was for long the largest mall in the world. *West Edmonton Mall*

peasant in Europe who shot the king's deer!), through the democratization of amenity in the nineteenth century—public parks, public baths, public libraries—to the present day, when the amenities available are more numerous than ever before, but increasingly are privately enjoyed—in the home, garden, and swimming pool. But whatever form the amenities take, private or public, there is likely to be an increasing emphasis on quality of life and an effort to end overcrowding.

2. *Pay increasing attention to preservation and restoration.* The ruthless removal of the old to make way for the new, which for so long marked out North American cities, can be expected to slacken off in coming years. A society with such large productive surpluses can afford not to demand the utmost use of every building plot in the city; can afford, in other words, to preserve some of its past and even find new uses for old structures.

This trend has been very marked in the past two decades. Sometimes it is the city that has taken the lead, by intervening to save old landmarks; sometimes private enterprise has found it profitable to convert old buildings. At Lowell, Massachusetts, for example, some of the continent's earliest and largest textile mills now house computer assembly plants, and in old market areas of Boston and San Francisco dozens of new businesses flourish in what were once warehouses. There is scarcely a city across the continent that does not have its band of restoration enthusiasts and its project; scarcely an old waterfront area that has not had new life breathed into it, now that its former usefulness as a port is at an end.

So far, the economic success of these projects has

been generally encouraging. If success continues, so too will the trend. Such projects do not, of course, actually solve urban problems, but they are welcome evidence that city folk are willing to stop and think before bringing in the bulldozers.

3. *Possess an increasing ethnic variety.* As we have already noted, recent immigrants (especially those from Asia and Latin America) head for the cities. Canadian immigration since World War II has been very largely a matter of making for Toronto; almost one out of every two newcomers has made it his or her initial destination. Vancouver is the gateway from, as well as Canada's gateway to, the Orient. In the United States, Los Angeles has been called the new Ellis Island: it serves today, as New York's immigration terminal served for so long, as the entry point for a bewildering variety of peoples. The cities will, it is tacitly assumed, continue to absorb and employ them all.

This may lead to heightened tensions between ethnic groups, but it need not. It would be hard to deny, for example, that Toronto today, with its great ethnic variety, is a much livelier place than it was forty years ago when it was dominated by people of British origin. The critical factor seems to be the presence or absence of what we can call a permanent underclass; that is, an ethnic group stuck at the bottom of the socioeconomic ladder and prevented, by whatever circumstance, from climbing up it in the way that, say, the American Irish or Italians have climbed. Whichever may be the group that forms the underclass, if there is one, then social unrest seems to be inevitable.

Of those three trends we can be fairly certain. The following three questions, on the other hand, admit as yet of no sure answers:

1. *Will the movement of people and business out from the central city to the suburbs continue into the years ahead?* Are there more, and ever larger, suburban malls and office blocks still to come? Will the bulk of morning commuters eventually travel outward from their homes to work in still remoter suburbs, rather than inward to the CBD?

There is some fragmentary evidence that a turning point of sorts may have been reached. A central city, after all, represents a tremendous capital investment: it makes no sense for a million people to turn their backs on it, write off the investment, and build a new city outside it. The Carter administration of 1976–80 made federal funds available for city-center retail projects, and sought to discourage the further growth of suburban shopping centers. Will reversal work?

2. *Is there a future for "gentrification"?* One of the interesting features of urban North America over the past two decades has been that a considerable number of urban residents, instead of following the general movement out to the suburbs, have reversed the tendency by taking over inner-city houses and renovating them. Most of them have been young and active, and some of the results, if small in scale when judged against urban dereliction generally, have been locally dramatic. The long decay of these buildings has been halted: they have now become sought-after properties, and the social status of the neighborhood has risen.

Not everyone approves of *gentrification*, which is the name given to this process.[8] Sought-after properties command high prices, and the renewal operation is likely to displace original occupants to whom the area was affordable and was home. Where are the displaced to go? To this objection, it can be answered that there is nothing to stop those original inhabitants from doing their own gentrifying (and some of them have succeeded), and that it must surely be better to increase the value and improve the quality of property than to let it decline still further.

The argument is unresolved. Gentrification may, as some observers have predicted, prove to have been no more than a fad of the seventies and eighties, with no lasting effect. Or it may be the spearhead of a movement to turn the city right side up again, and bring back life to the dead or dying.

3. *What is the optimum size for a city?* There has always been a general assumption—and nowhere more so than in North America—that once a city made a successful start, its population would inevitably grow larger with time. Indeed, most American cities record with pride their growth rates as a measure of their success.

But there are presumably limits to this growth—desirable limits, and not merely the constraints of geography. There comes a point where mere size—the miles that have to be traveled within the city—be-

---

[8]See J. J. Palen and B. London, eds., *Gentrification, Displacement and Neighborhood Revitalization* (Albany, 1984); and N. Smith and P. Williams, eds., *The Gentrification of the City* (London, 1986).

Jill R. Schuler and others, in "Neighbourhood Gentrification," *Urban Geography*, vol. 13 (1992), 49–67, a study made in Cleveland, found a high correlation between gentrification and (1) percentage of college-educated residents, (2) percentage of whites, (3) percentage of residents in the 25–40 age range, and (4) percentage with high median incomes.

comes an obstacle and when, with modern electronic communications, dispersal is no problem to commerce or industry, and commuter journeys reach their practical limit. There will be, in other words, a disincentive to further growth.

The fastest-growing cities in North America today are not the giants but the medium-sized cities. New York CMSA, with its more than 18 million inhabitants, is still the largest urban area in the two countries, but it is interesting to note that, in *Metropolis 1985*, a predictive study made in 1960, the population forecast for New York in 1985 was not 18 million but nearly 24 million.

This makes it probable that, among the current crop of half-million-plus cities, there are the metropolises of tomorrow, where new factors shape urban life, and new forms of communication make much of what goes on in today's CBDs unnecessary. As to which of these cities will emerge as the new leaders—Calgary or Charlotte, Dallas or Phoenix or San Jose—it is interesting to speculate and impossible to predict.

4
Government, National and Local

Parliament Hill, Ottawa. *National Capital Commission. Photo by Terry Atkinson*

The patterns of a country's human geography are shaped not only by the physical factors that govern land use or routeways, but also by the political conditions under which settlement and development take place. In North America, there are a number of such conditions whose effects on the distribution of economic activity have been of great importance, and it is the purpose of the next three chapters to call attention to them, as a necessary preliminary to the study of the continent's regional geography.

The Nation-States

The first and most obvious political condition is that in America north of the Rio Grande there are two separate nation-states. They share a common border from the Atlantic to the Pacific, with Alaska as a detached part of the southern nation "behind the lines." Although the two countries are justly proud of their good relations, which today make it possible for them to think in terms of military cooperation rather than military rivalry, the situation has not always been so happy, nor is the relationship free from periodic tensions. Some of these concern the exploitation of resources the two countries share or the environmental problems caused by exploitation. Others arise because each country has given shelter to refugees from the other—the British turned French settlers out of Nova Scotia, many of whom then moved to Louisiana; Loyalists moved north into Canada at the end of the Revolutionary War; American draft resisters found shelter in Canada in the 1960s.

But in a deeper sense, the very existence of two nations in this one continent can only be explained in terms of a longstanding aversion; an aversion that has kept Canada, with its smaller population and slower industrial development, doggedly unmoved by the apparent advantages it might gain if it joined the United States. And these are quite substantial. Not only would Canadians be able to buy more cheaply those American goods on which they now pay customs duties, but they would be linking themselves with a country where the Gross National Product per capita is more than 20 percent greater than in their own. In other words, for their independence Canadians are willing to pay an economic penalty that may amount to the renunciation of 20 percent of their potential wealth. The Canadians are, in the words of H. Hardin, "the world's oldest and continuing anti-Americans."[1]

The existence of these two nations, therefore, raises the primary question: Why is there a Canada at all? The existence of any nation presupposes two sets of factors—*negative* factors, which deter it from linking up with a neighbor, and *positive* factors, which give it, as an independent unit, a reason for its separateness. In the Canadian case, however, the negative factors have been, and still are, much stronger than the positive. Most Canadians are sure that they *do not* want to be Americans; what they *do* want to be or to do is much less clear.

Canada came into existence as a political unit in 1867. The confederation born in that year grew out of two considerations: (1) the practical need for a working compromise between the French in Lower Canada (Québec) and the British in Upper Canada (Ontario), and (2) fear of the United States, where the Civil War had ended in 1865, leaving the country with the world's largest army and a vocal section of the public suggesting the piecemeal takeover of Canadian territory.

Thus, fear played a part in the forming of the Canadian union, and, although the military fear of the 1860s soon passed, the early economic fears that led

[1]H. Hardin, *A Nation Unaware: The Canadian Economic Culture* (Vancouver, 1974), p. 3.

84

Canada to build canal and rail routes to the West simply to compete with the American routes have never evaporated; rather, they have grown with the threat of economic and cultural domination by the United States. To justify these fears, Canadians could point to their growing dependence on the United States as a trading partner, to the impact of American culture at the popular level, and to the enormous total of American investment in Canadian industry as described in Chapter 8. But regardless of the validity of these particular arguments, a defensive stance of Canada's government and people over against the United States has become second nature. In the words of a wise Canadian:

> It must be recognized that the nature of the relations between Canada and the United States is such that explosions of Canadian resentment will occur periodically. The intimacy of the relations of the two peoples and the disparity of power between the two states make it inevitable that American indifference and power will provoke Canadian resentment from time to time. The United States, it is to be hoped, will remove such grievances as can be removed and will suffer the outbursts philosophically, will avert the accumulation of grievances by dealing firmly and responsibly with grievances as they arise, and so keep the explosions at as long intervals as possible.[2]

In a country the size of Canada, the physical obstacles to unity are tremendous. One would, therefore, expect that the negative grounds of separateness and the physical barriers to unity would need to be balanced by very strong positive motives for political coherence if the nation were to survive. Yet the striking fact is that in Canada, the positive forces of nationalism—of Canadian nationalism, at least—are hard to discover. The French and British elements (who together make up two-thirds of the population) agree that they do not want to be Americans, but they agree on virtually nothing else, and certainly not on an emotional attachment to the British crown.

This failure to establish national goals and achieve agreement about them is the more striking when we realize that Canada is now playing a most responsible role at the international level. Before World War II, Canada was a young country with a small population, and little was asked or expected of her except within the British Commonwealth. But the new generations of nations born since 1945 have made Canada "middle-aged" and certainly middle-sized. It is now one of the relatively large and mature nation-states, and acts as such on the international scene. But it has done so without ever defining clear goals for the nation at home.

The failure is not for want of trying. Many individual attempts have been made to give expression to what it means to be Canadian. The country has been active enough on the international stage, supplying programs of relief, and troops for United Nations forces, that it must seem as if outsiders have a clearer view of Canada's role than her own citizens do. But the fact is that all such attempts at defining a national role must start by confronting two facts of geography. One is that Canada consists of four or five separate areas of settlement, all of which have easier ties on the ground with the United States, south of the border, than they do with each other. The second is that there is a very obvious role for Canadians, but it is not on the international scene: it is the development of the huge Northland which is theirs and theirs alone.

One of these facts of geography makes it very difficult to conceive of Canada as a whole, and the other means that, if Canada withdrew today from the outside world, Canadians could spend the rest of their lives, and their children's lives, in ceaseless efforts to make the Northland a truly integral part of their country, without completing the task. In practice, however, they show little enthusiasm for this task: as Trevor Lloyd once wrote: "Canadians may be a venturesome, northern people at heart, but when it comes to the things that matter they continue to live as far south as their citizenship permits" (see Fig. 4-1.)[3] It is easier to send out aid and peacekeeping troops.

We can only ask: Is there no middle way between a role which is wholly inward-looking, and one which consists of Canadian actions in remote corners of the world? This is a question raised by Hempson and Maule in their 1991 book and answered, at least by one of their contributors, under the title "Canada Discovers Its Vocation as a Nation of the Americas." From this standpoint, Canada has been distracted, as it has come to maturity, by its own strong, traditional

---

[2]W. L. Morton, *The Canadian Identity*, (Madison, Wis., 1968), p. 82. For another, more recent and worthwhile view of Canada's national origins, see G. Wynn, "Forging a Canadian Nation," in R. D. Mitchell and P. A. Groves, eds., *North America: The Historical Geography of a Changing Continent* (Totowa, N.J., 1987).

[3]In J. Warkentin, ed., *Canada: A Geographical Appraisal* (Toronto and London, 1967), p. 585. In this same connection, it is intriguing to find one of Canada's leading geographers, L-E. Hamelin, publishing in 1978 a book which, in its English translation, bears the appealing title, *Canadian Nordicity: It's Your North Too* (Montréal).

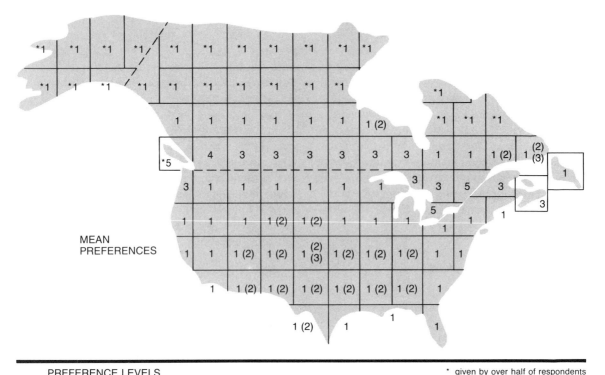

MEAN
PREFERENCES

PREFERENCE LEVELS

5. Would like very much to live there.
4. Would like to live there.
3. Neutral about living there.
2. Would prefer not to live there.
1. Would really dislike living there.

* given by over half of respondents

( ) together with major response,
given by over two–thirds of
respondents

Fig. 4-1. Canadian residential preferences: The result of a nation-wide survey of Canadian attitudes, and one which seems to bear out Trevor Lloyd's view, quoted in the adjoining text, of Canadian preferences for living "as far south as their citizenship permits." (Repro-duced from "Canadian Residential Preferences," a map by H. A. Whitney and G. W. Brown, presented to the Canadian Association of Geographers in 1975.)

links with Europe, and by the cold war, which made it—because of its position—the watchman at the gate of North America against possible attack over the North Pole. It is now time for the nation to recognize that things have changed, that where it belongs is in the Americas, and that there it should work to build up its future relationships.

It has to be said that the Canadian government has itself not been particularly helpful in presenting national goals, because of its habit of delegating some of its own powers to the individual provinces. The result has been that, in matters of policy, the provinces tend to grow apart, so that the idea of a truly national policy or goal becomes more and more remote. As we shall see in the next section of this chapter, this process of "growing apart" reached a critical level in the past decade. But the Canadian government has itself contributed to the process by adopting measures over the years that do little more than help one province

or region at a time. Legislation helping the Prairie wheat farmers is followed by subsidies for the Atlantic Provinces or government investment in British Columbia, in a process that has been described as "political self-bribery." Within Canada, therefore, what counts is not so much a national policy as the play of regional interests. But systems that promote regional values and consciousness are working against national consciousness, not contributing to it. Only if the nation has something to offer as a nation can feeling on the national level transcend regional sentiment.

We have considered the geographical factors which make nation forming difficult in Canada. But we cannot leave the subject without referring to the historical factor which, for the Canadians, can never be left out of consideration. This is the French–English divide within the nation.

In recent years, Canada has gone through an identity crisis unprecedented in its history as a nation. It had no sooner celebrated its centennial in 1967 than the separatists in French Québec stepped up their campaign to take the province out of the federation—to set up an independent nation. It is the same part-against-the-whole issue as the United States faced in the years before 1861; even the arguments and threats of the two sides sounded familiar. In an era when communities of less than a million people are sovereign states and members of the United Nations, there is nothing bizarre about the 6.5 million inhabitants of Québec seeking a separate identity; the questions at issue were, and are, rather how many of the 6.5 million actually want this, and given their particular position and resources, how likely would they be to create a viable state, and not merely to create it, but also to maintain it without lowering their standard of living.

We shall return to this question of viability in Chapter 13. In the context of the present section, however, there is one other matter that must be raised: what would be the future of a Canada *without* Québec? What would happen—what *could* happen—to this remainder? How would the Atlantic Provinces survive as part of the nation, given the enormous geographical separation between Gaspé and the Ottawa River (not to speak of the considerable French-Canadian population in these provinces, which might well prefer adherence to Québec)? And would this rump of Canada-without-Québec command the loyalty of the newly rich provinces of the West? If Québec left the confederation, would others not follow?

We can only wait and see. In the meantime, and whatever its justification as a line of separation, the United States–Canadian border exists. Its effect on the movement of goods and people has been somewhat reduced by sensible arrangements made by the

Québec City: The Château Frontenac from the Place Royale in the lower town. *Communauté Urbaine de Québec, Office du Tourisme et des Congrès*

two governments, and today it is crossed by oil pipes and power lines, while international agreements cover the movement of tourists and the flow of rivers. But it still has its effects, as we shall see in later chapters, in dividing into two some regions whose interests are in one, and in acting as a kind of strainer on the flow of ideas and even of money from south to north across the continent.

A Federal Constitution

Both Canada and the United States have federal constitutions; that is, the powers of government are divided between the national and the state or provincial governments. But there are interesting differences between the two constitutions, both in the principles that guided the original division of powers and in developments since then.

The Constitution of the United States was written in the year 1787 as a statement of the principles, laws, and rules by which the new nation should be governed. Its framers, having gained their freedom from what they considered an oppressive central government (in Britain), deliberately specified and restricted the powers to be given the new national government, and reserved all unspecified powers to "the States respectively, or to the people."

The British North America Act of 1867, the act that created a Canadian federation out of the four then-existing provinces of Québec, Ontario, New Brunswick, and Nova Scotia (by 1873 they had been joined by Prince Edward Island, Manitoba, and British Columbia) was, by contrast, an act of the British Parliament. It was not designed as an exhaustive statement of the rules of government, but rather it served as the basic constitutional document of Canada. The awkward fact was, therefore, that Canada's government existed by virtue of the action of Britain, not of Canada.

In 1982, the government of Canada brought about the "patriation" of its constitution: that is, the transfer of the ultimate lawmaking power from the British Parliament in Westminster, where it had been held since British Canada was organized, to the Canadian Parliament in Ottawa. This might seem simple enough, for Britain was quite willing to enact the necessary releasing legislation. But in Canada, the proposal merely raised and intensified the old problem of the division of powers between the nation and its provinces. Only after long and difficult negotiations within Canada was the constitution patriated to a country still in dispute over the powers of the national government. But at least, from then on, the dispute was purely Canadian.

Canada achieved its federation in 1867, in the aftermath of the American Civil War (1861–65), in which strong state governments had defied national authority, and its constitution was a kind of mirror image of that of the United States, assigning only specified powers to the provinces and reserving all others to federal authority. But in the years since, the balance of power in both countries has tended to tilt in the opposite direction from that codified in the constitution.

In Canada, on one hand, the provinces have tended to *gain* power at the expense of the central government. This can be explained partly by the realization that it was one way—perhaps the only way—in which the central government could hope to reconcile the conflicting interests of the provinces, especially those provinces that were either strongly distinctive in their culture (like Québec) or rich in their resources (like Alberta, with its oil and gas). Partly, too, it resulted from the fact that the British North America Act assigned to the provinces powers over "property and civil rights," a catchall clause that proved to be capable of almost boundlesss enlargement. And boundless, too, has been the scope for federal–provincial disagreements over minerals, lands, fisheries, and other forms of "property."

In the United States, on the other hand, the federal government markedly increased its size, scope, and powers as the years went by. But this has not been a straightforward process. In order to intervene in any field of government action, the federal government must first prove (in the courts, if necessary) its right to do so by reference to the terms of the constitution. Whenever it attempts a fresh initiative, it can count on the dogged opposition of the "states' righters," who will argue that it is acting beyond its powers.

A good example of this constitutional conflict between the federal and state levels of government concerns what we nowadays know as regional planning. The most rational or "natural" unit of planning may well not be a state as such; it may be, as in a river basin, an area covering parts of several states. The federal government has no powers to plan the economy or land of any state, let alone several states. If it is to act at all, it must therefore find a justification for doing so—some authority specifically granted to it under the constitution.

In the case of a river basin (and the Tennessee

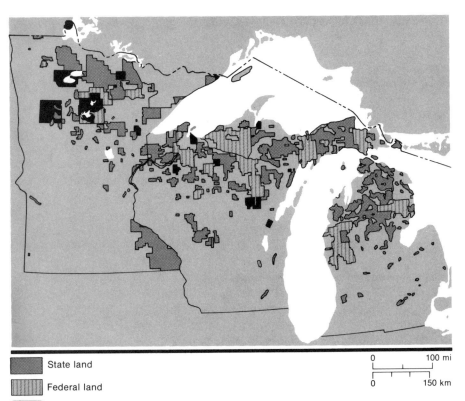

Fig. 4-2. The United States: Public and special land ownership. The northern Great Lakes region of the United States, where there is a characteristic intermixture of federal, state, and special owners; here the latter category is represented by the Indian reservations.

State land

Federal land

Indian land

0   100 mi

0   150 km

River comes immediately to mind) the federal government has found that authority, rather ingeniously, in the clause of the constitution that gives it the task of overseeing interstate commerce. A river that cannot be navigated because of natural obstacles or flooding (as was true of the Tennessee River in the early 1930s) is certainly a barrier to interstate commerce. The federal government—or so it argued—may therefore intervene with a plan to remove the barrier, and if along the way other aspects of the valley's life are affected, then that is purely incidental.

In this kind of backhanded way, the federal government tries to do good by carrying out functions of government that its opponents claim it does not constitutionally possess. The truth of the matter is that a federal authority has two ways, and only two, of overcoming the legal limitation of its powers: example and bribery. Under the first heading, "example," it can create model legislation or codes of practice—for planning, labor relations, conservation, or health. It can then make available this information, and it can also apply its model programs to any areas it owns or controls itself, in the hope that lower levels of government will follow suit—that they will find

the example worth imitating. And since the U.S. government owns 763 million acres (309 million ha) of public land and controls the District of Columbia, in which the federal capital stands, it has plenty of space, and different kinds of space, in which to follow its own advice (Fig. 4-2). The federal government of Canada owns almost 1 billion acres (405 million ha). The management of these federal lands can and should serve as an example to the states, to local governments, and to individuals.

The second method available to a federal government tied by its constitutional limitations is a kind of benign bribery. It is to offer money to a state or local authority to follow a particular course or program, and to withhold it if specified standards are not attained. It pays with federal money to have its will done in spheres where it is powerless to insist. Today the number of such federal programs is enormous. In many of them, the principle of "matching funds" is used to enforce standards—the federal government will only share in the cost if it approves the program. The local government, or the individual, in order to obtain the funds, meets the federal standard.

It was by these methods, as we shall see in Chapter

The lower levels of government: The state capitol. With its fifty states and federal structure, the United States possesses fifty capitol buildings, most of them resembling either this one or the fifty-first— the federal capitol in Washington, D.C. The size and dignity of the building, however, has nothing to do with the size of the state; this is the capitol at Providence, Rhode Island, smallest of all the states, with a land area of 1212 sq mi (3140 sq km). *Greater Providence Convention and Visitors Bureau,* © *Ron Hagerman*

5, that the federal government introduced, in the 1970s, measures for safeguarding the environment—necessary, overdue, and combining our two principles. By way of example, the U.S. government bound itself to issue for every project it proposed an Environmental Impact Statement, to describe the project's effect, list the alternative strategies available, and alert the public to possible environmental damage before that damage is done. By way of payment, the government set aside several hundred million dollars for a program of cleaning up polluted lakes and of continuing the work begun in clearing out the pollution of slum areas in cities a decade or more ago. By methods such as these, a federal government can gradually standardize practices and improve the quality of life, without ever possessing the power to intervene directly.

Differences in law and tax structure between states and between provinces have always been important; they are a natural consequence of the federal division of powers. Some of these differences are sufficiently large to affect economic activity: low taxes attract firms, whereas high levels of compulsory welfare spending drive them away. States realize this: Nevada has built a flourishing economy and a huge tourist trade precisely on a relative *absence* of legislation and the slogan "Anything Goes."

If federal funds are withdrawn, states will have to look for new sources of revenue, and each will be on its own in what has been called a "State Eat State" situation. Already the share of personal income taken by state and local taxes is almost twice as high in some states as in others. Still more uneven, however, is the distribution of taxable resources. A state that possesses oil or coal can impose a *severance tax* on its extraction; it is, in effect, charging for the use of its resources, and users in other states who buy these resources are contributing revenue to the owner state, which can then afford to tax its own residents less heavily. Montana charges about $2.00 on every ton of coal extracted (almost all of which is sold outside the state): Kentucky charges $1.25. Alaska receives an enormous revenue from oil royalties and has only 500,000 inhabitants to help the state spend it.

But many states have no significant mineral wealth. Their products are agricultural or industrial. To obtain comparable revenues, they would have to place a severance tax on corn or cars. In the future, they may well do so, if only in retaliation against the followers of the severance tax principle. But the prospect is not a happy one.

The Lower Levels of Government

Frustrating as it must be for the federal authorities, most of the powers they would like to exercise—and indeed try to exercise by the methods mentioned— are possessed instead by the states and provinces. It is at this level of government that the reserves of

power are to be found. In particular, the all-important police power rests with the states; that is, "the general authority of a government or sovereign entity to take action and legislate in the public interest or the general health, safety, welfare, and morals of the people"; it has to do with "the *regulatory* power of the state."[4]

These powers the states can and do delegate to their subunits: counties, cities, or townships. What they will not do is to surrender them willingly to the federal government. But the particular difficulties here are that (1) some states have delegated too much of their power to the lower levels—to governments that have neither sufficient resources nor sufficient scope to exercise them properly, and (2) some states exercise the police, or regulatory, powers much less than others. This is true of all aspects of government

[4]R. R. Linowes and D. T. Allensworth, *The States and Land-Use Control* (New York, 1975), p. 40.

The lower levels of government: The county. This is the Monroe County Court House Bloomington, Indiana, a state noted for its imposing court houses. Rivalry among early settlements to become the county seat was intense, devious, and often downright criminal, but it was a competition well worth entering in view of the business this status assured the successful town. *Bloomington/Monroe County Convention and Visitors' Bureau*

activity, but it is especially true in two areas of particular interest to geographers—the economy and the environment. One of the underlying causes of economic weakness and misuse of land and resources is the reluctance of state governments to enforce regulations that, from a national point of view, are desirable; that the federal government would enforce if it had the power to do so, but that lie outside its constitutional power.

A study[5] of the performance of state governments in the field of planning revealed in the mid-1970s how wide were the discrepancies from state to state. The results showed that the state scoring highest marks in this field was Hawaii. Second was Wisconsin, a state long noted for its concern with issues of environmental management and the "think tank" for much of today's planning legislation. The Middle Atlantic states scored high, no doubt because the pressure of population and the demands on space have obliged them to face planning issues squarely and so, too, did Oregon and California. At the other end of the scale, Alabama, Arkansas, and Mississippi had done little in this field; the idea of the state exercising planning powers found only very limited public acceptance, as it did in Nevada.

A number of states have assigned their powers to local governments where special interests can too easily block efforts to regulate: county interests block state plans, and recalcitrant townships block county plans. But in the 1980s, with environmental issues in the news and environmentalist groups doing their best to keep them there, the states seem to have resumed some of the regulatory powers they too readily handed over. As Bosselman and Callies comment, "states, not local governments, are the only existing political entities capable of devising innovative techniques and governmental structures to solve problems such as pollution, destruction of fragile natural resources, and the shortage of decent housing that are now widely recognized as simply beyond the capacity of local governments acting alone."[6]

The state governments, in other words, should govern; they should exercise the powers they have fought so hard to keep in their own hands. If they do not, nobody else can.

---

[5] A. J. Catanese, "Reflections on State Planning Evaluation," *State Planning Issues 1973* (Lexington, Ky., 1973), p. 27.

[6] F. Bosselman and D. Callies, *The Quiet Revolution in Land Use Control* (Washington, D.C., 1972), p. 3.

5
Government,
Land, and Water

Government and land: One of the U.S. federal land offices through which the Public Domain was disposed of. This one was located at Hollister, Idaho, in the early 1900s. *Denver Public Library, Western Collection*

Public Land and Private Land

One of the earliest problems to confront the infant republic of the United States of America after its formation during the war of 1776–83 was the creation of a land policy. No branch of the small federal government was more active than its Committee on Public Lands; none contributed more documents to those early records of government activity, which today fill the volumes entitled *Statutes at Large*.

There were excellent reasons why this should be so. There was, in the first place, the undeniable political fact that the United States, though now independent, shared the continent with four other powers. To the north lay Canada, retained by the British when thirteen, but only thirteen, of the American colonies opted for independence. To the west lay the French territory of Louisiana, claimed for France in the most grandiloquent, but vaguest terms by seventeenth-century explorers who had little idea where they were, let alone what they were claiming. Immediately south of Georgia the Spanish Empire began, an empire that stretched west to the Pacific and, for that matter, south to Cape Horn. Far away in the Northwest the Russians were active, small in numbers, but feeling their way down the coast into what is now northern California. Whatever decisions were taken about land, they were going to have international repercussions.

Then there was the economic consideration. It was one thing to set up, even on a modest scale, the machinery of a federal government; it was quite another to pay for it. The War of Independence had cost enormous sums, and the only sources of revenue available to the new government were (1) customs and excise duties, which, since most colonial trade had been with the ex-enemy Great Britain, might well take time to amount to anything, and (2) the sale of what-

ever assets it possessed. There was, in practice, only one such asset—land. The sale of land would have to see the government through its early years.

The third reason the land question occupied so much government time in those first two decades we may call ideological. For here was a situation without precedent. In Europe, from which most of America's colonists and institutions had come, there was no such thing as land that belonged to nobody. There were small areas of common land to which traditional rights of use existed, but even including these, all land ultimately had its owner, a landlord who held from an overlord, who in turn held from the king. The concept of land owned by a government rather than by an individual was a novelty. The Spanish Empire, for example, was the personal estate of the king of Spain. All grants of land originated with him. But now a republican government had taken possession, in the name of the people, of a huge area not included in the original thirteen colonies (now states)—an area calculated in 1790 at 888,685 sq mi (2.30 million sq km), to which were added by the Louisiana Purchase of 1803 a further 827,192 sq mi (2.14 million sq km), and at various dates between 1819 and 1853, over a million sq mi (about 2.60 million sq km) of the former Spanish Empire.

Apart from the precedents set by the land policies of the former colonies—policies that had been very diverse in character—there was nothing to guide the new government; no example to follow; nothing but the ideals of the Constitution and the practical constraints of money, counterclaims, and defense. It is not surprising that mistakes were made: what is surprising is that, in these circumstances, policy was formulated as quickly as in fact it was, mistakes and all.

Eighty years later, the land question also played an important role in the negotiations over Canadian federation. West of the developed areas in Upper Canada—the peninsula of southern Ontario—there was a great wilderness barely marked by the settlements on the Red River of Manitoba and the goldfields of British Columbia. This land had been the preserve of the Hudson Bay Company, whose interest in furs had done nothing to encourage human occupance of the area. At confederation or shortly afterward, most of the area passed into federal hands, a small government overseeing a very large area, but with one advantage—it could learn from the mistakes of the United States. The Canadian government could, and did, formulate a land policy marginally more suitable than that of its neighbor, while imitating it in so many ways, just as the American system was imitated, with local amendments, in the Australian settlements and the Argentinian colonies as the years went by.

Original land titles in British North America were European: the Crown granted land to individuals or to chartered companies, and they in turn made grants to settlers. There was no shortage of land: there were differences in the size of grants but, from the earliest days, there was a clear distinction between lands already granted and lands still vacant; a distinction that, in due course, became that between private and public land. Each colony evolved its own policies about size of grant, conditions of sale, pattern of holding or survey, and attitude to squatting (that is, the occupance of public land by private individuals who had no legal title, but who claimed title by right of priority).

The new federal government, steering a middle course between its belief that the public lands should be distributed to as many of the public as could benefit from them and the undeniable emptiness of the public purse, formulated an initial land policy that rested on two propositions: (1) land should be sold, at a fixed price, to raise revenue for the government; (2) no land should be opened for sale before (a) Indian title to it had been extinguished by treaty (by which it was hoped to put an end to the long series of Indian wars), and (b) it had been surveyed by the government. The latter principle of "prior survey" was something of a novelty; whereas it existed in an embryonic form in New England, in the South it was almost completely new. In the southern states, the land claimant normally went out, staked a claim, and then hired a surveyor to map it. The idea that emerged from the Committee on Public Lands was that not only the *location* of available lands would be

predetermined by the survey, but also the *shape* and *size* of the unit.

The problem was simply stated: how to carry a survey across unknown lands inhabited by often-hostile Indians, given a shortage of trained surveyors and a lack of money. The solution adopted was to become, in the following century, one of the most formative of all influences on the North American landscape. It was to survey the public lands in 6-mi (9.6-km) squares, forming a *township* of 36 sq mi (93.2 sq km), and later individual square miles (2.59 sq km) within the township, which were known as *sections*. Later still, the sections were divided into halves, quarters, eighths, and even sixteenths (Fig. 5-1).

The Committee's reasoning was clear and logical. The simplest and quickest operation a surveyor could be called upon to perform was to lay out a straight line, from east to west or north to south. This yielded a cheap survey, within the competence of the available surveyors.[1] The only serious problem was that caused by laying out a rectangular grid on a spherical surface, and the surveyors were given detailed instructions about making allowances for this.[2] Besides the benefits of speed and cheapness, the system also had the advantage that it could always be rechecked by astronomical fix. It was hoped, too, that its regularity might reduce the amount of litigation about boundaries, which arose wherever the casual southern system of "metes and bounds" was in force, and history has proved the expectation valid. Only where the gridiron of squares ran into patterns of previous French (Fig. 5-2) or Spanish landholdings did it cause problems. These exceptions apart, the pattern of survey initiated in 1785 in the lands immediately beyond the Pennsylvania state line marched from there (Fig. 5-3), westward across the continent, in an endless succession of squares that stretched all the way

[1]Anybody who imagines, however, that a straight line is something that even a child could lay out needs only to consult the 1:24,000 U.S. topographic maps today to see into how many different shapes a mile square can be missurveyed. In defense of the very first surveyors, however, it can be argued that they had to start work in the hilly, forested terrain of what is now southeastern Ohio, a peculiarly difficult area in which to see where they were going.

[2]The story of the survey is well told by W. Pattison, *Beginnings of the American Rectangular Survey System, 1784–1800,* Univ. of Chicago, Dept. of Geography Research Paper No. 50 (Chicago, 1957); and particularly by Hildegard B. Johnson, *Order Upon the Land* (New York, 1976).

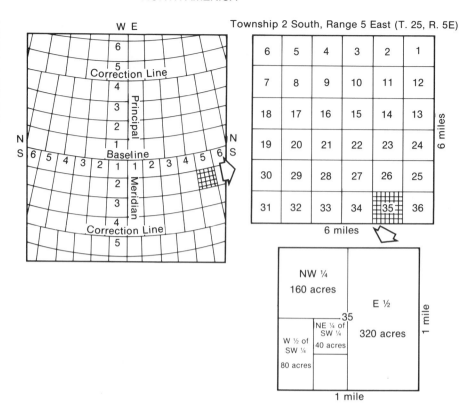

Fig. 5-1. The rectangular survey system of the United States, adopted in the last years of the eighteenth century and continued, in its essentials, to the present day on all U.S. land disposed of since. The same system was adopted by the Canadians for the survey of their own West.

to the Pacific Ocean, and northward in due course to cover the Canadian prairies in their turn. Eventually, it is calculated, gridiron survey covered 69 percent of the surface of the United States.

That the surveyor's checkerboard has imposed itself on the landscape of western North America is evident from one's first flight across the country. That it provided a framework for subsequent land policies of the American and Canadian governments we shall see in the next section. But one thing that has not been so immediately recognized by geographers or historians is its impact upon the growth of settlement. It turned it into a kind of lottery. For every one of those several million squares had exactly equal locational value, and four corners where it adjoined the neighboring sections. Which of those squares would prove to be the most important? Which of the section crossroads had the best hopes of becoming the focus of a metropolis? The questions are as unanswerable as if we were asking them about a chessboard. In practice, any one of several million crossroads might be the site of a future metropolis—and a good many of them, as we shall see again in Chapter 14, aspired to be just that. In other words, the survey system pro-

vided for virtually unlimited competition, since no one point had an advantage over any other.[3]

Whether or not the system worked to the benefit of the West would be hard to say. It may, for example, have encouraged the growth of too many central places in the formative years. What is clear is that the gridiron had other drawbacks. One was that it tended to maximize the length of roads needed to supply and serve a population settling the surveyed lands. Another was that travel was easy in north–south or east–west directions, but often, to this day, it is tedious to make diagonal journeys since all the roads run along the survey lines: one has to go around two sides of a square. A third drawback was that such a regular system left no gaps that could serve later as reserves of land; settlers were packed in, cheek by jowl, which was admirable in the early days when it prevented the frontier from fragment-

[3]In the humid eastern sections covered by the survey, this statement would need to be modified by admitting that river and lakeshore sites had a clear advantage over the other squares. But further west, the absence of navigable waterways removed even this element of choice.

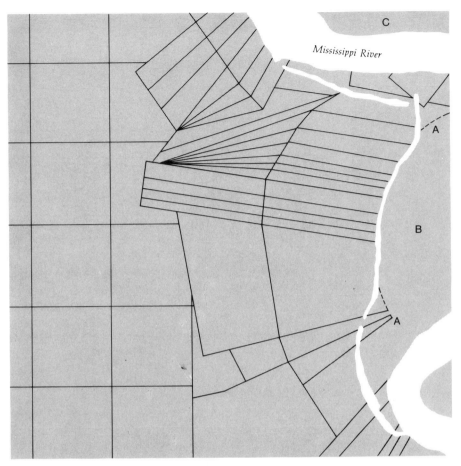

Fig. 5-2. A clash of survey patterns. In this area on the lower Mississippi, the old French survey pattern of "long lots" perpendicular to a river frontage antedates the American rectangular survey pattern. Lands disposed of by the French prior to the Louisiana Purchase of 1803 have been brought into the overall pattern; this legal maneuver occupied a great deal of time in the early years of the republic, before the two systems could be harmonized. (**A**) The dashed line represents a former course of the Mississippi (which has constantly shifted its bed in the past 300 years), and the property lines run down to this old course. (**B**) and (**C**) Two areas that have come into existence since the French survey was made, as the river changed course.

ing, but which poses immense problems today, when farmers are seeking extra acreage in order to maintain the efficiency of their farm operations. But this brings us to our next topic—distribution of land to the settler.

Land Disposal and Land Law

Driven by its poverty, the U.S. government was in the land business from its earliest days. Only by disposing of its holdings could it obtain revenue, and in the disposal process it was in competition with the individual states, most of which had lands within their own boundaries that they were anxious to have settled.

From the first, there was an idealistic hope that it might be possible to give away public lands in small quantities to penniless, worthy applicants who would be "actual cultivators" (the phrase recurs in early debates; clearly, the Americans had not forgotten the absentee landlords of Europe and the early

colonies). It was an ideal eventually realized some eighty years later in the Homestead Act of 1862. But for the moment, idealism had to bow to the necessity of placing a value on land, first, to raise revenue by selling it, and second, to enable the government to pay its ex-soldiers with land, since it could not pay them with money. If land had no value, the soldiers would receive nothing for their service.

The first federal land law offered land at $2.00 an acre, cash ($4.95 per ha). The blocks offered were large, for the surveyors had not progressed beyond some initial six-mile squares, and there were few buyers. Land in the states was also available and generally cheaper. With these beginnings, alterations in that first law of 1785 can best be understood as modifications to suit changing times and conditions.

1. *The government was obliged to offer smaller units for sale.* This implied, of course, an increasingly detailed—hence more expensive—survey. But it was the only way to create demand. In 1804, when what is now Indiana was being settled, the federal govern-

Fig. 5-3. Origins of the U.S. public lands survey. The system of rectangular survey began in what is now southeastern Ohio, in the so-called Seven Ranges. From there it spread across the continent to the Pacific.

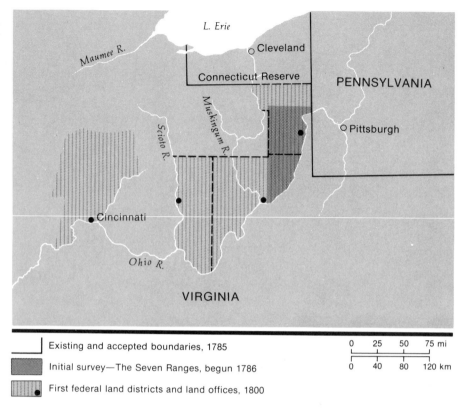

Existing and accepted boundaries, 1785

Initial survey—The Seven Ranges, begun 1786

First federal land districts and land offices, 1800

ment first began to deal in units that, for better or worse, came to figure in nearly all subsequent land laws—the quarter section of 160 acres (64.8 ha). Further subdivisions came later, but the idea of the quarter section achieved a kind of sanctity, which made it virtually impossible to dislodge it from congressmen's minds even decades later, when the frontier of settlement lay far out in the arid West and a quarter section had about the same economic value as a small field back in the humid East.[4]

[4]A quarter section was the area of free land granted by the Homestead Act of 1862. The present author has examined the *Congressional Record* for the seventeen years (1845–62) during which this act was under discussion, and is interested to find that, all through this period, it was the quarter section that both supporters and opponents of the homestead were arguing about and that, on the few occasions when some eccentric legislator proposed a homestead of another size—say 320 acres (129.5 ha) or a half section, the *Record* curtly reported: "Negatived without debate."

The size was, of course, important: a quarter section was recognized by both sides as being too small a unit to make worthwhile the introduction of slave labor—which is why most Southerners opposed the Homestead Act, and why it was finally passed only in 1862, after its opponents from the slave states had gone home to the South and taken up arms against the Union.

2. *The price had to be lowered and graduated.* To offer all land, regardless of quality, at $2.00 an acre was nonsensical, especially in competition with the states. The price quickly fell to $1.25 an acre ($3.10 per ha). There was later added the principle of graduation, whereby land that still had no buyers after years on the market could be offered at progressively lower prices down to 12½ cents an acre (31 cents per ha). Seen in this light, the Homestead Act of 1862 represented the ultimate price reduction—to zero.

3. *The original intention of selling land strictly for cash was modified to allow credit sales.* At that point the revenue consideration in U.S. land policy was, for practical purposes, lost. The credit terms were progressively weakened, and the revenue from land fell off. Fortunately for the government, its income from customs duties proved unexpectedly buoyant, so that early in the nineteenth century the revenue factor in land policy began to diminish in importance, giving way on the one hand to the original idealism about free distribution and, on the other, to the far-from-idealistic concern of the country to settle its public lands as quickly as possible to ward off claims by Britain, Spain (later Mexico), or Russia.

Saskatchewan: On the Regina Plains. Here, on one of the flattest and most fertile sections of the Prairie Provinces, the gridiron pattern of survey and land disposal shows up most clearly. Only the railroad line runs on a diagonal course across the checkerboard. *Saskatchewan Government Photo*

4. *The principle of "prior survey" survived until 1841.* In that year Congress passed the Preemption Act, the effect of which was to legalize squatting. A squatter who had gone ahead of the surveyors could, when they caught up with him, now lawfully purchase a quarter section of the land on which he had settled.

By the middle of the nineteenth century, a new principle effectively governed U.S. land policy, and was to govern that of Canada. Land was no longer seen primarily as a source of revenue, but rather as a truly public asset, which should benefit as many citizens as possible. Political and economic considerations alike favored the rapid settlement of empty lands. The federal government became increasingly generous with the public domain.It gave away millions of acres to newly formed states. At the other end of the scale, the 1862 Homestead Act (copied by the Canadians in 1872, but with slightly easier conditions) offered the individual settler title to a quarter section for the cost of the registration fee alone, if he would settle it, improve a part of it, and remain on it for five years. In between these two extremes, the main category of disposals consisted of those for "internal improvements." Recognizing the need for transport routes in opening up the continent, the U.S. government began, and the Canadian government continued, a system of granting land to companies that would construct canals, roads, and, especially, railroads. The government had not the funds to subsidize construction, so it granted land along the proposed right-of-way. The idea was that the construction company would sell this land (land whose value would be increased by the provision of transport

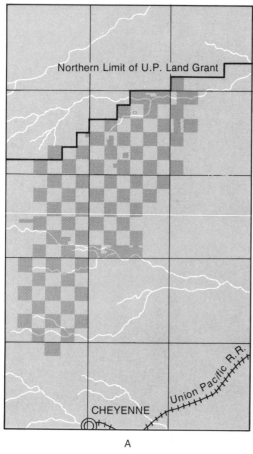

A

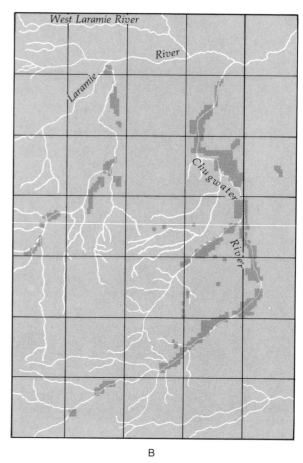

B

Fig. 5-4. The U.S. government land policy and its effects. The maps show (**A**) the effect of granting "alternate sections" on either side of its right-of-way to the Union Pacific Railroad: the shaded sections were bought to form a single ranch, effectively sterilizing the remaining (government-owned) sections, which must then be either sold cheaply to the ranch owner or used by him without formal purchase; (**B**) the acquisition of waterside land as a means of securing control of a huge waterless area of range: the Swan Land and Cattle Co. holdings in Laramie County, Wyoming. (From E. S. Osgood, *The Day of the Cattleman* [Chicago, 1929], pp. 212–14.)

routes) and use the proceeds to finance construction. It was, in theory, a sensible enough scheme; that it left enormous opportunities for fraud and oppression of individual settlers might have been foreseen, but was, for the most part, overlooked in the general haste to open up the territory. The big land grants under this head of disposal began with an award to the Illinois Central Railroad in the 1850s and culminated in the 1881 land grant to the transcontinental Canadian Pacific, which was to receive 25 million acres (10.1 million ha; later reduced to 18 million acres, or 7.3 million ha) as its subsidy for building across the empty West. Moreover, the company was free to pick and choose lands "fairly fit for settlement" as the U.S. railroads had never been able to do—the latter generally received alternate sections on either side of the line, no matter what the character of the land (Fig. 5-4).

By one means or another, therefore, the U.S. government divested itself of a great part of its public land. It reserved some areas, such as the forest and mineral lands of the West, but generally only after it saw the private fortunes that could be accumulated by obtaining title to these lands. One of the most farsighted reservations ever made, however, dated from the very beginnings of public land policy: that of reserving two sections out of every thirty-six in a township for educational purposes. But on the whole, the public domain, at the end of this period of disposal, consisted of the driest and most worthless parts of the original public lands—the parts nobody else wanted. Excluding for the moment Alaska, which occupies 375 million acres (152 million ha) and which became a state only in 1959, the public domain now administered by the Bureau of Land Management is just under 400 million acres (160 million ha) out of an original 1900 million acres (766 million ha). However, if one adds the areas administered by the U.S. Forest Service, the National Park Service, and other federal agencies, the area that never passed into private ownership is 678 million acres (271 million ha). And if one adds the lands these agencies purchased from owners, the total area of public land administered by federal agencies is nearly 740 million acres (296 million ha).

There were good reasons why the federal government was in such a hurry to divest itself of its public lands (and why the Canadian government acted more cautiously). But one side effect was important: the impression was created that what was so easily disposed of was of little value. If the nineteenth century saw, as it did, a huge wastage of natural resources in the United States with land, grass, and forest squandered on an unprecedented scale, then some of the blame attaches not only to the individuals who exploited the resources but also to the government, which seemed to care so little for their fate.

Carelessness can, in part, be explained by the inability of the federal agencies to keep a close watch over the whole enormous area of settlement. There were too few land officers to travel around verifying claims; consequently, many claims were fraudulent. Perhaps harder to excuse was the second weakness of federal policy—inflexibility. Despite mounting evidence that in the dry West a quarter section was a pathetically small unit, Congress moved at a snail's pace to modify the land laws. Clinging to a legacy of pre-Civil War belief that anything more than a quarter section was in effect a plantation (with all the connotations of slavery that term had once carried) the legislators condemned hopeful, but ignorant, homesteaders to slow bankruptcy by the thousand. In Canada, although the homestead of the period was of similar size, it was made as easy as possible for the homesteader to buy an adjoining quarter section (and so effectively to double his holding), and an alternative in the form of a pastoral lease was usually available in areas too rugged or too dry for cultivation.

GOVERNMENT LAND: TO DISPOSE OR NOT?

For all the lavish giveaway of earlier years, the federal government of the United States remains by far the nation's largest landowner. It owns, as we have seen, some of the worst of America's lands, both desert and ice cap, but this is not to say that there would be no buyers if government land were to be placed on the market. Periodically, the issue is raised as to whether the federal government should not dispose of the remainder of the public domain, either to the states (just as the Canadian government made over the Crown lands within their borders to the provinces), or simply to the highest bidder.

To dispose of this land to the state concerned has about it an air of natural justice, at least when we realize that nearly 90 percent of the territory of Alaska and Nevada is owned by the federal government, leaving the state only 10 percent on which it is properly master in its own house.[5] To dispose of the land to private buyers has always appealed, understandably, to western interests, which stand to gain most. But since the 1930s, the federal government has resisted all such urgings to dispose of this land: it has seen itself as the best custodian of an area, much of which is easily damaged and much, too, which is subject to competing and conflicting uses. In this stance it has been supported by the many environmentalist groups that have sprung up during the period.

It is true that, during the 1980s, some attempts were made to sell off parts of the public domain. But their scale was small, and most sales consisted of isolated blocks of land, or even single buildings, and certainly did not affect the huge areas upon which, no doubt, speculators were hoping to get their hands. The federal government has not persisted in this policy.

Much of the domain will never be sold—for example, the national parks, wilderness areas, and Indian Trust lands (see, for example, Fig. 4-2). But this still leaves substantial areas, and numerous fragments that once had some long-defunct military or government function. By selling these lands, the government might hope to make money without raising taxes. If it embarks on this course again, however, it can count on the opposition of environmentalists, and of some groups in the West that prefer the status quo, or that fear the environmental effects of development, and so want to limit the sales.

[5] Other figures are: Idaho and Utah, 63 percent federally owned; Oregon, 52; Wyoming, 48; and California, 46 percent. As can be seen, by far the largest part of the public domain is in the western states.

In practice, the question raised by these sales is only a rather special case of a much broader issue the North American nations have, on the whole, never really faced because they have never been forced (as have European nations) by the pressures on land to confront it. It is the issue of *control of land use*, and to it we must now turn.

The Control of Land Use

In the days when the young nation was creating its Constitution and its land policy, there was one simple criterion by which to decide whether land was being "used"; it was either being farmed or it was not. With the trifling exceptions represented by the few cities, all land use was agricultural in character. And apart from local restrictions that might be applied by the community or the town meeting, individual farmers decided how best to use their land.

If today we draw up a list of land uses, the difference is at once obvious. Apart from agriculture, land is required for housing, for industry, for means of transport (some of them, especially airports and freeways, very large space-users), and for recreation, to name only the major competitors. We might also add space needs for waste disposal, defense installations, water catchment, and mineral workings. Such is the competition for land, in fact, that it comes as something of a surprise to learn that the area in agriculture in North America today is about the same as it was in 1920, to such an extent have we come to think of farming as a kind of residuary legatee of land uses. The farms of today occupy the same amount of space they did sixty years ago, but not the same spaces; farms have been swallowed up by suburbs and parking lots but so far these losses have been made good by new lands elsewhere.

We shall consider American agriculture in Chapters 6 and 7. For the moment we need only note that, once two or more users are competing for land, the conflict can only be resolved in one of two ways: by the price mechanism, which secures the space for the user who will pay most for it, or by some form of regulation, which will impose a use regardless of price. Such regulations, by whomever they are imposed, constitute land use control.

The control of land use forms part of the police power that in North America, as we saw in the last chapter, reposes in the state or province. If arbitration among conflicting land uses is needed, then it is from the state that it must come. In practice, however, the states have generally been rather inactive about this power in the past. Insofar as anything has been attempted, it has been by delegating power to local units of government, like counties or cities—that is, to units often too small to plan comprehensively or to cover the most important problem areas.

The problem is actually more specific. It is that the control of land use in cities has been a feature of North American life almost from the first settlements, with *urban zoning ordinances* applying very generally in the twentieth century. But control of land use outside the city is not only in many areas nonexistent, but is also looked upon with genuine hostility by many Americans. In other words, land-use control is considered a necessary feature of urban life, but an undesirable one in rural areas—the very areas where, needless to say, so many of the present-day use conflicts are focused.

To explain this paradox would involve a book in itself, and we have space here for only a very brief summary of the position. Why should zoning, or control, be acceptable in the city, but not in the country?

THE POSITION IN THE CITY

Zoning, in its modern form, was introduced in American cities in the early 1900s. It prohibits particular land uses in a given area and as such provides sensible safeguards to health and amenity, like keeping the glue factory or the oil refinery away from residential areas. Yet this is only the beginning of the story. It can also be used as a discriminatory device to prevent particular kinds of residential development, even within a residential area. Low-cost housing, for example, may add little to a city's tax base, but a good deal to its school population. On the other hand, if the city zones an area for development in large, single-house lots of, say, 2.5 acres (1 ha) minimum, a city can ensure at a single stroke that it will attract wealthy residents, that they will put up expensive, high-tax houses, and that the need for new school places will be limited, at worst, to one family per lot. A block of apartments on the same site might produce 50 or 100 school pupils and wreak havoc with the educational budget.

It is such use of the zoning power that leads to complaints that zoning is basically a matter of politics and not of planning; that it is a discriminatory power rather than a proper application of the police power as we earlier defined it. Nevertheless, all the major cities in the United States employ it, except Houston,

Texas, which has steadfastly maintained its faith in a free land market, and which claims that by so doing it has benefited its less well-to-do citizens by providing more low-cost housing than would have been built under zoning laws.

THE POSITION IN THE COUNTRY

Once we cross the city boundary, however, a quite different situation exists. We may understand it best by visualizing a farm located just beyond the fringe of a built-up area. In the absence of any intervention or control, the farmer can confidently expect that, within one to five years, he will be approached by builders, supermarket chains, and restaurant owners, all of whom will bid up the price of his fields until he decides to sell out and retire a rich man. But now let us suppose that the county or township in which he lives declares that his land is zoned for agriculture alone or lies within a "green belt," which is to be preserved from built-up uses. Immediately the procession of would-be buyers to his door will stop, and immediately the farmer can claim that under the terms of the Constitution of the United States (Article 5) he has been deprived of his property without due process of law. For his right to the property (which in most cases he holds in fee simple) includes, as he sees it, the right to sell to the highest bidder; that is, he claims the right not only to own the land, but also to make the most of it. And if necessary, he will go to court to maintain this right. He will argue that a green belt, which is merely one of a number of planning devices that limit the range of uses to which land may be put, and which is in common use in Europe, is unconstitutional.

However, this is only one-half of a complex whole. The other half of the drama is being played out in the offices of the local tax assessor. It is the task of this official to set a value on all property, including the farm in question. But this value is normally based not on the farm's output or sale price as a farm; it is based on what the builders or supermarket chains would offer the farmer to sell out. As the fringe of the city creeps nearer, therefore, the farm produces what it has always produced, but the assessed value, and so the tax bill, rises sharply. At a certain moment, the farmer decides that he can no longer meet his tax obligations, even though he may wish to remain a farmer; high taxes have driven him to sell out. We thus have, on the urban fringe, the vicious circle shown in Figure 5-5. Whereas, originally, it may have been the farmer who opposed the idea of a

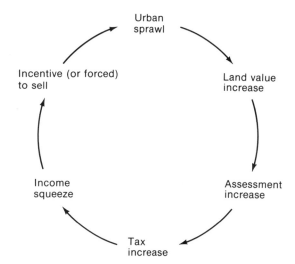

Fig. 5-5. Land use on the urban fringe. The diagram illustrates the vicious circle of land-use changes described in the text.

green belt because it limited his freedom of sale, he is now joined in his opposition by the taxing authority, which counts on increasing the tax revenue simply because the city's edge is approaching, and which will be equally opposed to any land-use control that has the effect of halting it. It is a simple and infallible recipe for urban sprawl and, to judge by present indications, it has been what, up till the recent past, a majority of American governments at all levels wanted.[6]

Before such reasoning, the non-American can only bow in awed silence. Nevertheless, there are steps that can be taken, if anybody wishes to do so, to break the vicious circle in both city and country, and there are some indications that this is occurring. In the city, an immediate reduction in the political use of zoning could be achieved if the cost of education could be transferred, to a substantial degree, from the local government to the state. Education, at present, accounts for so large a share of local taxes that the temptation to zone schoolchildren out is very strong. If the incidence of school costs on the local taxpayer could be reduced, the temptation might be removed.

In the country, there is an equally simple device for arresting urban sprawl, given only the will to do so. It is to assess farms for tax purposes as farms, so long as they continue in that use, and to increase the as-

[6]To justify this last statement, one might cite R. G. Healy, *Land Use and the States* (Baltimore, 1976), who reported that, out of 38,000 county and lower government units in the United States, only about 14,000 regulated land use in some way.

sessment only when their use changes. To assess them as potential supermarket sites is to ensure that, sooner or later, supermarkets are what they will become.

One after another, over the past quarter century, the American states have been drawn into action over this problem, and legislation dealing with it now exists in virtually all of them. Oregon passed a Greenbelt Law in 1961; California its Williamson Act in 1965. Today, there exists in most states some device for ensuring *either* that land can be zoned for agriculture just as much as for industry, commerce, and housing, *or* that a farmer who wants to continue farming can have his land assessed at agricultural values for as long as he does so. Some states have gone so far as to zone their agricultural land on the basis of soil surveys, which preserve the best soils for the farmer's use. Others have legislated minimum farm sizes to preserve commercial viability, although that is a measure to preserve farming as an industry rather than farm *land*, and we shall deal with that topic in Chapter 7.

But as so often happens with such legislation, its effect is not altogether what its framers hoped for. Landowners have realized that in order to obtain the low, agricultural level of tax assessment on their land, they do not actually have to farm it, so much as merely avoid doing anything else with it. They can, in fact, reduce their tax bill simply by doing nothing—by holding land idle.

This is merely another way of saying that, in general, there is nearly always some method of getting round planning law, and that for it to succeed there has to be not only the legal framework, but also the will to make it work in the way its designers intended. In any case, it is evident from what has been written here and in Chapter 4 that the burden of responsibility rests with the states. The local government unit is too small and self-centered; the federal government can do no more than exhort. The states have the necessary power and some, if not all, of the funds. We have already noted that the level of enforcement of land-use controls varies widely from state to state, but in general terms, there is a drift toward greater state involvement. It falls, at present, far short of marking a new American Revolution. But it certainly represents either a change of heart or a yielding to public pressure on the part of some American state legislators.

In Canada, a parallel movement toward government intervention in land-use decisions has been taking place. Here, the major initiative has lain with the provinces, in view of their substantial powers within the federal framework and their control of the Crown Lands within their borders. But from the start the Canadian government has shown greater awareness than that of the United States that not all lands are the same in quality or potential, and that land-use policies need to take account of these differences. The Prairie Farms Rehabilitation Administration of 1935 (see Chapter 6) recognized the importance of establishing wise land uses; so, too, did the Agricultural Rehabilitation and Development Act of 1961. The problem of loss of good farm land to other uses was addressed in the Food Land Guidelines of 1976, and provincial measures (especially in Ontario, the most highly urbanized of the ten provinces) have been introduced to separate farm and nonfarm activities on the urban fringes.

Government, Land, and Leisure

Defective as the United States' land policy was in some ways, there was at least an early recognition that some areas were of particular value because of their natural splendor or unique character. Out of this recognition grew the national and state park systems, a form of land-use reservation which today extends (if we include the huge areas of Alaska designated for the system but so far undeveloped) over some 85 million acres (34 million ha) of the United States.

The original intention was to preserve these areas from piecemeal seizure in the great land bonanza of the nineteenth century. The Yosemite Valley was originally secured by the state of California in the 1850s, and the Yellowstone Plateau by the federal government in 1872. These first acts of preservation by nation and state have proved to be entirely to the public's advantage. What could not, however, have been foreseen at that time was the way in which, in a society of growing prosperity, (1) personal mobility would one day place even the remotest of these preserved areas within reach of the bulk of the population, and (2) increasing leisure would heighten to such an extent the demand for recreational areas.

In 1985, there were in the United States rather more than 75 million acres (30 million ha) within the National Park system and, throughout the 1980s, these areas have received annually between 300 and 350 million visitors. (For 1970, by way of contrast, the figure was 172 million.) Other millions have visited National Forest lands. The 10 million acres (4 million ha) of state parks have recorded as many as 660 million visits annually, since they are more widely distributed and more readily accessible to centers of population than the national parks.

This colossal human movement is produced by leisure and the desire for recreation; by a search for solitude or natural beauty. Needless to say, solitude is the last thing the visitor will find around the "honeypots" of the natural world: in 1985, over 2 million people spent a night in Yosemite National Park and, at Yellowstone, most visitors in the season drive the park's roads in a convoy of cars which grinds to a halt every time a motorist sights a bear. This is only another way of saying that the 85 million acres (34 million ha) that the United States devotes to major recreational land uses are probably not enough to meet an escalating demand.

The heaviest pressures fall, naturally, on sites with spectacular scenery but limited access—the Grand Canyon is an example—and also on areas where

land recreation combines with water. Glen Canyon and Lake Mead recreational areas both recorded more than 1.5 million overnight stays in 1985, and this in an area that was, until quite recently, uninhabited desert.

Canada's national park system is steadily expanding, and now comprises thirty-one major parks, covering 34 million acres (14 million ha). Included among these are not only the Rocky Mountains' famous scenic areas of Banff, Jasper, and Waterton, but dramatic, barely accessible parks like Kluane in the Yukon and Auyuittuk on Baffin Island.

The recreational pressure on Canada's land is not, as yet, so acute as in the United States. But it, too, has its honeypots. In the context of this chapter, what this means is that, whereas once the only rural land uses for which any government or planner needed to legislate were farming and forestry—neither of which attracted much attention from the lawmakers—there is now a third rural land use in which the whole electorate is interested, which is constantly en-

Hydraulic mining in California: A destructive practice that persisted from the earliest days of mineral exploitation in the American West until the end of the nineteenth century. Before the arrival of the bulldozer, it was the quickest method of removing unconsolidated materials like river alluvia, but its erosive effect was appalling. *U.S. Geological Survey*

larging, and which forms in many areas a far larger and more stable source of income than the activities at present carried on there. It seems inevitable that government involvement with this type of rural land use will increase in the future.

Government and Water

The involvement of the North American governments with water resources has been slower to develop than that with land, but it nevertheless has a considerable history. It first became explicit when settlement penetrated into the dry areas of the West, and water became no longer a universal presence but, rather, a resource in short supply. Even then, the settlers might have resolved their difficulties without recourse to law had it not been for the emergence of two activities that used water on a far larger scale than hitherto and that led to bitter disputes among users. These activities were irrigation agriculture and placer mining.

Most of the early immigration into North America was from northern Europe—the British Isles, France, and Germany—where rivers flowed, for the most part, year around and the accepted basis of water use was that any property owner whose land included a water course had the right to reasonable withdrawals. There were no uses so large as appreciably to affect the stream flow. But southern Europe (the Mediterranean lands including Spain) knew of such uses; specifically, of irrigation. As we shall see in Chapter 7, the quantity of water involved in irrigation agriculture is so large that, in the dry West, it was possible for a settler beside a stream to enjoy a water supply one year and before the next growing season find that diversions for irrigation upstream had left him literally high and dry. Placer mining was less demanding of water in the sense of absolute withdrawals, but the stream was likely to be choked with sediment, because the whole idea was to wash the earth away to get at the gold.

The introduction of these two water uses forced upon the West a change in custom and eventually in law, so that today the subject is one of the most complex and regionally variable in the whole legal field. Water rights are a subject of endless litigation, up to and including the international level, as we shall later see. The governments of all the western states and provinces are involved.

The U.S. government became involved in water resources originally through its efforts to control floods and to improve navigation; it built the locks at Sault Ste. Marie and virtually rebuilt the Mississippi River south of Cairo through the efforts of its Army Corps of Engineers. From here, as we noted in Chapter 4, it moved into other aspects of river basin control, using the interstate commerce clause of the Constitution to justify its operations: it was removing barriers to commerce. As early as 1902, it entered the field of irrigation when the Bureau of Reclamation was set up, with the responsibility of bringing irrigation water to such sections of the public domain as could be assisted in this way.

The years between World War II and the late 1960s were a period of great government activity in water control, both in the United States and in Canada. They were years of dam building on a tremendous scale—in the West for irrigation and power, in the East for flood control, navigation, and water supply. The Central Valley Project of California, the Columbia Basin Project, the South Saskatchewan River scheme in Canada, and the damming of the Missouri River all date from this period, and all will be referred to in later chapters of this book. The costs have been astronomical, the value of the finished projects open to debate, and the burden on the taxpayer undeniable. Fortunately for the latter, the crest of this particular wave is now long past: it is a decade or more since Congress authorized a major project, although it is still every congressman's hope that he or she will obtain authorization for at least a small river widening or harbor deepening in their constituency by the Corps of Engineers.

Rather, attention has shifted to an increasing shortage of water in the continent as a whole, and this led to a U.S. presidential commission to examine the whole water resource pattern. Meanwhile, treaties with Mexico and Canada and joint action on the rivers concerned involved the Colorado and the Rio Grande, the St. Lawrence, and the Columbia.

The 1960s and 1970s brought Americans to the realization that water is not to be taken for granted, but rather is a valuable resource. Before we consider, however, the conservation of resources in the next section, let us be clear as to the nature of the problem. Taking the continent as a whole there is, of course, no absolute shortage of water. There are local shortages, as in the lower Colorado Basin, or west Texas, and there are important shortages of clean water, especially in the eastern states. There are problems caused by competition *among* states for water, since once again, it is to the states that the main legal powers over resources belong, and states today are sen-

Western land and water: Glen Canyon Dam, on the Colorado River, in Arizona. It holds back the river to form Lake Powell, creating a number of recreational areas in the process. *U.S. Bureau of Reclamation*

sitive about water as they have not always been about land.[7] We shall later refer to interstate rivalries for the waters of the Delaware in the East and the Colorado in the West. And then there are, as we have already noted, international overtones to the water problem—the expectation of Mexico that the United States will not drain the Rio Grande[8] or Colorado dry

[7]Since water is a fluid resource, it is sometimes possible for one user or one state to "poach" water from beneath the surface of another state and deprive its rightful owner of it. A number of states have laws specifically protecting their groundwater supplies against such action.

[8]Along the Rio Grande, a good deal of water is bought and sold within the overall flow. F. A. Schoolmaster, "Water Marketing and Water Rights Transfers in the Lower Rio Grande Valley, Texas," *Professional Geographer*, vol. 43 (1991), 292–304, writes that there is a flourishing market in water, especially along the lower Rio Grande, but that in any case several other U.S. states participate in marketing their water.

before they reach the border, and the expectation of Canada that works built on the lower Columbia will not affect the upper, or Canadian, section except in accordance with treaty.

If these are the three levels—local, interstate, and international—at which there is a water resource problem, is there a level for solution? So far as the government is or might be involved, there are two:

1. *The river basin approach.* The natural unit of water planning is the river basin. In almost no case does a major basin fall within a single state, so that interstate rivalry must be overcome before there can be a practical solution. This is, of course, precisely what the federal government achieved in the Tennessee Valley (see Chapter 16) and might, given the opportunity, have achieved elsewhere. But there has been no second Tennessee. In the other major basins, the states have preferred a more informal arrange-

ment and have kept the federal government at arm's length.

2. *The continental approach.* Far transcending all other projects for harnessing the water resources of North America is a plan that has been in existence for almost thirty years, which was designed not by government, but rather by private enterprise, although its fulfillment would certainly involve the continent's governments.

The most water-conscious area of North America is the Pacific Southwest; specifically, southern California and Arizona. And it was precisely from this area that the scheme emerged. It was based on the fact that within the continent there is one area of abundant water supply where pure water, in an economic sense, is running to waste. It is the Canadian Northwest. The plan, which was known by the name of North America Water and Power Alliance (NAWAPA), called for the "wasted" supplies of northern water to be tapped, stored, and reversed. By a series of canals, old riverbeds, and tunnels they could then be brought southward to the thirsty areas of the Southwest, with a subsidiary feeder to the Great Plains and something left over for Mexico. It would be a great north–south transfer forming, in effect, a water grid for the continent like that for electricity. Nor is that all. Northern water could also be sent eastward, to flush out pollution from the Great Lakes (which for all their size have a very small catchment area) and so help to restore life to the lower lakes.

Technically, the scheme appears feasible. But politically, it seems even more unlikely to succeed than when it was first mooted, in the dam-building, ditch-digging 1960s. Not only is it impossible to think of anything the Canadians would accept from the United States in exchange for their water; not only is Canadian hostility to making over national resources to the Americans a strong and abiding sentiment,[9] but the American states themselves have proved unwilling to collaborate where water is concerned. Washington and Oregon have consistently refused to share the huge flow of the Columbia River with water-thirsty California, on the grounds that they might one day want the whole Columbia flow for

themselves. And as recently as 1985, by the Great Lakes Compact, the Lake States of the United States and the Province of Ontario pledged themselves to oppose all efforts to divert any water south to the Sunbelt States.

On the whole, it is simpler to keep on with existing technologies for conserving water than to attempt the miracles of turning Canadian water into American, or northern water into southern.

Conservation of Resources

There is one further topic to be considered in this chapter on the relationship of government to land and water resources, but it might be argued by some enthusiasts that it does not really belong here. The conservation of resources in North America, they might argue, owes little to the government, and much more to group initiatives, which have goaded it into action or taken it to court; which have carried out the educational processes necessary to alert the nations to the true realization of what they possess and what they must do.

It is true that private groups have played an important part in the conservation movement and continue to do so. But the record of government is by no means negligible, and for any program to have teeth, it must ultimately have the force of law. We shall, therefore, consider this subject before going on in the next chapter to some of the specific issues, such as energy policy.

One of the major decisions facing the population of any area is the rate at which its resources shall be used. Many resources are irreplaceable; others are replaceable only by slow processes, such as forest regeneration. To use these assets more rapidly than they are replaced, or to use more of them than current need justifies, is wasteful exploitation; a rate of use that is adjusted to speed of replacement and to current need represents conservation. Experience on almost every new frontier of settlement shows that waste and exploitation accompanied the first settlers, and that regulation of use is the only alternative to a rapid deterioration of the area's natural resources. Upon a community's willingness to accept such regulation depends the future of its supplies of minerals, timber, water, or food. Of the existence or absence of these policies of regulation the landscape will often contain visible evidence; there is a landscape of conservation and a landscape of waste.

Throughout the first three centuries of European

---

[9]Lest anyone should miss this point, Canada's federal Northern Inland Waters Act of 1970 proclaims it with all the splendor of a royal command: "3(1) . . . the property in and the right to the use and flow of all waters are for all purposes vested in Her Majesty in right of Canada." See also D. Pharand, *Canada's Arctic Waters in International Law,* (Cambridge, Eng., 1988).

The threat of forest fires in the western United States: Damage caused by the Oakland, California, fire of 1991. *James Wilson/Woodfin Camp*

settlement in North America, the continuing impression made upon the settlers was one of inexhaustible natural riches. In a land where space was unlimited, where the forest seemed to go on forever, where fish and game abounded, the early generations of Americans can be excused for feeling that they need never fear a permanent shortage of the materials necessary for a livelihood. If they exhausted the possibilities of one area, they could always move on. Imbued with this spirit, they did move on, until there stretched across the continent a series of frontiers littered with the debris of exploitation and abandonment. In the north, the trappers

decimated the population of fur-bearing, forest animals. In the Great Lakes area, lumbermen passed through with all the devastating effect of a forest fire, cutting out only the species of timber they could market, and leaving behind the tangled remnants of the forest cover as useless, fire-prone slash. In the South, farmers grew cotton or tobacco until the soil had lost its virtue, and then abandoned their farms and moved west to repeat the process on newly cleared lands.

Every nineteenth-century frontier saw some kind of bonanza—the wasteful consumption of resources far beyond any scale of need. But in North America, the technical means of spoliation were more available, and the consequent waste probably greater, than anywhere else. The government's land disposal policies cannot have helped: what was so freely handed over could be assumed to possess little intrinsic value.

But already, before the end of the century, there were signs of change. Individuals and groups fought to preserve particular treasures—Yosemite Valley, or the Yellowstone Plateau. In the last years of the century, the government moved over to a policy of land reservation; late in the day, it began to conserve what was left of its lands. This policy meant the withdrawal from further public entry of lands whose resources—in the first place, timber—made them of special value to the nation. National forests were set up and in them the federal government undertook to regulate land use. Later, control was extended also to the open grazing lands of the public domain.[10]

In the meantime, the situation had been deteriorating in the eastern states. The pressure of an increasing population on land and resources used with traditional American freedom had made deforestation, soil erosion, and diminishing crop yields common. The vicissitudes of business during and after World War I, which culminated in the Great Depression of 1929–33, made it clear that a new approach to the problem of resource conservation was required. As if to underline this, the droughts of 1934 and 1936 produced the Dust Bowl in the southern Great Plains. In 1936, therefore, the Soil Conservation Service (SCS) was established, with the task of organizing farmers into district groups for the development of anti-erosion measures, and of instructing them in more scientific methods of land use.

Since then, it is calculated that the United States has spent over $15 billion on soil conservation measures. This being the case, it is regrettable to have to go on to report that, $15 billion later, soil erosion is currently reducing productivity on an estimated one acre in every three of U.S. cropland—including some of the most productive of that land, for example, in Iowa and Illinois. The drive for greater output and higher income and the use of machines and chemicals to achieve it have overridden the care and caution the SCS strove so hard to instill.

In Canada, government control of national resources has been somewhat closer than in the United States. For this fact the decision to hand over control of the public lands to the provinces may have been responsible. In certain respects the contrast is marked. In the northwestern forest states of Washington and Oregon, for example, the rate of timber cutting has until recently far exceeded the rate of growth. In British Columbia, however, where some 94 percent of the forest is owned or administered by the province, it has been possible to keep the annual cut to a figure lower than the total annual increment. Again, when the big Alberta oil fields came in after 1947, the government took immediate steps to regulate development and, profiting by the unhappy experience of some of the oil regions south of the border, to avoid wastage by restricting competitive drilling. On the whole, then, the time lag between the development of the United States' West and that of Canada gave the latter time to profit by the former's mistakes, and the allocation of all crown lands to the provinces gave local governments an immediate stake in the orderly development of their resources.

If conservation has been a twentieth-century refrain in North American life, there is no question that it has grown much louder in the last two decades. Indeed, to read the literature of the early 1970s, one might get the impression that the North American nations, having been the world's wealthiest, had been bankrupted of their resources and become poverty stricken almost overnight. At the same time, the focus of attention has shifted somewhat from the familiar problems of the last half-century, of which the best known was that of soil erosion, to a range of new and apparently more pressing problems. For the Canadians, one of these is acid rain, generated by industrial plants and power stations (most of them in the United States), and causing damage to Canada's forests and lakes. For both Canada and the United States, water pollution has become a major problem.

This pollution arises from the discharge of waste, domestic or industrial, into rivers, and from an enormous increase in the use of chemical fertilizers on American farms: the chemicals are washed out of the soil and fill the rivers with compounds of nitrogen, phosphorus, or potassium. Closed bathing beaches and polluted oyster beds were the warning signs and the focus of concern (unbelievably as it would have seemed to the earliest settlers) was the Great Lakes—the largest body of inland water in the world—and especially Lake Erie, with Detroit at one end, Buffalo

---

[10]Although individual reserves were set aside earlier, a National Forest Service was established in 1905, a National Park Service in 1916, and a Grazing Service not until 1934.

at the other, and Cleveland midway along its shore. But the problem is a general one and, like that of soil erosion or forest destruction, was a long time in the making before it forced itself upon the attention of the public.

It is, of course, possible to reduce pollution levels greatly by insisting on the treatment of wastes before they are discharged, but treatment costs money, and the question is, Who pays for it? Unless, for example, a very strong law made treatment of a particular waste obligatory in every state or province, the manufacturer who treated his waste merely out of public conscience would simply add to his costs and place himself at a competitive disadvantage against his less scrupulous competitors.

If such measures were to apply nationwide, in either the United States or Canada, they would have to originate with the federal government, constitutional limitations notwithstanding. Following its normal practice, the federal authority would have to bribe or cajole its corporations into complying with the standards it set in its federal codes. Firing the first shot in the campaign, the National Environmental Policy Act of 1969 opened with the words, "The Congress . . . declares that it is the continuing policy of the Federal Government, in cooperation with state and local governments, . . . to use all practicable means and measures, including financial and technical assistance, in a manner calculated to foster and promote the general welfare." The tone was, generally speaking, one of pious hope. However, the Environmental Quality Improvement Act of 1970 then followed, and after it came a series of acts in 1972 and 1973 aimed at producing clean water: "it is the national goal that the discharge of pollutants into the navigable waters be eliminated by 1985." Since then, 1985 has come and gone without the problem being eliminated, but there has at least been considerable progress toward the goal.

Other legislation set standards for noise abatement (1972), coastal zone management (1972), and clean air (1973).[11] In the meantime, and in the realm where practical measures can actually be enforced—that is, by the action of the states—these years brought some

notable achievements: the California Coast Commission of 1973, given a popular mandate to control coastal development up to 1000 yd (1 km) inland, which paved the way for the state's Coastal Act of 1976; the Vermont Environmental Control Act of 1970, which enabled local governments to control certain types of development and which specified standards (in the first ten years over 3000 land-use permits were issued, many of them with stringent development conditions attached); the Environmental Land and Water Management Act of 1972 in Florida, a state whose natural resources were under the heaviest pressure of all as its population increased faster than anywhere else.

The early 1970s were equally productive of legislation in Canada. In 1970 alone, four important measures were placed on the statute books: the Arctic Waters Pollution Prevention Act, the Canada Water Act, the Northern Inland Waters Act, and the Oil and Gas Production and Conservation Act covering the federal northern territories and offshore areas. The year 1972 brought the Great Lakes Water Quality Agreement with the United States, an agreement reiterated by both countries and strengthened in 1978. The Environmental Protection Ordinance of 1974 rounded off this avalanche of legislation. Suddenly, environmental safeguards had become politically fashionable. Even the water in Lake Erie improved as pollution levels dropped.

We might summarize the change in outlook in this way: when the conservation movement began in the late nineteenth century, its first objective was to halt the attacks on American nature that had been so numerous in the preceding decades; in other words, it was a backward-looking effort to put an end to the gross carelessness of the past. By the time that the 1930s brought the Dust Bowl and soil erosion to public attention, the conservation movement was dealing with contemporary, urgent problems, reflecting concern about damage occurring before the eyes of farmers and ranchers. Then, as time passed, the focus of interest shifted to the future. If it is sensible to arrest damage to the environment as it occurs, it is even more sensible to forestall it.

It is this latest phase of the movement that finds its expression in the U.S. government's acceptance of the *Environmental Impact Statement* (EIS). This is a commitment to weigh the effects of every federal project or proposal and to consider alternative strategies *in advance.* It offers the opportunity for public appraisal of the side effects of building a dam, or in-

[11]In J. W. House, ed., *United States Public Policy,* (Oxford, 1983), p. 162, the authors list thirty-eight public acts of the U.S. government, up to 1978, related to environmental quality. These range from the Historic Sites Act of 1936 to the Noise Control Act of 1972 and the Safe Drinking Water Act of 1974. The Canadian government's record is similar: it is calculated that there are at the present time some twenty environmental acts in force.

stalling a nuclear plant, or creating a military base, and at that point it can count on the Statement being examined under a magnifying glass by the powerful environmentalist lobby, that cluster of groups that represents the interest of thousands of Americans in their continent's future. Court actions and counter-proposals may then hold up the project for months or years, until an acceptable scheme has been evolved. But it is the federal government itself that has voluntarily tied its own hands in this way.

What all this environmental legislation in fact does, together with the EISs, is to put a brake on de-velopment—the sort of brake that North Americans would in the past have rejected out of hand. Devel-opments can be held up for months or even years, and have their costs raised by the environmental safeguards demanded in their construction. In fact, a 1992 estimate published by Resources for the Future (*Resources,* 106, Winter 1992, p. 7) suggests that just two of the legislative measures of the 1970s—the Clean Air Act and the Clean Water Act, together with the controls they imposed—caused the Gross Na-tional Product of the United States to be, by 1990, as much as 5 percent lower than it would have been without these controls.

The rate of use of the world's natural resources has greatly increased in the twentieth century. This is a result partly of the tremendous increase in world population and partly of the technical revolution, which, particularly in the Western world, has so markedly enlarged the productive capacity of the in-dividual. Crudely put, this means that we can now overwork our soil and strip our forest land more swiftly and thoroughly than ever before. What is at least encouraging in these conditions is that aware-ness of environmental damage, hazard, and possible disaster is also now at a higher level in North Amer-ica than it has ever been before. *Ecology, ecosystem, environment* have become household words. Pres-sure groups favoring conservation measures have multiplied, and every legislative action is reviewed and criticized by those who fear further destruction. In a remarkably short time, in fact, what was once the world's most wasteful society has been converted to enthusiastic salvage and conservation.

So strong, in fact, has been this growth of environ-mental awareness in North America that it has pro-duced something of a backlash. All too easily, opposition to particular projects becomes obstruc-tionism against all development; at its limit, life is brought to a standstill. So we find that, in a number of instances such as those we have mentioned in Cal-ifornia and Florida, the early momentum of the con-servation movement has been lost and, in particular, the overall plan, which should have followed the ini-tial legislation, has either not appeared or not won approval. The prejudice against land-use planning as such—"that un-American activity," as it has been called—remains strong. But at least in the future, in-dividual activities and particular violations of the en-vironment are less likely to pass unnoticed.

6

Government and Economic Activity

Government and research: Part of the U.S. Department of Energy's
Hanford research site in the state of Washington. *U.S. Department
of Energy*

The involvement of the North American governments in the national economy has been developing gradually over a long period. Some of this involvement was explicit in the U.S. Constitution, which assigned to the federal government such tasks as levying customs duties, establishing post roads, or regulating foreign commerce. Some of it was implicit, such as the responsibility "to promote the progress of Science and Useful Arts," a phrase that may be taken to have covered, say, the setting up of the U.S. Department of Agriculture in 1862. Some of it was, to say the least, unexpected, such as the way in which the Interstate Commerce Commission, a regulatory body with considerable powers, emerged in 1887 out of the apparently innocent wording of Article 1, Section 8 of the Constitution, authorizing the federal government "To regulate Commerce with foreign nations, *and among the several States.*" Today, this involvement is taken for granted: government and economy have grown to overlap each other to such an extent that their separation is impossible.[1]

This chapter will review some of the principal economic areas in which the federal governments of both North American nations are active. Expressed in geographical terms, they represent the distortions in spatial patterns of production brought about by the intervention of government in a theoretically free economy.

Government, Transport, and Freight Rates

One of the areas of the domestic economy in which both North American governments earliest became involved was the regulation of transport and, espe-

cially, of the freight rates charged by canal and riverboat lines and railroads. Not only was this an area of the economy that cried out, in the mid-nineteenth century, for the services of a referee or policeman but it was one which, as the frontier moved west, became increasingly important to the nation. In all economic activity, the cost of transport is a factor. But the greater the distances involved, the more important it becomes—and here the North Americans were soon to be dealing with the whole length and breadth of a continent.

Each transport company, whether dealing in land, water, or air freight, wants to offer acceptable terms to shippers of goods, in order to capture their business. A century ago, when canals and railroads were being built in North America by the score, it was often a matter of life and death to the carrier to capture the traffic of a city or area from rivals. In these circumstances, savage "rate wars" developed, which frequently led to the ruin of all those involved; and the freight rate "structure" became simply a series of special arrangements between carriers and shippers, reflecting their relative bargaining strengths. Already in the 1850s a number of states had set up government bodies to safeguard the public interest in efficient transport service, and inevitably the federal governments were drawn in to check what had become a scandal. In 1887, the United States established the Interstate Commerce Commission (ICC) and, in 1903, the Canadian government provided for a Board of Railway Commissioners (later called the Canadian Transportation Commission and now subsumed in the National Transportation Agency).

But, although these controls checked the worst abuses of the system, the freight rate structure remained basically unchanged, a mass of millions of separate agreements between carriers and shippers. The commissions merely acted as referees, and by enforcing certain rules of the business, they saw to it

---

[1]J. W. House, ed., *United States Public Policy* (Oxford, 1983), opens with a good discussion of the "Policy Arena" of government action. The rest of the book is also useful although, unfortunately, it covers neither of the topics dealt with in the first two sections of the present chapter—freight rates and agriculture.

that a reasonable degree of competition was maintained. Indeed, if a local market was in the undisturbed possession of one set of producers, the commissions might create competition by allowing producers from elsewhere an exceptionally low rate to enable them to enter the market. The commissions' basic aim was to see that the public was provided with efficient service and safeguarded against costly monopoly. They supported the bargaining power of the shipper against the carrier and of the carriers against each other.

Some of the most influential special freight rates ever approved by the ICC were the transcontinental commodity rates[2] agreed between the railroads and the fruit growers of California. Oranges, for example, are grown in both California and Florida. But California is more than twice as far as Florida from the urban market areas of the East, and there was little chance of western oranges competing in these markets unless they could be cheaply transported across the country. For the railroads, the alternatives were low freight rates or no freight at all, and they set a special commodity rate from California, which tapered so sharply from Chicago eastward that virtually the whole eastern market could be covered for the same freight charge, and competition with Florida oranges became possible.

On a much broader scale, the Maritime Provinces of Canada put forward the claim that, owing to their geographical isolation from the rest of the country, they were entitled to a lower level of rates than other sections in order to be able to place their produce on the home market on competitive terms. The Canadian government accepted this argument in principle, and by an act of 1927 reduced the railroad rate level for goods moving either within or westbound out of the Maritimes by 20 percent. This subsidy to the region was extended to road haulage in 1968.[3]

With the passing of time, the nature of the competition among the carriers changed. Under increasing pressure from truckers, the railroads stopped

fighting each other and instead banded together to preserve their interests. In certain limited spheres, water routes also provided a challenge that forced a special reduction in freight rates by rail. Instead of bidding separately for traffic, therefore, the railroads negotiated their rates through regional committees of their representatives, watched over by the ICC, the CTC, and the state commissions.

But the functions of the commissions were not purely preventive. It was early realized that the freight rate structure could have an important positive influence on the development of a country or region: in economic terms, that the difference between cost of service and value of service could be manipulated by either the carrier or the government in a highly creative manner. The Canadian government has been very sensitive to this fact, particularly since it became the owner of one of Canada's two great railroad systems. Its chief concern was to assist the outward movement of Canadian raw material exports and the forward movement of settlers' necessities into the frontier zones. Thus in 1897, it concluded the Crow's Nest Pass agreement with the Canadian Pacific Railway, whereby, in exchange for a subsidy, the railway agreed to fix low rates on a number of commodities from the western frontiers, particularly grain. This favorable rate on grain remained, justified on the ground that, as the 1951 Royal Commission on Transport stressed, production of grain for export is for Canada "an industry requiring special consideration as in the national interest." In economic terms, therefore, this low rate, secured by the government, acted as a subsidy to the grain producers.

But by 1983 the effect upon the railroads of offering this low rate had become a heavy burden, especially since it applied to one of the roads' chief forms of traffic. The railroads were losing money, and their equipment was in poor condition. In that year, therefore, a new deal was struck: the Western Grain Transportation Act was passed, under which both higher rates on grain and government subsidies were authorized. After a hundred years, the grain traffic was at last carrying something like its fair share of the railroads' costs.

Once the goverment starts to subsidize freight rates, however, it is not easy to stop. The western grain farmers were helped in the way just described: it was now the turn of their customers. The Canadian government offered Feed Freight Assistance to livestock farmers in eastern provinces—a system of sub-

[2]A *commodity rate* is a freight rate offered on a specific—usually large—quantity of freight of a single kind moving between two named points or areas. For obvious reasons, such a shipment, especially one dispatched regularly between the same points, is much cheaper for the carrier to handle than a small, irregular movement of goods in half-empty vehicles, to which the ordinary *class rate* applies. The carrier can therefore offer a commodity rate to the shipper that is lower than the class rate.

[3]Since the road distance between Halifax, Nova Scotia, and Montréal is about 800 mi (1300 km), this is the equivalent of moving the two cities more than 150 mi (260 km) closer together.

sidies planned to make sure that the freight cost of obtaining animal feeds outside the grain-growing areas would be held down to an acceptable level. Meanwhile, the railroads themselves were in need of assistance, especially in crossing the great empty, unproductive stretch of the Shield. To help in financing the cost of transport across these "bridge" lines, a so-called Bridge Subsidy was paid to the two companies involved.

The Canadian government has been careful to maintain a general level of rates comparable with that of the United States, since over many routes, especially the transcontinental ones, Canadian and U.S. services compete with each other and with the sea routes through Panama. But in any case the Canadian government has been so sensitive to what has been called "the tyranny of distance" over national development that all through the nation's early years, and at least up to World War II, Canada's general level of freight rates per ton-mile were the lowest, or next-to-lowest, in the developed world.

As time went by, government surveillance was extended to new modes of transport—pipelines and airways—to the allocation of routes, and to the withdrawal or retention of services. But the passage of time has brought other changes, too; in particular, a change in the power structure among carriers. The railroads, whose stranglehold on producers, shippers and regions led governments to intervene in the first place, are no longer in a position to exert this kind of pressure; they have too many competitors and are in need of government help much more than government restraint—those, at least, that have not long since gone bankrupt. Many of the survivors are merely parts of financial conglomerates that make their profits elsewhere.[4]

With this change in power structure has gone a process of *deregulation* of transport; that is, the government has withdrawn from many of its earlier functions, to leave carriers to compete with one another. The carriers themselves discover by trial and error on which routes, and at what rates, they can

profitably operate. A considerable part of the railroads' financial troubles, for example, was caused by their obligation as public carriers to maintain particular services, which lost money but which could not be withdrawn without government permission (see also Chapter 9). Deregulation means a free, or freer, play of market forces within the transport economy. Since the danger of gross abuse of its position by a carrier is now remote, the government can afford to relax the control it formerly exercised on behalf of the public.

This deregulation, where the U.S. railroads were concerned, was brought about by the Staggers Act of 1980. It freed the railroads from some of the ICC's controls, and where these remain it laid down that the ICC should regulate rates in such a way that the railroad had a reasonable chance of earning revenue from the traffic. Much more remarkable, however, in the light of history, was a provision of the act that allowed carriers and shippers to agree individual, confidential rates, known only to themselves and the ICC, for particular customers. It was this very arrangement of private rebates to favored shippers that was the foundation of the monopoly achieved by John D. Rockefeller over carriage of oil in the early days of the petroleum industry (see Chapter 8). The wheel has indeed come full circle.

So this particular era of government regulation of a sector of the national economy seems to be coming to an end. But the aftereffects remain: the effects of the millions of separate bargains that together made up the freight-rate structure of Canada and the United States, and that, therefore, guided locational decisions and choices in the past.

Government and Agriculture

We have already considered the government's attitude to land in North America, and this section deals with an extension of the same topic: its influence upon the agricultural use of land. This is a sphere where the government has gradually moved from a peripheral to a central role, as its involvement has increased.

In the nineteenth century, the federal government was involved in agriculture in two main ways: in an advisory capacity and as landlord and user of the public domain. In the United States, the year 1862 saw the founding, not only of a federal Department of Agriculture, but also of a program for land-grant colleges in the West, where the farm population

---

[4]The most spectacular bankruptcy of an American railroad in the past twenty years followed the merger, in 1968, of the Pennsylvania and the New York Central, two old and powerful lines that formed the Penn Central, and that passed into receivership only twenty-eight months later. The corporation as such still exists, however, and has paid off its debts. It deals in property and, at a recent date, owned among other things an amusement park. The one thing with which, quite deliberately, it has nothing to do is running trains.

could receive further education. From these beginnings, the advisory functions of the government grew to cover experimental farms, on-the-farm counseling, and a huge range of publications. The public lands, for their part, were in due course placed under the care of various government services, which became responsible for controlling their use: the National Forest Service, the Bureau of Land Management, the Bureau of Reclamation, and in Canada, the Department of Northern Lands.

Then in the 1930s, both the Canadian and U.S. governments abruptly found themselves in the position of having to rescue farmers from depression, shrinking markets, and bankruptcy; mere advice or example were no longer sufficient. They organized programs to save the farmers, and have kept those programs going for several decades, during which the original causes of disaster were lost to sight and farm fortunes rose and fell. In this way, government involvement in agriculture became direct; it continued to advise (those years saw, for example, the founding of the U.S. Soil Conservation Service), but it also paid farmers to plant, or *not* to plant, particular crops.

In the period since World War II, it has been influential in agricultural planning; its policies help to decide which lands shall be used for agriculture and which types of farm shall flourish. Without giving up its old functions, the federal government has come to possess a series of new ones.

The turning point in this development came in 1929–33. After World War I and throughout the 1920s, agricultural demand and prices had fallen off, only to be struck a catastrophic blow by the onset of the Great Depression. At the same time, decades of carefree land use and unwise cultivation had produced serious soil erosion by water in humid areas and dust bowl conditions in dry areas. With the farmers desperately trying to raise unwanted crops on worn lands and sell them to nonexistent buyers, the government stepped in.

In the United States, the New Deal legislation of 1933 brought in the new program. Its object was to raise both demand and farm prices, and to do so it had first to reduce supply—to get rid of the unmarketable surpluses that had built up. The method adopted was to pay farmers a subsidy based on the number of acres planted to the particular crop in surplus. Each farmer was given a quota acreage, and assured of a fixed minimum price for the produce of those acres, but no others. In that way, one set of

measures simultaneously reduced the area under the crop—cotton, wheat, tobacco—and boosted the price received by the farmer. By adjusting the price, which it guaranteed, to the farmers (and which it expressed as a stated percentage of a "parity" price each year), the government could, theoretically at least, fine-tune the economy and divert production from crops in surplus to crops in shorter supply.

The system had its advantages and its disadvantages. One of the advantages was that the land on which the farmers did *not* obtain a quota was taken out of the traditional crops—crops that, incidentally, may well have been grown for generations and have caused serious soil depletion—and used for other plantings, such as pasture or peanuts. As various crops were placed on acreage quotas, shifts took place to nonquota crops, generally to the benefit of the land. One of the disadvantages was that a quota system that limited acreage rather than output was undermined by the constant rise in agricultural productivity. This simply meant that the surpluses (which the scheme was designed to eliminate) were produced from a smaller acreage than before, but were produced just the same.

Another disadvantage was that, once the farmers had become accustomed to the idea of government support prices, they were most reluctant to forego them, even when times were good. Surpluses continued to pile up, only this time the problem of disposing of them was the government's, not the farmer's. The 1950s were a period of embarrassing overproduction, and saw the introduction of another type of restraint: the "Soil Bank." Under this scheme, the government made payments to farmers to take their land out of cultivation altogether and allow it to rest. The land was benefited and the area under crops reduced, both without loss to the farmer, whose problem has always been that of carrying the initial costs of land conservation measures over the period when production is reduced.

Since the 1950s, there have been a large number of new congressional measures and policy initiatives, but they all relate to the same problems and the same three or four types of solutions. The problems are those of cutting down surpluses, keeping up farm incomes, and—increasingly—saving and improving the land. The solutions are to pay farmers not to plant, to support prices, and to limit imports. Unfortunately, such schemes have unwanted side effects: one is that they make U.S. prices for farm produce much higher than world prices (and let us notice that more than 20 percent of U.S. production has nowadays to be marketed overseas), and the other is that they cost the taxpayer—who is also the domestic consumer—huge sums to finance. Beyond that, of course, consumers outside the United States, who may well have a choice of suppliers, are paying premium prices for any produce imported from the United States.

Reviews of these policies take place every five years or so. The early 1980s, for example, saw a new and ingenious scheme called Payment in Kind (PIK). Under it, the farmer was invited to take out of production land normally planted to cotton, wheat, rice, and coarse grains. For doing this, the government then offered him a quantity of those actual products from its stores, equal to 80–95 percent of what, in an average year, the idle land would be expected to produce.[5]

It was only one more scheme, but its take-up and its effect were both impressive. Farmers opted for the scheme on 82 million acres (33.3 million ha), or 36 percent of the nation's eligible cropland at that date. Since 1983 was a good harvest year, the fall in output of the crops in question was nothing like 36 percent, but the U.S. government estimated that the cost of its agricultural support and storage program was reduced by $9 billion and most, if not all, farmers seem to have benefited by PIK. The ultimate beneficiary of this scheme it was hoped, as of all the others, was of course the land.

The 1990 Farm Bill continued the now-traditional policies of support and subsidy, together with an increasing emphasis on protecting land from overcultivation; this latter effort is embodied in the Cropland Reserve Program, which is more or less the Soil Bank of the 1950s revived. Within the program's first few months, 36 million acres (14.6 million ha) had been entered for it by farmers.

In a changing world, why did the U.S. goverment persevere for so long with agricultural policies that were basically those of the depression? One clear reason, which has already been mentioned, was the reluctance of the farmers to see those policies changed: they represented a guarantee of continuing government commitment to their economic well-being, and the farmers used their political influence to see that the system was perpetuated. Another factor may

have been that the government supported farm incomes to slow down the drift of Americans from farming into other occupations and to improve the prospects of the smaller operator. This it did not so much because it disapproved of the drift on economic grounds as because of its commitment—a commitment that in the United States goes beyond mere party politics—to the ideal of the family farm and its preservation. To keep small farmers—"actual cultivators" (see Chapter 5)—on the land has been a goal for two centuries of idealism and it is a bold politician, except perhaps in California, who will be heard challenging it. The support program may have done something to maintain the family farm.

Perhaps it has. But although the idea was to encourage the survival of the small farmer with his marginal income, the effect was to benefit most the large farmer. Support prices naturally do most for those who have most to sell—bushels of corn or bales of cotton. The government was, therefore, keeping a certain number of small farmers in business by helping to make a few large ones rich.[6]

In Canada, as in the United States, the 1930s marked a turning point. The Prairies provided a dramatic focus for the troubles of agriculture: as a region, they were suffering from drought, and dependent as they were on exports for their livelihood, the Prairie farmers were in an even worse situation than the Americans, many of whose markets were at home. So it was in the Prairies that the principal relief measures were instituted. There were two. One was the establishment in 1935 of the Canadian Wheat Board. Ever since 1912, there had been a Board of Grain Commissioners, whose powers included the control of grain prices and quality, but the 1935 Act gave the Board control over all grain exports from Canada; it relieved the Canadian farmers of at least some of the problems of finding markets for their crop, and in doing so it set a price for each year's crop, which acted as a guaranteed floor for grain prices (although a very low one).

But the Board had first to find its markets, and in this respect we may notice a difference between the Canadian and U.S. systems of support. Canada, in the 1930s, was a nation in whose economy agricul-

---

[5]When PIK was introduced, U.S. government stocks stood at 1.5 billion bushels of wheat, 3.5 billion bushels of corn, and 7.9 million bales of cotton, all of which had to be stored somewhere by the government.

[6]Typical calculations in support of these statements are: (1) that at a recent date, 70 percent of all U.S. farmers—the smaller ones—received in aggregate only 22 percent of the benefits from the government's farm programs, and (2) that under the PIK scheme, fifty farms benefited to the extent of $1 million or more. Half of these were in California.

ture played a very large part, and the rest of that economy was not strong enough to subsidize the farmers on, say, the American scale. If everybody is a farmer, nobody is left to subsidize farmers. The Wheat Board only paid for what it could sell; it did not buy up surplus crops in the American fashion. It was not, in fact, until 1958 that the Canadian economy had matured enough to sustain agricultural support on something approaching the U.S. scale. In that year, the Stabilization Act introduced a system of guranteed prices similar to that of the United States, applying to other products besides grain, although the difficulty of disposing of the great Prairie crops year after year continued to be the focus of the problem.

The second Canadian measure of 1935 was the passing of the Prairie Farms Rehabilitation Act (PFRA). Under it, the government took wide powers to restore prosperity to the region. Soil erosion was checked, marginal lands were taken out of cultivation, scientific farming was encouraged, irrigation schemes were assisted or developed to bring security to farmers on the dry margins, and grazing reserves were gradually acquired to support stock in emergencies. Over the years, the PFRA has been used in so many different ways that it has become a kind of governmental Aladdin's lamp for the Prairies.

Governments all over the developed world have been sensitive in recent years to the well-being of their farmers; in this respect, the Canadian and U.S. governments are following along some distance behind, say, their counterparts in the European Economic Community. But with wheat nowadays an instrument of foreign policy and farmers a loud-voiced minority in national legislatures, there is little danger of agriculture being overlooked by government in the future.

Government and Industry

Ever since the beginnings of manufacturing in North America, governments have been involved with industry in two ways. One has been to protect it by imposing tariffs or restrictions on imports of foreign manufactures. The heyday of protectionist policy lies well in the past now but, even today, there are periodic calls for protection from supporters of particular industries that see themselves threatened by overseas competitors. The second has been to control it, when it threatened to become too powerful or ruthless. When it neglected the basic rules of health or

safety, it was subjected to government inspection. When monopolies developed, they became the target of antitrust legislation.

But over and above these two basic connections between government and industry, there has long been a third: the government itself has become one of industry's major customers—in some industries, the only customer. This has been the case principally during periods of war or national emergency. Two world wars saw North American industry swiftly adapting itself to the production of armaments. Korea and Vietnam maintained the connection. In between, the conquest of space and the need for military preparedness have both kept government spending high.

Plants exist today—steel mills, aluminum smelters—in locations that, had it not been for defense requirements, would never have attracted them. At the end of World War II, in addition, defense planners encouraged the dispersal of industry away from the traditional, crowded areas of the Northeast where, even in that period of relatively unsophisticated weaponry, something like one-third of all U.S. industry was within target range of missiles deployed in submarines.

But the major impact of government on industry in recent years has certainly been through its role as customer. In 1985, the U.S. Department of Defense alone purchased goods and services to the value of nearly $260 billion. Using an alternative measure of this customer's importance, over 250,000 manufacturing workers in California, out of a statewide total of some 2 million, are employed in "defense-orientated industries." Government demand may take the form of supplying capital and plant for an installation that it owns, or it may simply mean that industry is producing for the government. In either case, government money is likely to play a significant part in regional development, particularly in areas where the accumulation of investment capital would otherwise be difficult.

A great deal, therefore, depends on how and where the government awards its contracts. In wartime, governments are usually in a hurry to obtain results, and rely on existing plants or industrial regions for the necessary added production. But in peacetime, when there is more opportunity for long-term planning, the regional balance of government investment may well favor previously nonindustrial regions, and it will be the task of politicians from those regions to see that it does so.

Table 6-1. Distribution of Military Prime Contract Awards, Fiscal Years 1941–45, 1951–53, 1961, 1971, 1980, 1985, and 1989 by Regions.

Region	Annual Averages		Fiscal Year 1961	Fiscal Year 1971	Fiscal Year 1980	Fiscal Year 1985	Fiscal Year 1989
	1941–45 (World War II)	1951–53 (Korea)					
New England	8.9	8.1	10.5	9.5	13.2	11.0	13.6
Middle Atlantic	23.6	25.1	19.9	18.5	14.1	12.9	10.6
East North Central	32.4	27.4	11.8	10.4	8.9	9.5	8.7
West North Central	5.6	6.8	5.8	5.9	8.3	9.4	8.2
South Atlantic	7.2	7.6	10.6	15.8	14.9	16.1	16.1
South Central	8.8	6.4	8.2	14.7	12.7	13.0	13.9
Mountain	1.2	0.7	5.7	2.6	2.9	3.7	6.5
Pacific	12.3	17.9	26.9	21.2	24.3	23.5	22.8
Alaska and Hawaii	—	—	0.6	0.8	0.9	0.9	0.9
	100.0	100.0	100.0	100.0	100.0	100.0	100.0

Consequently, we find that, over the years, the distribution of industrial contracts to regions by the U.S. government has changed considerably (Table 6-1). In World War II, 56 percent of the government's defense contracts were awarded within the Middle Atlantic and East North Central regions—roughly speaking, the existing industrial core which was known as the Manufacturing Belt. In 1989, these regions received only 19 percent of the contract value. California had become the runaway leader; this one state received nearly 20 percent of all defense contracts. While this switch was in part due to the fact that the last two wars in which the United States was involved—Korea and Vietnam—have both been Pacific Coast wars, there were other factors: available skills, research facilities, the growth on the West Coast of the aircraft/space industry on which so much of U.S. military spending has been concentrated.

What is true of government contracts for industrial production is perhaps even more clearly true of the related activity of research, on which, in 1989, some $142 billion was spent, almost one-half of it by the federal government. Industrial research and development (or R & D) is taking place predominantly on the Pacific Coast, with a secondary concentration in New England and the Middle Atlantic states. Much of this R & D is carried on by industry itself *for* the government, but a part also—some 9 percent—by universities; in this respect, institutions in California and New England are outstanding. R & D breeds new industry so that, with the government as paymaster, some areas have a huge advantage over others in industrializing. In view of the size of the expenditures

involved, it would be surprising if they did not become the object of political maneuvering, in an effort by the "have-not" states to attract away from the "haves" what is, after all, taxpayers' money from the nation as a whole.

The relationship of government and industry in Canada includes, on a much smaller scale, the same military–industrial linkup as in the United States. The Canadian government has, however, its own unique concerns. One of them, to which we shall need to return in Chapter 8, is the problem of control of a large part of Canadian industry by corporations based in the United States. This has meant, over the years, not only that many of the profits of Canadian manufacturing were transferred south of the border, but also that many of the actual decisions about which products, or parts of products, Canadian plants were to manufacture were taken at corporate headquarters in the United States. This subservience showed up, for example, in a low Canadian expenditure on R & D; the industrial research was done elsewhere, and the purely mechanical operations were given to the Canadian plants.

The Canadian government has been concerned about this situation for years, and done its best to remedy it by insisting on a fair Canadian share in all industrial corporations. Both federal and provincial governments have been ready to intervene, directly where necessary, in industrial enterprises—in taking over bankrupt railways, in bailing out steelworks, in controlling oilfield development, or exploiting the tar sands of the West (see Chapters 18 and 22). Since 1946, when it was founded by the Canadian Parliament, the Canadian Commercial Corporation has served (like the Wheat Board) to promote Canada's exports to other continents, and to find worthwhile investment opportunities for Canadian capital overseas.

Government and Energy

The need for government action in the field of energy has been a product of the growing variety of energy sources available to man in the developed world. Societies using only wood or coal or water power face no problems other than those of supply and demand, but where petroleum, natural gas, hydroelectricity, and nuclear power are also available, the principal problem becomes one of *choice*. The question then arises as to whether users and customers are to be allowed to make that choice on the basis of market forces, or whether a government has a responsibility to intervene, to influence use or encourage conservation.

It is a question that has been answered hesitantly by the North American governments, sometimes with one answer and sometimes with another. The result is that, while both the United States and Canada have plenty of policies about particular forms of energy, neither of them, in the late 1980s, actually had an *energy policy*. It is true that, under the Trudeau government, Canada had such a national policy but, in 1985, it was formally abolished.

The reason for this hesitancy is not far to find. Since the main forms of energy generation that have brought us this far into the twentieth century—coal, petroleum, and natural gas—are all based on finite and nonrenewable resources, it would demand remarkable powers of prophecy to predict whether it will be better to use those resources now, and let the future take care of itself, or to hoard the existing supplies against an emergency, and buy in present requirements from more spendthrift producers elsewhere. Since there is no definitive answer to that question, it can be argued that a national energy policy is an illusion. And in a country where some states or provinces are large energy producers while others are consumers, it can be further guaranteed that to formulate any national energy policy is a sure way to split the country.[7]

It is not difficult to specify the principles that should underlie such a policy: they can be expressed as efficiency, equity, environmental integrity, and national security.[8] But those principles do little more than identify the problem, which is how to combine them: (1) How to encourage exploration for new sources of oil or gas at the same time as discouraging overproduction and (2) How to benefit nationally from the present fuel supplies of a state or province without impoverishing its future. On the whole the federal governments have preferred not to grasp what must look to them like a whole bed of nettles. This leaves the state and provincial governments free to legislate for their own needs.

There was, however, a short period when the U.S. government had a national energy policy, although even then it did not go by that name. It happened when, in 1973, a cartel of oil-exporting countries in the Middle East and Africa decided to raise the price of oil, a decision that led to turmoil in the petroleum market. In that year, petroleum and natural gas accounted for 77 percent of energy consumption in the United States. More than one-third of these fuels were imported and, in the years that followed, this proportion increased.

The federal government was faced with the need for an immediate reappraisal of the energy position. The United States, for so long the world's largest producer as well as consumer of petroleum products, was unaccustomed to the feeling of being dependent upon the whim of foreign suppliers, particularly since most foreign oil fields had been developed by American companies. Oil could usually be exploited more cheaply overseas than in the United States, and since the depreciation allowances granted to the companies by the government were the same in both cases, the oil companies had tended to concentrate on overseas fields and to neglect home exploration. With one-third of the world's petroleum consumption, the United States had, by 1974, only 6 percent of the world's known reserves. And overseas, the oil companies were faced by demanding governments, which no longer treated them as equals or looked on them as benefactors, but rather as foreigners exporting national resources "on the cheap."

The response of the U.S. government was to launch a national energy policy concealed behind the title of Project Independence. It was a five-year program to cut American dependence on foreign energy

---

[7]House, *U.S. Public Policy,* Chap. 5, has a useful discussion of the difficulties involved in formulating energy policy and quotes (p. 177) the *Washington Post* as declaring, "Energy policy is now the most divisive regional issue to afflict this country since civil rights."

[8]House, *U.S. Public Policy,* p. 178.

Energy and the future: The alternatives. The photographs show (left) the Tennessee Valley Authority's Paradise steam-generating plant on the Green River near Greenville, Kentucky, the largest thermal power station in North America (2,558,200 kw capacity). Around the coal-burning plant, with its tall smoke stacks, are areas of strip mining from which coal is fed to the plant. (Right) Brown's Ferry nuclear power plant near Athens, Alabama. Both types of power plant have their opponents, but once available hydroelectricity potential has been fully exploited, the choice for the future appears to lie between these two. *TVA*

supplies. It had two aspects. One was to cut down the consumption of energy in America, and the other was to find new energy sources. Some of these sources would represent technical breakthroughs, such as more efficiently harnessing the energy of sun or wind. It was also hoped to find new reserves of conventional fuels: more oil in the Arctic or on the continental shelf; or to exploit more fully the mid-continental oil shales, like those of Colorado (see Chapter 8). A major role was assigned to nuclear power. By 1989 there were 108 nuclear units on line, with a capacity of 94.7 million kw. The project represented a serious attempt to think of energy production for the first time as a whole rather than a series of separate power-producing industries.

To judge by the figures, there has been some modest degree of success under the first aspect of the policy—a drop in energy consumption. In reducing dependence on overseas energy supplies there has also been some limited success. The North Slope oil fields in Alaska were developed during the project period, and there was a burst of activity (see Chapter 8) in producing oil from shales, although this was frustrated by a subsequent fall in oil prices. As for the important dependency index of imported oil supplies, by 1985, the proportion of crude oil input to refiner-

ies represented by imports was down to 25 percent: by 1989, unfortunately, it had risen again to 43 percent.

There are two major obstacles to the success of energy-saving and energy-converting policies. One is the reluctance of people, who have become accustomed to using huge amounts of cheap energy, to change their habits, even when energy is no longer cheap, and especially when (as happened in the 1980s) there is a world surplus of oil, so that warnings about shortages and lines of cars at gas stations are seen simply as alarmist. The other is that plans for replacing imported oil with domestic oil have run into strong opposition from environmentalist groups.

The quickest ways to reduce dependence on imported oil are (1) to press ahead as rapidly as possible with nuclear development and (2) to revert to the use of the other hydrocarbon source, which is available

in quantity but has for decades now been overshadowed by petroleum; that is, coal.

Neither way is easy. The construction of a new generation of nuclear plants is fought at every step by groups hostile to the use of radioactive materials, or fearful of accidents at the power plants. The new plants should already be under construction if they are going to contribute their planned share of future energy requirements, but they are not. On the other hand, the use of coal has fallen foul of so much new legislation that it, too, is of limited usefulness. Clean air is the order of the day. In California, for example, coal-fired power stations have been outlawed, and oil-burning plants must use a specified grade of fuel. Where coal is concerned, much of that produced in Appalachia, for so long the major supplier to the continent, has a high sulfur content, and such coal is banned in many states and areas on grounds of air pollution. Meanwhile, environmentalist pressure has been building up against strip mining, at least in the form that a coal state like Kentucky has known it in the past (see Chapter 16); yet strip mining accounts for about 60 percent of total U.S. coal production.

We can best summarize all these developments in a few simple sentences: (1) A national economy, the world's leading consumer of energy, has become highly dependent upon one energy source—petroleum products. (2) This energy source is now in short or diminishing domestic supply. (3) Alternative sources exist, given freedom of action on the part of the government and the power producers. (4) But at this point in time, freedom of action is denied by a newly vocal section of the American community, which is highly critical of what such "freedom" has meant in the past. (5) Stalemate ensues.

Meanwhile, the U.S. government has been moving steadily toward "deregulation" of the oil industry. Although it would be wrong simply to interpret deregulation as a lack of energy policy—for it is obviously a policy decision itself—it does mean that the trend toward more federal involvement in energy matters is being reversed, and that the government is committed to reaching its energy goals by letting the prices of oil and gas find their own level—in fact, to rise. Indirectly, then, new energy sources may be opened up as they become economical to exploit. But indirect the policy method certainly is.

Meanwhile, the Canadians have wrestled with their own energy problems, from 1980 to 1985 under the terms of a formal National Energy Policy (NEP), since 1985 without one. Canada's problem is not, in the first instance, shortage of energy; in this respect the nation is favorably placed. The problems are, to Canadians, familiar ones: (1) how to reconcile the conflicting interests of energy-producing and energy-consuming provinces, and (2) how to keep a firm Canadian hand on the nation's resources and, at the same time, attract outside investment to supplement Canada's own limited capital supplies.

1. The NEP of the Trudeau government in the early 1980s assumed direct federal participation in the petroleum industry, restraint on prices, and a federal share in the revenues yielded. This being the case, the Province of Alberta, as the main petroleum producer, led the opposition to the NEP. Summarizing the position during the Trudeau years, the Economic Council of Canada put the case as follows: "For Alberta, the single most important issue—and it is hardly a new one—boils down to retaining control

over its natural resources. The province perceived the NEP as a takeover [by the federal government]."[9]

In this opposition, the oil-producing provinces were supported, rather surprisingly, by Québec, an oil consumer, because the province feared that, if the federal government entrenched its position as a price setter or controller in the oil or gas industries, there would be nothing to prevent it from transferring its attention, later, to electrical energy, Québec's prime power source. Ontario, on the other hand, as the most populous and most industrialized province, supported the NEP's assumption that the Canadian price of oil should be managed in favor of Canadian consumers.

Without an NEP, however, the Canadian government has since 1985 continued its usual balancing act between federal pressure and provincial rights. This is best seen in the series of agreements struck between Ottawa and the provinces after the NEP was laid to rest. In 1985 two "accords" were signed. The first, the Atlantic Accord, was made between Newfoundland and the federal government for the joint management and revenue sharing of the offshore oil fields of the Grand Banks, where exploration has been in progress since 1966 and two fields, Hibernia and Ventura, are in the process of development. The second, the Western Accord, with Alberta, British Columbia, and Saskatchewan, eliminated much of the pricing machinery of the NEP and made over to oil producers a large amount of revenue previously claimed by Ottawa. A third step was taken in the same direction in 1987, with the passage through Parliament of a bill "to implement an agreement between the Government of Canada and the Government of Nova Scotia on offshore petroleum resource management and revenue sharing."

2. "We reject," said the Standing Senate Committee of the Canadian Parliament on Energy and Natural Resources, "the premise that energy, in its various manifestations, is nothing more than an article of commerce; at times environmental, social, strategic or political considerations hold sway over market forces."[10] Conscious of its natural wealth and of the

---

[9]Economic Council of Canada, *Connections: An Energy Strategy for the Future* (Ottawa, 1985), p. 8. The whole document is one of the best short summaries of the Canadian energy position available.

[10]Senate of Canada, Proceedings of the Standing Senate Committee on Energy and Natural Resources, Proceedings of August 21, 1985, p. 3.

need to conserve it, the government has the problem of deregulating the oil industry and satisfying the provinces without at the same time allowing Canada's grip on its resources to be weakened to the point where it loses effective control of them.

This it must do, also, without weakening its trade position, for energy exports—crude and refined petroleum, natural gas, coal, and electricity—account for several billion dollars annually on the credit side of its balance of trade.

In the area about which it is most concerned—the oil industry—the government already has a stake through its Crown Corporation, Petro-Canada. A Canada Petroleum Resources Act has now been passed which requires that, at the production stage, there shall be at least a 50 percent Canadian participation in all oil field development. In financial control, then, Canadian interests are to have at least a half-share and, in the actual supply of petroleum products and natural gas, a policy of "Canada First" is to be followed.

7
Agriculture

American agriculture: Center-pivot irrigation. Watering fields by means of a rotating arm, a system that can irrigate about 135 acres (54.6 ha) of a 160-acre (65 ha) quarter section. *U.S. Department of Agriculture, Earl Otis*

The Formative Factors

In the United States, about 57 percent of the land area is put to agricultural use of some kind. If Alaska is excluded, the proportion rises to 67 percent, but the area actually under crops is only 15 percent. For Canada, the comparable figures are area in farms— 7 percent, area under crops—3 percent. For Canada, however, neither of these low figures is as important as a third—that only 11 percent of the nation's land is considered to have potential for agriculture at all.

Yet Canada's 7 percent of land in farms was responsible for sustaining the nation's export trade for many decades before its growth as a manufacturer transferred the main responsibility for exports to other sectors, and the farms still provide about 10 percent by value of Canadian export shipments. In the United States, agricultural exports have gone some way toward offsetting the massive foreign trade deficits of the 1970s and 1980s: during that period they have run at $20–30 billion a year, or 10–12 percent of total U.S. exports.

One generalization about North American agriculture, which it was for many years possible to make, was that its patterns were dominated by a series of belts—Corn Belt, Cotton Belt, Wheat Belt— and every geography book contained a map of these. In human terms, this simply meant that all the farmers in a particular region followed the same practices and grew the same principal crops, and there was nothing surprising about that. In a continental area where most farm produce, moving to either domestic or export markets, had to travel hundreds, if not thousands, of miles[1] there was every incentive to produce for shipment whatever the neighbors pro-

duced; it simplified the transport problem, as well as enabling isolated farmers to give and receive technical help with well-practiced operations.

But we can now see that those belts or regions were not permanent divisions: some of them have already disappeared. They represented simply a stage in the development of a pattern which has been evolving for over 300 years, and which has, in fact, changed significantly since World War II. So we must ask: what are the forces that have brought into being a pattern such as this, and that are now changing it once again? For the sake of convenience we may divide them into three groups: historical, environmental, and economic.

HISTORICAL DEVELOPMENT

Much of the story[2] of American agriculture takes its character from the fact that the continent was settled by agriculturalists spreading westward from the East Coast. The earliest settlers on the Atlantic Seaboard had to be as nearly self-supporting as possible. They brought with them wheat and livestock, learned the use of corn (maize) and squash from the Indians, and imported little but sugar. Hemmed in along the coast and the St. Lawrence by mountains, forests, and Indian opposition, they had to make the best of the available lands, unsuitable for cultivation though they often were, since the state of communications made it virtually impossible to transport food beyond the confines of the individual community.

As time went on, the problem was modified in two ways. First, the gradual improvement of communications did away with the need to produce every re-

---

[1]Someone has gone to the trouble of calculating that, for every $2 worth of agricultural goods placed on the North American market, $1 has been spent in getting it there.

[2]Some examples of the literature on American agricultural history are: J. T. Schlebecker, *Whereby We Thrive: A History of American Farming, 1790–1840* (Ames, Iowa, 1975); D. P. Kelsey, ed., *Farming in the New Nation: Interpreting American Agriculture, 1790–1840* (Washington, D.C., 1972); and G. C. Fite, *The Farmer's Frontier, 1865–1900* (New York, 1966).

quirement in each settlement, and made possible a regional division of labor. Areas particularly suited to the production of one crop began to specialize in it. The new situation also meant that intensive forms of farming could be practiced nearest to the settlements, where land was in greatest demand, and that extensive forms, such as the growing of cereals and the raising of sheep, were relegated to the fringes of the settled area, where space was no problem. Provided that transport facilities were satisfactory, this represented the most economical form of land use.

The second modifying factor was the discovery, made as the frontier of settlement moved westward, that beyond the mountain barrier were wide expanses of the continent which were more fertile and easier to farm than the original agricultural lands along the East Coast. To the hill farmer of New England or the western Piedmont, the level grassland of Kentucky and the forest openings of Ohio or southern Ontario seemed a paradise indeed. Once the interior was settled and linked by river, canal, or rail with the Atlantic Seaboard, the high-cost, marginal farming which had been forced upon the East Coast states could not hope to compete with that of the favored interior.

The effects of these two processes have been most marked.

1. *Geographically*, the effect has been that the centers of production have moved westward, their position dictated by the availability of transport facilities. The clearest example is that afforded by wheat and is illustrated by the maps in Figure 7-1. Before 1825 most of North America's wheat crop grew in the Atlantic coast states, between Vermont in the north and Virginia in the south. In 1825, however, the Erie Canal was opened, linking the Hudson River with the eastern Great Lakes region. This event brought about such a revolution in transport costs that the center of wheat production promptly moved from New England to the open lands on the Lake Erie shore of western New York. Through the period 1830–40 the wheatlands spread further westward, this time into Ohio and southwestern Ontario. The next decisive step was the development of a railroad network throughout the Great Lakes region. This occurred during the 1850s—New York and Chicago were linked in 1852—and made available for wheat growing the grasslands of Illinois. Through the 1860s and 1870s the process continued, with the centers of production moving on westward across Wisconsin and Iowa. The 1880s, marked as they were by the

construction of the Northern Pacific and Canadian Pacific railways and a number of shorter lines, saw the partial end of the process—the establishment of the wheat regions, as we shall later see in Figure 7-3, eventually established in Kansas, the Dakotas, and the Prairie Provinces. In those areas, aridity brought to a halt a movement that had covered 1200 mi (1900 km) in seventy years. But in the period 1870–90, following the completion of a transcontinental railroad, California's Great Central Valley enjoyed a wheat boom reminiscent of those that had occurred earlier in the East. In the 1880s, also, the Palouse country of southeastern Washington utilized its new railroad connections to establish a wheat-growing specialization, which, like that of the Great Plains, has endured.

In much the same way, further south, wheat growing spread from Maryland and Virginia to Tennessee and Kentucky, and from there across the Mississippi into Kansas and northern Texas. The southern crop, however, did not figure prominently in the nation's markets until Kansas began to produce; until that time much of it was consumed within the subsistence farming economy of the southern states.

2. *Economically*, the effects of these processes have been no less striking, and it is interesting to notice that there is a close parallel with the agricultural situation of Great Britain over the same period. For eastern North America, as for Britain, the opening up of fertile and extensive agricultural lands meant severe competition for the farmers of the older areas. At the same time, moreover, as the railway network was spreading westward, there occurred a parallel technical advance in the manufacture of agricultural machinery, which brought further advantages to the farmers of the level interior lands and further increased the handicap of the less-favored eastern regions.

The succession of westward-shifting wheat booms represented the operation of the principle of comparative advantage. In each newly opened center, superior climate and soils conferred a comparative advantage in producing wheat for sale—enough so as to offset the cost of longer distance to eastern markets. But the advantage was often temporary, holding only until a newly settled area or a change in technology changed the equation.

Nor did the principle always operate so simply, particularly in the last of the shifts. In Iowa, wheat yields are generally higher per unit area than those in Kansas or North Dakota. Similarly, the irrigated soils

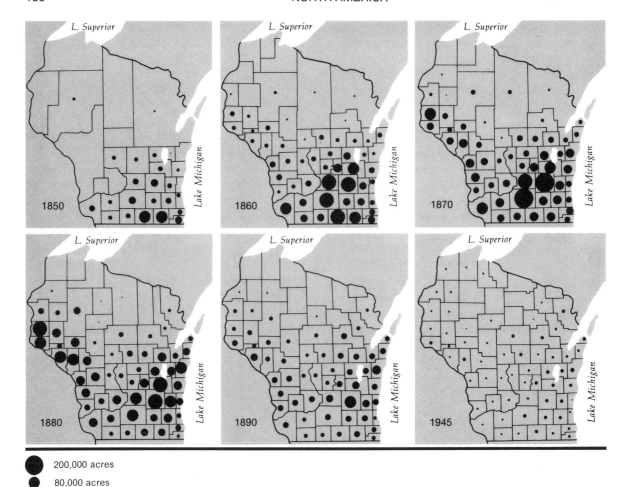

200,000 acres
80,000 acres
25,000 acres

Fig. 7-1. The Wheat Belt moves west: A series of maps of Wisconsin, showing how the Belt traversed the state during the period 1850–90, as it moved toward its present location in the Dakotas.

of California's Central Valley far out-yield those in the unirrigated Washington Palouse. Like all farmers, those in Iowa and the Central Valley apply the principle of comparative advantage, specializing in corn–soybeans–hogs–beef or in fruits and vegetables, respectively, because not only do they enjoy a comparative advantage in those specialities, but also they make more money than they would raising wheat or, presumably, any other crop. Wheat yields per unit area in Kansas, North Dakota, and the Palouse are lower than in Iowa or California's Central Valley, but these areas would be more at a disadvantage were they to grow corn and soybeans or fruits and vegetables. It could be said that, other things being equal, the best-endowed areas have first choice

among specialities; the lesser-endowed areas choose from what remains.

What happened, therefore, was that the Easterners were forced to retreat from one form of agricultural production to another. Except in a few specially favored areas, the eastern farmers could profitably produce only what could not easily be shipped into the region from outside. Their farming had to fulfill two conditions: that production should be intensive, and that it should be concentrated on commodities for whose sale closeness to market, the Easterner's one solid advantage, was of prime importance. In these circumstances, a concentration on dairy products and fresh vegetables could safely be predicted, and this prediction is largely borne out by facts. Yet even with this limited range of output, the eastern farmer depends on first-class communications for his ability to sell. Without them, as we shall later see, he is likely to be deprived of a market altogether.

These last comments apply mainly to the farmers of the Northeast who, coming to the New World from Europe, brought with them European crops, livestock, and farming practices. But in the Southeast the story was rather different. In the region that stretched southward from Chesapeake Bay, the crops of Europe were poorly adapted to the hot, humid growing season. They might supply local needs, but could hardly compete in national or export markets.

New crops were tried. One, the Indian crop, tobacco, was an almost instant success, both in the productivity of the Chesapeake lands and in the ready and growing acceptance by the European market. For the first time, wealthy and venturesome European entrepreneurs had an "American" crop in whose production they could invest on a relatively large scale. But tobacco was a labor-intensive crop; that is, labor (plowing, planting, cultivating, weeding, suckering, harvesting, and curing) was by far the major factor in its production—and it was hard, bending labor under a hot sun. Large-scale production could not be achieved by a settler family on a small farm. Initially, there were family-sized tobacco farms, but production eventually was dominated by large plantations whose immense needs for labor were met by slaves captured and transported like cattle, directly or indirectly from Africa, by European entrepreneurs. Thus was established the slave plantation system that came to dominate the economy and society of the American Southeast. It was used first for tobacco growing in Virginia and Maryland, for rice in the Carolinas, for sugar in the Mississippi Delta, and, after 1800, most widely for cotton. These were crops that the farms of Europe could not produce, but that (except for rice) the markets of Europe sought avidly, and so the plantation system grew and prospered.

The early settlers on the Pacific coast found themselves confronted by a situation not unlike that of the seventeenth-century immigrants to the Atlantic Seaboard. In the 1840s their supply route lay round Cape Horn, or at best crossed the Isthmus of Panama; their penetration inland was shallow, and the settlements had necessarily to be self-supporting. So there occurred in the isolation of the West an agricultural development on much the same lines as that already described in the older settlements of the East. Local self-sufficiency gradually changed to regional specialization. The West Coast developed its Wheat Belt in eastern Washington, and its Hay and Dairy Belt in the Willamette–Puget Sound Lowland. Since World War II it has even developed its Cotton Belt in central California, although it bears little resemblance to its older eastern counterpart. Only the midwestern Corn Belt, product of a unique combination of physical and economic circumstances, has no replica in the West.

As the mining booms of the 1840s and 1850s brought prospectors and camp followers to the West Coast, the early farmers enjoyed a seller's market so profitable as to lure many an unsuccessful fortyniner away from mining to the safer business of feeding his more persistent partners. So long as western agriculture fulfilled this limited function of local supply, its problems were few. But with the passage of time two changes took place. First, in spite of the region's phenomenal population growth, the total agricultural output quickly grew to exceed western demand. Second, the Pacific coastlands developed a series of specialty crops that are produced not for the local but the national market.

The combined effect of these two changes is to make today's western farmer dangerously dependent on outside markets, and so also on his means of transport to those markets. Here his isolation beyond the mountains counts against him. Before they can begin to sell in the main markets of the continent, Calfornia oranges must travel 1600 mi (2550 km) from Los Angeles, and Washington apples a similar distance from the valleys of the Cascades. British Columbia's farmers are, if anything, worse off in this respect than those further south. For many western products the Panama Canal offers a better route to market, even to the North American market, than do the transcontinental roads and railways.

In the meantime, what is to be said of the great intermediate region that divides East from West—the Agricultural Interior? Between the Ohio and the Missouri rivers and the Canadian Shield are to be found some of North America's most favored farmers. They possess all the advantages of their fellow farmers to both east and west of them, and yet escape most of their problems. Farmers in the East have the advantage of a large urban market close at hand, but lack wide stretches of fertile farmland. Farmers further west have space and fertility in plenty, but have the problem of getting their produce to distant markets, as well as the hovering threat of climatic variability. The Midwesterners, on the other hand, have both fertile drift plains to cultivate and excellent markets in the cities of the Manufacturing Belt; they have at their disposal, moreover, a communication network that is probably unrivaled throughout the world.

The greatest expanse of land in the United States with an almost ideal combination of terrain, soil, and climate for agriculture is the region that came to be called the Corn Belt. We have suggested that its rank as the best-endowed region gives it first choice among a variety of possible items to produce. Its great productivity in corn (partly achieved by genetically designing the crop to fit the region), coupled with the desire of an increasingly large and affluent American population to eat meat, resulted in the initial choice being predominantly corn-fed beef cattle and hogs. However, changing world markets and needs during recent decades are shifting the emphasis from corn-fed meat to corn and soybeans for export.

West and south of this region of what might be called agricultural equilibrium, the situation of the farmers becomes less favorable. Westward across the Great Plains, this decline can be explained in terms of increasing distances to market and of climatic hazards. Prosperity in a single year, or in a series of as many as ten years, may be equal to that of the Midwest, but there can be little security. Drought, dust, or grasshoppers will redress the balance. Southward, beyond the Ohio, there are neither the urban markets which are basic to the midwestern equilibrium nor the same midwestern fertility in the soils. Furthermore, the pattern and, in the long run, the prosperity of southern agriculture was distorted by its historic loyalty to King Cotton. Only since World War II, as Chapter 15 describes, has any semblance of a midwestern type of equilibrium emerged.

ENVIRONMENTAL FACTORS

The preceding paragraphs have outlined some of the historical circumstances responsible for the present pattern of North American agriculture. They have been presented first because all too often they disappear behind the physical factors of climate, relief, and soil, which are much more accessible to the geographer in search of explanations. It must now be recognized, however, that the ultimate control of this

pattern is not human but natural; and, on the continental scale we are considering, the principal environmental influence is climatic. Thus the migration of the Wheat Belt from the Atlantic coast to the western plains, which has already been described, was the outcome of economic change, but it was confined within limits set by climatic factors—on the north by temperature and on the west by aridity. Again, in the Southwest, the spread of slavery as an institution was checked not merely by political force (for it was legal in areas it never touched) but by the climatic limits of the region in which cotton, its economic accompaniment, could be grown. Where the Cotton Belt could not reach, slave owning lost much of its purpose.

In general, where the boundaries of the agricultural regions run from east to west, they are determined by temperature, and where they run from north to south, by rainfall. The northern limit of successful cotton cultivation is governed by the diminishing length of the frost-free season. The northern limit of the Corn Belt depends upon the amount of summer heat required to ripen the corn, and its boundary with the Winter Wheat Belt in the Southwest upon the prevalence in that quarter of dry summer winds, which parch the corn. The northern limit of the Spring Wheat Belt is a product not only of poor soils and distance to market but also of diminishing amounts of summer sunshine and increasing frost hazards.

The rainfall control of the longitudinal boundaries is most clearly seen on the Great Plains. The cultivation of cotton, corn, and hay in the south, center, and north respectively of the Agricultural Interior, give way along the 20- to 25-in (500–625-mm) isohyet to wheat growing, and this in turn is abandoned in favor of range livestock farming further west, where the precipitation drops to 12–15 in (300–375 mm) annually.

In the dry regions of the West, irrigation is necessary for raising crops and feeding stock on anything other than range grasses. The Indians of the Southwest had long farmed small areas by diverting water from streams. Spanish settlers, familiar in their Mediterranean homeland with this same technique, carried on the practice. But it has mainly been in the twentieth century that levels of technology and investment have been high enough to make possible the large-scale projects which have raised the irrigated area of the United States to 45–50 million acres

(18–20 million ha). In much of the Mountain West, the proportion of irrigated cropland exceeds 70 percent.

Thus the major agricultural divisions reflect climatic influences. On a smaller scale, relief and soil account for local differences. The ruggedness of the Mountain West, the lack of soil over much of the Canadian Shield, the sandy character of the southern Pine Barrens, and the inadequate natural drainage of the low-lying Florida Everglades all serve to illustrate the limiting effects of the natural environment on the North American farmers. So, too, do the ravages of various crop and animal diseases. Of these perhaps the best known is the cotton boll weevil; a U.S. Department of Agriculture publication commented that "it has encouraged diversification of crops in the Cotton Belt more than has any other single factor" since its appearance in the 1890s.

A further comment in this section must be devoted to the blank parts of the map of agricultural regions. A combination of the factors we have been considering, both natural and economic, makes them unsuitable for farming under present conditions. In the United States these empty areas appear to be relatively small and scattered. The largest of them are in the western states: the desert of southern California and Arizona, where average rainfall is below 5 in (125 mm), and the rugged and in part snow-covered terrain of the Sierra Nevada–Cascade chain and the central Rockies. But the map of this western region is misleading, for over much of the Range Livestock Belt farming is very extensive indeed. The difference between the apparently "farmed" and "unfarmed" areas may be no more than the difference between complete emptiness on the one hand and stocking at the rate of one head of cattle to every 100, or even 200, acres (40–80 ha) on the other. Furthermore, in the mountains and on the desert margins such stocking is only seasonal, and the sum total of agricultural activity in any area may be represented by the presence of a few sheep that find summer grazing in the open forest or above the treeline. Thus the map gives an exaggerated impression of the "agricultural" West.

In the East, nonagricultural lands are found in the swampy and sandy areas of the Southeast and in the forested uplands of northern New England. Here as in the Canadian provinces across the border, remoteness from market combines with poor soils and the cost of forest clearance to militate against the devel-

opment of agriculture. Indeed, as we shall later see, the farming frontier has tended to withdraw rather than advance in this area over the past fifty to seventy years.

In Canada the nonagricultural areas cover the greater part of the country. Over much of the empty north, physical conditions of climate and soil are too harsh for successful farming, but even where some form of cultivation is possible, distance from market limits expansion along the agricultural frontier. There are, in fact, some stretches of these empty spaces that could be farmed, given the necessary demand for their products. But with the present Canadian population, the need to compete with farmers located far closer to markets and ports, and the cost of clearing or draining these lands, their development must await, for the most part, an unpredictable future.

ECONOMIC FACTORS

We have considered the factors of historical geography and environment that have gone into the making of North America's agricultural patterns. But farming is a business, and we must now turn to the third set of factors that influence those patterns—the economic considerations.

All farming represents the input of three factors of production: land, labor, and capital. And European settlement of North America took place, for the most part, under a conjunction of these factors which was, and remains, unique—unique to the period of overseas expansion that began with the colonization of the Americas and continued with the settlement of southern Africa and Australasia.

In Europe, it had for centuries been land that, of the three factors, was in shortest supply: indeed, it was shortage of living space that lent a stimulus to the overseas expansion. Labor was plentiful; if anything, too plentiful. Apart from a few critical times in Europe's history—the Black Death in the fourteenth century, or the potato famines of the nineteenth—Europe's problem was one of too much labor on too little land, and capital concentrated in the hands of a wealthy few.

And then, quite abruptly, the historic alignment of these three factors of production changed. As the nineteenth century opened, with an Industrial Revolution in progress, new means of transport (especially steam powered) available, and a rapid increase in information for would-be migrants, ordinary settlers found themselves for the first time in a position to occupy huge new areas of the earth's surface, many of them sparsely settled and virtually untouched. Never before, and certainly never since, has there been in a few short decades such a glut of lands available: never before had the frontiers of settlement advanced with such speed. It was, in areal terms, roughly as if the twentieth-century landing on the moon had opened up to us a surface ready for immediate settlement: to the same extent was the useful area of the planet's surface enlarged within a generation or two. For a brief period, this single factor of production—land—ran out of alignment with the other factors and, in the nature of things, it *could* only happen once. By 1914, the episode was virtually over; 40 million or more Europeans had left for other continents; already, in 1890, the U.S. Bureau of the Census had declared the nation's frontier closed.

Here, then, was the initial condition of North American agriculture—abundant cheap land (and, as we saw in Chapter 5, much of it actually becoming cheaper as time went on). The corollary of this abundance, however, was that where any man could become owner and operator of his own farm, no one was obliged to resign himself to being a hired laborer. Indentured servants and ex-soldiers could become landowners when they had served their term just as, later, ex-convicts could turn farmer in Australia. Where land was plentiful, *labor was scarce.*

This left the third factor of production—capital. Most of the men and women who actually settled the North American lands (as opposed to speculating in them) had little enough. Although they were usually obliged to start by clearing their land of trees, they had few resources left over for improvements such as drainage; they simply avoided the wettest areas. These "actual cultivators" used the labor of their own families and invested their meager capital in the simple tools of the farmer's trade. Unless and until they had something to sell, and a market to send it to, there was no way of enlarging that capital supply.

This being the case, the efforts of the early frontier farmers to reach a market were little short of heroic. They converted corn into whiskey and carried it across the Appalachians on muleback. They drove cattle, and even hogs, across these same mountains, all the way to the eastern cities. They rafted wheat flour and salt down the Mississippi to New Orleans and so, by oceangoing ship, east and north to Philadelphia and New York in the years before the Erie Canal opened (1825), or steam power enabled them to haul these goods north against the current.

Their nearer markets grew with the spread of an urban-industrial society across the Interior. Beginning in Ohio and southwestern Ontario in the 1820s, there developed that close tie between farm producer and city market which, as we have seen, characterizes the Interior to the present day. As population increased and the market expanded so, too, did the farmer's capital.

We can represent this relationship between the input factors and the output to market by a simple table (Table 7-1). This will enable us to compare the economic circumstances of the frontier farmer in middle North America with the experience of his later counterpart on the drier grainlands of the West

Table 7-1. Sample Types of World Agriculture: Inputs, Outputs, Population, and Surplus for Sale

Sample Type	Inputs			Output per Acre	Population Density Supported	Surplus Left for Sale
	Land	Labor per Acre	Capital per Acre			
1. Early American settler farming	A–B	B–C	D–E	D–E	C–D	E
2. Contemporary Midwest agriculture	C–D	C–D	A	A–B	B–C	A–B
3. Great Plains/Prairies wheat farming	A–B	D–E	B	C–D	E	A–B
4. Asian paddy rice farming	E	A	D*	A–B	A	E
5. West European mixed farming	D–E	B–C	B–C	A–B	A–B	B
6. Tropical plantation agriculture**	B–C	A–B	A–B	A–B	0***	A

Note: In this table, a rating of A or B indicates a large quantity or high value; E or 0 represents small or nil quantities.

*Capital mainly in the form of irrigation channels and water works.

**Antebellum plantations of cotton and sugar producers in the southern states would, in the main, fall within this same category.

***Plantation crops usually not consumed by labor force or their families: food crops must be grown elsewhere.

and with the input–output situation of the midwestern farmer today. The next section of this chapter will then suggest how the factors of production have changed in importance and why.

Without assigning exact quantities to the factors involved, the table indicates their typical combination in a number of forms of world agriculture. The first three are North American; the other three are added by way of offering a comparison with other world farming systems.

In most systems, the amount of land available is not open to discussion; it is dictated by political geography, or population pressure, or physical constraints. In this regard, the occupance of the new lands of the nineteenth century was truly unique: there was as much land as anyone could use, and more (although *how* to use it was still to be discovered, by often painful trial and error). On this unlimited supply of land, it was then necessary to decide what combination of labor and capital inputs would best serve the settlers' purpose. On the same land, one can practice low input–low output farming (which in economic terms is *extensive* land use), or high input–high output farming (which is *intensive*).

The point to note about labor and capital in farming is that they are largely interchangeable. In our table of farming systems, the Asian rice farmer cultivates his paddy fields with a maximum of backbreaking labor and a minimum of equipment, because he seldom possesses any capital to invest in easing the task. In Texas and Louisiana, on the other hand, rice is cultivated on an American version of the paddy field by farmers who can cope with fifty or one hundred times the acreage because they are equipped with aircraft, tractors with balloon tires, and, to quote from a Louisiana Agricultural Experiment Station bulletin of recommended equipment, a truck, two breaking ploughs, a disc harrow, a section harrow, a grader, a roller, a drill, a combine, a grain cart, and 70–150 kg of chemical fertilzers per hectare (60–130 lb per acre). It hardly sounds as if they could be farming the same crop.

It is the contrast in inputs and outputs between lines 1 and 2 of Table 7-1 that reveals how North American farming has changed in the course of the past century and a half. In addition, it has diversified into a large number of specializations, from East Coast truck farming to West Coast fruit growing, each of which would require a separate line in our table. How and why this has happened we must now inquire.

Agriculture and the Present

The original situation of the North American farmer, as we have described it, has been modified out of all recognition over the past century or so. To explain what has caused these changes, and what have been their effects, is a highly complicated business, for cause and effect are endlessly interwoven. What, therefore, we shall do in this section is to consider three particular elements in the changes, which are clearly critical in importance, and try to see how they have been responsible for present-day North American farming. Between them, the three have produced what can be described, without exaggeration, as a technical revolution. They are: (1) mechanization; (2) science, technology, and education; and (3) market openings.

MECHANIZATION

The mechanization of agriculture grew out of both positive and negative conditions. On the positive side, the initial abundance of cheap land in North America encouraged farmers to develop techniques for exploiting larger areas. The land was there: the only limitation was the amount of it that an operator could cultivate in a single season, and mechanization enabled the farmer to increase that amount. On the negative side, the perennial shortage of farm labor, to which we have already referred, forced the farmer to search for labor-saving methods. And once the urban–industrial economy began to grow, the city and the factory offered far more job opportunities than the farm, further reducing the incentives, not only for hired laborers, but for farmers themselves, to continue to work in agriculture.

The process of mechanization was gradual. Colonial farmers used methods and tools little different from those of biblical times. The first changes came with the use of horse-drawn machinery. Between 1820 and 1850, steel ploughs and horse-drawn mowers, reapers, and grain threshers came into use.

A later stage of development saw the replacement of horses and mules by tractors. Although steam-powered tractors had been used as early as the 1870s, it was not until about 1920 that draft animals began to lose their importance. The results were far reaching. As gasoline-driven tractors improved, the implements they drew grew larger and more sophisticated, and so covered a wider range of tasks, and more rapidly. The 25 percent of cropland that had, in 1910, produced feed and pasture for horses and mules now

Mechanization of farm operations: A tomato-picking machine. The plants are raised by the prow of the machine onto a shaker, which breaks the tomatoes off the vines. A team of four to six workers, vis- ible on either side of the picker, then sorts and examines the crop and removes dirt and debris. *U.S. Department of Agriculture*

became available for other purposes. The area that one farm worker could cultivate multiplied, and the number of workers needed fell.

Mechanization did not everywhere proceed at the same rate. It was successfully applied first to close-planted (drilled) crops, such as small grains and hay. Implements were then developed to plant, cultivate, and harvest the row crop, corn, because of its preeminent position in midwestern agriculture. Most difficult to mechanize was the harvesting of fragile fruits and vegetables, especially those that ripen progressively or unevenly. The South's two major crops, cotton and tobacco, were labor-intensive: cotton especially because of the need to thin the plants in the rows ("chopping") and to harvest bolls without including leaves and other plant debris; and tobacco because of the need for careful planting, removal of unwanted suckers during growth, and care in the harvest and curing of the fragile leaves. Cotton-harvesting machines were not commonly used until after 1950; the machine harvesting of tobacco developed only during the 1970s.

Now, however, that a virtually full range of machines is available, not only to harvest cotton and tobacco, vegetables and grapes, but also to aid in livestock raising, the impact on labor inputs has been very striking. To take only the period since 1960, Table 7-2 shows how the input of labor per unit of production fell in just twenty years.

Mechanization had other effects. One of them was that, the more sophisticated—and therefore expensive—that machinery became, the greater was the need to employ it fully. To the farmer, this meant following one of two courses: he must either increase the acreage on which his machinery was used—that is, enlarge the farm to obtain the economies of scale—or he must specialize in the crop or crops for whose production he had equipped himself. Both these courses have been followed; so, too, has a third—by equipment contractors who control the

Table 7-2. United States: Labor Inputs in Hours per Unit of Production, 1960–64 and 1980–84

	Hours	
	1960–64	1980–84
Cotton, per bale	47	5
Corn, per 100 bu	11	3
Tobacco, per 100 lb	26	11
Milk, per cwt	1.2	0.2
Hogs, per cwt liveweight	1.9	0.3

Cotton-picking machine, Texas. *U.S. Department of Agriculture*

machinery and move around carrying out operations for farmers, who are thus saved the expense of owning and maintaining the equipment themselves.

SCIENCE, TECHNOLOGY, AND EDUCATION

Mechanization represented the application of technology to farming, but only in one, limited sense. Our second element of change, like the first, grew out of causes both positive and negative. On the positive side, the nineteenth century saw a general rise in the level of education and expertise of the North American farmer. The Committee on Public Lands in the United States gave the process a splendid start in 1785, by reserving two sections out of thirty-six in every township for financing education. In 1862, the establishment by the U.S. Congress of the land-grant colleges was another long step forward in training young Americans in the proper use of agricultural lands. The same year saw the foundation of the U.S. Department of Agriculture. States, provinces, and railroad companies set up agricultural advisory services. Some of the hardest lessons were learned in the midst of disaster—dust bowl or crop pest. To the development of scientific agriculture, the United States and Canada have made a full contribution.

The other, negative, side of the matter must also be mentioned. North American agriculture needed the help of scientists and technicians because much of it was of very poor quality. Land use was wasteful, forest clearance was careless, soil protection was disregarded, and yields were low. The abundance of land had no doubt contributed to this carelessness and so, too, had the breakneck speed with which the land had been occupied: it was cheaper and easier to abandon and move on than to conserve.

In humid North America—that is, in the East and the Interior—the main applications of science and technology have been made in three directions:

1. Drainage of wetlands
2. Use of chemicals to fertilize the soil and check crop or animal diseases
3. Improvements in plant and animal breeding

Once settlement had stabilized, and the illusion of boundless, empty land had been dispelled, it became apparent that humid North America contained many areas that it would pay to reclaim from marsh or swamp. First among these was the Mississippi bottomland, with its highly fertile soils. The Midwest, also, contained many former lake beds and river val-

North Carolina: Tobacco farming. *North Carolina Department of Economic and Community Development*

leys, such as the Maumee in Ohio and Indiana, or the Saginaw in Michigan.[3] Later on, the old lake beds of the Québec and Ontario Clay Belts would receive the same treatment. Once suitable equipment was available for large-scale drainage work, these areas soon proved their value.

The use of chemicals by the agricultural scientist has certainly transformed North American agriculture even if, in certain respects, this may now be looked on as a mixed blessing. It may be difficult now to recapture a sense of the heavy burden of risk run by the early farmers, as their westward movement led them into increasingly unfamiliar environments, or the despair that racked a farm family as it watched helplessly while a plague of grasshoppers wiped out its crops. Out on the Prairies of Canada, it was not only drought and low prices that ruined farmers in

the depression years of the 1930s, but all manner of pests that assailed the wheat crop, while a small band of research workers fought them to a standstill.

Some chemicals poison other flora and fauna than those at which they are aimed: their very success is a threat to the ecological balance. All we can do is to consider the alternative—that of *not* combating the natural enemies of the farmer—and try to agree, farmer and nonfarmer together, where the balance of advantage or safety lies.

By far the largest use of chemicals in agriculture is, of course, as fertilizers. The fertilizer habit is a recent one in North America: one has only to look at the line in Table 7-3 labeled ''agricultural chemicals'' to notice how inputs have risen over the past twenty-five years. In part, this input has been forced upon farmers, to replace the soil nutrients washed away in decades of erosion. In part, it belongs to the process of intensification that farmers are obliged to follow when there is no more cheap land to be had, but their mounting capital commitments force them to increase their turnover to match their obligations.

[3]Not only did brick and tile factories develop in these areas to supply the builders of the drainage network, but also most of the major manufacturers of earth-moving equipment (originally steam shovels and ditchdiggers) had their origins there; for example, in Lima and Bucyrus, Ohio, or in Peoria, Illinois.

Table 7-3. United States: Volume of Farm Inputs. Indices for 1960, 1970, 1980, and 1989 (1977 = 100)

	1960	1970	1980	1989
Farm labor	206	112	96	76
Farm real estate	103	105	103	93
Mechanical power & machinery	83	85	101	73
Agricultural chemicals	32	75	123	122
Feed, seed, & livestock purchases	77	96	114	119
Taxes & interest	95	102	100	94
Miscellaneous	77	89	96	123
All inputs	98	96	103	88

This leads us on to the third set of scientific measures that have helped to transform North American farming: the development of hybrid plants and the improvement of breeding stock. There is no better example than the history of North America's native crop—corn. In 1935, the average yield per acre of corn was 30 bushels (75 bu per ha). But in that year began the use of hybrid corn, bred for planting and since then developed on specialized farms staffed by geneticists and equipped with laboratories. The high-yield, short-season varieties of the 1980s gave national average yields in the United States of 116 bushels per acre (286 bu per ha) in 1989, and 118 bushels (291 bu per ha) in 1990. For the central Corn Belt states, the figures ranged in 1989 from 123 to 135 bushels (304–333 bu per ha).

On the northern fringe of the Corn Belt, the dairy state of Wisconsin experienced similar changes: annual milk yield per cow rose from less than 7000 lb in 1950 to more than 12,000 in 1980 (3180 to 5454 kg). Putting that in national terms, between 1960 and 1990 the nation's milk yield rose by 20 percent, but this yield was produced in 1990 by only half the number of dairy cows being milked in 1960. The increase is largely attributable to upgrading of herds through artificial insemination. Sires are kept on specialized farms—farms that serve the Dairy Region as the seed-corn farms serve the Corn Belt.

It goes without saying that all the measures we have so far considered are available to farmers in all parts of the continent, but that their importance varies: the development of frost-resistant grains is important to the northern Prairies, and the breeding of disease-resistant cattle to the humid Gulf Coast. But there is one technical development of particular importance to the drylands—the spread of irrigation.

Early irrigation schemes usually consisted of simple stream diversions carried out by individuals or communities. But in the twentieth century the national governments came to recognize that irrigation had become a matter too big and complex for individuals, or even local or state governments, to handle, especially as interstate and even international claims and transfers of water were involved. In 1902, there was founded within the U.S. Department of the Interior a Bureau of Reclamation, charged with bringing into use suitable drylands in the federal domain. Although most of the irrigated acreage in the United States is still in private, municipal, or cooperative projects, it is the huge Bureau of Reclamation schemes of recent years that have caught the public eye. In Canada, the federal government was drawn into the field of irrigation finance by the distress of the Prairie farmers in the depression and drought years of the early 1930s, and in 1935, the Prairie Farms Rehabilitation Act was enacted. As new and large-scale irrigation schemes were developed, they incorporated the smaller, older projects.

Questions of rights and ownership were, of course, involved, leading to a long and complex evolution of water law. As we saw in Chapter 5, two different sets of legal and cultural concepts and practices came into conflict in the West. The one, from humid northern Europe and eastern America, viewed streams as important for navigation—hence water could be diverted and used but should be returned to the stream so as not to deny use to others. This concept has been called the "right to undiminished flow." The other, from the Mediterranean lands and Spanish Mexico, viewed streams as a resource for irrigation. Little of the water appropriated could be returned. Conflict arose if water being depended upon downstream was appropriated further upstream. The concept of "prior appropriation" evolved—that the rights of first users take precedence over would-be later users.

Irrigation is used today in three quite different situations: (1) in areas of arid climate (accounting for, in the United States, a large part of the total irrigated area), (2) in areas of unreliable rainfall, where drought is a frequent menace (the Great Plains), and (3) where rainfall is adequate for normal farming, but where supplemental watering makes heavy fertilization and maximum yields profitable (increasingly applied in parts of the humid East since 1960).

Various movable watering devices have been developed, especially for use in areas where irrigation supplements rainfall. Most common are huge sprinklers on wheels. Some support high-pressure rotating gun-valves that shoot the water over a wide circle. Some are quarter-mile-long pipes, with regularly spaced valves. Radial sprinklers are most popular; they consist of a quarter-mile-long pipe with water spigots that rotates around an electrically powered turbine pump. Clusters of circular fields, each one-half mile in diameter, are becoming a common sight to travelers flying over such areas as northeastern Colorado, southeastern Nebraska, or central Wisconsin. Also used are lightweight portable aluminum or plastic pipes or tubing. One of the most innovative devices has valves in the tubing which release water in trickles. The system, called "trickle irrigation," helps to conserve water and to prevent soil erosion.

A major advantage of these devices is that a field need not be level. They also have made possible the use of areas formerly considered unfit or poorly suited for agriculture, for example, the central sand plain of Wisconsin, which is a glacial lake bed with sandy soils, a high water table, and poor drainage. Radial irrigation and heavy fertilization have given it a new name, the "Central Golden Sands"; it is now a major producer of potatoes and canning crops.

That irrigation helps productivity can hardly be disputed, given figures like those offered by Frederick and Hanson (Table 7-4).[4] Not only is each acre producing more, but the acreage required for a given output is strikingly reduced—according to Frederick and Hanson, by as much as 7 million acres (2.8 million ha) for these four crops alone in the dry West. Worn-out land can be rested; marginal land can be abandoned to rough grazing.

The decision to irrigate involves a number of questions, some physical and some economic. On the physical side, the major factor is water supply. Irrigated acreages were considerably enlarged in the 1970s, especially on the Great Plains; yet there are few Plains rivers capable of supporting this acreage. Increasingly, irrigation depends on tapping groundwater and, on the Plains, this groundwater level is sinking deeper and deeper. In the 1980s, in fact, some areas of the Plains—the Texas Panhandle and parts of Nebraska and western Kansas—have seen a *reduction* in irrigated area as the water supply has sunk to levels that make it too costly to pump up.

Abundant water, however, is not the automatic remedy. To pour water on desert soils may simply

[4]K. D. Frederick and J. C. Hanson, *Water for Western Agriculture* (Washington, D.C., 1982). Figures in Table 7-4 are taken from p. 27.

Table 7-4. United States: Effect of Irrigation on Crop Yields (all figures in percentages)

	Corn	Sorghum	Wheat	Cotton
Western U.S.A.: Irrigated crop exceeds dryland yield by	135	70	45	135
Eastern U.S.A.: Irrigated crop exceeds nonirrigated yield by	30	26	3	37

draw up to the surface all the soluble salts which have accumulated over millennia in the dry climate, and turn the soils into salt pans. One must then *either* use a lot more water to flush the salts out, *or* abandon the area for agricultural purposes.

The important economic questions about irrigation or other land improvement projects are Who benefits? and Who pays? As the projects develop, from small and local to huge and national or even international, these questions become extremely complex. People in eastern cities, for example, wondering how the hungry people of the world are to be fed, may favor costly plans—funded by the federal government—to reclaim lands in a western valley by irrigation. On the other hand, an Iowa or Ontario farmer trying to market surplus corn or beef in order to stay solvent may question subsidizing a project in the West that may result in more corn or beef coming to market. A farmer working irrigated fields in Colorado or California expects city dwellers to help bear the costs of irrigation projects: they are the beneficiaries of power produced by dams and, more often than not, the vacationers who frequent campsites or reservoirs. Most large projects in the dam-building years of the 1940s to 1960s were planned on a multipurpose basis; that is, they were designed to provide irrigation *and* flood control *and* recreation *and* wildlife protection *and* hydroelectricity. The federal government provided the funds and then recouped its outlay by charging the cost to those users who stood to make money from the scheme. Social goods like flood control and recreational values were paid for, in this way, by users of electric power.

The question then became: Who pays for the irrigation? Irrigation farming is not a social good; it is a business, out of which farmers are hoping to make a living. But to the extent that these farmers did not bear the true cost of bringing water to their land, they were being subsidized—either directly by the taxpayer, or indirectly by power users—who were taxpayers in their turn. And what a subsidy! In many cases, on Bureau of Reclamation schemes, it is calculated that farmers have been charged a mere 3–3.5 percent of the total cost of the project; the other 97 percent is being paid for by electricity users (more than half) and general tax revenues. It is small wonder that Congress grew tired of financing projects such as these, especially since, even when the project was complete, no one could force farmers to buy the irrigation water on offer: they could simply decline to take part if they considered the water too expensive. Meanwhile, on private irrigation schemes, many western farmers enjoy rights to cheap, or even free, water inherited from the earliest days of settlement, and protected by law. If these farmers had to buy water at anything like an economic price, they would turn off the taps tomorrow.

Canada has about 1.8 million acres (730,000 ha) of land under irrigation, the bulk of it in Alberta. In the United States, the figure is some 48 million acres (19 million ha), of which the Bureau of Reclamation's projects account for between 9 and 10 million acres (3.6–4.0 million ha) (Fig. 7-2).

MARKET OPENINGS

The transformation of subsistence agriculture to commercial agriculture has taken place everywhere in the world to the extent that a market exists for any surpluses the farmer produces: only then is it worth his while to produce them. In the early days in North America, as we have already seen, access to market over such great distances was very difficult: ingenuity was needed to gain that access and, even so, only a limited range of products could be shipped—either nonperishables like grain or whiskey, or products that walked themselves to market, like cattle or hogs.

The key to the commercialization of American agriculture was transport—transport to the farm frontier of tools and equipment, and transport from the frontier to market of products. Whether that market was in the cities of the East, or in Europe, made little difference to the farmer; either way, it was hundreds of miles away, and until he had means of shipment, he could not reach it.

But once the Industrial Revolution got into its stride, and the two federal governments recognized and addressed the importance of transport by making land grants to canal and railroad companies, the means of transport spread very rapidly across the continent, not merely up to the frontier of settlement, but often in advance of it. The railroads were at least

as anxious to find freight to haul as the farmers were to ship it, and a network was laid out in a few short decades far in excess of the needs of anyone but a farmer with produce to ship. It was a network designed precisely to move agricultural produce to collecting centers—Chicago, Winnipeg, Kansas City, lake and river ports—as directly as possible. By the beginning of the twentieth century no area in the world, except perhaps the Argentinian Pampas, had a network so well adapted to moving the farmer's produce to distant markets. The farmers themselves complained incessantly about high freight rates, but they could have little complaint about the density of the network.

At the same time, the range of products that could be shipped was widening. Canning became common practice in the first quarter of the nineteenth century; refrigerated ships and railroad cars were available from the 1870s onward, and the past fifty years have seen the introduction of many new techniques of freezing, drying, and packaging. With road service from the farm gate to the supermarket, and air transport to carry the most perishable products, the farmer can today market virtually anything he can raise, even in a continent the size of North America.

To assist them in their marketing, farmers have formed a wide range of cooperatives and associations among themselves. But nowadays they, as the producers, represent only one rather small segment of a total enterprise generally known as *agribusiness.* This term covers (1) industries supplying farmers' requirements for equipment, chemicals, or energy; (2) the farmer-producers; and (3) processors and sellers of farm products. It has been estimated that the farm sector proper is by far the smallest of the three, and that it accounts for only 12 percent of the value of domestically produced food bought by U.S. consumers.

The effect of these developments off the farm has been to constrain considerably the farm operator's freedom of action. While he can, naturally, grow what he wants, or raise any kind of stock, if he wishes to market his product he must conform to the market's demands, which in practice are usually set by the great food processors and retail outlets. In many cases, this dominance of the market over the producer is overt: the farmer is contracted to a large corporation to produce solely for it, and to its precise specifications. Even when he is not, however, he is bound to feel the shadow of public tastes or public demand falling across his operations. The urban–in-

Fig. 7-2. Western North America: Principal irrigated areas.

Areas containing irrigated lands, existing or projected

| 0 | 100 | 200 | 300 | 400 | 500 mi |
| 0 | 200 | 400 | 600 | 800 km |

dustrial consumer is, in this sense, the ultimate arbiter of his fortunes.

FEATURES OF TODAY'S AGRICULTURE

These, then, are the processes that have produced the North American agriculture of today, the statistical outlines of which are shown in Tables 7-5, and 7-6, and 7-7. The features of this agriculture are:

1. *Reduced labor inputs.* Table 7-3 shows, for the United States, a very substantial reduction in labor inputs since 1960. Farming has been shedding labor at the same time as the nonfarming economy has been absorbing it. With reductions in the man-hours needed for farm operations such as those recorded in Table 7-2, however, the agricultural labor force, small as it has become, must even so be substantially underemployed.

Table 7-5. Canada: Number of Farms and Area in Farms, 1951–86

	1951	1961	1971	1976	1981	1986
Number of farms	623,091	480,903	366,128	338,578	318,361	293,089
Area in farms (million ac)	174.0	172.5	169.7	169.1	162.8	167.5
Area in farms (million ha)	70.4	69.8	68.7	68.4	65.9	67.8

Table 7-6. Canada: Area in Farms, Area Under Crops, and Unimproved Farmland, 1986

	Nfld	PEI	NS	NB	PQ	Ont	Man	Sask	Alta	BC
Number of farms	651	2833	4783	3554	41,448	72,713	27,336	63,431	57,777	19,063
Area in farms (million ha)	0.04	0.27	0.41	0.41	3.64	5.65	7.74	26,599	20,655	2441
Representing % of provincial area	1	45	8	6	3	6	14	47	32	3
Area under crops (million ha)	0.004	0.16	0.10	0.13	1.7	3.46	4.52	13.3	9.2	0.57
Representing % of farm area	10	59	24	32	47	61	58	50	44	24
Area on farms unimproved (million ha)	0.026	0.09	0.26	0.24	1.5	1.55	2.34	6.55	7.75	1.51
Representing % of farm area	65	33	63	58	41	27	30	24	37	62

Note: To convert to acres, multiply hectares by 2.47.

Table 7-7. United States: Number and Acreage of Farms, by Regions and Selected States, 1974 and 1987

Region	Number of Farms (thousands)		All Land in Farms (million acres)		Average Acreage per Farm	
	1974	1987	1974	1987	1974	1987
New England	23	25	4.8	4.2	206	169
Vermont	6	6	1.7	1.4	282	240
Middle Atlantic	104	98	18.6	17.2	178	175
New York	44	38	9.4	8.4	215	229
East North Central	445	365	90.0	86.6	202	237
Illinois	111	89	29.1	28.5	262	321
West North Central	573	497	272.9	263.8	477	531
Nebraska	68	60	46.2	45.3	683	749
South Atlantic	296	240	60.9	51.2	206	214
North Carolina	91	59	11.2	9.4	123	159
East South Central	306	250	53.7	45.6	175	183
Kentucky	102	92	14.4	14.0	141	152
West South Central	326	335	191.0	184.4	582	551
Texas	174	189	134.2	130.5	771	691
Mountain	112	124	253.0	244.1	2262	1965
Montana	23	25	62.2	60.2	2665	2451
Pacific	127	154	72.0	67.3	567	460
California	68	83	33.4	30.6	493	368
U.S.	2314	2088	1017.0	964.5	440	462

Note: To convert to hectares, divide the number of acres by 2.47.

This means that, for many farmers, farming need occupy them only part-time; they can and do seek other employment. In Canada, the 1986 census showed that about 40 percent of farm operators were engaged in some off-farm work, and that 15 percent of operators actually worked full-time in other occupations. In the United States, figures for the mid-1980s revealed that only 38 percent of farm operators reported no off-farm work at all; only 55 percent gave farming as their principal occupation, and 35 percent of operators reported that they worked off-farm for 200 days or more in the year.[5] The part-time farmer has long been a familiar figure in areas where forestry or fishing offered seasonal employment. Today, he is much more likely to hold a city job, visiting the farm only infrequently.

Accompanying this trend, there has been a drop in the farm population both in absolute numbers and, even more, in proportional terms, as Table 7-8 shows. The U.S. farm population today is less than one-fifth of what it was in 1910, although the nation's total population has increased two and a half times since then.

The decline of the farm population has created another census category: that of the rural nonfarm population. With so many farmhouses abandoned by farmers, and the closure of buildings (like rural school houses) that serviced the farm population, the countryside has become the place to look for cheap housing. Road networks have improved, and so we have witnessed the movement out from the cities and "back to the land" which we noted in Chapter 3. The consequence is that, in the United States, the total rural population was 5.5 million larger in 1980 than in 1970. The disparity between the farm and the rural nonfarm components had strikingly increased during the decade.

2. *Increased capital investment.* The reductions in labor input necessary to produce a bale of cotton or a bushel of wheat (shown on Table 7-2) have been brought about, in large part, by the introduction of machinery. These machines form one part of the capital investment needed by the modern farmer. Another part is represented by the increased storage space he needs for fuel, chemicals, and spares and the cost of those items. A third is represented by the value of his land. During the 1970s and early 1980s,

Table 7-8. United States: Changes in Farm Population

Year	Total Population (Millions)	Farm Population	
		Millions	% of Total
1910	91.2	32.1	34.9
1916	102.0	32.5	32.0
1920	105.7	32.0	30.1
1930	122.8	30.5	24.9
1940	131.7	30.5	23.2
1950	150.7	25.1	16.6
1960	179.3	15.6	8.7
1970	203.2	9.7	4.8
1980	226.5	6.1	2.7
1990	248.7	4.8	1.9

this value sharply increased. In the late 1980s and early 1990s, however, it fell, and in falling added one more to the American farmer's assorted problems.

Some idea of the capital required to operate successfully in North American farming can be gained from the counties given as examples throughout the regional chapters of this book. The figures have fallen somewhat in the past few years, but they are still very high. Since nearly 90 percent of North American farms are operated by individuals or families, this means that the enormous capital sum required to succeed in farming today must be raised by these individuals. In practical terms, there is almost no way in which a young farmer can become an owner-operator unless he or she inherits from a relative. It also means in practice that the burden of debt carried by even the largest and most successful farmers is very heavy; so heavy that a bad harvest, or a sudden drop in land prices, can drive them out of business.

In the 1980s, this burden grew to a level where, for the first time in many decades, there was a falloff in the sales of farm machinery. In the United States, for example, the number of combine harvesters and corn pickers on farms in 1985 was only some 60 percent of the number in use in 1960: it was cheaper and less risky to put out the work to contractors than to run the machines on individual farms. Sales of tractors, too, fell off steeply in the 1980s, although the individual machines were admittedly far more powerful than their predecessors.[6] For some decades now,

[5]As a commentary on all these figures, readers may wish to consult J. F. Hart, "Nonfarm Farms," *Geographical Review*, vol. 82 (1992), 166–79.

[6]The farm machinery industry in the United States has been undergoing changes in its character as drastic as those in the demand for its products. According to *1988 U.S. Industrial Outlook* (Washington, D.C., 1988), p. 25-1, "In 1951, U.S. manufacturers pro-

however, capital investment per worker in agriculture has been far higher than in industry. It is all a far cry from the pioneer days of cheap land and dig-it-yourself operation.

3. *Increased intensity of operation.* We saw earlier that intensity of farm operation varies with the combined inputs of labor and capital per unit area of farmland. It is possible to make a living on a low input–low output basis, and this has the advantage that it minimizes risk: farmers often adopt a low-intensity policy in times of depression, or when they are nearing the end of their working lives. But a farmer carrying a heavy burden of debt, mortgages, or interest charges cannot afford to adopt this course: to intensify his operations—to make his ground yield more, acre for acre—is the only way of sustaining repayments. To offset his capital commitments, the North American farmer is obliged to maximize his production.

This is not, of course, the whole story. The advances of science have, as we already saw, contributed to higher yields. In the United States, the index of crop production per acre (1967 = 100) more than doubled, from 56 in 1910 to 129 in 1979: for livestock, the comparable figures were 45 in 1919 and 116 in 1980. But a good deal of this increase has to be paid for by the farmer, in buying seed and fertilizers. Investment per acre, in the long run, decides yield per acre.

A further factor contributing to more intensive farming has been the policy adopted consistently by the U.S. government since the 1930s of basing farm support payments for quotas on acreage rather than on produce. Provided that a farmer planted only his quota acreage, the government schemes have applied to whatever those acres could produce. Higher yields have meant more subsidy.

4. *Farm enlargement.* The farm is the unit of agricultural operations, and the size of the unit has grown larger with time. In Canada, for example, the area in farms in 1986 (167.5 million acres, or 67.8 million ha) was only slightly smaller than it had been in 1951 (Table 7-5). But the number of farms that made up that area had been more than halved—293,089 against 623,091 in 1951. There are, admit-

duced almost all the 567,446 farm tractors sold in the United States; in 1987, the U.S. industry produced less than 15 percent or about 15,000 units . . . there were over a dozen major tractor producers in 1951; today, two remain." Shipments by the whole farm machinery industry, in constant dollars, declined by 58 percent between 1981 and 1987.

tedly, a few areas where the reverse tendency shows up, but the general trend toward fewer and larger farms is clear.

Partly, this has represented the dropping out of small farmers, who came to realize that their few acres would never yield a satisfactory living and who, as alternative work became available in city or factory, gave up and left the land to larger operators. Partly, it has been a change forced upon larger farmers in order to pay for their investment in machinery, and to keep themselves fully employed. In the areas of heaviest farm mechanization, there has been a desperate scramble for more land, to preserve the economies of scale.

The fact is, however, that farmland has been scarce. Out of this fact has arisen a tendency to rent additional land where buying is out of the question. With its traditions of owner-operators and homesteads, North America has long regarded tenancy as a social evil. But today, under the name of *part-owners*, nearly 30 percent of U.S. farmers, and over 30 percent of Canadian, are glad to be tenants of some land, even land at a considerable distance from their home base, in order to increase their scale of operation.

Farm enlargement, in other words, has not yet gone far enough to make use of the full benefits of modern techniques, or create optimum-sized units for operators still in business. And the 1980s were a period of very mixed fortunes for agriculture as a whole, so that several of the long-standing trends were halted or reversed. The price of land rose and then fell; commodity prices seesawed; farm enlargement faltered, for example, in New England. In a sense, it was a decade of waiting-to-see; no one seemed to know what next to expect.

In Canada, farm enlargement continued, if only because the government has, for some years, adopted a policy of positive encouragement to elderly and marginal farmers to sell up, leave farming, and make their land available for younger successors. At the same time, it still has, on the Peace River, a frontier where new farms are being brought in (see Chapter 22).

It is out of all these conditions that the present pattern of agriculture, as represented by Figure 7-3, has emerged. Obviously, it bears a certain resemblance to the once-familiar map of farming "belts" that dominated the continent's agriculture for so long, and that have left their legacy behind. Those specialized belts were produced partly by lack of technical

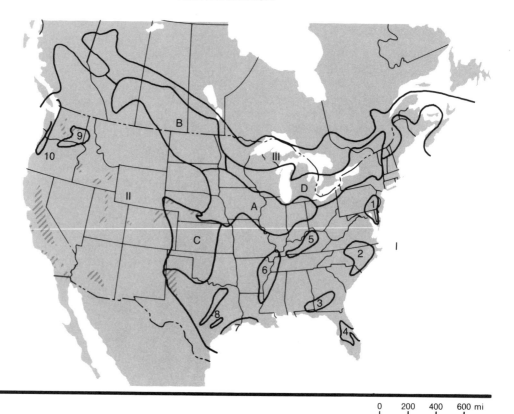

Fig. 7-3. North America: Agricultural regions:

MAJOR REGIONS

A Corn Belt (cash: beef cattle, hogs, corn, soybeans; feed: corn, hay, oats)
B Spring wheat region (with barley, oats, flaxseed, rapeseed, cattle)
C Winter wheat region (with grain sorghum, cattle)
D Dairy region (with other livestock, specialty crops)

SECONDARY REGIONS

1 Southeast Pennsylvania–Delmarva (dairy, poultry, hogs, vegetables)
2 Carolina–Virginia tobacco region (with hogs, corn, soybeans, cotton)
3 Georgia–Alabama Coastal Plain peanuts region (with hogs, pecans, cotton, corn)
4 Florida citrus region (with vegetables, cattle)
5 Kentucky Bluegrass and Pennyroyal tobacco regions (with cattle, hogs, horses, wheat, corn)

6 Mississippi Alluvial Valley (soybeans, cotton, rice, cattle)
7 Texas–Louisiana Gulf Coast (rice and cattle)
8 Texas Black Prairies sorghum region (with hogs, cattle, cotton, corn)
9 Columbia Basin wheat region (with barley, peas, lentils)
10 Willamette Valley–Puget Sound (dairy, fruit, vegetables)

OTHER AGRICULTURAL AREAS

I South and East (crop and livestock farming interspersed with wooded areas, some of which are pastured)
II Western Livestock Ranching region (cropland limited mainly to small irrigated areas. Surrounding areas of grassland, open forest, brush, and semidesert grazed by beef cattle and some sheep)
III Northern Fringe (minor areas of marginal agriculture advancing and retreating)

The principal irrigated areas cross-hatched.

knowledge about alternatives, partly by transport difficulties over long distances, which made it advantageous to grow and ship the same products as one's neighbors.

Today, these considerations are not so important. There are, of course, climatic limitations on crop choice, but today's specialization is more likely to be that of an individual than of a region. A single farmer is not likely to be able to maintain the equipment nec-

essary for a wide variety of crops; he will equip himself to produce one or two and so, almost by default, he becomes a specialist.

Agriculture and the Future

We have already noted that, since the 1980s began, the future of agriculture has been clouded with uncertainty. We must, nevertheless, try in concluding this chapter to find answers to three questions about the shape of things to come: (1) Upon which lands will farming be concentrated? (2) Which products will be in demand? (3) Which farms or farmers will produce them?

WHICH LANDS?

Change in the kinds and location of land used for agriculture in the United States and Canada has occurred on two different scales and at two different periods of time. First was a large-scale expansion and spread of agriculture, mainly westward, but also northward, until it reached margins imposed primarily by aridity or shortness of growing season. This expansion ended about 1890 in the United States and in the 1920s in Canada. Second has been a rearrangement of agricultural patterns on a regional and local scale, as land quality was reappraised, technological advances were made, and economic conditions changed. Rearrangement has occurred, in some instances, over relatively large areas; for example, the retreat from marginal lands in hilly and mountainous parts of the eastern United States and Canada and on the northern fringes; the irrigation of valley and piedmont lands in the arid West; and the drainage of lands in the Mississippi floodplain and the coastal plain in Florida and of glacial lake beds in the Midwest and Great Lakes areas. Although some rearrangements have been on a small scale, they are cumulatively important. For example, crops may be concentrated on the best fields of a farm in response to government programs of acreage limitation and price support.

Locational changes occur largely in response to changes in market demand and pose no problem so long as the quality of the land is maintained. During the period 1920–50, the total cropland harvested in the United States ranged from 350 to 370 million acres (142 to 150 million ha). From 1962 through 1972, the cropland area fell below 300 million acres (121 million ha). Increased productivity per unit area allowed the national and world market demand to be met, despite the withdrawal from cropland use, through the Soil Bank and other government programs, of more than 50 million acres (20 million ha). Following the world food crisis of 1972, those 20 million hectares were again planted to crops, only to be withdrawn once more in 1983 in response to the federal Payment in Kind program to reduce crop surpluses.

In Canada, similar variations in the area under crops have taken place. Between 1981 and 1986, for example, the area of the nation's farms rose by 4.7 million acres (1.9 million ha.) The area under crops on those farmlands rose by rather more: 5.4 million acres (2.2 million ha).

The size of these fluctuations, and the ease with which they apparently occur, alerts us to the existence of two schools of thought about the future of farmland. One school stresses the serious and continuing losses of good land to such nonfarm uses as urban growth, highway and airport construction, or recreation. Those who take their stand here can point to an estimated annual loss of about 1 million acres (400,000 ha) of farmland to these other purposes in the United States, and perhaps 50,000 acres (20,000 ha) in Canada lost each year to urban uses alone.

The other school of thought will challenge these statistics but will, in any case, argue (1) that the elasticity of the continental area under crops shows that these losses of farmland to supermarkets or parks can easily be made good in the future, (2) that with the huge agricultural surpluses that have built up in North America, what is needed is not more farmland but less, and (3) that the time has long passed when agriculture should assume an automatic priority in all areas not actually built over. In particular, the huge and ever-growing demand for recreational space should be given greater prominence, if only because it normally generates more income per acre than the primary activity it displaces.

There is about this more than a touch of the Mad Hatter's Tea Party:

"No room! No room!" they cried out when they saw Alice coming.

"There's *plenty* of room!" said Alice indignantly.

The argument continues.[7] But one thing upon which both sides can readily agree is that quality of land—and the preservation of that quality—is more immediately important than quantity. North America can only continue to produce as it does if the quality of existing farmland is stabilized.

In the United States, the Soil Conservation Service (SCS) has been in existence since 1935. Half a cen-

[7]On land-use competition, see O. Furuseth and J. Pierce, *Agricultural Land in an Urban Society* (Washington, D.C., 1982); R. Platt and G. Macinko, eds., *Beyond the Urban Fringe: Land Use Issues of Nonmetropolitan America* (Minneapolis, 1983); A. M. Woodruff, ed., *The Farm and the City: Rivals or Allies?* (Englewood Cliffs, N.J., 1980); and all or any of the writings of Marion Clawson, long-time analyst of North American land and its uses.

tury later—and half a century of patient educative effort by the SCS—soil erosion losses of 3 tons per acre (6672 kg per ha) annually are considered tolerable. But it is estimated that the average loss over the United States is four times that rate, and that one-third of the nation's topsoil has been lost by wind and water erosion.

It seems clear that the question Which lands will be used? must be joined to another, What will their quality be? Further losses in quality may lead these nations, one day in the future, to a frantic search for new farmlands, and to bitter regret that they did not, much earlier, adopt a policy of safeguarding their best soils.

WHICH PRODUCTS?

Important locational changes in the production of various crop and livestock products have resulted from regional specialization, as indicated, and from changing technology and market demands, such as changes in diet or rising demand for more elaborately packaged and ready-to-eat foods. Three of the most significant changes since 1940 have been: (1) a great increase in poultry production, especially of broiler chickens, and its shift from small, diversified farms to large contract farming operations in specialized districts mainly in the Southeast; (2) the rise of soybeans, especially in the Corn Belt and Mississippi River floodplain, from a minor, primarily forage crop to a major oilseed export crop; and (3) the rise of grain sorghum production as a major crop in the central and southern Great Plains, and its attraction to those areas of large-scale cattle feeding operations. Such changes, although few are of such magnitude, will very likely continue. The application of capital and ever-developing technology, which have given North American agriculture its most distinctive characteristics, will also continue.

WHICH FARMS?

Ever since the founding of the republic, one of the ideals to which the United States has been committed is that of a rural population of actual cultivators (to use the eighteenth-century phrase once again), living as owners on the farms they work. Europe, with its serfdom and its landlords, lay behind them; the future for these idealists was a vision of free yeomen holding their land in fee simple; that is, possessing the maximum of legal rights over their farms, safe from eviction or rack-renting. It was an ideal that, years later, came to be enshrined in the Homestead

Acts—1862 in the United States, and 1872 in Canada. It created one of the pillars of American society—or one of the sacred cows, according to taste—the family farm.

A family farm is one where the family itself owns the land, makes the operating decisions, and provides all, or nearly all, the labor. Furthermore, a series of legislative acts in the nineteenth century gave it a normal, or standard, size: that quarter section of 160 acres (64.8 ha) to which we referred in Chapter 5. In the United States, the Preemption Act of 1841, the Homestead Act of 1862, and, much later, the Reclamation Act of 1902, were all based on the quarter section as the "proper" size for a family farm—proper because it was too small for speculation; proper because it was too small to be a plantation, and therefore would discourage slavery; proper because it was right that the public lands should be disposed of to as many members of the public as possible.

Today, family farms come in all sizes, without breaching our initial definition. The nineteenth-century legislation lies far in the past, although only under the Reagan administration were the terms of the Reclamation Act (limiting the number of acres of irrigated land to which federal water might be supplied for an individual) formally eased: the Act had not been strictly enforced for years. But whatever North American agriculture has achieved, it has achieved while organized largely on the basis of family farms: as we have already seen, almost 90 percent of all farm units conform to the family farm definition.

Will this continue to be the case in the future? Granted that the legislators of both countries have gone out of their way—too far out of their way, it might be argued—to favor the survival of the family farm, is this the best pattern for the future?

Supporters of the family farm, and of the government measures which, they believe, help to sustain it, need to be realistic about the value of these measures. Small farmers in North America, as in other parts of the world, welcome government subsidies, on whatever grounds they are given, but the fact is that subsidies enrich most those who produce most, and who have most acres on which to produce it. In other words, the corollary of helping small farmers to survive is making large farmers rich. But many of the smallest family farmers have so little output that they barely qualify, if at all, for the kind of subsidies and supports which, meantime, are turning their larger neighbors into millionaires. To the question we raised, Is this the best pattern for the future? the answer would clearly seem to be no.

Competition for the family farm certainly seems to be formidable, at least for the smallest category. Industrial farms in the United States account for up to one-fifth of the total value of production, and have a particularly strong grip on such activities as poultry raising, cattle feeding, and production of fruits and vegetables for freezing or canning. Larger-than-family farms may constitute 6–8 percent of the total number of farms, and account for another one-fifth of production. At the other end of the scale, many family farms are very small, and the number of these small enterprises is growing, as city people move out onto the land, for health reasons, or hobbies, or to grow their own food, or all three reasons. Commercially, these contribute very little.

But this leaves the bulk of family farms, and there seems no reason why they should not continue to operate efficiently and economically, *provided that they can find space to enlarge their area when necessary.* The evidence suggests that, given this important proviso, the family farm can operate as economically as a larger unit; that is, a family farm can achieve full economies of scale without changing character. North America's family farms have for decades fed a world in need and, although there will continue to be a certain mortality rate among those that are smallest, or whose owners give up the struggle, or where unwise investment has occurred, there should still be a central role for the family farmer to play into the twenty-first century. It does not seem necessary, in this case, to choose between idealism and efficiency. Some sacred cows give milk, too.

8
Industry

Silicon Valley: Rapid expansion in the electronics industry has created new industrial areas, such as Silicon Valley, west of San Jose, California, brought into being by the silicon chip and computer development. *Air Flight Service, Santa Clara, California*

In 1961, according to a United Nations estimate, the United States and Canada produced industrial goods that, in aggregate, accounted for almost one-half (47.7 percent) of the value added by manufacture in the world outside the USSR. If that was so, there can be no doubt that the figure for, say, 1951 was much higher since, at that time, so soon after World War II, the industrial production of Germany and Japan was close to zero, and the production of most other industrial nations had yet to recover from the destruction and dislocation of war.

There can be equally little doubt, however, that by 1971, North America's contribution to world production had been considerably reduced, and that by 1981, it was far lower. Not only had recovery taken place in war-ruined countries in Europe, but also a whole set of newly industrialized nations were sharing in world manufacturing. Some of them had grown so rapidly that they were already industrial giants in their own right—Brazil and India, South Korea and Taiwan, even tiny Hong Kong and Singapore; size of territory evidently is no serious handicap to industrial advance and sophistication. North America had challengers around the globe to its industrial pre-eminence.

The rise and decline of that pre-eminence have followed a pattern with which the long-industrialized countries of western Europe have become all too familiar, but there has also been a distinctively American character about it all. The Industrial Revolution originated in Europe; more exactly, in Great Britain, from where development spread to such continental areas as the Meuse Valley in Belgium and the coal fields of France and Germany. The revolution passed through several phases, all of which had their counterparts in North America. First, the invention of machines to replace hand operations; then the harnessing of many machines to a water-driven power source; then the assembly of machines in mills and factories and the consequent movement of workers to these large industrial units; then the general application of coal–steam power to replace water power. In the twentieth century, coal and steam have given way, in their turn, to newer power sources, and so the location of industry and the structure of the industrial unit—the factory—have been altered.

This story of industrialization begins in the late eighteenth century—by coincidence, at about the same time as that of the new American republic. Only in parts of the continent settled at that time, therefore, has it been possible for industrialization to pass through all its stages and, in practice, only in southern New England and the Middle Atlantic states has it done so. But there, at least, a very fair replica of a European mill landscape evolved, early in the nineteenth century, with mills clustered along the rivers, tall and compact around the waterwheel (or, later, the boiler-house) in New England, just as in Old England.[1]

From this starting point, the character of the industrializing process varied with the region's date of entry into it. The old Northeast lived through all the

[1] In these industrial beginnings there were similarities, but there were also differences. The English novelist Charles Dickens visited Lowell, Massachusetts, in 1842. Contrasting what he saw there with what he knew of industry at home, he was impressed by many features of Lowell's factory life, and that on a trip where very little impressed him favorably, to judge by his *American Notes*. Having remarked upon the cleanliness of the mill girls, the flowering plants in their workrooms, and the care with which the managers of their hostels were selected, he concluded: "I am now going to state three facts, which will startle a large class of readers on this [British] side of the Atlantic, very much.

"Firstly, there is a . . . piano in a great many of the boarding houses. Secondly, nearly all these young ladies subscribe to circulating libraries. Thirdly, they have got up among themselves a periodical called THE LOWELL OFFERING . . . whereof I brought away from Lowell four hundred good solid pages, which I have read from beginning to end."

stages, whereas the new Southwest entered during the latest stage, when the need for closeness to the power source has long gone, the purpose of building tall factories has been lost, and the advantage now lies with ground-floor spread. Between the two industrial landscapes, the old and the new, there is no resemblance.

There is, however, another similarity to the European situation. The first country to industrialize, Great Britain, saw itself for a time as "the workshop of the world"; that is, Britain would specialize in supplying the world's needs for manufactured goods while other nations, in their turn, provided her with food and raw materials. In rather the same way, in the years of westward expansion across North America, the Northeast saw its role as that of "the workshop of the continent," supplying the needs of an advancing frontier for manufactures.

The geographical expression of this specialized role came to be known as the Manufacturing Belt. It was an area that eventually stretched from New England west to Chicago, and from the middle St. Lawrence Valley southward roughly to Baltimore. It achieved its clearest expression about the time of World War I, when it contained virtually all the significant industry of the continent.

But what both Europe and, now, the Manufacturing Belt have learned is that to be early in the field is always to run the risk of being overtaken by later imitators and improvers. The first main challenge to the industrial hegemony of the Northeast came in the 1920s, with the rise of a cotton textile industry in the South, usurping the role of that of New England (see Chapter 15). Since World War II, the challenges to the old Manufacturing Belt have multiplied; today, the area where the highest proportion of the work force is employed in industry is not in the Belt, but rather in the Carolinas. In the process, the older areas of the Belt suffer; wages rise and taxes increase, and they no longer offer attractive conditions for new industries. The effects of industrial obsolescence outweigh the advantages of an early start. Some parts of the old Manufacturing Belt are now better known as the Rust Belt.

In these respects, the North American experience has paralleled that of Europe. But we referred earlier to a particularly American dimension of industrial development, and we must now identify what this is. What has been the truly American contribution to the manufacturer's art?

Know-how and spirit of enterprise we can take for granted. North America has them in plenty, but so have all the other nations that have successfully industrialized—that is what made them successful. It is rather the *increase in scale* North America brought to world manufacturing that has been most distinctive. Europe knew how to make steel, for example, but when the first reports were received in Sheffield, England, of the production figures from Carnegie's Pittsburgh mills, they were greeted with total disbelief—the English ironmasters could not even visualize output on such a scale. With a continent to occupy, and a rapidly enlarging population to supply, North America excelled at the techniques of mass production, which it has since taught the rest of the world.

The rest of the world has proved apt. American production figures no longer appear miraculous; they are within the range of numerous rivals. The seemingly unattainable has been attained, first and foremost by Japan, a country that less than fifty years ago lay in ruins. Some American industries have suffered by the competition; others have reacted by fighting back. North America may not today occupy the exclusive position it held in the 1950s, but it is still an industrial giant, and still capable of great, and often unexpected, industrial achievements.

These last comments enable us immediately to point to a moral of this chapter on industry. It is that manufacturing is a process that is constantly changing not only its location—we shall deal with that later—nor its character, but its role in the economy of a nation or region, and the way in which it is perceived by inhabitants of that area.

The Industrial Revolution introduced manufacturing as the first major alternative in Europe to agriculture, hunting, or fishing. The craftsmen of the past had formed closed bodies; their object was to keep out newcomers rather than provide employment. With the coming of the Industrial Revolution, processes were involved which in size, scale, and technique lay beyond the capacity of craftsmen, and which called into being a new economy based, in large part, on new materials.

These processes were available to only a few societies, which quickly became wealthy on the proceeds. Industry became the symbol of rich and progressive nations; their goods crossed the world; they supplied railway locomotives and steamships to countries which never built a locomotive of their own. Possession of industry was a matter of pride—not merely in terms of national prestige, but even in those of the individual plant. Successful industralists had pictures of their architect-designed factories printed on their letterheads; their works were built to look like Egyptian temples or Gothic cathedrals. Military intelligence became preoccupied with other states' industrial capacity for war. The South lost the war of 1861–65 for lack of it: the North, and subsequently the Union, became the greatest industrial power on earth.

And then everything changed again. Neither in employment nor in output does industry now represent the largest sector of the North American economy. The architect-designed factory is derelict, its successor an anonymous shell built on a greenfield site alongside other, identical shells. (Today's architect-designed buildings house finance companies or research institutes.) Offered a choice between an industrial plant and a finance house to provide new employment, most communities will opt without delay for the finance house. Industry is something best left to others, probably in Third World countries where no one will complain about its environmental impact.

Today, the nations that brought in the Industrial Revolution have service-based economies. Agriculture, industry, and its attendant mining and construction account for perhaps 30 percent of employment: the rest is service. Whether such a situation can be, or should be, sustained into the future, only time will tell: for the moment it merely looks dangerously top-heavy.[2]

The Distribution of Industry

As industry has spread from the Manufacturing Belt to other, newer locations, it has become more difficult to identify particular manufacturing regions. Today they are widespread because, as has already been suggested, there is no technical reason why they should not be. Whereas the older regions owed their existence to coalfields, or ports of shipment, or supplies of suitable labor, the newer industries are, for the most part, footloose. In a great free-trade area like Canada or the United States, there may be a dozen locations where costs of production are roughly equal, and each of these locations may be ideal for

serving a particular regional market. Once the tie with the coalfield or the raw material source is gone, industry is free to select its location on other, quite different criteria: the size of the market or, just as probably, the pleasantness of the environment for its workers or the richness of community life in a given city.

Equally obviously, footloose industry can be lured to particular areas where it will provide employment, investment, and local taxes. State with state and city with city, the communities of North America compete with each other to attract new plants. Most communities want to expand and, in doing so, to keep a balanced employment structure (one that does not depend too heavily on a single employer) and to attract desirable types of industry—those that are clean and that bring in skilled workers with high wages. A manufacturer with this type of industry will probably be wooed by the local chamber of commerce; he may perhaps be offered bargain tax rates as an added inducement. If his coming would harm the social environment of the community, he will be discouraged.

This form of competition has been particularly evident where the federal government has been building industrial plants for military purposes. Government investment of this kind is usually a valuable addition to the assets of an area, and competition to secure contracts is keen. What atomic power has meant to Tennessee, or missiles to California and Florida, can never be fully assessed, but it amounts to hundreds of millions of dollars of additional income for these states.

As a result of all the changes time has brought, it is probably easier today to identify the few regions of North America that do not have industry than the many that do. Concentration of manufacturing has been replaced by dispersion—away from large centers to small ones, and away from the city to rural locations. If we nevertheless attempt to classify industrial regions, it is perhaps simplest to categorize them by their relative ages: old, intermediate, and recent.

1. *The old industrial regions.* These lie almost entirely within the historic Manufacturing Belt and constitute, in fact, a series of subdivisions of the Belt.

(a) Southern New England, the oldest industrial area of the continent, with its original specializations, in textiles and leather goods, now largely replaced by light metal goods of all kinds, and by sophisticated, electronic-based products of recent research.

(b) The Middle Atlantic area, which includes New

[2]Basic to the study of U.S. industry is Walter Adams, *The Structure of American Industry,* 7th ed. (Toronto, 1985). For Canada, see D. F. Walter, *Canada's Industrial Space-Economy* (London, 1980). This may also be the point to introduce another concept; that represented by D. Clark, *Post-Industrial America: A Geographical Perspective* (London, 1985).

York, Philadelphia, Baltimore, and the cities between and to the west of them. As a "funnel" for most of the pre-1960 immigration into the United States, and for much of the nation's import–export traffic, this area has the widest range of industry, heavy and light, of any of the continent.

(c) The coal and iron area of Pennsylvania and eastern Ohio, the core area of U.S. heavy industry. Its historic focus has been Pittsburgh, but the area extends north to Lake Erie and includes such lake cities as Cleveland and Buffalo, as well as Ohio River towns and those of the Kanawha Valley in West Virginia.

(d) The Montréal area of the middle St. Lawrence Valley, the original heartland of Canadian industry.

(e) The Chicago–Lake Michigan area. Chicago was already a major manufacturing center with a big steel mill when, in 1906, the U.S. Steel Corporation established Gary, a city that became the focus of the world's largest iron and steel complex. Subsequently, industry spread for 100 mi (160 km) and more along the shore of Lake Michigan, and Chicago and Gary formed the nucleus of the western end of the historic Manufacturing Belt. From these focal areas, industry spread into the intervening spaces in the Belt, most obviously into southern Michigan with the rise of the automobile industry, but it also blanketed much of the eastern interior of the United States, and began to spread outward from Toronto through southwestern Ontario in Canada, and up to the power sites on rivers flowing into the St. Lawrence.

2. *The intermediate industrial regions.* The rise of these regions came later than the growth just de-scribed, but still belongs to the period before 1945. By then, industry had sprung up in the cities of the western interior—Winnipeg, Minneapolis–St. Paul, St. Louis, Kansas City—and had appeared on the West Coast, where fruit canning and lumber milling gave place during World War II to the production of ships and planes. On the Appalachian Piedmont, the textile industry first gained ascendancy over its New England rival in the late 1920s, and held on through the Depression to grow and diversify, while the provision of cheap power from the dams of the TVA began to attract industry from the late 1930s onward. On the Gulf Coast, the oil fields attracted petroleum-related industries, but outside these areas, much of the South remained nonindustrial and so, too, did virtually the whole of the western plains and mountains.

3. *The recent spread of industry.* Some of the most recent industrial developments represent dispersion of manufacturing plants from regions already moderately industrialized; the U.S. western interior and southwestern Ontario are examples of this process. Other areas, however, really are new to industrial growth. Such areas are the Deep South away from the earlier growths on the Piedmont or the Gulf Coast or in the Tennessee Valley, the Alberta Prairies, and the southwestern states like Arizona and Utah. On the roll of industrial centers new names are appearing: Houston, Dallas–Fort Worth, Calgary. In 1950, the regions constituting the Manufacturing Belt contained some 72 percent of U.S. industrial employment. By 1987, the figure had fallen to 52 percent (Table 8-1).

Table 8-1. United States Manufacturing: Percentage Distribution of Value Added and Employment, by Region, and Percentage Change in Regional Shares, 1950–87

	Value Added			Employment		
	1950	1987	Change in Share	1950	1987	Change in Share
New England	8.3	6.7	−1.6	9.8	7.1	−2.7
Middle Atlantic	26.2	15.4	−11.2	26.9	15.9	−11.0
East North Central	33.2	22.8	−10.4	30.0	22.1	−7.9
West North Central	5.7	7.3	+1.6	5.6	7.0	+1.4
South Atlantic	9.4	15.3	+5.9	11.1	16.4	+5.3
East South Central	3.8	6.4	+2.6	4.4	6.9	+2.5
West South Central	4.3	8.7	+4.4	4.0	7.5	+3.5
Mountain	1.2	3.1	+1.9	1.1	3.1	+2.0
Pacific	7.9	14.2	+6.3	7.0	14.0	+7.0
	100	100		100	100	

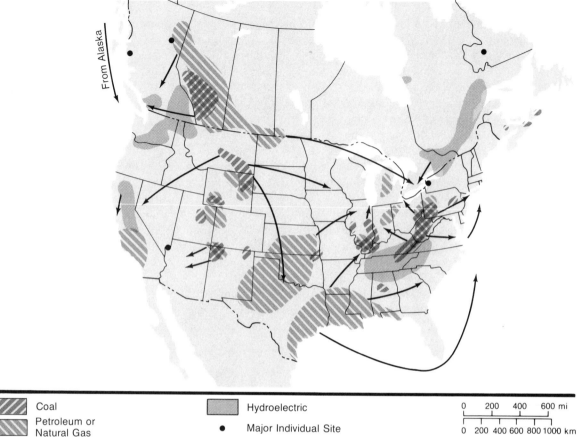

Coal

Petroleum or
Natural Gas

Hydroelectric

● Major Individual Site

0   200   400   600 mi

0   200 400 600 800 1000 km

Fig. 8-1. Energy in North America: Major producing areas and movements.

Coal, Iron, and Steel

COAL

For a century past, North America's industrial output has been undergirded by a continental coal production that rose from 440 million tons in 1962 to 1050 million tons in 1989,[3] and a steel-making capacity, at peak, of 150 million tons. These were the traditional sinews everywhere of the Industrial Revolution and, if today they have lost some of their importance as an industrial base, it would nevertheless be idle to pretend that the annual movement of a billion tons of materials around the continent could be easily overlooked (Fig. 8-1).

The chief source of coal during this industrial century has been the great Appalachian field that stretches from Pennsylvania south to Alabama. The

earliest development occurred in the northern end of the field; today, the center of production is in eastern Kentucky and West Virginia. Further south, the field was developed around Birmingham in Alabama, where the coal was conveniently partnered by iron ore deposits. The continent's secondary source of coal was formed by the Interior fields in Illinois and Indiana. Output in the western United States and in Canada was, compared with these two sources, negligible.

Within long-established patterns of supply, Appalachia met 80 percent of the continental needs for coal in the early 1960s. But since then there have been remarkable changes in the industry, one of which is that Appalachia's share of the market has fallen to little more than 50 percent. During this interval, methods of production, conditions of mining, size of market, location of producing fields—all have changed.

[3]Canada, 73 million tons; United States, 976 million tons.

Method of Production

While America's coal output has been increasing, the part of that production extracted by strip, or open-cast, mining has been increasing, too. For the United States as a whole, the proportion is now around 60 percent, while in several of the newer fields of the West, it is 100 percent.

Successful strip mining depends on the presence of coal in thick, level seams within easy reach of the surface. Where these conditions exist, as they do in much of Appalachia, in Indiana, and in the West, strip mining is cheaper and safer than shaft mining, and output per man-shift, the standard index of the coal industry's productivity, is many times higher in the open pits, with their huge mobile shovels, than it is underground. Even in such a traditional coal area as eastern Kentucky, today, the proportion of the output strip mined is well over one-half. The effect of this change is to eliminate the large, experienced underground labor force, and to favor mining wide

Canada: Strip mining of coal, with a drill rig in the foreground. *Energy, Mines and Resources, Canada*

and level spaces as against rugged terrain or built-over surfaces.

Conditions of Mining

The coal mining labor force has fluctuated considerably in size over recent decades. In 1950, it numbered 415,000. By 1969 it had fallen to 125,000. It rose again to reach 242,000 in 1978, but by 1987 was down once more to 163,000. Demand had fallen off in the 1950s and 1960s, and when the upturn of the 1970s came, the labor force began to grow once more, but in nothing like the same proportion as the output. Partly, this was accounted for by the switch to strip mining, which employs only small numbers of workers. But the switch itself was, in part, a response to legislation designed to make mining, and especially underground work, safer and less hazardous to health, which in turn made labor more expensive. The Coal Mine Health and Safety Act of 1969 laid down conditions that were costly for employers to implement: the effect was that they employed fewer workers under better conditions.

This was, however, only the start of a program of all-around control of the industry. The Clean Air Act of 1970 reflected public concern about the atmospheric pollution generated by coal, especially with regard to the burning of the high-sulfur coal that produces sulfur dioxide and, subsequently, acid rain. It so happens that some Appalachian coal has a high sulfur content; in consequence, western coal, which is low in sulfur, has taken its place in a number of internal markets.

Nor was this all. In 1977, the federal Strip Mining Control and Reclamation Act came into force. Strip mining, as we have already seen, is technically faster, cheaper, simpler, and safer than underground mining, but, unless mine operators clean up behind themselves, it leaves a legacy of spoiled landscape and unusable surfaces—which is exactly what, in earlier decades, many operators did leave. The object of present legislation is to prevent this: to make mine operators responsible for all the effects of mining, and for remaking the landscape.

The Market for Coal

The events of 1973, when world oil prices rose dramatically, forced the North American nations to reappraise their energy resources. Years of cheap petroleum and gas had led to a dwindling demand for coal; now the reverse process began, with the big oil companies themselves in some cases buying their

Newport News, Virginia: Coal terminal. Appalachian coal is brought down to tidewater at Hampton Roads by rail and is then loaded for sea transport at terminals like the one shown. There is ground storage behind the pier for stockpiles of 1.3 million tons. *Pier IX Terminal Company*

way into the coal industry to diversify their energy holdings, and in the expectation that, in any case, by the end of the century coal might well be providing the raw material for a synthetic oil and gas industry. And since American coal has generally been highly competitive on world markets, the universal increase in oil prices stimulated demand for coal not only at home but also overseas, and the export business boomed.

With new mines, and new overseas customers like Japan, the most sober estimates put production by the end of the century at well above 1000 million tons, with exports likely to rise by then to between 100 and 200 million. What that will mean—what it already means—for shipping lines and for near-bankrupt railroads in terms of extra tonnage to haul needs no stressing.

The Location of Coal Mining

The impact of all these forces upon the location of the industry is best summarized by saying that, whereas, twenty years ago, 80 percent of North American coal

came from the eastern fields (almost all from the Appalachians), and 16 percent from the central (or Eastern Interior) region, leaving a mere 4 percent to be accounted for by the West, today's proportions are nearer 50:10:40. Most of the factors we have reviewed confer an advantage on new, western coal-fields over older, eastern fields. Today's leading producer-state is Wyoming (1989 production, 171.5 million tons), a state whose production was separately listed in the industry's tabulations only in 1970. Texas is also high on the list (first appearance, 1972). Kentucky and West Virginia, among the traditional coal states, rank after Wyoming. In Canada, national production in 1969 was 9.7 million tons; in 1989 it was up to 73 million. Apart from some 10 million tons of lignite mined in Saskatchewan, the bulk of this production came from British Columbia and Alberta.

This is not to say, however, that the output figures for Appalachia declined; the reverse is the case. What we have witnessed is a decline in the number of mines and a relative shift in the East–West balance.

But there is one important difference between eastern and western production: hardly any of the coal mined in the mountain and plains states is used locally. In the Appalachian field there is, of course, a huge local market; proximity to coal accounts for the presence of the heavy industries of Pennsylvania and Ohio. There is an intermediate-range market in the whole industrial Northeast, including eastern Canada. And there is a heavy movement of coal by rail, the short distance to the ports; eastward to Hampton Roads, Virginia, for export, or northward to Toledo, Sandusky, Conneaut, and Ashtabula, for shipment throughout the Great Lakes. From the new fields of the West, by contrast, the coal must move to market over thousands of miles, generally by rail to the nearest navigable water, whether that is the Pacific coast, the Mississippi system, or the head of the Great Lakes.

IRON ORE

Over the past hundred years, North America's iron ore production has been just as markedly dominated by a single source as has its coal production. This source is near the shores of Lake Superior—vast ore bodies first made accessible to industry by the opening of the Soo Canal in 1855, and reaching full development with the opening up of the Mesabi Range in the 1890s. From this wilderness area, where the main ore bodies are almost ideally accessible for working, there has flowed a century's supply of high-grade ore for Canada and the United States alike.

But just as there have been changes in the mining of coal in Appalachia, which we considered in the last section, so the mining of iron ore has undergone a transformation. Like coal mining, iron mining has come into the open; ore is no longer mined underground on Mesabi, and the number of individual mines has been greatly reduced. Then the tremendous drain of the past century has removed the most accessible, and higher-grade, hematite ores. There is plenty of lower-grade, mainly taconite, ore still in the ranges, but it must be part-concentrated, or "beneficiated," before it is shipped, to prepare it for use in the furnace and to reduce the bulk of the waste that is carried down-lake in every shipload. Beneficiation

Minnesota iron ore: The Hull-Rust mine at Hibbing, Minnesota, is the world's largest open pit iron mine. *Minnesota Office of Tourism*

plants have been set up on and near the shores of Lake Superior, processing the low-grade ores into a product that is superior to the natural ore originally shipped. Pelletized ore requires so much less heating in the blast furnace that this more than makes up for the energy used in producing the pellets. Most of the ore now used in North American furnaces is pelletized.

Alternative sources of ore have nevertheless become important. Within the United States, there are small producing areas in the Adirondacks, Pennsylvania, Wyoming, Utah, and California. In Canada, the northern shores of Lake Superior and the extensive deposits of northeastern Québec and adjacent areas of Labrador have been yielding 30–40 million tons of ore in the 1980s. Indeed, the Québec–Labrador ores were an obvious replacement for the Superior ores when they were opened up in the 1950s, if only because they could be brought to the steel mills of the Atlantic Coast by water and to those of the Great Lakes region after the opening of the St. Lawrence Seaway (see Chapter 9) in 1958. Other ores have been imported from Venezuela and Brazil. The quantities involved have fluctuated with the fortunes of the steel industry, but the supply position has been as shown in Table 8-2.

THE STEEL INDUSTRY

Because of its role in supplying basic materials to so many other industries—car manufacture, construction, military hardware, railroads—the steel industry in any country occupies a special position, and nowhere more so than in North America. Because of its character, it has also always held a special interest for geographers, since the quantities involved in this heaviest of industries are so large, and the cost of transporting them plays so important a part in deciding its location. For the production of 100 million tons of steel (which the United States produced annually up to 1981) involves the movement of, probably, 350–400 million tons of commodities—coal, ore, and finished product. Only the petroleum industry handles comparable quantities of materials, and that largely by water or, invisibly, by pipeline.

This great American industry was situated in and around Pittsburgh, from where it spread down the Ohio Valley to other towns like Wheeling, and over to Youngstown. Then, with the principal ore supplies for its furnaces coming from Lake Superior and Upper Michigan, the industry spread back along this ore route, to Cleveland, Buffalo, Detroit, and Chicago, and to Hamilton and Sault Ste. Marie on the Canadian side—all places where the coal from Appalachia could meet the ore from Superior at a point of transshipment.

On the East Coast, the first major plant to be established was Bethlehem Steel's Sparrows Point works at Baltimore in 1916. In the early 1950s, U.S. Steel built its Fairless works near Trenton on the Delaware River. Both these plants were within easy reach of Appalachian coal and could receive their ore supplies by sea; they were also located in the heart of their market area. Meanwhile, far to the north, on Cape Breton Island, Nova Scotia, a Canadian steelworks was established, using local coal and iron ore shipped over, at least until their closure, from mines in Newfoundland.

In the interior of the continent, Chicago and Birmingham, Alabama, were for long the western and southern outposts of the industry, but for the solitary exception of Pueblo, a steel town in Colorado. Only during and after World War II did steelmaking spread to the West, to Fontana near Los Angeles and Provo near Salt Lake City, and with small, scrap-based plants in the major cities of the Pacific and Gulf coasts.

Before World War II, this pattern of distribution had been stable for almost forty years. It was maintained by a number of factors, among them the power of the great corporations, which found expression in the price system known as "Pittsburgh Plus." Under this system, all steel sold was, by agreement within the industry, priced as if it had been produced in Pittsburgh and shipped from there, even though the purchaser may have fetched it himself from a mill in his own city. The object of this device was to discourage production in outlying market areas and, thus, to protect the steelmakers' huge in-

Table 8-2. United States: Iron Ore Supply, 1970–88 (production figures in millions of tons)

	1970	1975	1980	1984	1988
Number of U.S. mines	74	67	37	20	21
Domestic production	89.8	78.9	68.6	51.3	57.0
Of which,					
Lake Superior	69.6	66.7	62.3	49.7	56.0
Imports	44.9	46.7	25.1	17.2	19.3
Of which,					
Canada	23.9	19.1	17.3	11.2	8.4
Venezuela	13.0	13.1	3.6	1.5	4.2
Brazil	2.0	7.5	2.0	2.5	5 1

vestments in the central region. It was a kind of stage-managed industrial inertia, enforced by the overwhelming power of the U.S. Steel Corporation.

But the pattern has now changed. In 1924, Pittsburgh Plus was declared illegal,[4] and in 1948 its successor, the basing point system, was also abolished, freeing the industry from artificial restraints upon relocation. At the same time, other, technological changes took place. The industry is using less iron ore and more scrap metal in its furnaces. To the extent that this occurs (and on the West Coast, 85 percent or more of the furnace material is scrap), it lessens the importance of access to ore and increases that of access to scrap—which is found in areas of greatest steel use; that is, in the market areas. The industry is using less coal, too, as plant efficiency increases, especially with the development of continuous flow systems, by which the metal is heated only once and goes through all the stages of production in one continuous process. To these changes in technology and pricing must be added the effects of replacing the high-grade hematite ore from Lake Superior by beneficiated taconite ore from the same area or by ore from Labrador and Latin America. All these influences must be weighed against the impact of changes in demand; in particular, the growth of the automobile industry as the steel industry's prime customer. The industry responded by increasing its capacity in Michigan and in the West, while the relative importance of the old steel heartland around Pittsburgh diminished.

The 1980s, however, saw change of another sort affecting the steel industry: a steady rise in competition from overseas producers. The figures speak for themselves: in 1970, imported steel made up 13.4 percent of U.S. consumption. For 1983 the figure was 20.5 percent, and for 1984, no less than 26.4 percent, with Japan and the E.E.C. as the major suppliers.

This is a situation that confronts all the long-established steel-producing nations of the world. There is today a global surplus of steelmaking capacity: too many members have joined the club, members whose costs (labor costs in particular) are far lower than those of the North American producers. In Britain, France, and West Germany, as well as in the United States, a large amount of capacity is idle or already demolished, and times are hard in steel towns.[5] It would be a strange irony if in this recession the U.S. steel industry found its competitiveness reduced by the legacy of Pittsburgh Plus—by the dispositions made in that period of success and expansion when the industry obliged its customers to come to it, rather than going to them.

Petroleum and Natural Gas

We must now move on to consider briefly North America's other sources of power for industry. The first of these are petroleum and natural gas. The United States is the world's greatest consumer of both these fuels. It was once also the largest producer, but those days are past. Today, it accounts for about 15 percent of world crude oil output, and imports some 30 percent of its requirements.

The rise of the U.S. petroleum industry was one of the great economic phenomena of the past century. It is about 130 years since the discovery of oil in northwest Pennsylvania ushered in the era of commercial exploitation.[6] The early days of the industry were marked by savage competition, from which John D. Rockefeller's Standard Oil Trust emerged in 1882 with a virtual monopoly of the continent's refining and pipeline facilities and, thus, a stranglehold upon the producers who depended on these facilities. The early Pennsylvania fields were soon eclipsed by those of California, which opened in the 1890s, and much more so by the southwestern fields, which came in after 1901. New fields were discovered and new wells were drilled in such numbers that production far outstripped demand and untold wastage occurred. Under the mining law of the United States, the holder of the mineral rights is entitled to anything he can extract from the area covered by his

---

[4]Legal niceties did not, however, prevent the oil industry from establishing a similar arrangement in 1928. It was called "Gulf Plus," taking the Gulf of Mexico coast as its Pittsburgh, and it covered the whole world except the USSR, as it then existed.

[5]It may or may not console the folk in these towns to know that the foreign competition, especially the Japanese, are buying into the American steel industry, as well as selling it foreign steel. But the facts remain that, according to *1988 U.S. Industrial Outlook* (Washington, D.C., 1988), nearly 22 million tons of annual raw steelmaking capacity was shut down in the two years 1986 and 1987, and that "more than 450 steel and steel-related facilities have been shut down since 1980" (p. 20-4).

[6]Canadians prefer to commemorate the discoveries at Oil Springs and Petrolia near Sarnia in 1858 as representing the start of their own petroleum era. The Ontario oil smelled so badly when brought up that it became known as "skunk oil." Until 1898, there was a refinery at Petrolia; it was then moved to nearby Sarnia, which has since become one of the major refining and petrochemical cities of Canada.

Oil field development in the United States: (1) The early years. This picture was taken in 1865 in Pennsylvania, not far from the site of the 1859 strike, which ushered in the modern era of the petroleum industry. Note the closely packed wooden derricks in the barely cleared forest valley. Compare this scene with the next. *Standard Oil Company of New Jersey Collection, Drake Well Museum*

lease, an arrangement suited to mining for gold or silver, but not to drilling for something so mobile as oil or gas. The result was that drilling became competitive; it paid to sink as many wells as possible. The resulting pattern is illustrated by Figure 8-2, which shows an area of the western suburbs of Los Angeles now adjoining the "Miracle Mile" along Wilshire Boulevard. Not only was such a large number of wells quite unnecessary to extract the oil, but under these conditions the natural pressure, which forces up the oil once it is tapped, is very quickly lost, so that it becomes necessary to pump, which increases the cost for all concerned. It is small wonder that the cost of producing oil on most of the American fields is many times higher than on those of the Middle East, where a single large concession enables the holder to drill the most rational and economic way.

The wastage and accompanying fluctuations in price gave both to the oil states and the oil companies an interest in controls, and in the 1930s, quota systems were introduced. In World War II, the federal government intervened in the national interest with a nationwide quota and rationing scheme, which the industry voluntarily applied after the war. After eighty years of competitive exploitation, a self-imposed sobriety had come to mark the industry. Today the leading producer-states are Texas, Alaska, Louisiana, California, Oklahoma, and Wyoming, while a number of other states play a smaller but still significant part.

As the petroleum industry has grown over the past century, so too has its market. In the early days, the product in demand was kerosene, for lighting and heating. Then, at the turn of the century, the internal combustion engine created an entirely new market, which was the prime cause of the great increase in exploration and production in the period 1900–40. Today 85 percent of the total petroleum output is converted into motor and aviation fuels and oils. The next development was the emergence of the petrochemical industries, which provided a market for the by-products that had formerly gone to waste in the processs of gasoline production. Demand for these enormously increased when, during World War II,

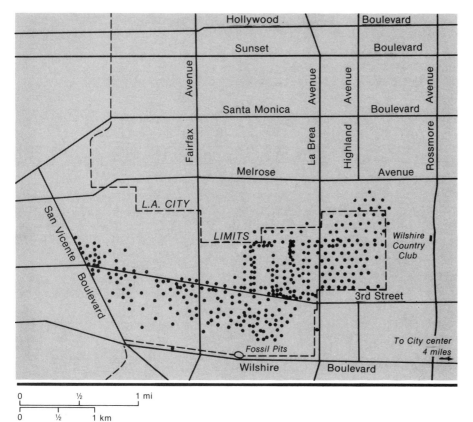

Fig. 8-2. The pattern of competitive exploitation on an American oil field. The map covers an area in western Los Angeles, which includes part of Wilshire Boulevard and the La Brea fossil pits as they were in the 1920s. Note the dense clusters of oil wells and their relation to the city limits. Such clusters result from conditions explained in the text and still survive on some American fields, although little trace of this particular field remains beneath the luxury apartments in the area today.

the "plastics revolution" began, starting with synthetic rubber.

Curiously, it was only as a relatively recent development that natural gas came to be used as a fuel. Early oilmen regarded the gas, which was often present in their wells, as a nuisance to be burned off or allowed to escape. Only from about 1940 onward did natural gas begin to compete nationally for a market alongside coal and oil, and gas from the gasworks.

In Canada, the industry's development after the discovery at Petrolia was very slow. As late as 1936, the country produced only 1.5 million barrels of oil.[7] A long period of intensive, but largely fruitless search was crowned by the discovery of oil at Leduc, near Edmonton, Alberta, in 1947. This was quickly followed by other discoveries in the western provinces, among which Alberta has, from the start, been the largest producer.

The petroleum industry has three phases: production, transport, and refining. The three can be, and sometimes are, organized separately. But, because of the nature of the product and the need either to store or to move the oil as soon as it reaches the surface, the producer requires a means of transport. This is especially the case when oil is flowing under natural pressure, and it was also especially true in the early days of the industry, when the oil was collected in anything available, such as empty whiskey casks, and the barrelmakers and carters held the producers to ransom by asking exorbitant prices. It was by appreciating the relative weakness of the oil producer vis-à-vis the transporter that Rockefeller scored his first success in the petroleum industry; he would have nothing to do with the producing end of the business. Today the big oil companies—the "majors," which include parts and branches of the empire Rockefeller himself founded—still adopt something of the same caution toward the actual production of oil, where their share of the operation is less than in refining or transport. But modern extraction is so complex and costly a business, particularly when carried out under the sea or in arctic wastes, that the major companies have been drawn into it because of its very sophistication.

[7]In 1979, Canada officially converted its statistics: barrels of oil and cubic feet of natural gas both became cubic meters. For the curious, the conversion factors are: 1 cu m of oil = 6.3 barrels; 1 cu m of natural gas = 35.3 cu ft.

Oil field development in the United States: (2) Oil rig at an offshore well in the Gulf of Mexico. Offshore drilling represents an extension of the continental oil fields beneath the adjacent shallow seas. *Gulf Oil*

The network of transmission pipelines for petroleum products that has grown up over the years now forms a vitally important element in the continental communication system. Although the trunk lines run from the Southwest to the Middle Atlantic region and the Chicago and Detroit areas, few settled parts of the continent are today isolated from the network. Alternative routes to markets are provided by pipelines especially to the Gulf Coast, from which a fleet of tankers carries both crude and refined products to great refining and market centers such as New York, Philadelphia, and Baltimore. In Canada, the Prairie fields are connected by pipeline with both the East and West coasts, although Montréal and Atlantic Canada are supplied mainly by imports. Edmonton possesses refineries, as do the Montréal, Toronto, and Sarnia areas, on the Atlantic–Great Lakes water route.

The most recent extensions of the continental oil–gas transport system have been the pipelines linking the fields in Alaska and the Canadian Northwest and their markets. When oil was discovered on the Arctic shore of Alaska (see Chapter 22), the question of its transport was crucial. The technical difficulty of moving it was complicated by a general uncertainty about where it, and neighboring Canadian fields' product, would be sold. This uncertainty continues; so, too, does oil prospecting in the Arctic. We may be sure that proposals for new pipelines, most of them running northwest to southeast, will be with us for many years to come.

The refinery may, in principle, be located anywhere between the well and the market, but for technical reasons refineries are usually either in the market areas or at the ports of shipment. Refinery towns are dotted along the coastlines of Texas, Louisiana, and California; important concentrations are also to be found on the New Jersey–Chesapeake Bay shore, adjacent to the great eastern cities, and at points along the Great Lakes and Mississippi waterways, such as Sarnia and St. Louis.

Since World War II, North America has been living through a technological era dominated by power supplies derived from petroleum and natural gas, which in the 1960s and 1970s together accounted for three-quarters of all energy consumed. This situation could be partly explained in terms of the genuine advantages of using petroleum rather than coal in powering vehicles or in heating buildings. But it could also be attributed to the cost advantage that oil and gas enjoyed over other forms of power, which encouraged their use and built up demand. These developments also, however, led to a very rapid depletion of reserves, which first alarmed the oil-producing states and provinces and then began to threaten the United States' future supplies. Quite suddenly, as it appeared, North America was facing an energy crisis.

Because so much has been said and written about this energy crisis in the past few years, we need to be clear about its character. In a sense it was merely coincidental that the crisis occurred during the oil–gas period of America's technical development: what produced it was the very sharp rise in energy consumption—on the order of 29–30 percent overall during the six years 1965–71, or over 20 percent per capita. The strain of providing for such a rapid increase falls naturally on whatever power source is supplying the largest share of energy at the time, in this case on oil and natural gas.[8] With an economy so thoroughly converted to the use of these fuels, it was little comfort to know that there were huge coal reserves remaining, or that nuclear power was steadily enlarging its share of the energy market: the question was whether there would be fuel for the family car tomorrow.

The danger of depending on a single type of energy source was clear. It was in these circumstances, as we saw in Chapter 6, that the U.S. government was confronted by the need for an energy policy. For one thing, the initiative in price setting in the petroleum industry had passed from its domestic oil companies (operating at home or abroad almost without price distinction) to foreign governments. For another, it would be suicidal for a superpower like the United States to depend too heavily upon imported fuel supplies. Project Independence (see Chapter 6) therefore proposed all the obvious measures: reduction in oil use, search for fresh reserves, switch to alternative energy sources. One predictable consequence of all these actions was to raise the price of energy, but that was not necessarily a bad thing: the whole difficulty had been caused by the relative cheapness and competitive marketability of petroleum in the past. It had been offered too cheaply for rational exploitation.

One of the consequences of the events of 1973 was to draw attention, in both Canada and the United States, to a well-known source of oil, which had been largely untapped for years: the oil-bearing sands and shales of the West. Recovery of oil from these was technically possible; it was simply a matter of cost—to be precise, of the price of oil. When the OPEC countries raised that price in 1973, and went on raising it, the processing of the Athabasca tar sands in northern Alberta and of the Colorado oil shales made immediate sense. In Canada, the federal government joined with the oil companies in setting up plants in the Fort McMurray area, while in the United States, a number of new towns appeared in western Colorado as the oilmen moved into that area. The scene was set for what, technically, was an impressive new development.

Then the price of oil began to fall. After the alarm of the 1970s, the 1980s brought a surplus of oil on the

[8]In 1990, petroleum and natural gas accounted for 65 percent of U.S. energy consumption. Comparable figures for 1970 and 1980 were 77 percent and 72 percent, respectively.

world scene: the North American countries can buy all they require, the Alberta developments have slowed, and those in Colorado have stuttered to a halt. On strategic grounds, these alternative oil sources should be exploited, to reduce the size of imports; that was the purpose of Project Independence. On economic grounds, and for the present, there is no reason to produce expensive oil at home, involving new technology and new settlements, when cheaper oil is available in huge quantities on the world market. In the absence of any specific national emergency, which would permit a government to act on strategic grounds, the only people less happy than the oil companies involved are the hotelkeepers and real estate agents in a dozen remote communities in the West who looked for profits, but have found desolation.

Canada's position as an oil and gas producer is rather different. Since the 1947 strikes in Alberta to which we have already referred, there have been abundant oil and gas supplies in the West, some of them available for export. Because of the distance, however, from the western fields to Ontario and Québec, it has been cheaper and simpler to continue to import oil into eastern Canada, as in pre-1947 days. This has had the effect of making the country both an exporter and an importer (see Table 8-3).

Now the oil supplies of central and southern Alberta are beginning to dwindle. This raises the question—which has in any case been exercising the Canadians for a long time—whether remaining supplies should be conserved for Canada's own use (the so-called Canada First policy), and exports reduced. Differences of opinion on this question, and on that of managing the price of Canadian oil, led to the formulation of a National Energy Policy and, equally, to its abandonment in 1985, as we saw in Chapter 6. The divergence of view between East and West, both politically and financially, has been sharp.

There are three possibilities for the future of Canada as a producer of hydrocarbons:

1. *To encourage a fuller use of natural gas* to ease the demand for petroleum. Canada has enormous reserves of natural gas, either untapped or capped awaiting an upsurge in demand. Yet the share of natural gas in the nation's consumption of primary energy has been lagging: in 1983 it was actually lower (20 percent) than it had been in 1973 (21 percent). As Table 8-3 shows, much of the present production is exported for lack of domestic demand, and this despite the existence of a gas pipeline link between Alberta and Montréal. The main recipient is the United States.

Table 8-3. Canada: Trade in Energy-Related Products, 1973–87 (in millions of C$)

	1973	1975	1980	1982	1983	1987
Exports						
Crude petroleum	1482	3052	2899	2729	3457	4855
Natural gas	351	1092	3984	4755	3958	2527
Refined petroleum & coal products	477	1132	3258	3807	4129	3720
Electricity	109	104	773	1120	1228	1200
	2483	5427	11,145	12,769	12,835	13,188
Imports						
Crude petroleum	941	3304	6919	4979	3274	3179
Natural gas	8	8	3	5	2	0
Refined petroleum & coal products	375	852	1499	1794	1887	2443
Electricity	6	13	3	5	2	9
	1331	4176	8489	6906	5276	5649
Balances						
Crude petroleum	541	−252	−4020	−2251	183	1676
Natural gas	343	1084	3984	4754	3958	3956
Refined petroleum & coal products	102	280	1759	2013	2241	1277
Electricity	103	92	770	1114	1226	1191
	1153	1251	2656	5863	7559	7539

Note: The items listed do not add up to the totals given; the difference is accounted for by trade in radioactive ores.

2. *To revive the exploitation of the tar sands.* This is an expedient that can be undertaken at any time: the only question is one of price. At present, world oil prices are too low to make operation worthwhile, but the tar sands represent a resource which is there, so to speak, on standby.

3. *To drill for new oil supplies.* Today's search for oil is being carried out not only on Canadian land but also from artificial islands built up in the daunting environment of the Beaufort Sea. A gas pipeline from the Boothia Peninsula, or even Melville Island, to southern Canada may one day exist, to parallel a pipeline from the Mackenzie Delta to Edmonton.

Electric Power and Nuclear Energy

The rivers of North America are a valuable potential source of power. The United States and Canada rank with the Russian subcontinent as leaders among the countries of the world in terms of installed hydroelectric generating capacity. The role of this capacity is, however, rather different in the two countries. In the United States, hydroelectric plants account for 9.5 percent of generating capacity and 10 percent of production (a fall, incidentally, from 20 percent in 1955). In Canada, 68 percent of all electricity used is hydro-generated, and for the province of Québec, the figure is 97 percent (Fig. 8-3).

The United States is favored by huge power resources—especially in the remote Mountain West and on the Pacific Coast. Canada's development to date is concentrated in two areas—the southern half of the Laurentian Shield and the mountains of British Columbia. In the first of these areas, it is well placed to supply the industries of the St. Lawrence Valley and has played a vital role in their development. A good deal of the potential, however, is in remote areas, and its use involves either long-distance transmission or the location of manufacturing plants at the power sites, as in the case of Kitimat (see Chapter 22).

Step by step, the province of Québec has been harnessing these remoter power sources along the rivers flowing south to the St. Lawrence. In the 1980s, the focus of attention had switched much farther north, across the Height of Land, to the east coast of Hudson Bay and the basin of the La Grande River. In an area with only a scattering of Indians (whose interests must nevertheless be considered), a work force of many thousands had transformed the whole basin

into a complex of dams and reservoirs, and laid lines to carry the power generated there 600 miles (1000 km) southward to the national distribution grid.

The two largest units in the project—La Grande 2 and 3—were already in production. Progress on the remainder of this vast scheme depended on finding markets for the power generated. Apart from Québec's own consumption, Hydro-Québec regularly exports 15–20 percent of its power production, a half of this to other provinces and a half to the United States.

Canadians have been building power plants of all kinds for years in the hope that electricity exports to the United States will justify their construction. As Curtis and Carroll put it, "Canadians have long since learned that they can build on a large scale and thus enjoy sizable economies if they build for *both* national markets simultaneously. The ideal is to overbuild at today's costs in the hope that ultimately the Canadian market will expand to absorb the excess." While penetration of the U.S. market is certainly possible, it is perhaps unfortunate for Hydro-Québec that the part of the United States that lies within economic delivery range of its power is an area of industrial loss and little population growth.

Almost every major river in the United States has been put to work, though not all of them as spectacularly as the now-famous Tennessee. Of the potential remaining undeveloped—some 100 million kilowatts—over half is in the Mountain and Pacific regions, with about one-sixth in the state of Washington alone. Here it is the Columbia River system, already harnessed by such giant works as the Grand Coulee and Bonneville dams, that represents the principal source for future development.

The three areas of the continent in which hydroelectricity has been particularly important in industrial development are the Pacific Northwest, the Tennessee Valley, and eastern Canada. The first of these, remote from supplies of either coal or oil, could never have responded as it did to the demand created by war in the Pacific without drawing upon its one local source of power. The Tennessee Valley, although close to the Appalachian coalfields, found an altogether new prosperity in power development and was the scene of a now-familiar type of rural industrialization, made possible by the use of electrical energy. In the St. Lawrence Valley and southern Ontario, another area situated none too favorably in relation to coal and oil, electric power supplies have been sufficient not only to support Canada's indus-

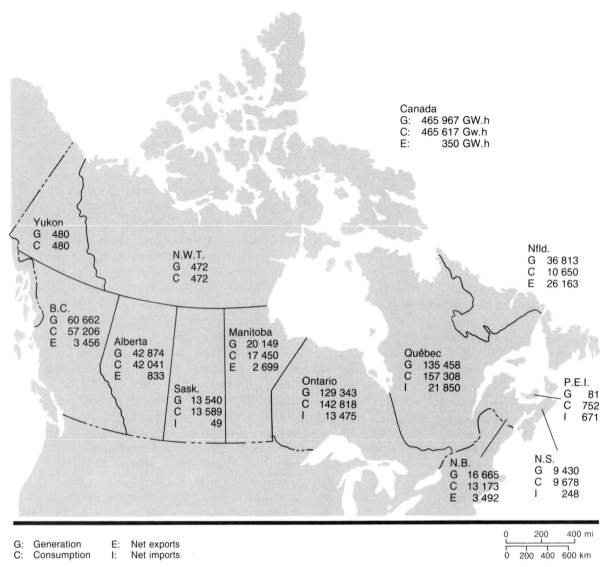

Canada
G: 465 967 GW.h
C: 465 617 Gw.h
E: 350 GW.h

Yukon
G 480
C 480

N.W.T.
G 472
C 472

Nfld.
G 36 813
C 10 650
E 26 163

B.C.
G 60 662
C 57 206
E 3 456

Alberta
G 42 874
C 42 041
E 833

Sask.
G 13 540
C 13 589
I 49

Manitoba
G 20 149
C 17 450
E 2 699

Ontario
G 129 343
C 142 818
I 13 475

Québec
G 135 458
C 157 308
I 21 850

P.E.I.
G 81
C 752
I 671

N.B.
G 16 665
C 13 173
E 3 492

N.S.
G 9 430
C 9 678
I 248

G: Generation   E: Net exports
C: Consumption   I: Net imports

0   200   400 mi
0   200 400 600 km

Fig. 8-3. Canada: Generation, consumption, and import/export of electric power, 1990. The two main generating provinces, Québec and Ontario, produced roughly equal amounts of electricity but, whereas Québec's production was almost entirely hydroelectricity, Ontario produced roughly similar amounts from hyrdo, thermal, and nuclear stations. All totals in gigawatt-hours. (Source: Energy, Mines and Resources, Ottawa, 1991.)

trial core, but also to permit a considerable export of power across the border to the northeastern United States, as we have already seen.

Despite the rapid increases in generating capacity brought about by the construction of large numbers of dams in recent years, many parts of the continent are suffering from shortage of power. Even the Pacific Northwest, for all its Columbia Basin development, has power shortages, and the TVA now produces far more electricity from thermal power stations it has built than from the turbines at the dams for which it is famous.

Apart from the material problems to be met in increasing capacity to meet present demand, there has been another, political, hindrance to expansion in the United States. The question is whether the new power schemes should be in government or in private hands. The emergence of the TVA, followed by the Bonneville Power Administration, as federal agencies selling electricity in competition with private companies, touched off a political dispute that is still a burning issue wherever new construction is projected, be it on the Snake, the Colorado, or the Missouri. Although the coordinating role of the gov-

Hydroelectricity in Québec: Part of the power scheme on the La Grande River. *Hydro-Québec*

ernment is generally accepted, there seems no need for the government to assume responsibility for all construction, nor indeed any likelihood that Congress would vote the funds necessary.

A new dimension was given to the production of electricity when it became feasible to replace waterpower or steam by nuclear energy. Considering, however, the lead the United States took in nuclear development, its application of the new energy source to power production was rather slow.[9]

Since one of the principal locational requirements of a nuclear station is a very large supply of water for

[9]In evidence of this fact, as recently as 1967, the *Statistical Abstract of the United States,* in classifying prime movers for generating electricity, included nuclear fuel only in a footnote, along with "wood and waste." By 1976, this charming anachronism had sadly disappeared, or rather, had slipped sideways: the "wood and waste" was now lumped together with coal.

cooling, the new stations are sited in such places as the shores of the Great Lakes and the banks of major rivers, including the Tennessee, which becomes the first region to produce electricity by all three of the main methods so far introduced—hydro, thermal, and nuclear. Since nuclear plants require neither coal nor waterpower, there has been a tendency also to build them in areas lacking these two resources.

Canada produces about 13 percent of its electricity from nuclear plants, nearly all of them in Ontario. For the United States the figure is about 19 percent.

A further generation of nuclear plants has reached the planning stage. That they do not already exist is due to the hostility the use of nuclear power has provoked, not only in North America, but in Europe also, on the grounds of environmental hazard. This opposition has increased rather than diminished as time has gone by, and it was powerfully confirmed and reinforced in its stand by an accident at Three Mile Island, a nuclear power station near Harrisburg, Pennsylvania, coming as it did in the wake of a series of leaks and "near misses." Thanks to the wave of environmental legislation to which we referred in Chapter 5, the ability of groups opposed to such a development to thwart it has grown almost to the point of veto, whereas bad design or planning of the plants has given a number of justified grounds for exercising it. Meanwhile, the question of where and how new demand for electric power is to be met remains unanswered.

Change, Growth, and Decline in North American Industry

With all their size and strength, industries in the United States and Canada are constantly changing in character. This need not surprise us; rather, it is surprising that, for so long, we have tended to regard industry as a fixed quantity—as if, once it was established, it was as immutable as the features of the physical landscape. All industry everywhere is, in fact, ephemeral. Most of the industries with which we are familiar are, in their present forms, less than a century old, and some of those industries that brought in the Industrial Revolution have since then either disappeared or been changed out of all recognition.

These changes in the industrial scene are of several types. We can recognize changes in

1. The character of inputs
2. The nature of markets
3. The nature of the product
4. The location

Let us consider each of these, in turn, drawing upon North American examples.

CHANGES IN THE CHARACTER
OF INDUSTRIAL INPUTS

The principal industrial inputs are raw materials, energy, and labor. We can recognize the impact of changes in each of these on American industry. An example of the first, a change in raw material inputs, has occurred in the steel industry, with the substitution of scrap metal for iron ore or pig iron in the furnaces, which has reduced the "pull" of the ore supply, and even of coal, on the industry's location. In the synthetic fibers industry, too, there have been

changes: the early synthetics were derived from cellulose—in practice, usually from trees—whereas more recent synthetics are petroleum based. The first group were, in effect, an offshoot of the forest products industry, the second of the oil industry.

Changes in energy supply have seen industries switching from a coal–steam base to electricity, oil, or gas. The general effect of these energy sources on the location of industry, as we have already noticed, is liberating: they enable areas like the Tennessee Valley, the Gulf Coast, and the Alberta Prairies to attract industry to places where it was previously absent; at the same time, they remove the old incentive for industry to build near coalfields.

Changes in labor inputs occur as the answer changes to the question: What is the labor needed *for?* In the early days of factory industrialization, the answer was usually that industry required a large number of strong hands to lift or push, and of unskilled docile workers willing to accept long hours and appalling conditions in exchange for low wages. But times have changed; modern industry has replaced many of the hands and much of the brute force by machines, so that fewer workers are needed, while those who remain carry out, in much pleasanter conditions, less physical tasks. For one thing, the staff employed in planning, or sales, or services form a larger proportion of the whole: the ratio of white-collar to blue-collar workers has steadily increased. The U.S. census, for its part, distinguishes "production workers"—those actually making the goods—from other employees. In 1929, the ratio of "others" to production workers was 1:6.5. In 1947, it was 1:5.0; in 1975, it was 1:2.2; and by 1988, it had fallen to 1:1.54. Among other things, this change means that job opportunities in modern industry increasingly favor the better educated. There are no longer the "lift and push" jobs for unskilled workers that once offered employment to new immigrants and those at the bottom of the educational ladder. In North America, this has had the general result of placing many members of minority groups—black, Hispanic, Indian—at a disadvantage in their search for work.

The other great change in the labor force has been the increasing participation of women. To take once again the figures for the United States: in 1970, the ratio of women to men in the paid labor force was 1:1.65. By 1989, this ratio had been reduced to 1:1.21; in other words, the United States is steadily approaching the time when women are as likely as men to be wage earners. This change has come about partly through women's own efforts to assert their ability to hold jobs traditionally carried out by men; partly, too, because so many physically demanding jobs, as we have just seen, have been eliminated from the labor agenda. This being accepted, attention then shifts to the pay and prospects of the women workers, as compared with their male counterparts.

CHANGES IN THE NATURE OF MARKETS
FOR INDUSTRY

In any country, the size of the internal market for manufactured goods is governed by the life-style and the purchasing power of its citizens, whereas the external market depends on international competition, and also on political circumstances—tariffs, import restrictions, wars—which increase or decrease the sale of goods.

In North America, the internal market has been affected by a general rise in standard of living, and by particular regional improvements in that standard, which have lifted the areas concerned—notably the Deep South and Appalachia—out of economic backwardness. These areas have become, for the first time, worthwhile market areas for the manufacturers of cars, refrigerators, and the paraphernalia of modern living. It has also been affected by the shift in population from north to south, cold to warm, Snow Belt to Sun Belt, which we described in Chapter 2. In its simplest terms, such movement means to the manufacturer an increased demand for sunwear, garden furniture, and air conditioners, and a reduced demand for snowshoes and heaters. When thousands of homes are involved, such changes have had a measurable effect on the market. So, too, has the steady shift of population from rural to urban areas over the past century.

International markets may change even more rapidly, opened and closed overnight by a political decision to permit or forbid trade. On the international scene, the past twenty years have been marked by a steady erosion of North America's industrial position through the rise of competitors whose costs, especially labor costs, are much lower, and whose products sometimes imitate those of the older industrial countries. American industrialists expect their governments to protect them from this kind of competition. But protection is a two-edged sword, and foreign manufactures have captured an increasing share of the American market, for cars and cameras, computers and steel.

Protection, in any case, is of limited value in a world where capital moves freely, even though goods may not. There is little or nothing to prevent foreign businesses buying into North America: after all, U.S. businesses have been buying into Canada for decades. If imports of foreign cars, for example, are curbed, then the foreign car manufacturer can establish a plant inside the tariff or quota barrier. Indeed, if the manufacturer creates a demand for labor in the new plant, it will be more than welcome. For years, it has been American corporations that have operated overseas in this way; now the return traffic has built up. By 1989, foreign direct investment in the United States was over $400 billion, $160 billion of it in manufacturing alone. Some 65 percent of this investment was by Europeans; 20 percent of it was Japanese.

CHANGES IN THE NATURE OF THE PRODUCT

At any particular moment, certain industries are expanding and others are declining. New industries arise as a result of research and development; in this respect, North America's newest products are derived from space exploration and electronics, to create entirely new industrial areas, like Silicon Valley south of San Francisco, brought into being by the silicon chip and computer development. A rather larger number of industries, however, are in decline. Among them are some for which the demand for their product has simply run out; railroad passenger car manufacturing is an example of a once-important industry that has now virtually died, and men's hat manufacturing is another. Other industries are in difficulties because of competition. The most obvious example here is the automobile industry, year after year America's largest. There were some bad years in the 1980s for Detroit and Michigan, with the industry fighting back from periods of depression. Table 8-4 gives the two sets of figures at the root of the problem: retail sales of domestic production, and imports.

There have been other changes in the nature of the

Table 8-4. United States: Retail Sales and Imports of Cars, Selected Years, 1970–89 (in thousands of cars)

	1970	1975	1980	1984	1985	1989
New cars						
Domestic production	7116	7050	6581	7952	8205	6823
Imported	2013	2075	3248	4880	6406	2805
Of which, from Japan	381	696	1992	2692	3404	2052

industrial product. One of them has been the re-placement of heavy materials by light: aluminum or magnesium replacing other metals; plastic replacing steel, even in car bodies and parts, which for years have afforded the steel makers their largest market. And while the size and weight of a few manufac-tured objects, such as aircraft, continue to increase, many more have been reduced by better design, or even miniaturized. Reduction in bulk means reduc-tion in transport costs and that, in turn, relaxes the constraints on locations of industry—which leads us naturally into our fourth and last group of industrial changes.

CHANGES IN THE LOCATION OF INDUSTRY

With all these factors of change operating within the manufacturing sector, we need not be surprised that the location of industry has altered. But these inter-nal factors, as we may call them, represent even so only half the story. The other half is to be read in con-ditions outside manufacturing; in particular, in the changes in population distribution that we noted in Chapter 2, and in the gradual reduction of regional disparities in standard of living (which, to a manu-facturer, means increase in demand). With a popu-lation steadily dispersing across the continent, the concentration of manufacturing in the old Belt of the Northeast was clearly uneconomical.

Let us consider, first, three examples of changing location among the older industries of North Amer-ica. We shall then turn to the glamour industries of today, the so-called high-tech industries, and exam-ine the factors likely to have influenced their location.

Our first example is that of an industry—meat-packing—that, for more than a century, seems to have been pursuing its raw material across the con-tinent. We first encounter it about 1840 in Cincinnati on the Ohio River (see Chapter 14), from where it moved west to Chicago, the city of the great stock-yards and of Upton Sinclair's famous novel of the in-dustry, *The Jungle*. From Chicago it moved west again to other major midwestern cities, then on to the western plains, to smaller centers; ironically, in the most recent phase, to the very towns—like Dodge City and Garden City in Kansas—from which, in the days of the great cattle drives, trainloads of live cattle had been shipped east for slaughter in the distant packing houses of Chicago. The industry has been, in this sense, closing in on its raw material and has fi-nally caught up with it.

Our second example is that of an industry that has moved to exploit particular regional advantages; in this case, mainly the advantages of raw material sup-ply and labor quality. The cotton and woolen textile industries of the United States grew up in southern New England, on a base of local water power, abun-dant immigrant labor, capital from the shipping in-dustry, and cotton and wool brought in from other regions. But in the southern states, where the cotton was grown, a local textile industry came into being about 1880; using low-cost labor, especially women and children. By 1920, it had grown to rival the New England industry. There was, after all, no particular reason why cotton should make the double journey from the South to New England and back again in the form of cloth: an industry within the South would benefit from considerable savings in transport costs.

But this was not all. The southern states, once launched upon textile production, proved to have other advantages, such as (1) more up-to-date ma-chinery, which made their workers more efficient, (2) less conservative mill owners, (3) less restrictive leg-islation covering employment and shift work, (4) lower property taxes, and (5) less unionized, lower-cost labor. The cotton textile industry moved south, followed, after an interval, by woolen manufactur-ing. The spinning and weaving of synthetic fibers (which now account for three-quarters of all textiles) was from the first mainly a southern industry. Today, the textile industry has largely deserted New En-gland and is concentrated very firmly in the Pied-mont region of the Southeast.[10]

Our third example is that of an industry that has not so much moved its base as dispersed from, and contracted back to, that base. This is the automobile industry. According to J. M. Rubenstein,[11] it has gone

---

[10]Because the cotton textile industry is a labor-intensive one, in which the cost of semi-skilled labor constitutes a large fraction of the value of product, it has tended to be the lead industry in the industrialization of regions. Thus, the progression from Lancashire to New England to the southern Piedmont and now to developing countries in eastern and southern Asia and Latin America. As each new area assumes precedence, the older one turns to higher-qual-ity, more expensive goods or to other kinds of manufacture, as New England has done and the Piedmont is now attempting to do.

[11]J. M. Rubenstein, "Changing Distribution of the American Au-tomobile Industry," *Geographical Review*, vol. 76 (1986), 288–300. Since 1986, Rubenstein has become the chronicler of the automo-bile industry's fortunes and misfortunes: see his further articles in *Geographical Review*, vol. 77 (1987), 359–62, and vol. 78 (1988), 288–98.

through four phases. From its beginnings about 1895, and until 1905, it was dispersed in workshops and small plants in the Northeast. Then, from 1905 to 1915, it came together in the area with which it will always be associated: southern Michigan. However, between 1915 and 1965, it decentralized. The headquarters and base were in Detroit. Around the city a ring of factories grew up to supply the thousands of parts that go into car manufacture.

But as firms were by now supplying a national market, it became obvious that to distribute complete cars to the nation from one small area in Michigan was entirely uneconomical. The manufacturers therefore set up assembly plants (with their attendant parts suppliers) in outlying market areas and shipped only the essentials from Detroit. At peak, American automobile manufacturers operated over 200 dispersed plants, including at least one in virtually every state east of the Mississippi.[12]

But since 1965, a fourth phase has become apparent: a return to base. Rubenstein reports that, in this period, a number of car plants outside the Midwest have closed while, of thirteen new plants opened by the industry since 1979, all thirteen are *within* that region. These new plants are on fresh, mainly greenfield sites, but the locational emphasis is clear.

Each of the three long-established industries we have briefly considered has reacted differently to changing times. But now we must ask the question: Where is *new* industry likely to locate? A 1984 survey revealed that the most popular states among firms seeking new locations were North Carolina, Texas, California, South Carolina, Georgia, and Florida. With one or two possible exceptions, that is a list that few people could have guessed—none of the states is in the former Manufacturing Belt, and four of them are in the Old South.

Their listing perhaps gives us a clue to what today's industrialist is looking for. He is not interested in coalfield sites, like his nineteenth-century predecessor. He is deliberately choosing to avoid older industrial areas with their aging infrastructure, and

doubtless crowded conditions: word of that Rust Belt has evidently leaked out. He is opting for space and an environment of small to medium-sized communities, with a work force that may well be new to industrial work and weakly unionized. He may be attracted by the presence of research or science parks—North Carolina has an outstanding one—to back up his efforts. He is choosing—as the American population is choosing—warm areas over cold ones.

If our industrialist is in high-tech industry, we can be still more specific. It is one of the characteristics of these industries that they tend to group themselves together, just like the textile mills of an earlier age. Their groupings tend to occur where the pool of their specialized labor requirements is greatest and under the shadow of a strong group of research institutions. A survey of some hundreds of these firms revealed that their first four locational requirements, in order, were labor availability, labor productivity, tax climate, and "academic institutions." (Considering the small bulk of much of their output, it is no surprise to find that access to raw materials came last on the list.)

Where are these conditions to be found? In the United States, the largest centers of high-tech employment are (1) Los Angeles; (2) the San Francisco area, especially San Jose and Silicon Valley; (3) Boston; (4) New York; (5) Chicago; and (6) Dallas–Fort Worth. Such centers are created by a combination of research, personnel, and either venture capital or government money. Any area of North America today that does not have such a center wishes that it did.

Canadian Industry

A separate comment must be made about Canada and its industries. We have already seen in Chapter 4 the importance to the continent of Canada's insistence on her separate identity. In the industrial sphere, it has led to the development of plants and firms that serve only the small Canadian market, merely duplicating much larger industries in the United States, which have ample capacity for both countries. From this point of view, it is easy to argue that Canada has become overindustrialized: in practical terms, the manufacture of automobiles in Detroit makes it unnecessary to manufacture them also in Windsor, Ontario, a mile or two away across the river.

In developing her own industries, Canada is only following the example of many other nations, in de-

---

[12]In some cases, the manufacturer may elect to concentrate the production of particular models in particular plants. If then the new model proves to be unsuccessful, which is not unheard of in the industry, the local impact of that failure may be very serious indeed, although the company, as a whole, is still prosperous. In North America, such a situation would be particularly serious if the plant whose model failed was in Canada and the company was based in the United States. The customer's choice of a new car then turns into an international incident.

clining to depend on a foreign power for manufactured goods. Even so, however, the Canadian government found cause for alarm in the fact that so much of the industry on its side of the border was either owned or controlled by non-Canadians. Either by share purchase or by the establishment of branch plants inside the country, individuals and corporations from outside Canada had come to control, in recent years, enterprises representing about one-half of all Canadian manufacturing industry. In the automobile, rubber, and electrical goods industries, the proportion was much higher. At the same time, over three-quarters of Canadian oil and gas production was foreign controlled. A large proportion of this control was exercised by firms in the United States. The situation was hardly new, but the amount of foreign control in the newest and most promising sectors of manufacturing was striking.

This question of foreign control came to a head in the 1970s and led to the introduction of measures to ensure that the Canadian government and people would receive a reasonable share of the profits from their own labor and resources. In 1974, the Foreign Investment Review Agency began work, examining all proposals for new investment in Canadian industry, to ensure that such investment would be of "significant benefit" to Canada. The impact of the new policies (some of which we have already referred to in Chapter 6) was considerable. Between 1971 and 1981, the share of non-Canadian control of the assets of sample Canadian industries was reduced as follows: mining, from 62 to 30 percent; petroleum and gas, from 90 to 49 percent; all manufacturing, from 53 to 41 percent.

It seems as if the problem has been brought under control. The dilemma for Canada is this: its long-term problem is that of finding the means to develop its huge territories with the limited capital resources a nation of 27 million can generate. Understandably, Canadians would prefer to accomplish the task themselves, but if they cannot, they presumably have to call upon capital from outside. In this respect, there is little difference between the so-called "underdeveloped" or "developing" countries of the world and a developed country like Canada with an underdeveloped hinterland. Both need development capital and if, for whatever reason, they interfere with its inflow, then it is their own economy that is likely to suffer.

Scarcely, however, had this problem been con-

fronted than another, associated worry for Canadian manufacturers loomed up. Following the general election in Canada in 1988, the government entered into a free-trade agreement with the United States, aimed to remove, over a period of years, the barriers to trade between the two countries. Present arrangements are that this free-trade area shall, in the near future, be extended to Mexico as well.

It is still rather early to try to judge whether this agreement is working to Canada's advantage; in particular, whether it is working to the advantage of individual Canadian workers or consumers. On the face of it, it would be surprising if it did. Consumers may gain some relief over prices, but workers may well lose jobs. Canadian industry may find it easier to sell in U.S. markets, but Canadian manufacturers may find themselves treated simply as managers of places to assemble goods planned, researched, and designed elsewhere. On the balance of these factors the agreement—and the government which negotiated it—must stand or fall.

But we must not give the impression that Canadian industry in all its forms is weak, protected, or dominated by foreigners. On the contrary, it employs 1.7 million workers and in some of its branches, notably pulp and paper production, is a world leader. Small though its industry may be by contrast with its giant neighbor, it has achieved a high degree of sophistication in certain branches, such as aircraft production and telecommunications.

In concluding this chapter, let us return briefly to a subject mentioned earlier—the role and importance of industry within the economy as a whole. A century ago, the amount and type of manufacturing which a country possessed were regarded as the truest indicators both of its standard of living and of its status as a modern power. Even those industries which made wage slaves of their workers and pollution traps of their surroundings were accepted as contributing to the nation's wealth, at least by those who lived far enough away from the plants. Industrial strength—or, in the case of the North American nations, the potential for rapid increase in that strength in time of emergency—was what counted.

That all this has changed can be seen in a number of ways. It can be seen in the virtual standstill in industrial employment in the two North American nations in recent years. There have been, certainly, *regional shifts* in employment; in the United States notably from north to south, but the total employed in industry has risen only very slightly since, say, 1977, a period during which U.S. employment has risen overall by 30 million.

It can be seen, too, in the way in which American industry has shifted some of its manufacturing operations overseas, to areas of cheap labor, from where parts are brought back for assembly and finishing. Some of these areas, as we shall see in Chapter 20, are not strictly overseas at all: they are the *maquila* plants situated a few hundred yards into Mexico, across the border. One writer goes so far as to say that: "The decline of 'American' manufacturing . . . has not been a decline in the performance of 'American manufacturing firms' so much as a decline in the role of the United States as a geographical location of production. The firms have held their own, but they are not attached to America in the way they used to be."[13]

It can be seen, thirdly, by the way in which environmental concerns have grown stronger; strong enough, in fact, to act as a controlling factor in industrial development. Manufacturers themselves have responded by spending more on pollution controls and amenities; governments have enacted laws

[13]G. Wright, *Old South, New South: Revolutions in the Southern Economy Since the Civil War,* (New York, 1986). Quotation from p. 273.

to spur on those whose response was too slow. The quest for the cheapest possible construction and the last possible dollar of profit has become, first, unpopular and, secondly, illegal.

So now geographers have turned much of their attention to the service industries—to tertiary and quaternary types of employment. But this has made their task harder, not easier. Not only is the range of services much wider than that of industries, but while some services—a large hospital, say, or a state university—create a site and a physical plant larger than most industries, many other services require only a few square feet of space, and an increasing number are carried on from people's homes. To speak of them as having a geography is therefore difficult. Only in a very few cases can we describe a service center in the same way that we could link Detroit with automobiles or the old Pittsburgh with steel. We can, perhaps, say that the state capitals are government-service cities, that Hartford in Connecticut and Des Moines in Iowa are insurance towns, or that places in Florida and Arizona are centers for services to the elderly, or to vacationers. But that is all. Many of these services are too universal to be precisely located; a telephone and a computer screen locate them wherever the operator happens to be.

With steel mills, at least you always knew where you were.

9
Transport

American railroads: The Northeast Corridor route in the United States provides fast service for Amtrak's Washington, D.C.–New York–Boston commuter trains, one of which is seen here in northern New Jersey. *Amtrak*

The most important early routes into the North American interior were along the waterways. It was by means of the St. Lawrence, the Great Lakes, and the Mississippi that the French made the swift westward penetration in the years before 1763 that almost succeeded in encircling the English colonists further south and east. It was by the water routes that the English thrust more slowly west through the mountains until, crossing the portages, they reached the Ohio River and made of it the main road to the West. And it was largely by water that Mackenzie made the first transcontinental journey, from Upper Canada to the shores of the Pacific, in 1793. There were few land routes, and those that existed—the Wilderness Road to Kentucky via the Cumberland Gap, and Zane's Trace through Ohio, for example—were so hazardous that, until the opening of the Erie Canal (1825), the safest and most economical freight route to and from the interior was a one-way circuit, whereby goods traveled down the Ohio with the current and returned by way of the Mississippi, the Gulf, and the Atlantic to East Coast cities.

Only in the lands beyond the Mississippi, where the rivers are mostly unsuitable for navigation, were land routes, from the first, of comparable importance. Here the historic trails struck out across the plains and mountains—the Sante Fe, linking the East with the Spanish settlements in New Mexico; the Gila, the Spanish route to California; the Oregon Trail that brought early emigrants to the Northwest, and its branch, the California Trail, leading (after 1848) to the goldfields. Later, when the first railroads came to the West, they followed much the same routes.

If the opening of the Erie Canal as a two-way route for freight revolutionized travel in 1825, the changes brought about by the railroads were even greater. In determining the regional balance of power within the continent and in the settlement of the West, the role of the railroads was of the utmost importance. It was in the late 1840s that a crude network of lines reached as far west as the Mississippi. Before that time, the all-important river traffic on the Mississippi had linked the new Midwest with the South. With the coming of the railroads, this link was replaced by a far stronger one between the Midwest and the Northeast, whose significance was quickly demonstrated. As J. T. Adams wrote: "When civil war at last came, and the South counted on the West joining with it on account of their being bound together by the arms of 'Ol' Man River' and an outlet on the Gulf, it was the newly completed railways between West and North that enabled those sections to hold together, instead of South and West."[1]

But it was west of the Mississippi and on the Canadian Prairies that the railroads had their maximum influence. We have already seen in Chapter 5 how both the U.S. and Canadian governments made the railroad companies their agencies for settling the West, giving them land grants with which to finance their operations. Equipped with millions of acres for disposal, the railroads set out to colonize the lands along their right-of-way, and in doing so became the greatest single factor in fashioning the cultural landscape of the West.[2]

[1]J. T. Adams, *The Epic of America* (New York, 1931), p. 244.

[2]By 1914, Canada had more miles of railroad line per capita than any other country in the world. In 1961 the Royal Commission on Transportation was moved to comment (Vol. II of its report, p. 93): "Historically, the transportation system in Canada was used so extensively as an instrument for the pursuit of broad national policy objectives that the character of the system as a system tended to become a matter of secondary concern. As a result, national transportation policy has often been a great deal more preoccupied with the question of how effectively the transport system was functioning as an instrument to fulfill national policy objectives, than with the question of how well it was functioning as an economic enterprise."

For the western railroad was so much more than just tracks. It was a whole economic system in itself. It had first to find its settlers. To do this, it conducted recruiting campaigns throughout Europe and eastern America and indulged in propaganda sufficiently fanciful to earn the name of "the Banana Belt" for one company's lands. It then carried the settlers and their goods, free of charge, to their new homes and established them with tools and grain or stock. In years of drought it gave out supplies to tide the farmers over, and it maintained agricultural advisory services to increase output. It was buyer and shopkeeper to the settlers, and sometimes it abused its monopoly. Finally, the railroad was responsible for the location and layout of the towns, itself deciding which cluster of huts and tents should become a town, and which should not. Along the western railroads today, it is common to find that, although half of the settlements may have Spanish or Indian names, the other half are either named after railroad engineers, or were named by them.

In eastern North America the heyday of rail transport lasted until World War I; in the West until the 1920s, for there the network of all-weather roads is of very recent date. (In the North, there are areas where transport has from the first been based on the airplane, where neither railroad nor road penetrated.) Not long after road transport came to challenge the railroad's monopoly of heavy freight haulage, the commercial airlines began to capture business at the other end of the scale—the conveyance of letters and parcels. In the field of passenger traffic, buses provided alternative transport for the less well-to-do traveler and airlines catered to the more wealthy, both at the expense of the railroad.

From 6.4 percent of the passenger-miles traveled in the United States in 1950, the railroad's share fell below 1 percent in 1970, and has never recovered. Buses, with 2 percent of the traffic, have not fared much better. Only the airlines have increased their share—but always within the limitation that 85–90 percent of passenger-miles, year in and year out, are covered by private car.

The position regarding freight movement is rather different. Although, as Table 9-1 illustrates, the railroads' share of goods traffic has fallen considerably since 1940, the actual volume of freight movement by rail is in most years double what it was in 1940. The population has increased, the whole economy has expanded, and the average length of haul is much greater. Although a considerable part of the track mileage is derelict, the remainder is busier than before; 40 percent of rail freight tonnage is coal, and the railroads handle two-thirds of all U.S. coal moving to market. However, the percentage increase for all the other modes shown in the table is much larger than for the railroads.

Intermodal competition for freight traffic has been a fact of life in North America almost from the start of settlement: overland routes versus coastal traffic, or the route round the Horn, or the Panama route; water versus rail; road versus water. In the field of bulk haulage, the 1980s witnessed a new phase of this competition between modes—created by the increase, which we have already noted in the last chapter, in coal production in western North America. If, as is foreseen, production rises in the West from its present level of 300 million tons to, say 500 million, and if coal exports rise to 150 or 200 million tons, this huge quantity of freight will have to be moved across the center of the continent to the West, South, or East Coast. There is already competition for this lucrative traffic between rail and water routes and a number of projects have been suggested for moving coal by

Table 9-1. United States: Domestic Freight Traffic, Selected Years, 1940–89 (in billion ton-miles [T-M])

	Railroads		Motor Vehicles		Inland Waterways		Oil Pipelines		Airways	
	T-M	%	T-M	%	T-M	%	T-M	%	T-M	%
1940	411	63.2	62	9.5	118	18.1	59	9.1	0.014	0.002
1950	628	57.4	172	15.8	163	14.9	129	11.8	0.318	0.029
1960	594	44.7	285	21.4	220	16.5	228	17.1	0.778	0.058
1970	768	40.0	412	21.4	307	16.0	431	22.4	3.400	0.176
1979	927	35.9	628	24.3	420	16.3	605	23.4	5.0	0.19
1985	898	37.2	600	24.9	348	14.4	562	23.3	6.4	0.26
1989	1048	37.3	712	25.3	444	15.8	597	21.2	9.8	0.35

Table 9-2. United States: National Traffic in Coal; Actual and Projected Coal Transport Mode (in million tons), 1975 and 2000

Mode	1975 Actual	2000 EEI	2000 BoM
Rail	418	608	1023
Water	69	101	170
Truck	79	115	193
Mine head use	74	107	181
Pipeline and other	8	11	19
	648	942	1586

Source: J. F. Davis, "Coal Transportation in the 1980s and 1990s," *The Environmental Professional*, vol. 4 (1982), p. 54.
EEI—Edison Electric Institute; *BoM*—Bureau of Mines.

Table 9-3(a). United States: Freight Movement on Inland Waterways, 1960, 1975, 1984, and 1988 (in billion ton-miles)

	1960	1975	1984	1988
Atlantic Coast	28.6	31.8	24.7	28.1
Gulf Coast	16.9	30.8	36.7	44.6
Pacific Coast	6.0	9.7	20.5	24.5
Mississippi River System	69.3	170.7	234.6	257.8
Great Lakes System	99.5	99.2	82.5	83.1
	220.3	342.2	399.0	438.2

Table 9-3(b). United States: Freight Movement on the Great Lakes System, 1960–84 (in billion ton-miles)

	1960	1970	1980	1984
Domestic	80.6	80.1	62.3	50.2
Foreign	18.9	34.3	33.7	32.2
With Canadian ports	13.8	23.8	23.6	19.0
With overseas ports	5.1	10.5	10.1	13.2
Total	99.5	114.5	96.0	82.4

pipeline, in the form of slurry. The present predictions for this future traffic are as seen in Table 9-2.

Waterways and Ports

North America possesses a system of extensive and well-integrated waterways (Table 9-3), thanks to a century of engineering improvements. In 1988 some 438,000 million ton-mi (700,000 million ton-km) of freight were carried over this system. Its two major components are the Great Lakes and Mississippi River with its tributaries, which together account for some 80 percent of the freight movement mentioned.

Although its relative importance in the nation's transport system has declined since the riverboat era before the Civil War, the Mississippi System has since been much extended and improved. From the ocean traffic terminals at New Orleans and Baton Rouge, 9-ft (3-m) channels now extend up the main stream as far as Minneapolis, up the Ohio and Monongahela beyond Pittsburgh, and up the Tennessee to Knoxville. The Missouri, perennially a navigator's nightmare, has a 9-ft (3-m) channel as far as Sioux City, and this can be extended in the future. The Arkansas River has a 9-ft (3-m) channel into eastern Oklahoma—450 mi (720 km) of river, with eighteen dams and locks.

Then in 1985, the system was given a kind of cutoff route when work was completed on linking the deepened Tombigbee River in Alabama with the Tennessee waterway, so halving the distance from the Gulf port of Mobile to Tennessee ports like Knoxville.

The whole system connects at New Orleans with the Intracoastal Waterway and, by way of the Illinois River and a canal across the low divide south of Lake Michigan, with the Great Lakes at Chicago.

The St. Lawrence and Great Lakes provide a natural route to the heart of the continent, but one whose present usefulness is based on improvements made by the governments of both the United States and Canada. To the west is the Sault Ste. Marie (Soo) Canal, opened in 1855 to clear the way from Lake Superior to Lake Huron and now the continent's busiest artificial waterway. Although significant amounts of western grain, coal, and timber are shipped through the canal, the bulk of this tremendous movement is represented by iron ore on its way from the Superior mines to the steel regions. To the east, the Welland Canal, completed in 1933, serves to bypass the Niagara Falls on the Canadian side and shares the strategic importance of the Soo Canal, although at present it carries less cargo.

As a result of these improvements, most of the Great Lakes route was open for ships 600 ft (182 m) and more in length, with a draught up to 30 ft (9.1 m). The weak link in the chain, however, was the section of the St. Lawrence itself between Kingston, Ontario, and Montréal, where there were still only a shallow channel and short locks. To open the deepwater system further inland to oceangoing vessels

The canal era: Building the Erie Canal. The scene is at Lockport near the western terminus. Lithograph from *Memoir of the Completion of the New York Canals, 1825,* by C. Colden. New York Public Library

was the object of the St. Lawrence Seaway, opened in 1959. Begun jointly by the Canadian and U.S. governments in 1955, the scheme provided, besides hydroelectricity for both countries, a 27-ft (8.2-m) channel from the lakes to the sea, thus making Chicago and Duluth as truly ocean ports as Montréal or New York (Fig. 9-1).

There is no doubt that, when it was completed three decades ago, the Seaway was a fine token of international effort and an important stimulus to trade and traffic on the Great Lakes. With the advantage of hindsight, however, one can only regret now that it was not built on an even larger scale. Its locks are too small to admit much of the world's oceangoing merchant fleet, in particular the ships in the bulk-carrying trades—trades that have always formed the staple traffic on the Lakes and that use the largest ships. In the west, the Soo Locks can now handle vessels up to 1000 ft (303 m) in length, with cargoes of 50–

60,000 tons.[3] The economics of lake transport being what they are, there is every incentive for the carriers to operate the largest ships possible. But these big ships cannot pass through the Seaway or, for that matter, the Welland Canal. They must spend their whole life between Superior and Buffalo.

The St. Lawrence–Great Lakes route is open for only about nine months of the year, owing to winter ice in the canals and along the lakeshores.[4] However, this navigation season is gradually being extended through improvements to the channels and the ships

[3]Up to the end of 1990, the record cargoes carried on the Great Lakes system were as follows: (1) Through the Soo Locks—iron ore, 72,117 net tons; coal, 70,706 net tons. (2) On Lake Michigan—iron ore, 81,033 net tons. By contrast, the record for the Seaway at the same date was iron ore, 31,922 net tons; coal, 31,028 net tons. (Source: Annual Report, Lake Carriers Association, 1990.)

[4]In 1990–91, the season on the Great Lakes lasted from March 20, 1990 to January 31, 1991.

Fig. 9-1. The St. Lawrence Seaway.

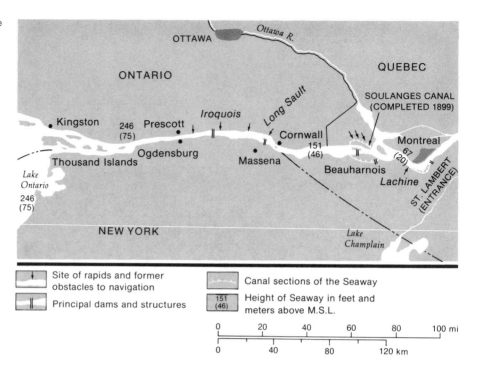

Great Lakes shipping: A bulk carrier negotiates the Poe Lock at The Soo, where shipping from Lake Superior must bypass the Falls of St. Mary to reach Lake Huron. The bridges link the United States shore (foreground) with Canada (background), crossing the shipping channel and the St. Marys River. *U.S. Army Corps of Engineers, Detroit District, Sault Ste. Marie Area Office*

Transport on the Mississippi River: Forty barges and a single pusher. *American Commercial Barge Lines, Inc.*

using them. The main effect of seasonal use has, in the past, been felt by the steel companies and other processing industries on the lakeshores, which have been obliged to stockpile iron ore or other lake-borne raw materials in order to continue operating in winter. Freedom from the necessity to stockpile would, among other things, release a large area of lake frontage in each of the port cities where the stocks are now kept. But keeping the lakes open for an extra day or an extra week is very costly and, up to the present, most of this cost has been borne by the taxpayer—who may well feel that it should fall to the user instead.

In terms of tonnage of goods carried, no other section of the waterway system can compare with the Great Lakes and the Mississippi (Table 9-3). Never-

theless, some other rivers and canals are of great local importance, especially those of the Gulf Coast. A number of the manufacturing and refining cities of the coastlands—Houston, Beaumont, Lake Charles—are linked with the sea by ship canals and cross-linked, in turn, by the Intracoastal Waterway, a route designed, when complete, to provide a sheltered 12-ft (3.7-m) channel from New Jersey to the Mexican border. Some parts of this waterway are canals in the accepted sense, whereas others make use of the lagoons and sheltered water behind the long line of offshore sandbars. The Intracoastal has become a lifeline for the industries of the "golden crescent" on the Gulf Coast; it now carries as much traffic as the Ohio River, and the number of industrial plants choosing a location on its banks is increasing.

In the Northeast, water transport plays an important part in the movement of goods between the great manufacturing cities. This water movement, however, is now essentially coastal, making use of such "shortcuts" as the Cape Cod Canal or the Chesapeake and Delaware Canal between Baltimore and Philadelphia. The historic canals of the interior, such as that from the Hudson to Lakes Ontario and Erie, whose opening revolutionized transport more than a century and a half ago, are now of little importance in comparison with the railroads and roads that parallel their course.

Elsewhere in the continent, it is interesting to find that water routes which played an important part in the movement of the earliest settlers and their goods toward their new homes, and which then lost significance as routeways, are now coming back into their own. This is true, for example, of the Kentucky River. In the West, a growing export trade in agricultural products to Japan and other Asian countries has brought cargoes of Great Plains and Palouse wheat to the Columbia–Snake system, via loading facilities at Lewiston, Idaho, and Portland at the mouth of the river.

Occasionally, however, the trend has been in the opposite direction. The Missouri, for example, has known fluctuating fortunes. Although it served many early traders as a routeway, its shifts and its sandbanks made it difficult to navigate for the larger craft of the riverboat days. Then it was regularized in the great era of dam building and channel improvement in the 1950s and 1960s (see Chapter 18), and became available for the development of a sizeable traffic. But this has not happened; despite enormous investment in these engineering works, the traffic has remained so small as to defy cost–benefit analysis, and there is talk now of cutting costs by converting the river from a commercial to a purely recreational stem.

Along the great inland water routes of North America, port cities have grown up to handle the river and lake traffic. The largest of them, however, are those that combine internal with external traffic, or river craft with oceangoing ships. As we have seen, the St. Lawrence Seaway has made foreign-trade terminals out of the Great Lakes ports, and a river port like Baton Rouge, 150 mi (240 km) above the mouth of the Mississippi, is accessible to foreign craft and handles millions of tons of overseas trade a year.

The foreign commerce of the North American countries was once directed overwhelmingly toward Europe, the source of both their immigrants and their capital for investment, and of the market for their primary produce. New York was by far the greatest port in the continent, with Philadelphia, Baltimore, and Montréal playing subordinate roles. New Orleans, with its river–sea connections, had a great past, but had also known lean times, while the ports of the Pacific were few in number, and trade with Asia was then spasmodic and unreliable.

Much has changed since the early part of the twentieth century. By tonnage handled, the largest foreign trade ports today are New York Harbor, the Delaware River (that is, the Philadelphia area), New Orleans, Norfolk–Newport News in Virginia, and Houston. But several of these totals depend on traffic in a single, bulk commodity (petroleum at Houston and coal at Newport News). However, both the direction and the nature of the traffic have changed. Canada's leading trade partner was once the United Kingdom; now it is the United States. For its part, the United States still exports more, by value, to Western Europe than to any other country, but imports more from Canada. However, the big changes have come with the shift of U.S. trade toward Latin America and Asia (Table 9-4). The first of these has greatly increased activity in the ports of the Southeast, led by New Orleans. The second has strengthened the links between Japan and Los Angeles, San Francisco, Seattle, and Vancouver; Japan now supplies more U.S. imports than the whole of Latin America. In Canada, trade with the United States, both imports and exports, eclipses all other foreign commerce, but Japan is now Canada's second-largest trading partner, ahead of Great Britain.

Throughout North America, the fast-growing ports are those that have made the swiftest adaptation to the new character of ocean traffic and, especially, to the ships that serve it—container ships and bulk carriers. Old port facilities are no longer suitable; what is needed nowadays is space to handle containers rather than warehouses, and deep water and rapid unloading facilities for the big bulk carriers of petroleum, coal, or iron ore. This revolution in cargo handling has given the opportunity to some ports to break back into business they had lost to the old, dominant ports.[5] Boston, for example, is experi-

[5]Two common reasons for transferring business to smaller ports have been (1) intensive and restrictive unionization of labor in the big old ports and (2) easier control of pilfering and damage within a smaller area.

Table 9-4. United States: Waterborne Imports and Exports, 1960–85

	Cargo (million tons)				Value (billion$)			
	1960	1970	1980	1985	1960	1970	1980	1985
Imports								
Atlantic Coast	135	219	183	190	7.8	15.0	71.5	94.4
Gulf Coast	27	47	243	141	1.3	2.9	56.4	32.8
Pacific Coast	23	31	56	51	1.6	5.5	45.0	90.4
Great Lakes System	13	26	16	17	0.5	1.3	1.9	2.8
Exports								
Atlantic Coast	45	79	117	93	7.0	11.9	51.0	35.2
Gulf Coast	37	78	163	144	3.7	6.9	41.5	31.8
Pacific Coast	19	48	78	81	1.8	4.1	25.2	25.8
Great Lakes System	23	36	45	34	0.7	1.4	4.6	2.4

Note: Values given are in current, not constant dollars.

encing a new lease on life and so, too, are some smaller ports that for long were overshadowed by New York. Along the Gulf Coast, a relatively new chain of ports handles a bulk traffic in petroleum and much else besides; the ports along the Gulf between Tampa and the Mexican border handle considerably more freight tonnage than all the ports from Boston south to Baltimore.

Railroads

In 1981, the United States had a railroad mileage of some 168,000 (269,000 km), and Canada of 57,000 mi (92,000 km). These figures were well below their peak at about the time of World War II, and have continued to fall, as little-used lines have been closed. The standard (4 ft 8.5 in or 1.43 m) gauge prevails over almost the whole of this system. The only important exception used to be 700 miles (1100 km) of 3 ft 6 in (1.06 m) gauge track in Newfoundland, but this has now been closed.

Although the system in the two countries is virtually identical, and traffic movement takes place freely across the border, subject only to customs check, the patterns of railroad ownership are quite different. In the United States, a large number of individual companies operate the system, of which 15 were categorized in 1989 as Class I railroads[6]—a number that, owing to numerous mergers and takeovers, had

fallen from 71 in 1970. In Canada, on the other hand, the bulk of the system is controlled by two great companies—the Canadian Pacific and Canadian National. The Canadian Pacific is the product of private investment backed by a government land grant (see Chapter 5); the Canadian National represents a consolidation, in 1923, of a number of earlier companies and is operated by the government. Since the Canadian National was formed to save a number of financial lost causes, it has been considerably handicapped in competing with its privately owned rival.

The railroad network as it appears today is a product of intense competition in the great railroad-building era of the second half of the nineteenth century. On the one hand, cities fought to secure railroad connections and, having succeeded, to use them to extend their spheres of economic influence. On the other hand, the railroads fought among themselves to attract the traffic. A large part of the struggle was financial rather than technical, and by the time that the scandals of the Railroad Era had attracted the attention of federal and state governments, the continent had been crisscrossed with lines, many of which could never be economically justified or, even if they could, had been built under conditions that crippled the line's working, wasted the investor's money, and

[6]The definition of a Class I railroad changes with the times. The class contains, as might be expected, the largest rail systems in the country. In 1987, a Class I company was one with annual operating revenues of $88.5 million or more. This group of railroads ac-

counted for over 90 percent of the freight tonnage moving by rail and employed about 270,000 staff.

While mergers and takeovers have greatly reduced the number of Class I railroads, they have paradoxically increased even more rapidly the number of small or shortline railroads. Most of these are abandoned branches of the main lines, which have been taken over and are operated by private interests for the benefit of local shippers, whose traffic feeds into the main lines.

enriched no one but the promoter.[7] It has been suggested[8] that, by the 1880s, the United States had twice as many railroads as the economy of that period could support. It is known that in 1876, two-fifths of all railroad bonds were in default, and it is estimated that between 1873 and 1879 investors—mainly in Europe—may have lost as much as $600 million through bankruptcy and fraud.

In North America, as in Europe, the railroads have suffered from the competition of other transport services—road, air, and water. And, as in Europe, they have met this situation with three measures: to get rid of services that do not pay, to improve those that do, and to reorganize themselves into larger groups or systems.

For most North American railroads, the least profitable part of their operations was their passenger service. In the ten years after World War II, the railroads fought hard to lure and to keep their passengers: they produced some superb trains with a wide range of facilities, but the cost was high in both staff and equipment. Gradually they lost customers, and then they were forced to reverse their policy and, where possible, to *dissuade* passengers from using these trains. The ICC, which had once been their scourge, now proved helpful in authorizing discontinuance of their passenger services. The railroads responded to the authorization with a kind of pernicious enthusiasm. Soon there were large American cities devoid of all passenger services.

So far did this decline reach from the great days of the passenger train that the North American governments felt obliged to intervene, to preserve at least a skeleton service between major cities. In the United States, a National Railroad Passenger Corporation was brought into being in 1970; it offered to relieve each railroad of its remaining obligations to run passenger trains. Most of the railroads quickly accepted and, under the name of Amtrak, the corporation now

provides virtually all the medium- and long-distance trains in the United States. Many of these services, admittedly, operate only once a day but, in areas of denser settlement (particularly in the Northeast Corridor between Boston and Washington, and between other pairs of cities like Los Angeles and San Diego), a frequent and regular service operates. Amtrak has been making progress, toward both covering its costs and attracting passengers away from overcrowded highways: one can only wish it well.

In Canada, the railroads have not given up their passengers without a struggle, and efforts have been particularly focused on providing a good, fast service along the eastern corridor route, from Québec City, through Montréal and Toronto, to London and Windsor. The government stepped in with an equivalent to Amtrak called VIA Rail Canada, but the transcontinental service, once the very epitome of railroading in its heroic era, seems doomed to extinction. In any case, the Canadian government has been chided by one of its own commissions for trying to keep rail passenger services alive by subsidy: "The study team believes that, except for a very small number of remote services (e.g., Churchill) there is no transportation need for subsidized rail passenger service because adequate lower-cost alternatives exist."

It goes on to add, severely, "Subsidies for rail passenger services as a means of achieving other, nontransport objectives have never been evaluated in terms of opportunities foregone in other social and economic sectors.[9] British Columbia and Ontario operate provincial lines and, in the densely settled area of Ontario centered upon Toronto, there is an effective suburban transit system known as the GO (Government of Ontario) Train.

A by-product of the reduction in number of passenger trains has been to make many of the great stations of a past era redundant, and to raise questions as to what is to be done with them. Many of them are architectural symbols of civic or engineering pride, and not to be lightly demolished. Perhaps the grandest of them all, Union Station in Washington, D.C., now houses a splendid shopping mall, with the trains out of sight at the back. Some, like the great station in St. Paul, Minnesota, are shopping areas without trains. Some, like Kansas City, have an elegant Union Station with only two or three operating

---

[7]We saw, in Chapter 6, how the competition between these railroads, expressed in the rate wars of earlier days, provoked government intervention in order to bring some order out of the chaos. In recent years, deregulation has been introduced, and we find that the railway wheel has come full cycle. The *Canada Yearbook* for 1990, reported (p. 13-2) that under the 1987 Canadian transport act mentioned earlier, "shippers can now negotiate confidential contracts with individual railways. Essentially, they can shop for the rates and conditions of service which suit their needs"—the very thing which, a hundred years ago, led to government intervention.

[8]By J. Moody, early historian of the American railroads, quoted in M. Josephson, *The Robber Barons* (New York, 1962), p. 292. Josephson also provided the other figures.

[9]*Economic Growth. Transportation:* A Study Team Report to the Task Force on Program Review (Ottawa, 1985). Quotations from p. 224.

tracks, over whose future the locals are still disputing. And some are like La Salle Street in Chicago, a station which once saw the daily departure of a dozen trains to New York, including the Twentieth Century Limited, but which is now confined to sending suburban trains out to Joliet, a mere 48 miles (77 km) away.

To keep their freight services, by contrast, the railroads have fought hard and well. They have accepted the fact that road and pipeline transport have come to stay, and have produced integrated services—as in the piggyback trains that carry semitrailers and containers, to be interchanged with trucks. They switched from steam to diesel for greater speed and efficiency (although a few sections of line are electrified). They have closed local lines and services as rapidly as the Interstate Commerce Commission, over the protests from small towns and businesses along the routes to be abandoned, would allow. And the railroad companies themselves have sought—by mergers for greater efficiency and by financing their own road transport and pipeline operations—to beat their competitors by joining them. Beginning in the 1960s, they began to diversify their services into other forms of transport and operation, just as the Canadian Pacific, long before, had branched out into transatlantic shipping and then into airlines.

The density of the railroad network corresponds very fairly with the distribution of population, with the same marked change, about the 98th meridian, from a closely covered East to a sparsely covered West, that we have already noted in other connections. Crossing the emptier West there were in their heyday nine transcontinental routes, two in Canada and seven in the United States (Fig. 9-2). The focus of the U.S. routes and the unchallenged railway center of the country is Chicago. Although other cities may be more exclusively concerned with railroading, none can rival the tremendous concentration of major railroads and track that still characterize the Chicago "terminal district."

From both Canada and the United States, railroad ownership and operation extend across the border. Conrail, whose main line runs south of the Great Lakes, from Chicago via Toledo and Cleveland, has an alternative, Canadian route via Detroit and the northern lakeshore to Buffalo, over which its trains pass in bond. From the Canadian side, in order to reach an ice-free port for winter use, the Canadian National Railways owns a line that cuts across the

United States to the Atlantic coast at Portland, Maine.

Despite the fluctuating fortunes of the American railroads, business has remained fairly steady on the commuting services around the few larger cities which possess them. Not only this, but rail technology has contributed to a whole series of newly built rapid transit systems. City after city has been pressing ahead with services which generally run underground in the city center, and above ground in the suburbs. The object is, of course, to ease road traffic problems. In New York City, with its islands and rivers, this is how the parts of the city have for years been linked together. On the Pacific coast, the city that most resembles New York in being divided up by water and linked with bridges—San Francisco—has developed the Bay Area Rapid Transit (BART) network. And now Montréal and Toronto, Washington, D.C. and Atlanta, none of which had the same problems of topography, but all of which had too much road traffic, have followed suit.

Roads

As North America took the lead in the production of motor vehicles, so it had become also the first continent to create a network of roads designed specifically to take advantage of this new form of transport. Since Canada and the United States between them account for one in three of the world's vehicle registrations, the task of providing for and controlling this huge volume of traffic becomes a problem of the greatest national importance.

Responsibility for this network and its maintenance is divided, as in Europe, among various bodies. In Canada, virtually the whole responsibility lies with the provinces, or, in the case of city streets, with the municipalities. In the United States, however, the federal government has established a basic network of principal routes for which it is responsible—some 200,000 mi (320,000 km) in all. The states develop their own systems within this framework, and, as in Canada, the municipalities are responsible for the upkeep of streets.

The establishment of this network has been no easy task, considering the natural hazards involved—lakes and muskegs in the North, mountains and deserts in the West, swamps along the southern coasts, and extremes of heat and cold that damage road surfaces. There could be no better expression of the economic meaning of federalism than the hun-

Fig. 9-2. Transcontinental railroads of North America: The pre-World War II network in the West. Note the convergence of lines on port cities, such as Vancouver and Chicago; on the "Hinge" or linkage cities east of the Great Plains, such as Winnipeg or Kansas City; and on mountain pass cities, such as Spokane or El Paso.

dreds of miles of expensive roads that crisscross the West and the Great Plains, roads that have little local use, but that are justified by and built for the through traffic they carry across the nation.

The network is not yet complete. There are still some links missing in the United States, and in Canada there is much to be done. The Trans-Canada Highway was only completed in 1962, and in the North, there is a vast wilderness to conquer. Only a small proportion of northern Canadian roads are surfaced with asphalt or concrete.

But in a wider sense, this road network is incomplete because the volume of traffic is constantly overtaking the rate of construction. Highways built a decade ago have been superseded by superhighways with double the traffic capacity. Around such cities as

New York and Los Angeles, the congestion of arterial roads has become intense, and finding space for these broad highways is a problem that is often solved by carrying the new artery above or below the old street level. But to a visitor returning to America after an absence of ten or fifteen years, the roads that have been built during the interval are always a most striking feature of the landscape.

Recognizing this situation, the U.S. government undertook the building of a completely new strategic network of Interstate expressways, 41,000 mi (65,000 km) in length. It is all but complete; however, it has taken so long to build that many sections of the earlier Interstates are in serious need of repair, long before the final links are completed.

In the meantime, the car placed its mark on North

Transcontinental roads: The Trans-Canada Highway passes through Banff National Park. *Alberta Tourism*

American life, and especially urban life, to which we referred in Chapter 3. Its two main effects can best be labeled *destruction* and *reorientation*. In this context, destruction is the process whereby the centers of cities are increasingly given over to thruways and parking lots: more and more of the available area is pre-empted by motor vehicles and less space, in consequence, is available for all other urban functions. In Los Angeles, which is usually regarded as the most outstanding example of this process at work, almost one-half of the surface area of the city center is dedicated to motor vehicles, moving or parked. Not only spatially, but also socially, this type of land use is disruptive. The modern, limited-access freeway, which is barred to pedestrians and of no use for purely local movement, may represent a barrier as socially uncrossable as the Great Wall of China, isolating sectors of the city and blighting areas along its course. New York City is a prime example of this (Fig. 9-3).

At a certain point, in most cities, there occurs a re-

versal of this process—the reorientation of movement and of business toward the suburbs, as we saw in Chapter 3. A city center that is merely a crossroads is no center at all, so far as life and work are concerned. At the same time, without universal access to cars such a reorientation would not have been possible. In the most direct sense, the car has created the pattern of the suburb and its most distinctive feature, the commercial strip.

It used to be said of the gauchos in Argentina that when they got up in the morning, they first mounted their horse, and then thought what to do next. Visitors cannot help feeling that suburban America has merely substituted the car for the horse, and the strip for the Pampas. To adapt the layout of cities and the distribution of economic activity to the motor age has been, over the past four decades, one of the most formidable tasks confronting planners and councils in North America. Fortunately, the time when the automobile carried all before it, including historic build-

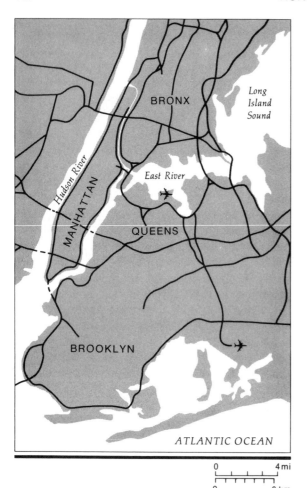

Fig. 9-3. The New York Metropolitan Region: The expressways, status in 1991.

ings and integrated communities, seems to be over and to have been replaced by the recognition that, even in the 1990s, there are other goals in life than the clearing of broader pathways for the motor vehicle.

Airlines

From the time air travel first became a commercial possibility, conditions in North America favored its development. This method of travel is well suited to two countries whose area is enormous and whose parts are separated from each other by considerable natural barriers, but whose administrative and commercial structure also makes rapid communication between regions essential. Once a network of principal routes had been created, no city of any size

could afford to be isolated from it, so that municipal airport construction was pushed ahead, and "feeder" lines to the main routes were developed.

North America today accounts for probably a third of the world's air traffic. The U.S. passenger volume has increased from 153 million passengers emplaned in 1970 to 453 million in 1989. But these figures mask the traumas endured by the major airlines in the interval. One of these was the huge increase in the price of fuel in the mid-1970s, and another was the deregulation of air services by the U.S. government, which threw the whole network of routes open to competition, trial and, inevitably, error. The third and continuing trauma is the problem of controlling this movement by air, especially around major hubs, where overcrowding occurs both in the air and on the ground.

Thanks largely to deregulation of the industry, however, the fundamental problem faced by the airlines˙is simply how to stay in business. Since 1978, with its abrupt flurry of competition, several of the oldest-established names in air travel have disappeared from the list of carriers. Their routes have been taken over by survivor companies, which themselves have in some cases survived only by amalgamation. Fewer flights are made, but at least those flights are commonly filled to the doors.

While the long-distance airlines have been wrestling with these problems, there has been a quiet, but real, revolution going on around them. This has been the rise of the regional or "commuter" airlines. Whatever overall increases the industry has achieved in carrying passengers, the commuter lines have set their own, higher marks. Between 1975 and 1985, the commuter lines of the United States increased their passenger load four times. They increased the total of their revenue passenger-miles six times.[10] They are evidently inculcating—successfully—the idea that it is better to fly than to drive to the major airport to make connections. In Canada, the experience of the regional air carriers is similar. Although, between 1981 and 1983, the number of passengers carried by the major lines actually fell, the regionals managed a modest increase in their business.

Within North America there has grown up a hierarchy of airline service centers, with a group of primary foci 500–1000 mi (800–1600 km) apart, and a

[10]Statistics for the commuter airlines for 1989 (1980 figures in parentheses for comparison) were: number of airlines, 151 (214); passengers emplaned, 37.4 (14.8) million; revenue passenger miles, 6.77 (1.92) billion.

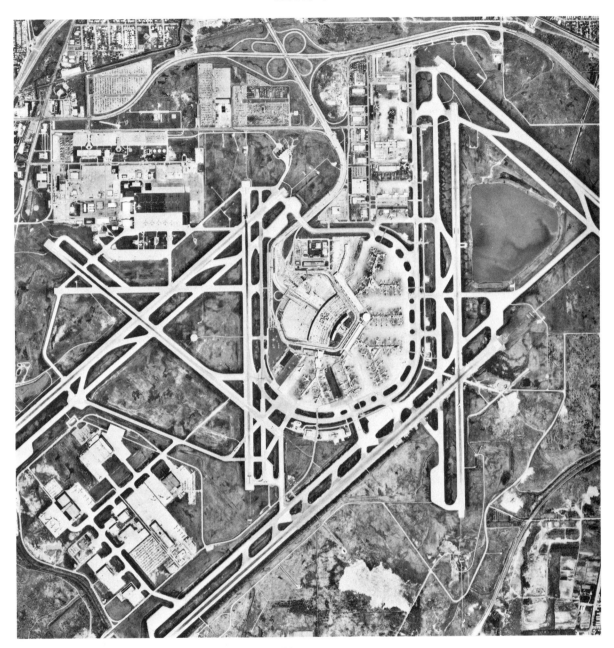

Chicago: O'Hare Airport, the world's busiest (east is at the top of the picture). Since this photo was taken, a large new international terminal has been added to the central complex at its eastern end.
O'Hare Associates

number of ranks of airports below them, which can be graded by frequency and direction of service. Of the dozen airports in the world that handle most passengers, nine are in North America: they include not only New York, Chicago, Los Angeles, and Washington, as one would expect, but also several of the other first-rank foci: Atlanta, Miami, and Dallas. If it does

nothing else (and even if it changes with time, as it is bound to do), this list shows that air travel has created a hierarchy of centers that is very different from that of the railway era.

The development of this hierarchy may draw our attention to a general phenomenon of recent North American geography—the growth of what are now

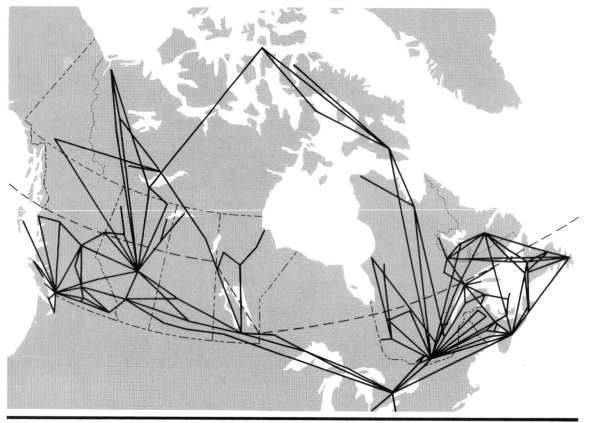

Fig. 9-4. Canada: Regional air carrier routes, 1981. It is the aim of the Canadian government to ensure that every community in the North-lands with more than 100 inhabitants shall be accessible by air. (Re-produced by permission from J. L. Courtney, "Regional Air Carriers in Canada: Network Evolution and Government Policy," *Canadian Geographer*, vol. 29 [1985], 4–16.)

called *load centers*, or transport foci at which goods are gathered and interchanged between air, road, and water carriers. A city like Atlanta, Georgia, is today not only one of the busiest air foci in the world; it is also home to a huge trucking fleet, and possesses rail and waterway connections into the bargain. Freight shippers working via the load center have never been so spoiled for choice as they are today in North America.[11]

Air transport, both passenger and freight, has become a factor of considerable importance in the location decisions of business and industry. For passenger movements, "several flights a day direct to Chicago"—or Toronto, or Dallas—has become a powerful argument for choosing a city as home base.

Thousands of business conferences in Chicago take place, not in the downtown area, the "Loop," but rather in one of the maze of offices, hotels, or restaurants that surround O'Hare Airport in an expanding zone shared by warehouses and light industries that depend upon airborne freight.

In the era of jet planes, when Chicago is less than two hours by air from New York (the train still takes 16–17 hours, overnight) and Denver a further two hours from there, the continent's dimensions have shrunk so remarkably that a whole new geography of economic and social activity has sprung up. Patterns of management, servicing, and supply have altered; air route centers have left their rivals behind; the concept of commuting has been revolutionized.

Apart from these general traffic movements, air transport plays a particularly important part in North American affairs in two special areas:

[11]See, for example, B. Slack, "Intermodal Transportation in North America and the Development of Inland Load Centers," *Professional Geographer*, vol. 42 (1990), 72–83.

1. In the opening up of areas of the remote Northland (Fig. 9-4), both the initial surveys of the region and the supply of pioneer settlements were entrusted to air services. Before the opening of the railway to Sept Iles, for example, the whole development of the Labrador iron ore field was based on air supply. And on the basis of per capita miles flown, Alaskans use air travel far more than does the average American in the lower forty-eight. As one indication of this habit, Alaska possesses well over 600 public airports.

2. In industry, use of company planes has become commonplace, especially when a firm has branch plants in several regions. And on the farm, aircraft are used not only to spray crops and to sow seed, but also to carry the farmer from point to point to control operations. To the increase in scheduled passenger traffic, therefore, must be added a comparable rise in private plane movements, and these latter more than double the total number of plane-hours with which American traffic control has to cope.

10
Regions and Regionalism

New York: In the financial district. *Steve Benbow/Woodfin Camp*

In every book on regional geography, there comes a point where the writer must commit himself or herself to a scheme of regional subdivision, and justify the choice. That point has now been reached, and this chapter serves as an introduction to the studies of individual regions that follow it. But let us begin by asking: Why is there regional variety in the first place? Is it inevitable and, if so, what kind of factors cause it?

Bases of Regional Diversity

First and most obviously, it is a product of physical variety: of latitude and aspect; of the uneven distribution of fertile soils or smooth relief; of minerals, forests, or water. It would be surprising if it were not so—if a continent that covers more than 8 million sq mi (more than 20 million sq km) of desert and ice cap, plain and mountain, had developed in all its parts at the same pace or, for that matter, under the hand of the same sort of people. In some areas, the occurrence of one resource compensates for the lack of another: the Canadian Shield has minerals and forest cover, but little soil, and the U.S. Southwest has sunshine, but little rain. There is, however, nothing assured about this compensation; nothing, from this point of view, to prevent some areas being disadvantaged beyond the power of any present technology to alter their status.

Once society has organized itself, another factor comes into play—the position of an individual region within the continental or the national unit. And the larger the unit, the more significant does position become, if only because, over longer distances, transfer costs from one region to another become an ever larger item in the cost budget. (We have already noted, in Chapter 6, a recognition of the importance of these transfer costs, to and from a peripheral region, when the Canadian government granted the Maritime Provinces a 20 percent subsidy on freight rates.)

Position must, of course, be judged not only with reference to the nation itself, and its core or capital, but also to other continents and nations. In North America, there are outward-looking and inward-looking regions. The eastern coast has, over the years, looked outward mainly toward Europe, the southern coast to Latin America, and the western coast now looks, to a far greater degree than ever before, toward Asia. The interior of the continent, by contrast, has tended to be inward-looking—isolationist is the politician's word—and still is, but only to the extent that most people's economic concerns there are represented by Chicago or Toronto, Winnipeg or St. Louis, rather than by overseas trading partners.

Positional factors alter with time and market relationships. When European settlement was confined to the Atlantic Seaboard, southern New England lay close to the center of colonial life. But, with the opening of the West, the nation's center of gravity moved away: the once central position became increasingly peripheral. On the other hand, the Pacific Northwest—Oregon, Washington, and British Columbia—once isolated beyond its mountain barriers, and with a permanent problem of transfer costs to and from the rest of the continent, has been steadily increasing its positional advantages as trade between North America and Japan has mounted to unprecedented levels in the last twenty years.

Next, there is the cultural factor to consider.[1] The Anglo-American realm was settled and developed by at least three major European powers, operating within a continent that already possessed distinctive

[1]See W. Zelinsky, *The Cultural Geography of the United States* (Englewood Cliffs, N.J., 1973); and R. D. Gastil, *Cultural Regions of the United States* (Seattle and London, 1975).

native cultures. From Africa was brought yet another group of totally different cultural background. It is not surprising that some of these cultural traits have persisted to form the basis of regional identities. Indeed, the past two decades have seen a conscious strengthening of these regional traits through the media of language, race, and politics: we have Black Power, Brown Power, and Red Power, as well as the movement for the liberation of Québec. The Spanish influence in architecture or the French influence in law and education are quite sufficient of themselves to generate diversity, even without their being consciously fostered by culture-based groups.

Cultural distinctiveness, in fact, can be either deliberate or enforced. It was enforced upon the American Indians by their being confined to the reservations and upon the blacks by a policy that segregated them as a racially distinct minority. But it is the commoner case today that cultural distinctiveness is sought and magnified, even by those groups on whom it was originally forced. Making a virtue out of past necessities, the minority cultures have asserted themselves: the last thing they desire is to merge and disappear into a uniform Canadian or American culture. Here, then, is a factor that tends to increase rather than to diminish cultural diversity with time.

A fourth factor in regional diversity is amenity, but this requires explanation. In a community living near the level of subsistence and threatened by periodic famine, there is little opportunity for individuals to choose either life-style or occupation; generally, there is only one way to stay alive, and this is forced upon the members of such a community. By the same token, the individuals usually have little choice as to where they may live. But in a community grown prosperous enough to have built a safe margin of surplus between itself and starvation, choice becomes possible: the productive capacity of only a fraction of the population is absorbed in providing the basic necessities of life, and a wide range of choices is possible for the remainder of the community as to what they will do and where they will do it. In practical terms, given such facilities as telephones, cars, and retirement pensions, a considerable proportion of all North Americans can choose where in the continent they will live. Even though they may not make the most of their earning power by their choice, they can make the most of amenity—by settling in a pleasant climate or in an uncrowded location or where they can best carry on their own interests.

In this way, in societies with a high standard of liv-

ing, high-amenity areas tend to attract and low-amenity areas to repel. Since World War II, as we have already noted in Chapter 2, millions of Americans have expressed their amenity preferences by moving to areas of sunshine and mild winters: Florida, Oregon, the Southwest, and coastal British Columbia are today high-amenity areas. More recently, amenity considerations have been leading people to settle near or in the mountains (growth of interest in winter sports has been phenomenal); for example, in Colorado. As it happens, the Colorado Rockies adjoin one of the areas of lowest amenity estimation in today's America—the Great Plains. In consequence, the Plains have been losing population steadily, and so offer us the curious paradox of empty plains and peopled mountains in close proximity.

Use of a term like "amenity estimation" or "amenity perception" reminds us, of course, that what we are considering are individual choices, made by people free to move on the basis of their assessment of where their values may best be satisfied. Consequently, we are measuring not objective, but perceived qualities: the perception, in fact, becomes more important than the reality. It may also vary with the passage of time.[2] Once a region has established a reputation for high amenity, it will attract individuals and may continue to do so (if southern California is anything to judge by) long after the influx of newcomers has gone far toward destroying the very amenity that was the original attraction. It is these perceptions, or "mental maps," that have been explored by geographers with increasing thoroughness in recent years.

So far, we have considered the physical, positional, cultural, and amenity bases of regional diversity. But these four factors are all, in a sense, preliminary to the fifth, which tends to reflect the influence of them all—the economic factor. Some regions advance economically, while others stagnate. Rich and poor regions exist alongside each other, and react upon one another. Income levels vary from region to region; so, too, do the cost and standard of living. In the two large countries of North America, both free from internal barriers to trade or movement, how does this come about? Why should per capita income in Alberta in 1980 have been 112 percent of the national average, and that of Newfoundland 64 per-

[2] For an example of such variation, see K. Thompson, "Insalubrious California: Perception and Reality," *Annals,* Assoc. Amer. Geog., vol. 59 (1969), 50–64.

cent? Why is per capita income in Alaska virtually double that of Mississippi?

The first step toward an answer is to recognize that incomes—which are, however, only one regional indicator out of many that might be used—vary significantly with (1) urban and, especially, metropolitan population (high regional incomes correlate with substantial urban growth); (2) percentage of non-white population (minority incomes are generally lower than white levels); and (3) the amount and type of industry present in the region. The second step is then to ask how disparities in these variables have come about.[3]

Any explanation should take into account two groups of factors, which we may call *legacies* and *stimuli*. Among the legacies that may be responsible for holding back regional prosperity are environmental poverty, which limits the range of options open to a region; past failure to invest sufficient social capital, especially in education; and a tradition of segregation or discrimination, which may artificially depress living standards for a minority group. In the Old South, as we shall see in Chapter 15, there was no environmental poverty (or if there was, it was induced only after settlement began), but the other two conditions were certainly present: educational spending was low and the black community was not exposed even to the little education the South could provide, nor was it given the opportunity to raise its own standard of living.

Yet, other regions with similarly unfavorable legacies from the past have succeeded in "taking off": indeed, the New South has done so. Such regions succeed because their economies receive a powerful stimulus—or, if the region is fortunate, a whole series of stimuli—from which they gain a regional momentum. The initial stimulus may be of several kinds. One kind would be an abrupt lowering of transfer costs within or across the region; for example, by the construction of an Erie Canal, or a railroad into Labrador. Or the stimulus may be the emergence of a single industry, but an industry of the kind some economists describe as "propulsive"; that is, an industry large in scale, fast growing, and related to a

number of other industries by the need to supply parts or to service the resultant product. In the North American experience, two great propulsive industries of the past, both of which fulfilled these conditions, have been the motor vehicle industry and the petroleum industry. What the latter has done for Texas and the Gulf Coast by way of regional "take-off" we shall consider in Chapter 17. Today, the aerospace industry is doing for some regions—particularly southern California—much the same thing that the automobile industry did for Michigan at the turn of the century. Or again, the stimulus may consist simply of the organized and large-scale export of a region's single resource—timber or coal or wheat—provided that capital is attracted to the region and reinvested in further exploitation of the resource.

There is, of course, nothing to prevent a government from creating an artificial stimulus. Within a region that lacks, say, a propulsive industry, it is possible to create a "pole of growth"—a center where investment is concentrated and around which new occupational or industrial opportunities may become available. In a sense this is a "propulsive place," rather than a propulsive industry, but the impact can and should be much the same in both cases.

We shall return in the next section to this topic of government policy toward regions. For the moment, let us note simply that regional diversity in North America, as elsewhere, is the result of many forces working upon the landscape, and amidst that diversity of landscapes we can look for the generalizations that we call regions. In the usually accepted definition, a region is an area homogeneous with respect to defined criteria. It is a form of general statement about landscapes and their natural or human features, and it is for valid statements of this kind that we are looking in Chapters 11–25 of this book.

So there is no shortage of variety among regions, but what is *regionalism?* It is the conscious subdivision, for whatever purpose, of a whole into parts; the identification of less-than-continental, or less-than-national, patterns that are clear enough, and significant enough, to be perceived. Out of those perceptions arise communities of interest—groups of producers, or consumers, or voters, who realize that their needs and problems are common within their group, or commonly threatened from outside it.

Regionalism, then, implies purpose—purposeful community of interest. That, in fact, is too weak a definition for the kind of regionalism that led to secession by the states of the Old South in 1861, and to

---

[3]On regional disparities of this kind, see R. L. Morrill and E. H. Wohlenberg, *The Geography of Poverty in the United States* (New York, 1971) and D. M. Smith, *The Geography of Social Well-Being* (New York, 1973). For a slightly different viewpoint on the same subject, see J. Agnew, *The United States in the World-Economy: A Regional Geography* (Cambridge, England, 1987), especially Chaps. 3 and 4.

four years of armed resistance to the extinction (as they saw it) of their regional interests. Regionalism, when it reaches the level of consciousness, can become a potent force, and there is generally somebody on hand with a vested interest in seeing that it *does* reach that level: that it is strong enough to change a law, or secure a subsidy, or overthrow a government. Regionalism may continue for years at the unconscious level and then, quite abruptly, be brought to the surface by a particular issue—by the sudden realization that there exists a common threat, or opportunity, which calls for a common response.

But this raises a problem. Unlike the concept of *region*, which emerges out of what we may call landscape and life—out of natural features, and the distinctive use made of them by man—and which, consequently, is a relatively static concept, *regionalism* may be quite transitory. A community of interest appears; then the opportunity or the threat that evoked it passes, and the consciousness of unity fades once more. Regionalism, in fact, is more than a simple act of subdivision. It belongs to those states of group consciousness to which nationalism also belongs. They wax and wane, usually strongest when most endangered and, in the meantime, taken for granted; sometimes presumed upon; sometimes proving illusory. At the beginning of Chapter 9, we noted an example of this illusory regional solidarity—the belief of Southerners that the Midwest of 1860 had more in common with the South than with the North. When the Civil War came, that illusion was quickly shattered. A regionalism that has not been recently tested may, in fact, have ceased to exist.

Of course, regionalism is not a matter of consciousness alone, but also of the organization of society and the circulation patterns of its life. Awareness of community of interest must depend on movement, contact, exchange of information, demand for goods. Just as nations may be bound together by transport ties—and the transcontinental railways in North America were avowedly built with this in mind—so a region may have its regional consciousness aroused by the spread of communications and enhanced by their improvement; by the impact of television or the dominance over its business life of a central city. Further extensions may lead to the enlargement of a particular regionalism, or to competition between two or more centers for the regional loyalty of unclaimed territory. In Canada, the frontiersmen of the Peace River Country stand between the regionalisms of the Prairies and the Pacific Coast, inhabitants of a kind of Debatable Land. In the United States, the Mountain and Desert West lies open, to be drawn into the fiefs of Denver or Phoenix, Salt Lake City or Los Angeles, according to the competing forces of metropolitan attraction.

It is a commonplace today to stress the importance in regional consciousness of the metropolitan center. The metropolis may indeed emerge as the embodiment of regional spirit, of its cultural aspirations or its pride. That this is true of the cities of Texas is hardly open to dispute. But although the development of the media has opened up new dimensions of metropolitan dominance, the idea of the region dependent upon the urban focus is not in itself new. We need only to recall the intense rivalries in the past between cities; the profound conviction that this city or that was the most impressive, or the most influential, in the nation; the belief that one particular city on the East Coast—New York or Boston, Charleston or Savannah—was destined to become the gateway to the interior. Regions of eighteenth-century America had their urban focus, no less than their twentieth-century counterparts.

What we have so far considered about regionalism would be true of any area sufficiently large to display regional variety in its makeup. It was true, in the earliest colonial days, of the division between Tidewater and Back Country in Virginia. It was true, on a far larger scale, of the contrast between the East and the frontier, as shown by the endless arguments in Washington and Ottawa between the representatives of each. Whether we label it regionalism or *sectionalism*—which, as it happens, has been the more usual term in North America—it draws attention to the same community of interest among the peoples of a particular part of this large continent.

It does so, especially, because both Canada and the United States are governed, under a federal consitution, within a system of provinces or states. The whole point of federalism is to allow *local* interests to influence life and welfare. And so they may—so long, at least, as the local units of government correspond to the interest groups affected: so long as the provinces or states have been created with this correspondence in mind. But in North America, for the most part, they have not.

Almost every province and state has at least some part of its boundary defined not by nature but by geometry—a line of latitude or longitude. Some of them—Wyoming, Utah, Colorado, Saskatchewan—have their entire boundaries drawn in this way; nature plays no part at all. The best-known piece of all North American geometry—the 49th parallel, which marks the western half of the United States' boundary with Canada—also provides the best example of this non-correspondence between geography and community. At its western end, it crosses the sea to chip off the tip of a peninsula (Point Roberts), which reaches southward just across the line, and so forms part of the United States, even though there is no way of reaching it except through Canada.

From a geographical point of view, in other words, most of these ten provinces and fifty states are artificial units. They do not correspond to the great realities of American life—river basins, plains, or mountain ranges. In many of them, two or more major types of environment exist side by side, with their economies and interests almost wholly opposed to one another. In Colorado, for example, the eastern half of the state could hardly be flatter. It has interests in irrigation and grain farming, a generally declining population, and a year-long fear of drought or blizzard. The western half could hardly be more rugged. It contains the principal ranges of the Rockies, backed by high grazing land; its interests are in forestry, ranching, mining, and increasingly, winter sports, and its population is growing. It is therefore quite possible to argue that the interests of these two regions of the state of Colorado would be better served by its partition between two other states: if the eastern, or Great Plains, section were attached to neighboring Kansas or Nebraska (with which it shares an interest in wheat prices and soil erosion), while the western, or mountain, section allied itself to Utah or Wyoming and made common cause over grazing permits or ski trails. Seldom a year goes by, in fact, when proposals of this kind do not appear. Not only the nongeographical shapes of the states, but the wider discrepancies in their areas and populations, fuel this discussion. Surely, it is felt, there must be some better way of governing a country than by according equal status to California (1990 population: 29.8 million) and Wyoming (453,000), or Prince Edward Island (130,000) and Ontario (10.1 million).

Few provinces or states, in fact, represent a single regional interest. Most of them are either too large or too small to fit the regionalisms of the day. Generations of geographers, economists, and politicians have therefore proposed scrapping the present framework, and reorganizing it along the lines of the nations' contemporary regionalism.[4] But this is not an easy game to play. For one thing, as we have already seen, regionalisms change as time goes by, and who is to say whether a reorganization that today seems sensible will so appear in ten or twenty years' time? For another, there is no reason to suppose that every part of North America can be associated with an identifiable region. We cannot demand this. And then there is a third thing: the political impossibility of carrying out such a reorganization over the ferocious opposition of the dispossessed (for it must be taken for granted that some of the smaller units would be absorbed in the larger). On the whole, it is simpler and safer to leave things as they are, and let the provincial and state governments make their own *ad hoc* arrangements for mergers or mutual aid on particular issues or occasions.

Government and Region

This last comment leads us on to our next topic: the interest or involvement of central government in regions. This interest is taken for granted in Europe but, in North America, it is a relatively recent development. In either case, it has usually been aroused through a growing awareness that some parts of the nation are faring badly compared with others; that a geography of poverty has become recognizable, as between region and region. Wherever a poor region has a political voice, it will demand attention. The central government, for its part, is likely to respond, partly out of concern for its own image when such backwardness is brought to public attention, and partly because there is a fear that a region of poverty may harm the well-being of a prosperous region nearby.

Over the past half-century, we have witnessed the North American governments moving into the field of regional aid programs, and then retreating again. One of the first of these programs that transcended the boundaries of a single state, and is probably still the best known, was the Tennessee Valley Authority of 1933, described in Chapter 16. What more natural regionalism than that of a river basin? But the TVA remains, of course, one of a kind. As a form of creative regionalism it seemed, in those far-off days, too

---

[4]See, as examples, G. Etzel Pearcy, *Thirty-Eight States U.S.A.* (Fullerton, Calif., 1973); and J. Garreau, *The Nine Nations of North America* (Boston, 1981).

avant garde to be politically acceptable, and the plans for a comparable authority on the Missouri or the Columbia never materialized. Partly this was due, no doubt, to the shocking spectacle (shocking, that is, to the eyes of conservative Americans in the 1930s) of the federal government, the government of Washington and Lincoln, brewing hydroelectricity in a tidy witch's cauldron called Norris Dam. But partly, also, it was due to a revolt against the idea that the state could in some way be superseded, or bypassed, by this new regional entity. In geography, the natural unit might be the river basin but, in politics, it was the state, and let no one forget it.

In those early modern days, then, state and region were in opposition. It was not, in fact, until the 1960s that regionalism surfaced once more but, when it did, its scale was altogether new and larger. The governments of both Canada and the United States brought in the new decade with a series of regional development programs (Fig. 10-1). In Canada, 1961 saw the passing of the Agriculture and Rural Development Act (ARDA); 1962, the Atlantic Development Board, to help the perennially lagging Atlantic Provinces; 1966, the Fund for Rural Economic Development (FRED); and 1969 the founding of a Department of Regional Economic Expansion (DREE).

In the United States, the legislative blizzard blew in the same years, to give the nation its Area Redevelopment Act (ARA) of 1961, the Economic Development Administration of 1965, and the Appalachian Regional Development Act of the same year. One effect of this blizzard was that, by 1973, there were 1818 designated Redevelopment Areas in the United States, and they contained 100 million people; that is, the ratio of those inside the area receiving aid to those outside paying for it was virtually one to one. It was a case of "everybody develop somebody."

The 1960s, then, were for both countries a decade of experiment in regional programs. Of these programs, much of the fabric remains in Canada: the Atlantic Provinces, in particular, have benefitted through the funding of many development projects, and in 1982 DREE was given new resources, to promote industrial development in all the regions of Canada. In the United States, however, much of the structure erected in the 1960s has disappeared, leaving the TVA, which has survived it all, and the Appalachian Regional Commission, which, after not one but several closure notices, was still receiving funds in 1992.[5] It is not that the regionalist idea has proved itself irrelevant on the American scene: "irrelevant" would be a poor word to describe the impact of several billion dollars of federal spending (Fig. 10-2). Nor was it opposition, this time, from the states that halted regionalism, for the very good reason that, in its 1960s legislation, the U.S. federal government took the states into partnership rather than making any attempt to bypass them and deal with the regions direct. (But this, as we shall see in Chapter 16, does not necessarily make for good regional planning.)

If there has been a retreat from regionalism, it can be traced to several causes:

1. *Genuine uncertainty as to whether regional aid programs are legitimate within a free-market economy.* There can be no doubt that to subsidize incomes, or production, or infrastructure, in a specific region *is* an interference with the free flow of goods and the free distribution of jobs and capital. It may benefit the region subsidized but, on the national level, a government pursuing such a policy may end up with nothing to show for it but a lot of high-cost production facilities at the end of a road to nowhere. At best, encouraging development in a poor region (where labor is presumably cheap) may create competition for a wealthier region next door, and so reduce its prosperity.

2. *Genuine uncertainty about the impact of rich and poor regions on one another* (Fig. 10-3). Will the rich region polarize growth and wealth, drawing off from the poor region what little enterprise, or capital, or skilled labor, it still possesses? Or will the growth in the rich "trickle down" to the poor: will activity overspill in time from the rich region, to bring employment to the underemployed?[6] On a smaller scale, is it better to choose, and subsidize, "growth poles" at the expense of the rest of a region, or to spread regional aid evenly across the whole area? Most of the evi-

[5]For details of the Canadian and U.S. regional initiatives during this period, two readily available sources are: P. Guinness and M. Bradshaw, *North America: A Human Geography* (London, 1985), Chap. 5; and J. W. House, ed., *United States Public Policy: A Geographical View* (Oxford, 1983), Chap. 2.

For further, up-to-date detail, see also M. Bradshaw, *The Appalachian Regional Commission* (Lexington, Ky., 1992); and J. B. Carmon, "Directions in Canadian Regional Policy," *The Canadian Geographer*, vol. 33 (1989), 230–39.

[6]The terms "polarize" and "trickle down" are borrowed from one of the protagonists in the long debate over regional economic theory, A. O. Hirschman, *The Strategy of Economic Development* (New Haven, Conn., 1958), Chap. 10.

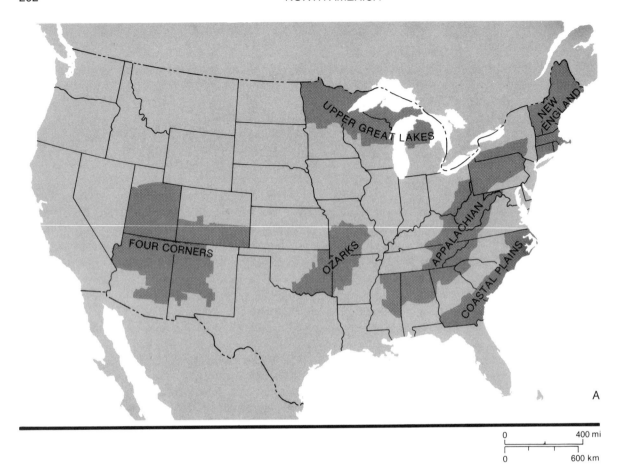

0   400 mi

0   600 km

Fig. 10-1. North American governmental regional intervention. Maps **A** and **B** show the situation in the United States and Canada in the late 1960s, when regional assistance programs were in fashion. The

U.S. program established the major areas of regional need shown in **A**: of these, only Appalachia (see later text) survives with a funded program. **B** shows the Canadian development region in 1969: the

dence upon which a huge superstructure of regional science has been built is to be found not in North America but in the Third World or in Europe: for example, in the relationship between northern and southern Italy, or Ireland and England.

3. *The structure of North American government and the financing of regional programs.* In Ottawa and Washington, where the money comes from, interest groups and lobbies were not organized on the necessary regional lines. Americans understand lobbying by states, or by businessmen; each lobby is identifiable; indeed, officially registered. But who was to lobby for, say, Appalachia? The road builders and coal miners working there? But Appalachia *as such* meant nothing to them: it was their jobs and their industry for which they would campaign. The states involved? But all those states, with the single exception of West Virginia, had a non-Appalachian section

whose interests must be considered. In the final analysis, their non-Appalachian population was being taxed to aid the Appalachian section, which was enough to deter any sensible politician from going overboard for regional development.

Whatever the reasons, then, the 1960s were a decade of political regionalism and the 1980s, on the whole, were not. The blizzard died down without any positive answer to the two basic questions confronting regional policy making everywhere: Should it be made at all, and What is the best way of making it?

A Regional Framework

The usual definition of a region, as we saw earlier, is that it is *an area homogeneous with respect to defined criteria.* But this definition raises at least three ques-

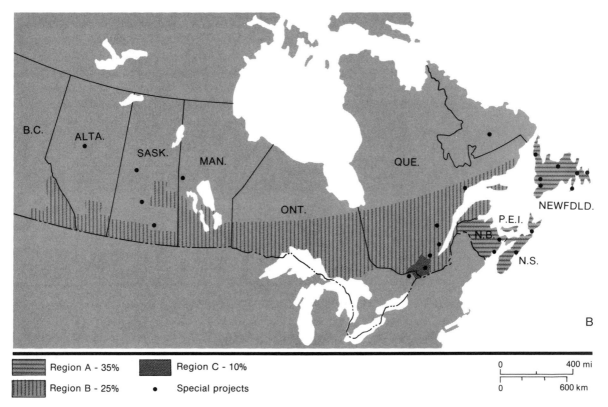

Region A - 35%   Region C - 10%

Region B - 25%   •   Special projects

0   400 mi

0   600 km

percentage figure shown in each case represents the maximum incentive grant payable there as a proportion of the cost of new plant

or fixtures. The program has gone, but the Canadian government is
especially sensitive to the regional needs of the Atlantic Provinces.

tions. The word *area* leads us to ask, "How large must an area be to be regarded as a region?" The word *defined* calls for clarification as to who is responsible for defining the criteria to be used. And the word *criteria* leaves us asking, "How many criteria, and of what sort, does this definition of a region involve?" The answers to these questions, and the regional divisions of North America to which they point us, will occupy the third section of this chapter.

The simplest of the three terms to explain is the second: the word *defined*. A region can be, and in practice is, defined by anyone, or any institution, with an interest in doing so. Governments and their agencies create regional divisions to improve the efficiency or control of their own operations; so, too, do commercial enterprises. This works well enough, although not so well as to eliminate the question: Could it be improved? Business firms spend time and money analyzing their operations in an attempt to answer this very question in its many forms: Is Atlanta or Dallas the better location for our southern regional headquarters? Should we move our Great

Plains agency from Chicago to Minneapolis (or Kansas City) to give our customers better service? This is a simple, practical exercise in identifying regions.

The activity of geographers is no different in principle: to choose and define the criteria that will form the basis of a regional subdivision. The implication is, however, clear: regions are not units that exist in nature. They are not natural subdivisions of the earth's surface like the perforations on a sheet of postage stamps. Everyone, geographer or government, defines them for themselves. Like beauty, they are in the eye of the beholder.

The next problem arises with the use of the word *criteria*. Here we are asking, Which indicator, or set of indicators, shall we use to define the region? With a single criterion, there is no problem. The government agency, or the public utility, simply subdivides its field of operations to give maximum ease and efficiency in supplying electricity or administering the law. Geographers can likewise produce maps of climatic regions, or poverty regions, based on a single criterion of temperature, rainfall, or per capita in-

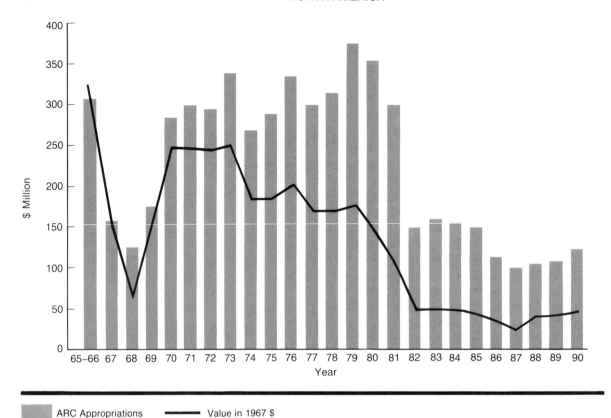

Fig. 10-2. Appalachia: Congressional appropriations, in million dollars, for the regional aid and development program there, 1965–90. The 1967 dollar value is added to the diagram to show the impact of inflation on the grants for this program. The point, however, is not so much that Congress's grants have fallen off as that they have continued so long: Appalachia is the sole survivor of the regional programs set up in the 1960s. (Reproduced by permission from Michael Bradshaw, *The Appalachian Regional Commission*, copyright © 1992 by the University Press of Kentucky, p. 109.) (Reproduced by permission of the publisher.)

come. It is when geographers attempt the use of multiple criteria, in order to define regions that will have a general validity, that the problems mainly arise.

As soon as we try to establish a regional boundary by the use of two or more criteria, we confront the problem of non-correspondence. "The South" and "The Midwest" are two of the best-known regional terms in the vocabulary of every American; yet no two users would agree as to exactly where they are, because no two criteria define exactly the same area.[7] We are then faced by two alternatives. *Either* we must plot a whole host of criterion-boundaries, and then take some kind of geometrical mean among them—simply divide this "boundary-girdle" into

two—*or* we must look for a single criterion, which in itself combines the effect of several, and use that one as a surrogate for the others. Without necessarily admitting it, that is what geographers often do. At the world scale, vegetation is a good single criterion, embodying as it does something of the effects of bedrock, relief, aspect, slope, soil, and climate. At the local scale, the choices among geographers have ranged over such criteria as newspaper coverage, commuting patterns, telephone calls, or bank deposits, all of which have the quality of combining the elements of regional life and its patterns in some way.

But the problem of regional definition is, of course, ultimately insoluble. What geographers are doing is looking for lines or boundaries in the real world that are a little more significant than the other lines that run across them or parallel with them. While governments owe it to their taxpayers, and businessmen owe it to their shareholders, to get their regional divisions right, geographers owe it to themselves to get

[7]To speak of "two or more" criteria here may give the impression that, at three or at four, we should give up the struggle. On the contrary: in his landmark study, *Southern Regions of the United States* (Chapel Hill, 1936), H. W. Odum worked his way through more than 700 criteria, on a state-by-state basis, to obtain a definition of what was, and what was not, a southern region.

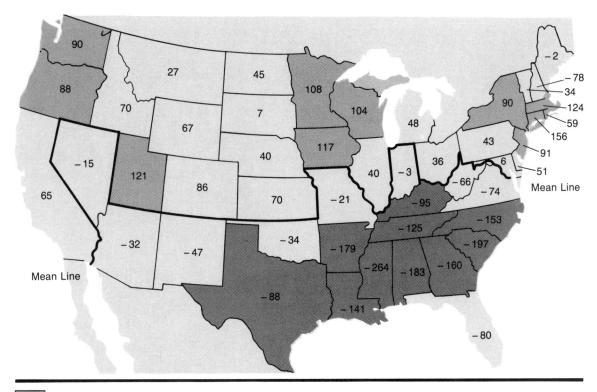

Fig. 10-3. The United States: Social well-being. The map shows state scores on a composite indicator of social factors. The use of a state basis, however, naturally conceals more local differentials in well-being. (Reproduced by permission from D. E. Smith, *Where the Grass Is Greener: Living in an Unequal World* [Baltimore and Middlesex, 1979].)

the best understanding they can of how things fit together.

This brings us to the third question—that posed by the word *area*. How large must an area be to qualify as a region? In other words, we must consider the question of *scale*. In reality, this need be no problem at all, if we can agree that a region is an area—any area—of a character sufficiently definite to mark it off from other areas adjacent to it. There are large regions and small ones, and their recognition depends solely on the scale chosen for their analysis.[8]

[8]P. Haggett, R. J. Chorley, and D. R. Stoddart proposed, in 1965 (*Nature*, vol. 205, 844–47), a scale of measurements (the *g*-scale) for regions of the earth's surface, based on the area of the whole globe, which they gave the value $g = 0$. On this scale, the United States has a value of $g = 1.82$, Texas one of 2.87, and Rhode Island 5.27. In these terms, a region may occupy any position on the *g*-scale from, say, 2 to 6; its treatment by geographers will be substantially the same at all points.

Let us at once place this in a North American context and, in doing so, lay the groundwork for the regional system adopted in this book. In a continent the size of North America there are hundreds, perhaps thousands, of distinctive small areas, with their own peculiarities of landscape or culture. Their names may be known only to the local inhabitants, and be preserved by them if they are to be preserved at all. These are regional divisions identified by men cultivating the land, and learning by experience the idiosyncrasies of terrain and soil. No one may hope to know them all and if, in fact, the knowledge of them survives it will be a small miracle in a continent where only 4 or 5 percent of the population has any direct contact with the soil. But the knowledge of them represents a part of the cumulative experience of North America, a part worth preserving in the face of the earthmover, the factory farm, and the blanketing anonymity of city or suburb.

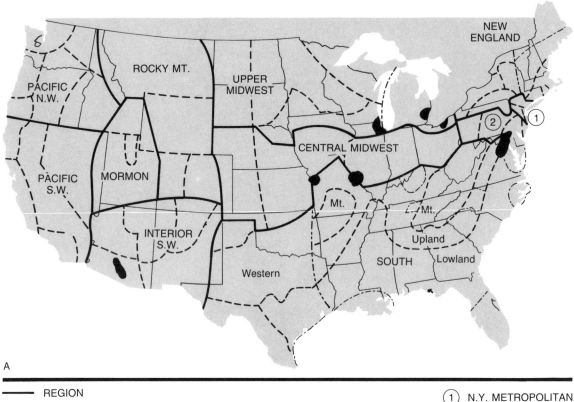

A

─────── REGION
─·──·── Subregion
─ ─ ─ ─ District
─··──·· Subdistricts or special areas
⬤ Nonconforming metropolitan
      areas

① N.Y. METROPOLITAN

② PENNSYLVANIAN

Fig. 10-4. Regions of the United States. The two maps offer two alternative schemes of regional subdivision. (**A**) is a map of cultural
regions, and is taken, by permission, from R. D. Gastil, *Cultural
Regions of the United States* (Seattle, 1975), Map 4. (**B**) is a map of
trade centers of the United States in three grades, with their trading
areas. On Gastil's map, the main metropolitan areas are described
as "non-conforming," while on the trading-area map these metropolises form the focus of the twenty-three regions.

At the other extreme of scale, we can propose a
subdivision of Canada or the United States into no
more than, say, five major regions each. Inhabited
Canada divides itself, with little dispute, into four
regions—Atlantic, St. Lawrence–Great Lakes, Prairie, and Pacific, with the great unoccupied North as
a fifth region.[9] In the United States, a division into
North, South, Midwest, Southwest, and West would
be accepted by most geographers, although there
would be plenty of scope for argument as to where
the boundaries between them should be drawn.

Even the boundary between North and South is
open to dispute, and that after the two fought a war
against each other. The boundary between them
should probably be drawn from St. Louis in the west
to the Potomac in the east, but for intermediate
points en route there is a wide choice—the Ohio Valley, the Kentucky–Tennessee state line, the Mason–
Dixon line (Fig. 10-4).[10]

For a book of the present kind, neither of these

---

[9]J. Lewis Robinson, writing in the official *Canada Handbook,* offers
a sixfold division which shows minor differences from that suggested here. His six divisions are: Atlantic Region, Great Lakes–St.
Lawrence Lowlands, Canadian Shield, Interior Plains, Northwest
Territories, and Cordillera.

[10]For example, in the two works on cultural geography referred to
in n. 1 of this chapter, Zelinksy draws this boundary as a straight
line from St. Louis to the Mason–Dixon line in south-central Pennsylvania. Gastil, by contrast, includes in the South most of the state
of Missouri, southern Illinois, southern Indiana, and the whole of
West Virginia before his boundary reaches the same point in
Pennsylvania.

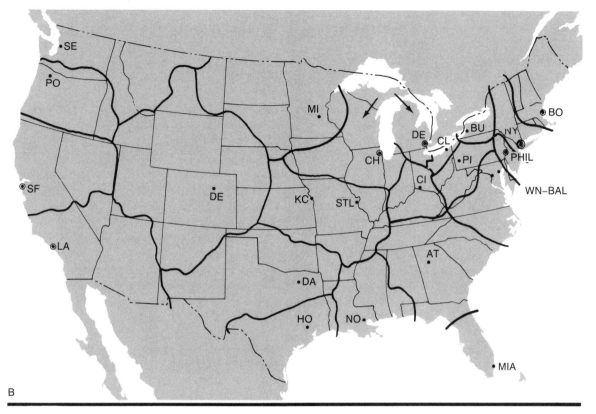

B

⊚ First–order trade center

⦿ Second–order trade center

• Third–order trade center

scales of division will do; one is too small and the other too large. The regions whose names head the remaining chapters are, for the most part, of inter-mediate size. But they are organized within the framework provided by a specific, economic crite-rion, one of those surrogate criteria that combines within itself the impact of many factors. This is the core–periphery concept.

Modern North America, the America of productiv-ity, infrastructure, and cultural life that we know today, was settled by a westward movement from the East Coast, the coast nearest Europe. On the whole, the occupation of the continent was carried out from bases in the East, through which moved both the immigrants who were to settle the frontier, and the tools and goods they needed to do so.

In due course, the ports of entry, from Montréal in the north to Baltimore in the south, became not only entrepôts for goods and people passing in and out, but manufacturing and commercial centers. Upon their activity and support depended the advance of the frontier. They came to form what was recogniz-ably an economic core of the continent and, as the frontier moved ever further west, this core area spread inland until it included not only the coastal cities but many of those in the northern interior—Chicago and Toronto, as well as New York and Montréal.

This provides us with a first breakdown of the con-tinent outside the Southwest into two: a core area from which settlement was organized, financed, and directed and a periphery. This periphery for long in-

Calgary, Alberta: The central business district. Since World War II, the economy of Alberta has grown to sustain two major urban centers, Edmonton and Calgary. Of these, Calgary, the former cattle town and home of the famous stampede, has developed as the financial and commercial center, as witnessed by the rise of its office blocks. During the past decade, it headed all other Canadian CMAs in rate of population increase. *Calgary Tourist and Convention Bureau*

cluded the Old South. Even before the Civil War dealt such a savage blow to the South's prosperity, the region showed few of the characteristics of a core area: it was content to remain agricultural and largely nonindustrialized; it attracted few of the immigrant millions who were flooding into the North (see Chapter 2); it was carrying its capital in the form of slaves and, while its political influence in Washington was considerable, its economic influence spread little beyond its own immediate hinterland. The Civil War reduced it swiftly and decisively to colonial status, and it was to spend many decades simply as part of the economic periphery over which the Northern core exercised control (see Chapter 15).

It was this twofold division to which Frederick Jackson Turner called attention in a famous paper in 1893, that made North America's core–frontier dichotomy probably the best-known regional division in the world, and gave rise to the so-called frontier hypothesis. Certainly, it has kept scholars arguing for ninety years over the question: What *was* the impact of core on frontier, or frontier on core?

We refer to a frontier "hypothesis," presumably to suggest that this core–frontier contrast was important, but we do not know exactly *how* important until the hypothesis has been further tested. In the years since 1893, a good deal of testing has gone on, and the hypothesis has become invested with elements of myth and mystery. The influence of the frontier was, surely, more than a myth but rather less than determining fate. It was, above all, a modifier. Policies evolved in seaboard comfort in Montréal, New York,

Los Angeles: View from the south over the city center, looking toward the San Gabriel Mountains. *California Office of Tourism*

or Philadelphia were increasingly challenged by the men of the frontier who were on the spot, facing the difficulties, pressing for policy change. But the economic core remained the core, larger in area as time went by, but still dominating the continent.

As the frontier advanced still further westward, there grew up, as a kind of secondary phenomenon, what is sometimes known to geographers as the "hinge": a line of cities lying on or beyond the westernmost edge of the core area, and through which the core's controls were exercised over the less-developed periphery beyond. These cities, to which we shall refer again in Chapter 14, formed a north–south line from Winnipeg, through Minneapolis–St. Paul, St. Louis and Kansas City, to Dallas and Fort Worth. They formed a hinge in the sense that they were connectors between the core and the western or northern peripheries with their agricultural and mineral production and their financial dependence on the East.

So the next four chapters of this book will cover sections of the economic core and its hinge within Canada and the United States. All the rest of the continent has, at one time or another, been periphery. Some parts of it manifestly still are—for example, the Mountain West and the Canadian North. But what is of interest to us in the present context is to ask whether this simple one core–one periphery model still applies and, if not, in what respect it is changing.

If we take as the characteristics of a core area a dense population, an economy with well-developed secondary, tertiary, and quaternary sectors, and an important role in providing capital goods and investment funds for the periphery, then those characteristics are today found not only in the historic core area of the Northeast, but in what we may call secondary cores elsewhere. These are areas where growth of population and generation of local capital supplies—in most cases through exploiting local resources or positional advantage—have given the region increasing control over its own development.

There is no doubt, in North America today, where the most important secondary cores are to be found: in California and Texas. Both have growing populations (indeed, California now has more people than the whole of Canada); both have accumulated capital by exploiting their resources—mainly oil and natural gas—and both lie at a considerable distance from the original core, in the heart of what has been the periphery. In Canada, Vancouver and the cities of Alberta have been developing wealth and business control independent of the Toronto–Montréal core. In the United States, there are other subsidiary core areas in the making—on the Colorado Piedmont around Denver; on the Carolina and Georgia Piedmont around Charlotte or Atlanta; on Puget Sound. Each of these will merit attention, as we try to trace and account for its rise.

Will the locations of core and periphery sooner or later change? Will the engine of the U.S. economy one day be situated in California, rather than in New York City or the Northeast? Will the present secondary cores at least grow to divide national functions with the primary core? After all, a diminishing proportion of the national populations of both the United States and Canada lives in the old core area: the Sun Belt and the Pacific Coast are home to ever-increasing numbers, not only of residents but of businesses. Corporations in rapidly developing areas, such as California, Texas, or Alberta, grow, become national or multinational, and so extend their control to distant areas. High-level decision makers are also human; they, too, are attracted by amenities. Today's communications networks make it possible for them to move their headquarters to attractive areas and to direct their business operations over long distances. To move the New York Stock Exchange to Los Angeles or Miami may at present seem inconceivable, but who is to say that, in the twenty-first century, this might not be a true reflection of future core function?

In the regional studies that occupy the remainder of this book, we therefore deal first with the historic core, and then with the periphery in its various parts, in the following sequence:

Regions of the Economic Core
1. The Middle Atlantic Region
2. Southern New England
3. The Middle St. Lawrence Valley
4. The Interior

Southern Regions
1. The South
2. Appalachia
3. The Southern Coasts and Texas

Regions of the Western Fringe—The Hinge and Beyond
1. The Great Plains and Prairies
2. Mountain and Desert
3. The Southwest

Regions of the Northern Fringe
1. The Northern Atlantic Coastlands
2. The Northlands

Regions of the Pacific Coast
1. California
2. Hawaii
3. The Pacific Northwest

11
The Middle
Atlantic Region

New York City: Looking south from Rockefeller Center toward the Empire State Building (center) and the World Trade Center (background), located at the southern end of Manhattan Island. *New York Convention and Visitors Bureau*

That a survey of the regions of Canada and the United States should begin with the Middle Atlantic Region is entirely appropriate, for it is the historic heart of the economic core. This is where, in the century of the immigrant (1815–1914), most of the new arrivals landed. It is where political and financial control of the westward expansion was vested. It became, in due course, the most industrialized, most urbanized section of the whole continent, containing a great urban concentration around New York, and four other metropolitan areas—Philadelphia, Washington, Baltimore, Newark—each with 2 million inhabitants or more. Its ports handle a high proportion of U.S. foreign trade and it has—partly as a result of this—a wide range of manufactures: heavy industry in Trenton and Baltimore, with their steel plants; light industry in New York City, with its clothing industry and luxury goods. The Middle Atlantic Region began as an assortment of English, Dutch, Swedish, and German settlements. In the three centuries that followed, it became the organizational center of the world's most powerful economy.

Today, the urbanized area that represents the Atlantic end of the economic core stretches virtually without a break from southern New Hampshire to northern Virginia. It has become known as *Megalopolis*, and it is more than 400 miles (640 km) long. This being so, there is good reason for treating it as a unit, since its present functions and problems (to which we shall later refer) unite its various parts. But historically, it grew from several different nuclei—originally, in fact, from *rival* nuclei. In recognition of this, we shall deal separately with its northern and southern extremities and concentrate, in this chapter, on the central section, on that part that lies in New York, New Jersey, Pennsylvania, Delaware, and Maryland.

The task of description of this region is complicated by the fact that its most important function within the nation—what we described in the last chapter as direction and control of the national economy—is, in a landscape sense, the least visible.[1] It is therefore necessary, as we describe in turn the physical features of the region, its agriculture and its industry, to recall that, important as those are, they are all economically overshadowed by that other regional function that is exercised by the core of the core, but which leaves behind it so little landscape evidence—the organizing, financing, insuring, controlling of the life of a subcontinent.

For present purposes, then, our region stretches from the mouth of the Hudson River at New York in the north to the Potomac in the south. Its inland edge is indefinite, for much of the region lies west of the Fall Line and some of it to the west of the Blue Ridge. Although we will be considering the Appalachian hinterland of the region again (in its appropriate context, in Chapter 16), it is useful to give it a place here in order to point out how the regional characteristics of the Middle Atlantic Region change as one moves westward.

Defined in this way, the region divides into several physical subsections. In the east lies the Coastal Plain, cut by the deep indentations of Delaware Bay and Chesapeake Bay, and diminishing rapidly in width as it stretches northward. In character, much of it is a sandy, infertile area, covered in large part by pine barrens. The long line of coastal sandbars and lagoons that extends north and south from Cape Hatteras and Pamlico Sound continues all the way to Sandy Hook, at the entrance of New York Harbor,

---

[1]This suggests a possible reason for the extreme shortage of geographical literature on this important and densely populated region. The only substantial work that can be recommended is the excellent John H. Thompson, ed., *The Geography of New York State* (Syracuse, 1966), and even that is nearly thirty years old. For the main cities, however, see n. 7.

and fringes the coast of Long Island; but to compensate the sailor for this inhospitable shore, the drowned valleys of Chesapeake and Delaware bays and the lower Hudson carry shipping far inland to magnificent natural harbors.

On the landward side, the sandy plains end at the Fall Line. Along this line the recent formations of the Coastal Plain meet the older rocks of the Piedmont, and there are falls in the rivers flowing off the Piedmont. Representing as it did both head of navigation and source of waterpower for early industries, the line became the site of a string of settlements stretching from Trenton, New Jersey, southward into Alabama.

West of the Fall Line lies the Appalachian System, with its four component parts (Fig. 1-3). But in this middle section of the system, altitudes generally decline from south to north, and indeed in southeastern Pennsylvania the Blue Ridge virtually disappears, to reappear in northern New Jersey. Elsewhere, the mountains are pierced by water gaps—notably those of the Potomac, Susquehanna, and Delaware—as well as by dry gaps like that at Manassas that presumably result from river capture. In peace as in war, to the railway engineers as to Robert E. Lee and Stonewall Jackson, the position of these gaps has been a matter of constant and decisive significance. Beyond them lies the Great Valley of the Appalachians, important both for its fertile farmland and as a route giving access, through further gaps, to the interior.

Between the Susquehanna and the Hudson the Piedmont also changes its character somewhat, thanks to the presence among its hard, crystalline components of a belt of softer Triassic formations (which incidentally reappear both in New England and in the Maritimes). The Triassic area lies, on the whole, well below the Piedmont surface level, but included in it are several bands of trap rock—igneous intrusions that form ridges in northern New Jersey.

Throughout this region, good soil is found only in limited areas. The climate, however, is favorable to farming. The presence of the ocean, here warmed by the Gulf Stream, and the deeply indented coastline give the area a long growing season and serve to moderate temperature extremes. On the other hand, the farmer's gain is the city dweller's loss, for in the region's great urban centers the hot, humid summer weather makes heavy demands on the workers.

Agriculture

Like the settlers of New England and of Virginia, the early inhabitants of the Middle Colonies had to look to the land for most of their needs. They were, however, more favored by natural conditions than the New Englanders and were, on the other hand, free from the commercial link that tied early Virginia's economy so firmly to tobacco production. There were Dutchmen on the Hudson, Swedes on the Delaware, and Germans in Pennsylvania, besides the British colonists who quickly became the dominant element. Each group developed its own agricultural methods, and farming was more mixed than in the colonies to the north and especially to the south.

The semi-feudal conditions of land tenure in some of the early settlements, and the pressure on the land of an increasing immigrant population soon produced a drift westward toward the mountains. Here in the rougher terrain of the upper Piedmont and later of the Appalachian valleys, independent farmers carved out their holdings, accepting the handicaps of infertility and remoteness in exchange for liberty of action. To the eighteenth-century farmer, the exchange seemed a reasonable one; his twentieth-century descendant, occupying the same hill farm, suffers the handicaps without the same compensation.

The present agricultural pattern of the Middle Atlantic Region is the product of one overwhelmingly important factor: the rise of the seaboard cities. Paradoxically, this factor accounts at one and the same time both for the presence of such agriculture as exists, and for its absence from large parts of the region, where it is squeezed out in the competition for space by cities, suburbs, parks, and country estates, which may once have been farmland but are certainly no longer farms.

In the general discussion on agriculture we saw how western competition has limited the range of products with which the eastern farmer can succeed and how, even so, he needs excellent transport facilities to market in order to compete. In the Middle Atlantic Region, these circumstances combine to create a dairying and truck-farming area based on supplying milk and vegetables to the huge urban populations of the region, and marked by wide differentials in farm prosperity between the areas adjacent to the cities and the remoter hill farms.

The importance of cattle rearing, especially in Pennsylvania and New York, is clearly revealed by the agricultural statistics. In Pennsylvania the crops of the classic "Pennsylvania Rotation"—corn, oats, wheat, and hay, most of which are fed to cattle—occupy almost 90 percent of the state's cropland. In 1985, 69 percent of Pennsylvania's farm marketings consisted of livestock—largely dairy—products. For New York State, the figure was 72 percent. It is mainly to the urban demand for fluid milk that the region caters.

On the sandy soils of the coastal plains, however, dairying is unimportant. Its place is taken by truck farming. In the Delmarva Peninsula[2] and in New Jersey west and north of the Pine Barrens, there has developed one of the world's largest concentrations of this intensive and highly specialized type of farming. The whole range of kitchen vegetables is grown, and one crop follows another on each plot throughout the long growing season.

Within the coastal region, potato growing has been concentrated in two areas: eastern Long Island and the southern tip of the Delmarva Peninsula. Here, the distinctive potato barns show that the farmers specialize in the crop. Potatoes have, in fact, become increasingly a specialist crop through the years. The days are long gone when every farmer in the East grew them; today, four or five specialist potato areas in the United States account for almost the whole crop.

This remarkable concentration of truck farming is due partly to a relatively mild climate and the suitability of the light soils for fruit and vegetable production, although use of chemical fertilizer and supplemental sprinkler irrigation compensates for deficient soil fertility and susceptibility to drought. However, it is the unrivaled location with respect to the huge markets provided by New York, Philadelphia, and Baltimore and lesser cities that underlies such specialization. An excellent road network supplements or replaces earlier rail and still earlier water connections. The region has developed a comparative advantage based on market location, soils, and climate; on early specialization and intensity; and on use of large-scale techniques of crop production, processing, packaging, and marketing. Processing of the crops in large plants nearby makes it possible not only to ship to other areas, but, especially, to extend the marketing period from a limited growing season to the entire year.

No other farm product has the region-wide importance of milk or truck crops, but several are of local significance. Poultry raising, like vegetable growing, is well adapted to poor soils and to the pockets of agricultural land that lie between the cities: it requires little space and caters for the urban markets, so that in Delaware, New Jersey, and the dwindling agricultural lands of Long Island it plays an important part in the farm economy. In Delaware, where the broiler-fowl industry began, no less than 40 percent of the farms are classified as poultry operations. Tobacco, the foundation of the economy in the coastal plains of Virginia and Maryland during the colonial period, is now important in only three districts: in southern Virginia mainly on the Piedmont; in southern Maryland between the lower Potomac and Chesapeake Bay; and in Lancaster County, southeastern Pennsylvania.

From a survey of the chief farm crops, we may now turn to the pattern of agricultural subdivisions within the region. They are listed in sequence as they would be noted by a traveler making a traverse from the coast to the Appalachian Ridge and Valley Province, and then either west to Pittsburgh or north to the shore of Lake Ontario (Fig. 11-1). On the coastal plains, as we have seen, truck farming occupies a large part of the agricultural land of New Jersey, Delaware, and eastern Maryland, together with the small areas of farmland remaining on Long Island. Inland across the Fall Line the lower Piedmont reveals a marked contrast. Here, with its focus in southeastern Pennsylvania and its heart in famous Lancaster County (which produces one-seventh of the entire agricultural output of Pennsylvania), is one of North America's most prosperous farm areas. The farms average only 75–100 acres (30–40 ha) in size, but their handling by successive generations of the Pennsylvania Dutch (the best farmers of whom were German Mennonites, by origin) has been scrupulously careful.[3] It is indeed the skill of the farmers, together with the proximity of the Philadelphia market, that underlies this remarkable prosperity; for, although tobacco, wheat, poultry, hog, beef, and dairy products are all major sources of income, it is corn and hay of the old four-crop rotation that form the basis of the farm operations (Table 11-1).

As the traverse continues across this prosperous

[2]The peninsula enclosing Chesapeake Bay and forming parts of Delaware, Maryland, and Virginia.

[3]An excellent account of these settlements is to be found in J. T. Lemon, *The Best Poor Man's Country* (Baltimore and London, 1972). See also J. W. Florin, *The Advance of Frontier Settlement in Pennsylvania, 1638–1850* (Philadelphia, 1977).

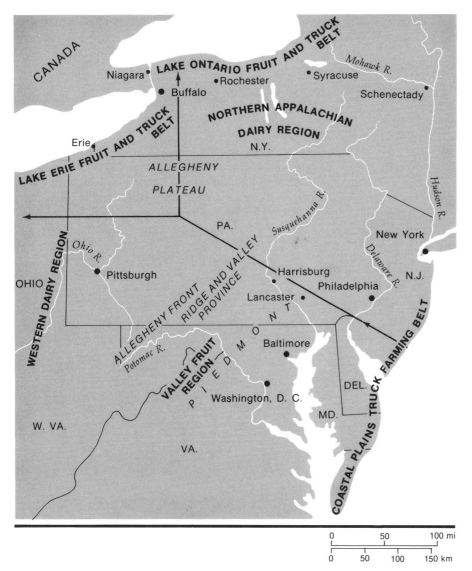

Fig. 11-1. Agriculture in the Middle Atlantic Region. The arrows pointing inland from the coast indicate the approximate line of the traverse of the various farming regions which is described in the accompanying text.

lower Piedmont and approaches the hills, the situation changes. Natural conditions become less favorable: the terrain rougher, the soils more patchy, and the growing season shorter. Distance to market increases. The effects are (1) the land in farms becomes more scattered; (2) the cash crops of the lower Piedmont disappear; (3) the farmers concentrate less on dairying, for their milk must be marketed in smaller centers like Harrisburg or Scranton; and (4) prosperity varies markedly with access to such local centers.

Beyond the Great Valley, with its fertile bottomlands and apple orchards, these effects become rapidly more pronounced, ending in the deeply dissected forest country of the Appalachian Plateau, in hill farms whose isolation and meager natural endowment prevent the practice of anything other than part-time or subsistence farming. Each year that passes now sees the abandonment of more of these plateau farms and the return of the land to the forest.

Beyond this agricultural "dead heart" of eastern North America (which lies at the heart of Appalachia also, and which will be considered again in Chapter 16), we can trace in reverse much the same sequence as that seen on the Piedmont, whether we go westward or northward. To the west, the market for milk recovers as we approach Pittsburgh and the coal and steel towns, whereas on the north side of the plateau, the industrial cities of the Mohawk Gap and the lake-

Table 11-1. Lancaster County, Pennsylvania: Agricultural Statistics, 1987

Population (1980)	362,000
Percentage change, 1970–80	+13.2
Area of county (ac/ha)	605,000/245,000
Percentage in farms	66.8
Percentage in cropland harvested	49.3
Value of agricultural produce sold	$601 million
Percentage from livestock products	90.3
Number of farms	4775
Average size (ac/ha)	85/34
Average value of land, buildings & equipment, per farm	$354,000
Average value of agricultural products sold, per farm	$125,800
Percentage of operators whose principal occupation is farming	73.4

shore provide a similar market. Thus in western Pennsylvania and New York State, there are dairy regions similar to, if less prosperous than, those of the Piedmont. On the north, indeed, there is even a counterpart of the coastal truck-farming belt, where market gardens, vineyards, and orchards extend along the southern shores of Lake Ontario and eastern Lake Erie.

In all the states of this region there has been a long history of decline in the number of farms and the extent of land cultivated (Fig. 11-2). Only in the latest period has this decline in farming apparently been reversed, and this may have more to do with the methods of census classification, and the answer to the question When is a farm not a farm? than with any long-term trend. We will have to wait to see whether the reversal is a temporary phenomenon or whether it does indeed mark a turning point. What is, however, fairly clear is the reason for the long decline that preceded the upturn in 1978.

Farming in Lancaster County, Pennsylvania. The county is famous for the quality of its farming and the density of its farm settlement. It is also well known for the Amish farmers who cultivate it, using traditional methods such as the horse-drawn tobacco planter seen here. *Pennsylvania Dutch Convention and Vistors Bureau*

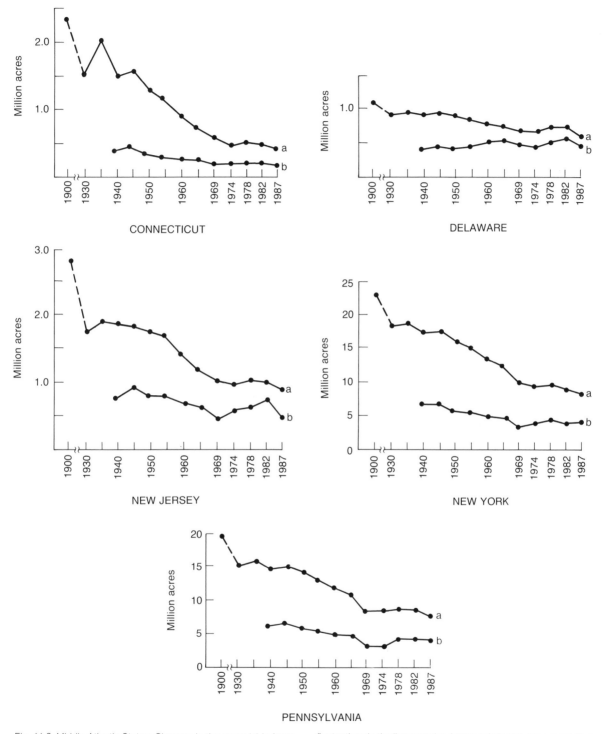

Fig. 11-2. Middle Atlantic States: Changes in the areas (a) in farms and (b) in cropland harvested, for census years since 1900. Minor fluctuations in the lines may be due to variations in census practice, but the general downward trend in size of the areas is obvious.

The first reason for the decrease in agricultural land has been the expansion of urban and industrial settlement over former farmland. This is a natural result of population growth, but it has been intensified by today's social concepts and ideals—the concepts of building "out" rather than "up," which have tended to spread the suburbs of cities over huge areas in low-density housing. American communities often have bylaws enforcing a maximum housing density of one house per one-half or one-quarter acre (0.2 or 0.1 ha). Small wonder, then, that the cities of the Atlantic Seaboard have overrun both agricultural land and smaller towns and villages many miles from their original centers.

The second reason, however, is due to the action of the farmers themselves in abandoning marginal lands. In some sections, such as the Tug Hill Plateau in upper New York State, and very generally above the 1000-ft (300-m) contour, farming has ceased altogether. In today's farming, the poorer and remoter lands simply will not yield a family a worthwhile living. So strongly is the trend toward farm abandonment running that the planners can forecast closure and replan land use even before the farmer has actually made his decision to give up. Only the better farmlands, such as the Pennsylvania Piedmont and the truck areas of the coastal plain, have held their own, agriculturally, in the national setting.

The landscape of farm abandonment is an interesting one and is actually far from uniform. Around the outer suburbs of the cities there is, characteristically (in both North America and Europe), a belt of total farm dereliction, where farmers, foreseeing the spread of the city and the arrival of developers, are simply waiting to sell out and retire on the proceeds. Platt calls it "a zone of disinvestment . . . characterized by rundown buildings, lack of visible activity, and a general 'air of anticipation.'"[4] Typically, the dairy herd, or other livestock, is sold; capital improvements cease, and farm buildings deteriorate, but cropland may be planted in cash crops, often by short-term renters, until urban development begins.

Outside this belt there is often a kind of agricultural aureole—an area of manicured lands, white rail

fences, and elegant paddocks. This is the domain of the city businessman and his country estate. Beyond this again, unless the land is of exceptional quality, prosperity and its appearances fall off with distance from the urban center. In none of these zones is stable commercial agriculture being carried on, yet in all of them the farm homes are occupied, at least part of the time.

What has happened (and again conditions here parallel those in Europe) is that in the vicinity of cities and in areas of particular rural quality, the farmhouses have been bought up by city people, who in many cases use them as second homes and occupy them only periodically. Farming as such is not their concern, and often they allow the land to become grown over and prefer it that way. But since they renovate the farmhouses and bring capital into rural areas, it seems likely that the practice will spread.

What of the future? To all appearances, all these trends will continue. The urban areas will increase their populations, although probably at a slower rate than in, say, the decades 1950–70. More of Long Island and the farmlands of southeastern Pennsylvania will disappear beneath suburban growth, and the rural areas themselves will gain in population through a strong inflow of exurbanites. The cities of the Hudson–Mohawk lowlands and the Lake Ontario shore will spread, overrunning some of the region's best remaining farmland.

From an economic point of view, as we saw in Chapter 5, there is little reason to regret the passing of low-value agricultural land. However, there must be some sensible limit imposed on the spread of built-up areas, and so there arises the question of other uses for abandoned farmland. In the Middle Atlantic Region, there are two such uses. One of these is forestry. The original farmers, when they cleared the land, left patches of forest that they used in relation to their fields—for fencing-timber or as shelter for their stock. Today, forests are planted for their own sake, and to control runoff and improve the catchment areas of the big city reservoirs, or to produce commercial pulp for the newsprint market. In this way, rather ironically, the land acquires a new value when new forests are planted, and the cycle that began when the old forests were cut down 200 years ago is completed.

The other use for these lands is recreational. The enormous urban population of the region must find

[4]R. H. Platt, *Land Use Control: Interface of Law and Geography*, Assoc. of Amer. Geog. Resource Paper No. 75-1 (Washington, D.C., 1976), p. 9. J. F. Hart, however, in "The Perimetropolitan Bow Wave," *Geographical Review*, vol. 81 (1991), 35–51, warns us that not all land on the urban fringe is waiting idly to be built over: *some* of it is intensively farmed.

an outlet, and as the land empties of farmers, much of it is enclosed in state forests and parks, and so is secured as open space in perpetuity. In this way, New York State has set up a planning agency to oversee the rational development of the Adirondack area, which includes 2 million acres (810,000 ha) of forest preserve. The Catskills include another 200,000 acres (81,000 ha) of preserve. In face of the advancing tide of urban sprawl, the states will be well advised to set up many more reservations, so that the cities of the future can provide their inhabitants with space in which to breathe and relax.

Industry

The growth of industries in the Middle Atlantic Region has been favored by circumstances of both location and history. When the revolt of the colonies threw the United States, in a new sense, upon its own resources, it gave a powerful stimulus to industrial production in the communities of the Atlantic Seaboard. Local deposits of iron ore in eastern Pennsylvania and charcoal from the Appalachian forests formed the basis of early ironworking, and skill in a host of crafts that later became factory industries was brought into the coastal cities by immigrants from Europe. This immigrant stream also ensured a constant supply of cheap labor when the factory phase of industry opened. Then, as the settlement of the West began in earnest, the East Coast states served as a supply base, providing the migrants with manufactured goods for their westward penetration—much as Europe had served as a supply base for the earliest colonists. Throughout three centuries, during which North America's most important connections, both cultural and economic, were with Europe, the Middle Atlantic Region profited by the constant passage of people and goods between the Old World and the expanding frontier of the New, to secure for itself a large share of the processing and shipping services required by both East and West.

Today, the strength of the region's industries depends partly on a continuing exploitation of this position in relation to foreign trade and partly on the huge size of the local market the position has, in turn, created. This has permitted the development of a full range of manufactures, and 95 percent of the 500 types of manufacture recognized by the census are represented in the state of New York alone.

In relation to industrial employment, however,

two materials have been of special significance. One of these is cloth, represented by the two industrial groups of textile mill products and apparel. The southern New England section of the developing economic core, during the nineteenth century, dominated the manufacture of cotton and woolen textiles. In the making of clothing, however, the Middle Atlantic Seaboard cities, especially New York, became and remained supreme. Although it is commonly found in any large city, the clothing industry has its focus in a remarkable area in New York, occupying only a few city blocks between 34th and 41st Streets, a "Garment Center" which contains over 60,000 workers in several thousand small establishments.[5]

The region's other principal industrial material is steel. Before the Civil War, the center of the nation's iron industry was in eastern Pennsylvania, where local ores were smelted, first by charcoal and later by anthracite from the Scranton area. After the war, however, a number of important changes occurred. A shift from the use of eastern Pennsylvania anthracite to western Pennsylvania bituminous coal as fuel and flux was related to technological developments in furnaces and to greater emphasis on production of steel. Newly discovered supplies of high-grade iron ore in the Lake Superior region began to replace the inadequate supplies of the East. Not only was there a tremendous expansion of the market for iron and steel, but also a shift, as the railroads and settlements expanded in the Old Northwest. The result of these changes was that the steel industry located and mushroomed in Pittsburgh, and then expanded to and along the southern Great Lakes. However, a large, concentrated market remained in the old economic core along the Atlantic Seaboard, and the steel plants of Bethlehem and Philadelphia supplied it. Still larger plants were built, by Bethlehem Steel at Sparrows Point near Baltimore in 1916, and by U.S. Steel (the Fairless Works) on the Delaware River opposite Trenton after World War II. At these tidewater locations, bituminous coal brought by rail from the Appalachian fields to the ports of Hampton Roads, and shipped downstream from there, meets ore from

[5]See A. Herod, "From Rag Trade to Real Estate in New York's Garment Center," *Urban Geography*, vol. 12 (1991), 324–38. Herod reports how, in spite of pressure from other business uses, the Garment Center has hung on to its space, and today even has a reserved "Special Garment District" between 35th and 40th Streets, from 7th to 9th Avenues.

South America. The finished steel is then moved by rail or water to other seaboard cities, to the southeastern United States, or overseas.

The finished steel is used by a host of other industries, from those making ships or railroad equipment to those making machine tools and fine instruments. Apart from these, two other groups of manufactures deserve mention. One is the chemical group, well represented in both its heavy and light branches, in both agricultural and industrial chemicals and in pharmaceuticals. Apart from numerous plants in the New York–New Jersey area, the region contains the

headquarters, though by no means all the branches, of the great Dupont concern at Wilmington, Delaware.

The other, and closely associated, industry of note is oil refining. By pipeline and by tanker, crude or semirefined products are brought from the Gulf of Mexico and overseas to the Middle Atlantic coast for processing and distribution. Since sea transport is involved in both import and export, most of the refinery sites are on tidewater, frequently on reclaimed marshland that affords the most suitable sites for this industry within the vicinity of cities.

Fig. 11-3. The Middle Atlantic and Southern New England States: Employment in Industry, selected dates, 1967–87.

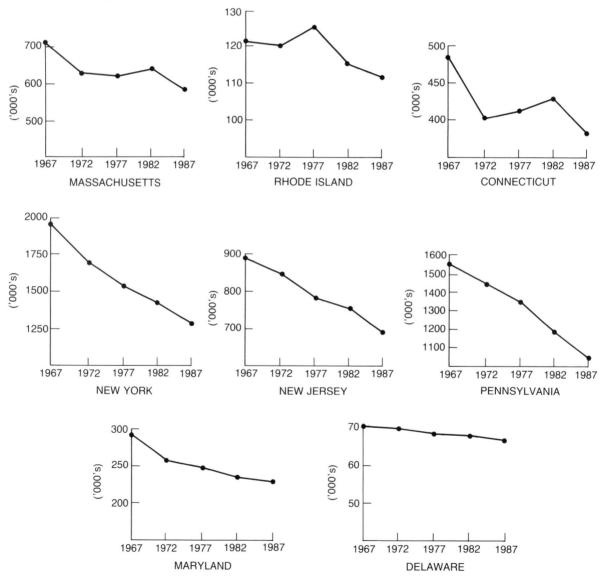

The Middle Atlantic Region is one of the great industrial areas of the continent. It remains so despite the fact that its status has changed markedly in the past thirty years. In 1950, this region contained more than one-quarter of the industrial employment of the entire United States. By 1987 the corresponding figure had fallen to 15.8 percent. It has lost ground not only relatively to newer industrial areas elsewhere, but also absolutely; about 1 million jobs in industry have disappeared from Megalopolis since 1969 (Fig. 11-3). Some of the reasons for this loss we have reviewed in Chapter 8. Overcrowding has been a factor; so, too, have been high levels of state and local taxes in the Northeast, and the high price of electric power. Newer markets have opened elsewhere, and newer plants to serve them.

Yet the Middle Atlantic Region still provides manufacturers with their most highly concentrated market. The early industries were grouped around the ports, along the waterways, and adjacent to deposits of coal or iron ore. Since then, industry has become much more widespread. It has moved up the valleys of the Hudson, the Delaware, and the Susquehanna, and particularly along historic routes to the interior. Most important is the old route of the Erie Canal and former New York Central Railroad, through the Mohawk Gap to the Great Lakes, with Albany, Troy, Schenectady, Syracuse, Rochester, and Buffalo as major industrial centers. The trans-Appalachian routes further south were more difficult and are less urbanized or industrialized. Harrisburg, Altoona, and Johnstown lie between Philadelphia and Pittsburgh, but no large centers have developed along Baltimore's routes to the interior.

Region and Nation

We come now to the primary role of the Middle Atlantic Region, the role we described in the last chapter—that of the earliest economic core region and the present heartland of the country. To suggest that the importance of this region could be measured today by its diminished farmlands or its remaining industries would be grotesquely to underestimate its significance. Only three American workers out of ten, after all, are engaged in agricultural or industrial production: they are clearly out-numbered by workers who serve, educate, finance, direct, or govern. And in these categories of employment, the Middle Atlantic Region is particularly rich. It has been commercial from the start and industrial for nearly two centuries,

but rather than thinking of it today as a trading region or as a part of the Manufacturing Belt (and it is both), it is more appropriate to consider it as a *control area*—a decision-making region.

Industry, transport, and enterprise elsewhere in the United States are directed from this core, which is a kind of general headquarters of Enterprise America, the national economy, even when production is carried on in other regions. This region makes decisions not only as to when and on what terms finance shall be provided, but equally what the nation shall read (for it dominates the publishing business) and how American goods shall be advertised or presented. Although much of this control is exercised from Washington, the seat of the federal government, or from New York, the financial capital of the nation, the rest of the region plays its part, too, as lack of space in the central cities forces offices, research laboratories, and government agencies to move to quieter out-of-town locations, where they are linked electronically with their headquarters. Not milk or steel or chemicals, but rather decisions are the chief output of the Middle Atlantic Region.[6]

The Seaboard Cities

This is, before all else, a region of great cities.[7] Over 18 million people live in the consolidated metropolitan area that sprawls across the mouth of the Hudson and has its dramatic focus on the island of Manhattan: New York. To the south, and dwarfing such intermediate centers as Trenton (326,000) and Wilmington (597,000), are Philadelphia, which has some 4.85 million inhabitants within its metropolitan area; Baltimore with 2.4 million; and Washington, D.C., the federal capital, with 3.9 million. Diverse in character, all four of these great centers have their

[6]J. O. Wheeler, in "Corporate Role of New York City in the metropolitan Hierarchy," *Geographical Review*, vol. 80 (1990), 370–81, writes: "New York City is the single most important source of economic power and decision-making in the United States and, indeed, the world. The city is the power center of the new information economy" (pp. 379–80).

[7]There is the same shortage of geographical literature about the great cities as about the region as a whole, but there are two rather elderly references to offer: John S. Adams, ed., *Contemporary Metropolitan America*, 4 vols. (Cambridge, Mass., 1976); and the Assoc. of Amer. Geog., *A Comparative Atlas of America's Great Cities* (Minneapolis, 1976). In the former work, the Middle Atlantic cities are dealt with as follows: New York–New Jersey, vol. 1, 139–216; Philadelphia, vol. 1, 217–90; Baltimore, vol. 2, 1–95; Washington, vol. 4, 297–344.

Table 11-2. Middle Atlantic Region: Percentage Shares in Nonagricultural Employment, by States, 1985

	Total Work Force (thousands)	Construction	Manufacturing	Transport, Public Utilities	Wholesale and Retail trade	Finance, Insurance, Real Estate	Services	Government
New York	7755	3.6	16.7	5.4	21.1	9.3	26.4	17.4
New Jersey	3419	4.1	21.0	6.6	23.8	5.6	23.2	15.6
Pennsylvania	4736	4.0	23.0	5.0	22.6	5.5	34.7	14.3
Delaware	293	6.1	24.6	4.1	22.2	6.8	21.2	15.4
Maryland	1885	6.8	11.5	4.8	25.1	5.8	25.0	20.8
District of Columbia	629	2.1	2.4	4.1	10.2	5.6	34.0	41.8
U.S., 50 states	97,614	4.8	19.8	5.4	23.7	6.1	22.5	16.8

distinctive functions, the first three as ports, manufacturing cities, and commercial centers, the fourth given over to the business of government (see Table 11-2).

When the end of British rule in the United States and, later, the Louisiana Purchase opened the era of widespread settlement beyond the Appalachians, there rapidly developed among the Atlantic coast ports an intense rivalry to secure a share in the lucrative supply and transit business with the West.[8] From the first, the decisive factor was transport; as communications developed, stage by stage, so the advantage swung from one city to another. In the first stage, Baltimore held a slight advantage. It was the newest of the ports, but it lay closest to the eastern end of the first main route across the Appalachians— the National Pike, authorized in 1806. Then, dramatically, the situation changed: in 1825 New York was linked by the Erie Canal with the Great Lakes. Both Philadelphia and Baltimore responded with canal projects, but it was one thing to cut through the low Mohawk–Lake Ontario divide and quite another to cross the main Appalachian barrier, with the Phi-

ladelphians reduced to hauling their canal boats up a cable incline to climb the Allegheny Front. Baltimore interests turned to other expedients, and, in 1828, began work on America's first railroad, the Baltimore and Ohio. A new phase in the struggle opened. Philadelphia responded with the Pennsylvania Railroad, New York with the New York Central and some lesser lines. The southern routes were shorter, but the northern were less heavily graded, and construction went ahead much faster. Competition was intense, and the last stage of the rivalry developed as a struggle for advantageous freight rates. This was won by Baltimore. With its shorter route to the interior, it secured, in 1877, an advantage over New York and Philadelphia, which it held almost up to the present time. Since the freight rates on most commodities being shipped to Europe were the same for all ports between Hampton Roads and Maine, Baltimore was able to use this rail differential to good effect in building up its export trade, especially in grain.

Yet New York outstripped its rivals in the end, for, although the canal era passed, the Hudson River proved a permanent and decisive advantage to the city at its mouth. To this advantage of strategic location, New York could add that of its splendid harbor, opening directly onto the Atlantic (both Philadelphia and Baltimore face southeastward down their bays) and its maritime and commercial links with Europe. Once New York established its leadership, it grew with increasing momentum, until today few American concerns doing business on a national scale can afford *not* to have a New York address.

To this story of intercity rivalry on the Atlantic Coast there is an addendum. In 1959 there appeared a new threat to the trade of all these cities—the St. Lawrence Seaway. Ever since 1825 the main trade

[8]Competition among the several seaboard cities had gone on even before they became involved in the race to establish transport routes to the West. In 1700, Boston was the largest city, with an estimated population of over 7000. By 1760, thanks to its superior agricultural hinterland, Philadelphia had forged ahead, with a population estimated at 32,000. But even before the Erie Canal opened in 1825, New York had capitalized on its central location within the developing economic core region and on its unsurpassed harbor to gain a lead that, in most respects, it has never relinquished.

Although in this section of the book we are concerned primarily with competition between cities within the United States, it should not be forgotten that the Canadian port city of Montréal also very actively competes with U.S. cities.

The rivalry between the seaboard cities in the canal era: Canal connections with their hinterland played an important part in this rivalry, and when New York secured access to the West by means of the Erie Canal, Philadelphia interests responded with a canal up the Juniata River. But to cross the Appalachians it was necessary to build a portage railway from Holidaysburg to Johnstown, and the southern cities were unable seriously to compete by water with the low-level Hudson–Mohawk route. The picture shows the portage railway in 1840.

route from the Great Lakes eastward had run overland to these Atlantic ports. Midwestern grain exports, in particular, were handled there, especially at Baltimore. But the Seaway cut very deeply into this transit trade, and grain-handling facilities at the Atlantic ports soon stood idle, in spite of reduced rail rates granted by the Interstate Commerce Commission. The Seaway did for Montréal, Toronto, and Chicago what the Erie Canal did for New York and Buffalo in the early days of the rivalry between the Atlantic Coast cities. These latter may well feel relieved, in hindsight, that the Seaway, as we saw in Chapter 9, was not built on an even grander scale, to challenge their position more seriously than it does.

NEW YORK

The mouth of the Hudson seems a place foreordained by nature as the site of a great city. As we have seen, its position here enabled New York to outdistance its nineteenth-century rivals. Yet paradoxically, it is far from ideal for the kind of city that has grown up there. Few of the world's major centers are so hemmed in by geographical restraints as New York. Its core is on an island—Manhattan, 12 mi (19.2 km) long, but in general only 2 mi (3.2 km) wide—which, although it more than served the pur-poses of the original Dutch settlers, is today hopelessly overcrowded. Its suburbs lie either on adjacent Long Island or on the mainland, and formidable water barriers interrupt the movement of workers and goods on all sides except the north. Furthermore, it is essential to New York's port that these waterways be unobstructed. This is especially true of the Hudson River west of Manhattan, which is navigable for oceangoing vessels as far north as Albany. But it is precisely from this western side that freight from the rest of the country moves into New York. Even beyond the Hudson, the roads and railways approaching New York from the west have to cross a series of inlets and marshes, and these form further constrictions, hampering the development of an adequate transport network.

Manhattan's links with east and west consist of six or seven bridges and four road tunnels. Of the half-dozen railroads that approach the Hudson from the west, only one penetrates (by tunnel) to Manhattan, and, even so, its main freight terminal is west of the Hudson. Terminals of other railroads line the New Jersey shore and, in the days when the city was served by the passenger trains of these companies, travelers began their journey by ferryboat. Only to the north, across the narrow Harlem River, is it rela-

New York City: The World Trade Center, located at the southern tip of Manhattan Island, looking south toward the open sea. Many of the buildings to the right occupy reclaimed land where the old piers of the port once stretched out into the Hudson River. *Port Authority of New York and New Jersey*

tively simple to leave New York. This fact of geography goes some way toward explaining the important part in the life of the city played by the subway network, which tunnels easily under the water barriers.

But the same water barriers that impede movement by land have contributed to New York's rise as a port. With waterfront available within the harbor area, with deep channels and a small tidal range, the port grew to dominate the Atlantic coast in terms of tonnage handled, but, even more markedly, in value of overseas trade. The immigrant traffic of the nineteenth and early twentieth centuries was funneled

through the notorious Ellis Island terminal, while in the heyday of the North Atlantic passenger ship, the world's largest liners tied up on the west side of mid-Manhattan at piers specially built to accommodate them.

In 1985, the port of New York handled about 152 million short tons of freight—103.7 million tons domestic and 48.9 million foreign (43.2 import and 5.8 export). Nevertheless, much of what one writes about the port must be written in the past tense, for underneath the statistics there have been many changes. Most visibly, the passenger liner has virtually disappeared from the North Atlantic route; the waters that surround Manhattan appear deserted by contrast with the scene they presented only a decade or two ago, and many of the island's piers have been removed or are in a state of disrepair. The reason for this last change, at least, is clear: it arises out of changes in the handling of cargoes and out of the demand for space in Manhattan. The piers on the island and the adjoining waterfronts offered no "backup" space for assembling cargoes and for storing them: all this had to be done on the pier itself. In particular, the advent of the freight container created a demand for assembly space occupying hundreds of acres rather than the narrow waterfrontage of the old harbor. Much of the activity of the port has therefore moved to outlying sections, particularly to such new facilities as the container terminal at Port Newark.

It is not too much to claim, in fact, that these facilities on the New Jersey side of the harbor have been the salvation of New York as a modern port. Only about a decade ago, traffic figures were declining; pilferage and strikes were discouraging shippers and driving business away to smaller rivals; Philadelphia and the Delaware harbors had overtaken New York in tonnage handled. It was one more phase in the maladjustment of facilities to traffic that had been overtaking the world's older ports for half a century, and a far cry from the 1870s, when it was estimated that nearly 60 percent by value of all U.S. foreign trade passed through New York.

At this point, the port saved itself by using the unused wetlands west of the Hudson, around Newark Bay. By excavation, dredging, and reclamation it created a new container port on the New Jersey shore, well placed for service by road and rail (since it was unnecessary for most traffic to cross the Hudson to get to it) so that it has been able to participate fully in the transport revolution the container has brought about.

Nevertheless, in the years when New York grew to dominate the continent as a port of entry, it did so also in relation to manufacturing. Within its metropolitan area there are over 1 million industrial employees. The pressure on space in the central districts of the city, of which the skyscraper is the expression, has driven all but the lightest industries (such as garment manufacturing, which has already been mentioned) down to the waterfronts and out to the fringes of the urban area. It has been calculated that, in 1910, 75 percent of the manufacturing employment in New York City was to be found in Manhattan and, of this amount, two-thirds was in establishments situated south of 14th Street; that is, in roughly the most southerly one-sixth of the island. Today, much of the heavy industry is to be found on the New Jersey shore, while for newer or expanded industries, sites have been provided by reclaiming the marshes and inlets west of the Hudson, in the Jersey City–Newark area. Space on Manhattan is a precious commodity. Extremes in the demand for it are reached in the two nodes of an extensive central business district (CBD), each distinguished by a remarkable cluster of towering skyscrapers. The Lower Manhattan node began as the original CBD of the eighteenth-century city at the southern tip of Manhattan, opposite the separate CBD of Brooklyn on Long Island. Both districts were clustered around the early hub of port activity in the entrance of the East River. As New York developed and expanded its role as control center of the national, continental, and world economies, so the great financial, insurance, and trading institutions of this Lower Manhattan node expanded their offices outward (onto infilled land along the island shores), but mainly upward. The twin towers of the World Trade Center are but a recent evidence of this function.

But New York's increase in economic functions demanded more space. In 1871, railroad access to Manhattan from Long Island and the north was focused, by the construction of Grand Central Terminal, at 42nd Street, almost 4 miles (6 km) north of the original core. Later, at 34th Street, Pennsylvania Station became the stopping point for the only passenger railroad (apart from rapid transit lines) that approaches from the west, tunnels under the Hudson, and then goes on to tunnel under the East River also, to Long Island. Today, this double-tunnel route forms a rather roundabout section of the trunk line from Boston to Washington; having made a swing across Long Island, it regains the mainland by way of a bridge into the Bronx. Offices, hotels, theaters, and other services were established nearby to cater for the incoming business traffic. The headquarters of national and multinational corporations expanded upward and outward (especially in a northerly direction) to create this second, larger and more diverse Midtown CBD node, now reaching from 34th Street to Central Park.

Within metropolitan New York, zonal differences in land use are a source of constant interest to visitors. Along a single street running for twelve blocks from east to west across Manhattan it is possible to find a complete range socially, from the poorest to

Port development in New York Harbor: Container terminals at Port Newark, on the western, or New Jersey, shore of the Hudson River. Such terminals require a huge backup space, which was never available on the Manhattan shore of the port. Much of the container development has been on reclaimed marshland. *The Port of New York Authority*

the richest, and economically, from port function on the West Side, through garment industry and transport terminals, to high-quality residential on the East Side. In culture and even language, too, New York is a series of cities within a city. For a century and a half it has been the funnel through which has flowed the greatest tide of immigration known in modern history, and it has acquired, in the process, a huge foreign-born population. Since these immigrants so often arrived with little knowledge of the ways of the New World, and without even the rudiments of its language, it is not surprising that groups of the same nationality have tended to congregate in specific areas of the city, in Little Greece or Little Italy, where they could do business in the language of their homeland and find the reassurance of at least some vestiges of a familiar culture. Within the narrow limits of Manhattan, these cultural divisions are clear and quite often very abrupt. As in most large American cities, the most clearly demarcated are the city areas in which the blacks live. The word *city* is used deliberately, because the black areas contain a range

of services, and a social spectrum, just as complete as that of New York as a whole. In New York, the focus of everything black today is in Harlem, north of Central Park. How this came about is a long and fascinating story,[9] the point of which, for present purposes, is that this has not always been so. New York City, perhaps because of its long history of immigration, shows with particular clarity the workings of the principle of residential succession. When blacks first began to arrive in New York in considerable numbers, early in the nineteenth century, they congregated in the area around the present City Hall. Then, between 1830 and 1860, a transfer occurred: the blacks congregated in what is now Greenwich Village and were replaced in their older quarters by Irish immigrants. In the 1890s, another shift took place: the blacks moved uptown to the area of the present Penn Station and beyond, and Greenwich Village became Italian. Meanwhile, Harlem was be-

[9] It is told with very ample documentation in G. Osofsky, *Harlem: The Making of a Ghetto* (New York, 1966).

coming the most fashionable residential suburb of the city. Linked with what was then the distant city of New York on lower Manhattan by the first electric railways, Harlem experienced a property boom that first inflated values and then abruptly lowered them. It was in the wake of the property speculators and in the years 1904–5 that housing in Harlem began to attract blacks. By 1914 there were 50,000 in a type of accommodation originally designed for wealthy white New Yorkers: "Negro tenants, offered decent living accommodations for the first time in the city's history, flocked to Harlem."[10] They started in the north of Harlem and moved south toward Central Park in a veritable "On to Harlem" movement, which drew blacks from all over the New York area. Indeed, they were drawn from far beyond the city and in particular from the South, where Harlem was looked upon as a promised land. And so the present Harlem of the blacks came into being.

Today there are two Harlems: the black area we have just described, and Spanish Harlem. This second Harlem is much more recent in its growth, but no different in its character. It grew in the 1960s with the arrival of Spanish-speaking Puerto Ricans and other peoples from the Caribbean. In a movement that New York City encouraged, but which then grew to totally unexpected proportions, the Puerto Ricans arrived looking for work and congregated, inevitably, where the cheapest available housing was to be found. Most of the processes that produced Black Harlem have operated again in Spanish Harlem, but with this difference: that these newcomers often spoke no English, and so found it even more difficult to obtain work in the city than did the blacks.

In keeping with the principle of residential succession mentioned earlier, however, we should not assume that these ghetto areas are, or need be, permanencies. Bad as Harlem has been as a home for New York's black and Hispanic populations, it is probably true to say, first, that today the Bronx, north of Harlem, is worse and, second, that there are signs of an upturn in Harlem itself. At least, experience elsewhere suggests that no area is beyond reclamation. The idea that, once an area becomes a ghetto, there is no other ultimate fate for it than ruin and the bulldozer has been controverted, if only on a small scale, in some American cities. In Harlem, there is a good deal of empty and semi-ruinous property on which a start could be made, if the city of New York

can persuade private developers to share in the task of restoration and private citizens to live in it and look after it.[11]

Around the Central Business District, southern Manhattan, stretch the suburbs, interrupted and elongated by the water barriers that surround the city. These suburbs increasingly contain not only the residential areas, but also the factories, offices, warehouses, and shopping centers of the metropolis, so that the former simple, periphery-to-center trek of the daily work force to Manhattan has been replaced by a much more complex pattern of movement; it has, in fact, been partially reversed as employment has migrated outward. In its outward spread, New York has overrun or overshadowed other cities, such as Newark (1,824,000) and Paterson (141,000), which, in a different setting, would be sizable themselves and which had their own origins.[12] The only way to slow down this outward spread is to build upward—to increase the number of skyscrapers and apartment buildings within the central areas, which, of course, increases the congestion there.

Although the canal link between the Hudson River and Lake Erie, which wrought such a revolution in transport in 1825, has now faded into insignificance, the Hudson–Mohawk route as such has retained, in the railroad and road eras, its importance both to New York City and also to the series of towns that lie along it. This route has been the main channel by which the output of the Midwest reaches its markets and is an admirable location for industries that tap this flow of goods to obtain their materials. At the eastern end of the corridor, where the Mohawk joins the Hudson, lies the Tri-City area of Albany, Schenectady, and Troy, with a population of 874,000.

[10]Osofsky, *Harlem*, p. 93.

[11]The size of the task is such as fully to warrant the question mark at the end of the following title: R. Schaffer and N. Smith, "The Gentrification of Harlem?" *Annals*, Assoc. Amer. Geog., vol. 76 (1986), 347–65.

[12]Those of Paterson are of particular interest: the town grew up at the falls of the Passaic River after Alexander Hamilton visited the site, was impressed by the waterpower potential of the falls, and promoted development through his Society for Establishment of Useful Manufactures. As a result, a mill town grew up and attracted a population of workers just as cosmopolitan as those of New York City across the Hudson. Much of the later growth of the city can be traced in William Carlos Williams's fascinating epic poem *Paterson*. For anyone who has been in Paterson lately, the most surprising details revealed by the poem are probably that (1) in 1917, the local lads captured a seven-and-one-half-foot sturgeon in the basin below the falls, and (2) one of the world's finest pearls was found in a mussel taken from a tributary of the Passaic about the year 1860.

Schenectady, "the electric city," is the home of General Electric Corporation, and the district also manufactures rail stock, machinery, and textiles. Then on the low Mohawk–Erie watershed and in the neighborhood of Oneida Lake are the three cities of Utica, Rome, and Syracuse, each with a considerable range of manufactures and with a combined population of almost 1 million. Further west again lies Rochester (1,002,000). Rochester is a canal town: even though it lies not far from the shore of Lake Ontario, it is not one of the cities of the Great Lakes (see Chapter 14) for it has no port function worth speaking of. It grew up, like Paterson, at a waterpower site—the falls of the Genesee—and the Erie Canal crossed the river just above the falls, giving the city what for the early nineteenth century was an ideal situation. Today, however, Rochester is better known as a cultural center and as the home of Eastman Kodak Corporation.

Besides these major cities there are a number of smaller ones strung out along the routeway. But in spite of their location, these are less prosperous. It seems as if the Hudson–Mohawk route has, in fact, encouraged an unfortunate type of urban development here—a linear development creating a series of towns, several of them dependent on a single industry, and none of them large enough to raise itself to the rank of a major service center or to achieve regional prominence. Although the larger cities on the routeway are relatively prosperous, the smaller ones are struggling. This phenomenon of metropolitan "shadow" is sufficiently noticeable in the Northeast to have prompted the comment that "in New York State it appears that there is a threshold size of approximately 100,000 for substantial manufacturing success.[13] Manufacturing alone, in other words, does little to guarantee the growth or prosperity of a city; that depends on the service functions it develops in relation to its hinterland.

PHILADELPHIA

Owing to the configuration of the river gaps to the north of the city, New York's sphere of influence extends far inland. Moreover, New York draws off, as we shall later see, a large part of the traffic of southern New England. By contrast, the hinterland of Philadelphia is more circumscribed. The city itself, carefully sited and planned by William Penn on his

arrival in 1682, lies between the Delaware and Schuylkill rivers just above their junction. This site gives the modern city advantages of location far beyond either the purposes or the imaginings of its founder: 40 mi (64 km) of navigable waterfront on the two rivers, deep channels, and room to expand. On the other hand, the lack of a natural route to the continental interior, of which Penn took little account, is a continuing problem. The Pennsylvania Railroad (now part of the Conrail system), Philadelphia's main link with the West, winds its tortuous way up the Susquehanna and the Juniata until, with a final contortion, it conquers the Alleghenies by means of the famous Horseshoe Curve. The Pennsylvania Turnpike, the first major trans-Appalachian expressway, led from Philadelphia to Pittsburgh and improved the situation of Philadelphia, but only temporarily: new Interstate highways (expressways) lead directly to New York. The pronounced southwest to northeast grain of the country behind Philadelphia, with its influence on the direction of railway routes, has the effect of bringing within the sphere of New York much of the industrial country of eastern Pennsylvania, which would otherwise be tributary to Philadelphia.

As a manufacturing center, the Philadelphia Consolidated MSA has almost half a million industrial employees and a wide range of industries. It is a smaller version of New York and, like New York, has suffered the loss of manufacturing plants and workers in recent times. It is a major publishing and printing center, it has 23,000 employees working in the clothing industry, and, in the east bank suburb of Camden, across the Delaware River, products from the truck farms of the coastal plains are canned or frozen. West of the city it has a high-tech industrial corridor, along Highway 202, resembling that along Boston's well-known Highway 128 (see Chapter 12).

As a port, the Delaware River handles a tonnage of goods comparable with that of New York, although here there is a good deal more bulk cargo making up the figures, especially iron ore and petroleum.[14] The Delaware terminals receive ore from eastern Canada and Latin America (see Chapter 8). Although some of the ore was carried up the Delaware to Trenton and

---

[13]Thompson, *Geography of New York State*, p. 250.

[14]In recent years, two of the United States' main suppliers of foreign oil have been Mexico and Venezuela. Therefore, apart from the domestic seaborne shipments from Texas and the Gulf, these imports are also likely to be handled at East Coast ports like Philadelphia.

the Fairless Works (the river had been deepened to permit the ore carriers to operate), this has also become an inlet for ore supplies to the Pittsburgh steel region.

It has been said that "with the exception of New York City, Philadelphia is probably the most socially heterogeneous city in the United States. It has a long tradition of immigrant settlement, and a diversified economy that attracted a diverse population."[15] The variety is great in this metropolitan area of nearly 5 million inhabitants, and the distances between disparate social groupings are short. The city has grown from the gridiron core laid out for William Penn in 1682, and which contains the renovated central area of the one-time capital of the United States, out through suburbs that show strong and varied ethnic concentrations, and dense black residential areas on the north, west, and south sides of the city center. In 1980, nearly 38 percent of the population within the city limits was black: for the metropolitan area as a whole, the proportion was 19 percent.

Along the Schuylkill Valley, northwest of the city center, is Fairmount Park which, with Central Park in New York and Prospect Park in Brooklyn, forms a trio of great eastern city parks associated with the name of Frederick Law Olmstead and the 1850s. Further out on the northwest side of the city, where the old Pennsylvania Railroad (now Conrail) ran toward the west, there grew up one of the most unashamedly exclusive series of suburban communities that North America ever produced—the so-called Main Line, served by the railroad, and insulated by every social sanction that its favored inhabitants could devise.

Behind the city, across the gently rolling Piedmont, stretches a virtually continuous spread of suburban settlement, from Trenton on the northeast to Wilmington on the southwest. Beyond the suburban ring, between 40 and 80 mi (65 and 130 km) from Philadelphia, lie a number of smaller industrial centers, such as Reading, Lancaster, York, and the Lehigh Valley towns of Allentown and Bethlehem,

each possessing significant manufactures, with steel goods and textile groups generally most prominent. All these cities lie within the orbit of Philadelphia. Scranton and Wilkes-Barre, in northeastern Pennsylvania, lie remote from the others, in the upper Susquehanna River valley. They owe their growth to the exploitation of the Pennsylvania anthracite coalfield, a field that, at its peak in 1916, yielded almost 100 million tons a year, but today is of negligible significance. After a long period of depression brought on by the declining demand for anthracite and the loss of Scranton's major steelworks to a site near Buffalo on the Lake Erie shore, these cities' fortunes are beginning to recover as improved communications move them effectively closer to Philadelphia and New York, enabling them to share in the overspill effects of industry and business activity generated by the two metropolitan cities. An important example of this is their participation in clothing manufacture under contract and control from the New York garment manufacturing district.

BALTIMORE

Baltimore, the third and smallest of the Middle Atlantic region's port cities, lies at the head of one of the branches of Chesapeake Bay, where the mouth of the Patapsco River provides a natural harbor.[16] Although founded only in 1729, Baltimore had, by 1800, become the third port of the United States. As "the most southern of the northern ports and the most northern of southern ports," it has always drawn upon a wide hinterland for its traffic, and suffered, in consequence, when the Civil War cut off its important southern trade area. Recovering rapidly after 1865, Baltimore exploited to the full, as we have seen, its connections with the interior in the era of railway competition, but its function as a port was nevertheless hampered until the 1930s by the necessity of routing all but the smallest ships down Chesapeake Bay and around the Virginia capes, 170 mi (272 km) to the south of the city. With the enlargement of the Chesapeake and Delaware Canal across the Maryland–Delaware Peninsula, however, a shortcut 35 ft (10 m) deep was created for northbound traffic out of Baltimore.

With its freight-rate advantage and good rail connections to the coal and steel regions, Baltimore's

[15]P. O. Muller, K. C. Meyer, and R. A. Cybriwsky, "Metropolitan Philadelphia," in Adams, *Contemporary Metropolitan America*, vol. 1, p. 219. Reference has, however, been made in Chap. 2 to the fact that, if present immigration trends continue, Los Angeles may be disputing this claim with Philadelphia. See also W. Cutler and H. Gillette, eds., *The Divided Metropolis: Social and Spatial Dimensions of Philadelphia, 1800–1975* (Westport, Conn., 1980); and R. A. Cybriwsky, ed., *The Philadelphia Region* (Washington, D. C., 1979).

[16]On Baltimore, see Sherry H. Olson, *Baltimore: The Building of an American City* (Baltimore, 1980).

chief port function has been the bulk handling of goods to and from the interior, with a particular interest in grain exports, although these suffered from competition through the opening of the St. Lawrence Seaway and improvements in river navigation. Imports are mainly ores and petroleum for local use, and the coastwise traffic consists largely of petroleum either from the Gulf ports or from the Pacific coast by way of the Panama Canal.

As an industrial center, Baltimore has always been regarded as the southeastern outpost of the Manufacturing Belt. Its long-established industries lie along its port frontage, dominated by Bethlehem Steel's Sparrows Point plant, 8 mi (13 km) direct-line distance southeast of the inner harbor. Adjoining the steel mills are shipyards for both construction and ship repair. Chemical, paint, and gypsum plants occupy other waterfront sites. The newer industries—electrical goods, household tools—are to be found inland of the city center, especially along the high-

ways that link Baltimore to Washington, 37 mi (59 km) to the south. The metropolitan area contains almost 150,000 industrial workers, widely distributed across the spectrum of trades.

The closeness of the nation's capital also contributes substantially to the city's economy, as a variety of federal activities have spilled out into Washington's suburbs, which, by a process of growth, have also now become the suburbs of Baltimore. The nation's Social Security Administration, for example, is an employer to rival Bethlehem Steel in importance, and some of NASA's installations lie within the metropolitan area.

As a city Baltimore, although the smallest of the Atlantic Coast giants, is highly individualistic, even in its architecture: indeed, considering the many parallels in the history of New York, Philadelphia, and Baltimore, and the basic similarity of the roles they all three now play, perhaps the most interesting observation to be made about them is that no one who

Baltimore, Maryland: The waterfront. Among the many efforts by U.S. port cities to find new uses for old waterfront areas which no longer conform to the requirements of modern freight handling, that of Baltimore has been particularly successful. *Bruce R. Weller/The Rouse Company*

knows them could possibly mistake any one of them for the others.

Baltimore's distinctive feature is its "row houses"—regular terraces of houses entered by flights of steps from the street; of a variety of ages, yet all conforming to basically the same design. Common enough in Europe, this design is much less so in North America; it has become a trademark feature of the city. To preserve or restore it has given Baltimore an extra stimulus in its urban renewal program. Few cities, it is widely agreed, have been more successful in this task.

This success has carried over into the port area. In seaports all over the world, changing technology has made waterfront structures unsuitable for today's traffic and, consequently, derelict. Boston, New York, and Seattle, and St. Louis on its river, have found this to be the case; in Europe, London, Bristol, and Liverpool have confronted the same problem. What is needed is to find new uses for the waterfront area.

Baltimore's Harborplace shopping and recreation area represents a spectacular response to this problem.

WASHINGTON, D.C.

There is a sharp contrast between Baltimore and its neighbor to the south, Washington. Baltimore is a largely industrial city. In Washington, the proportion of the metropolitan area's work force engaged in manufacturing is a bare 2.4 percent (Table 11-2), and this is made up of two principal categories—printing and publishing, and high-tech industry. The business of Washington is government; the city was founded for this purpose and government accounts for over 40 percent of the employment of the PMSA. Its only other important business, apart from providing services for the government employees, is to cater for the thousands of tourists who come each year to visit the city.

The site of Washington, on the banks of the Poto-

Sparrows Point, Maryland: Part of the steelworks. The works are situated some 10 miles southeast of the center of Baltimore, where the Patapsco River enters Chesapeake Bay. *Bethlehem Steel*

mac, was chosen for reasons of political equilibrium rather than for its locational advantages. Much of the area now covered by the city was marshland, unhealthy and liable to inundation.[17] On this unpromising base was imposed one of the most grandiose settlement plans of modern history, the fruits of which, after a century and a half of reclamation and construction, are to be seen in the broad vistas and carefully aligned buildings of the capital.

In order to free it from any pressure from the states, the seat of the federal government was established in an area—the District of Columbia—carved out of Maryland, which ceded the territory. For the small affair that was the eighteenth-century federal government, this was lavish provision. Its gigantic modern counterpart finds the District (which has an area of 69 sq mi [179 sq km]) all too small, and, especially in times of war overcrowding of both offices and workers has been a serious problem.

The modern expansion of Washington dates from the New Deal of 1933 when, under President Franklin D. Roosevelt's administration, the federal government enormously enlarged its field of operations. Events since that time have merely emphasized the trend. The District of Columbia itself has been losing population; it fell from 797,000 in 1965 to 607,000 in 1990. Like all these large cities, Washington has been emptying out from the center into the suburbs—suburbs that now stretch far out into Virginia, and into Maryland, where, as we have seen, they merge with the suburbs of Baltimore. Washington's spread has been encouraged by a height limitation on the city's buildings imposed by Congress; a city that cannot build upward must build out. Under these circumstances, such a city had better also provide itself with a rapid transit system, preferably underground, and this Washington has done. The present metropolitan area population is 3.9 million, eighth largest in the nation.

Apart from the presence of the federal government and its servants in overwhelming numbers, the other important social fact about the nation's capital is that it was the first large U.S. city to have a black majority in its population. The District of Columbia—that is, the core of the metropolitan area—is over 70 percent black in population: the metropolitan area as a whole

has 1.04 million blacks, which is nearly as many as Chicago, where the population is more than double that of Washington MSA. This is not altogether surprising: for one thing, Washington is the first big city on the road coming up from the South and, for another, the federal government's commitment to equal opportunities in employment means that, as an employer itself, it has provided job opportunities for many ethnic minority workers. By 1990, the MSA also housed a quarter of a million Hispanic people, many of whom, given their language difficulties, had taken over jobs formerly held by the blacks.

To keep it politically neutral, the government of the District of Columbia was for a long time in the hands of a committee of Congress, and not in those of its citizens. The latter now have a say in their own affairs, but the record of the years following this change has been an unhappy one. In spite of this, Washington leaves one in no doubt that it is an imperial city, and the tourists who swarm in its streets and avenues cannot fail to go away impressed.

The Problems of Megalopolis

With so great a concentration of city dwellers enduring the summer climate of the Atlantic Coast, escape from the city has become a major seasonal operation. The coast is lined with resorts, of which the most famous is Atlantic City, and a series of east–west routes through the pine barrens link the cities with their summer annexes by the sea. Some of these resorts are also fishing ports. The sheltered waters of Chesapeake and Delaware bays support extensive fisheries, and the region's shellfish catch is particularly valuable, although pollution from urban sewage is a problem.

This last comment brings us to the concluding point of our survey of the Middle Atlantic Region. The coastal section of the region has become well known to both geographers and general public in the past twenty years under the name given to it by Jean Gottmann—Megalopolis. This sprawl of urban areas along the coastal plain certainly deserves some distinctive name, if only because of its enormous length, its total population, and its distinctive problems. Furthermore, since Gottmann first publicized the name Megalopolis in 1961,[18] urbanization has continued and the built-up areas have grown appreciably. For

---

[17]The site of Washington has, from the first, had a multitude of critics. A geographer joined in when J. Valerie Fifer described Washington as "pegged into a vacant lot along the Fall Line." See "Washington, D. C.: The Political Geography of a Federal Capital," *Journ. Amer. Studs.*, vol. 15 (1981), 6.

[18]Jean Gottmann, *Megalopolis* (New York, 1961).

Atlantic City, New Jersey, with its famous boardwalk. Long a resort city for the urban populations of New York and Philadelphia, Atlantic City has in recent years done its best to become the Las Vegas of America's East Coast. *Atlantic City Convention and Visitors Bureau*

one thing, the interurban spaces—for example, that between Baltimore and Washington—are increasingly filled in by suburban development or by the kind of overspill activity, such as research and educational institutions, which are related to the cities, but actually do not need to be in the central areas. For another thing, Megalopolis is extending north and south of its former boundaries. We shall refer to the growth on the northern fringe in the next chapter. To the south, there is a good case nowadays for regarding both Richmond and the port cities of Norfolk and Newport News as part of the urbanized seaboard too.

Megalopolis has its special problems. Among those to which Gottmann drew attention three particularly should be mentioned. They are (1) water supply; (2) sewage disposal, and pollution in general;

and (3) political and administrative splintering.[19] The first and second of these are obviously linked. All of the major rivers of the region—Connecticut, Hudson,[20] Delaware, Susquehanna, Potomac—must provide for the needs of large urban populations and, what is more, for populations living in two or more states, so that plans for each of these catchment areas require interstate agreements—not the easiest thing to achieve in the United States, at least when a scarce resource is involved. The Delaware, in fact, is within measurable distance of becoming the Colorado River

[19]Volume 8 of the *New York Metropolitan Region Study,* published in 1961, was entitled *1400 Governments.*

[20]See R. W. Richardson, Jr., and J. Tauber, eds., *The Hudson River Basin: Environmental Problems and Institutional Responses,* 2 vols. (New York, 1979).

of the East Coast (see Chapter 20), as New York, New Jersey, and Pennsylvania bargain for its waters. In a conurbation of such unique shape and size, it is vital to coordinate public services and plans for the future. Megalopolis provides a test case that is probably the most pressing, and certainly the most complex, in the Western Hemisphere today.

From a geographical point of view, the problems of water supply, waste disposal, and pollution take on an added dimension because of the nature of the Middle Atlantic coastline. Behind the long line of sandbars that lie offshore are what amount to a series of lagoons, interrupted by long bays—the Chesapeake, the Delaware, and the Hudson River mouth. Behind the sandbars, tidal action is limited: the Atlantic storms and scours do not penetrate here. To all too large an extent, therefore, the outfalls of land water and urban waste are into slack, even stagnant, coastal waters; the shorelines are indented, island-lined, with a thousand piers and promontories to trap wastes and act as obstacles to a clean discharge.

The lesson of Megalopolis is perhaps best summarized in this way: for many centuries, we have thought of the distinction between town and country as consisting mainly of the fact that the space between the towns was used to grow food for the urban populations. We must now recognize that this is not the only or the most important function of interurban space. In Megalopolis, the chief function of space is simply to keep the cities apart—to provide not food, but water, air, elbowroom, and recreation for the huge concentration of people in the cities.

As Gottmann himself expressed it, "the area may be considered the cradle of a new order in the organization of inhabited space."[21]

---

[21]Gottmann, *Megalopolis*, p. 9.

12
New England

Industrial New England: The first industrial revolution. The old Slater Mill at Pawtucket, Rhode Island, now designated a historic site. *Slater Mill Historic Site*

North and east of the city of New York lies an area that forms one of the best-known regions, political and cultural, in American life: New England. For more than three and a half centuries, the six states of this northeastern corner of the United States have in many ways functioned as a unit. Yet in terms of regional geography, the existence of a single, distinctive New England region can be seriously questioned. Southeastern New England, from Portland, Maine, to the suburbs of New York City, is densely populated, urbanized, and industrialized. It is an integral part of Megalopolis, the oldest Atlantic sector of the economic core region. Northwestern New England is sparsely populated and rural, and has more in common with the Adirondack region of New York State or, in a landscape sense, with the Maritime Provinces of Canada than with southeast New England. In the terms we have been using, if the southeast is part of the Core, the northwest is part of the Fringe.

Before we pursue that distinction further, however, there are advantages to be obtained in our regional analysis by considering the two sectors together—in looking first at the total land area and the ways in which that land and its resources are used.

This northeastern corner of the United States, lying between the Canadian border, the sea, and the Hudson–Mohawk corridor, is occupied for the most part by a rolling upland, forested, lakestrewn, and agriculturally uninviting. The main mass of the New England Upland proper is divided by a north–south line of lowland, through which flows the Connecticut River. This is the same Triassic lowland we encountered in the Pennsylvania Piedmont and shall meet again in the Annapolis Valley of Nova Scotia, and nowhere is its significance for agriculture and industry greater than here in southern New England. West of the valley lie the Berkshires and the Green Mountains of Vermont; east of it are the White Mountains, which stretch away northeastward through Maine, to where the Upland is broken along the Canadian border by an area of softer rocks that underlie the valleys of the Aroostook and the St. John.

If glacial erosion has been the decisive natural influence in northern New England, smoothing relief and diverting drainage, glacial deposition has played a role of equal significance in the south. This is a region of outwash plains and moraines. At its seaward edge, the frontal moraine of the ice sheet, deposited on the barely submerged northern extremity of the Atlantic Coastal Plain, has given rise to a chain of islands—Long Island, Nantucket, Martha's Vineyard— and to the peninsula of Cape Cod. The presence of abundant loose, ice-borne material has made this a coast of sandbars and spits under the action of the waves. Inland spreads the drift cover, here sandy and there stony, and dotted with drumlins or marked by the lines of eskers. In the valleys, terraced clays and gravels remain as a product of the action of either ice or meltwater. Much of the lower Connecticut Valley was filled by an ice-front lake; its legacy takes the form of a clay bed into which the present river has cut a series of terraces. In these southern valleys, together with the lowlands around Lake Champlain and along the Maine–New Brunswick border, are found the region's most fertile farmlands.

New England has no coal, petroleum, or iron ore. To offset this near absence of minerals, however, the region does possess three types of resources that, especially in early years, brought it some prosperity— forests, fisheries, and building stones.

The splendid stands of New England timber early caught the attention of the British Admiralty, and throughout the colonial period the region's forests were one of the prime assets of England's transatlantic possession. Today, after three centuries of settlement and commercial use, although much of the for-

Vermont: Gathering sap for maple syrup. This photo shows the industry in its traditional form. Today snowmobiles have largely replaced horses. *Vermont Travel Division*

est cover remains (some 73 percent or 31 million acres [12.5 million ha] of New England are classified as commercial forestland, and forests cover the greater part of the Adirondacks), its value has greatly diminished. It is almost entirely second or later growth that has developed after the cutting and clearing of earlier years. The best of the timber has long since gone; the main forest product is now wood pulp. Maple syrup and sugar are local specialty products. For all its acres of forest, New England imports sawtimber from Oregon and Georgia.

The region's fisheries have fared somewhat better over the years. In 1989, New England's 30,000 fishermen were responsible for 7 percent by weight and 16 percent by value of the total U.S. catch. The principal species landed are cod, flounder, haddock, herring, pollock, and ocean perch. The shellfish catch, although small in weight, brings in over one-half the

returns by value. Among the fishing ports scattered along the coast, Gloucester, Boston, and New Bedford land by far the greatest share of the catch. Some of these places, such as New Bedford, are effectively living through their third career as ports: they began with the transatlantic trade in the great days of New England commerce in the eighteenth century, then they turned to whaling, and now, having abandoned that business, they are fishing ports for the coastal and Banks fisheries, as well as receiving ports for inbound shipments of coal and petroleum products.

The upland interior of New England yields a variety of valuable building materials. Vermont's granite and marble are famous, and the state is the leading U.S. producer of these stones, and of asbestos. Elsewhere in the region granite, marble, slate, and limestone are quarried, together with a variety of lesser-used materials, such as mica and quartz.

The New England Economy

No part of the North American economy has been so frequently and critically analyzed over the past fifty years as the regional economy of New England. Most of the statistical temperature-taking resulted in gloomy diagnosis, and indeed, to some observers outside the region, the patient's case seemed hopeless. Although the New Englanders themselves were far from accepting this diagnosis, they were nonetheless aware of the serious problems that confronted them.

The explanation of this flurry of interest is not far to seek. There was a time when, in terms of population, industry, or wealth, New England possessed a high proportion of the national totals. Its history went back to 1620; its people were industrious and turned early to commercial pursuits; it was here that factory industry in North America began. But the thirteen colonies on the Atlantic Coast gave way to a nation stretching across the continent, and inevitably the relative position of the East within the nation declined.

But this was not all. The relative decline applied to all parts of the East alike. What marked New England for special attention was that it was the first area to suffer, in some respects, an absolute economic decline. The well-publicized fact that textile firms left New England for the southern states was taken as a symbol of failure. For in the climate of U.S. opinion, and in the light of the nation's long history of successful expansion, failure to go on expanding is in itself bad enough; actually to lose ground is to incur a sort of social stigma.

Yet this absolute decline by no means applied to all aspects of the economy, and even if it did, it would still only represent a logical and, for the most part, beneficial redistribution of the nation's population and economic activities, as areas better situated or endowed for particular enterprises were settled and developed. The present size of southern New England's population, and the character of its economy, like those of Megalopolis of which it is a part, are not primarily the outcome of the region's natural attractions, but rather of historical chance—that it was from Europe that the great immigrant stream into the New World flowed. One of the main strengths of the American economy is that, within a great free-trade area possessing enormously variegated resources, the effects of such a historical coincidence can be adjusted with the passage of time. It is this process of adjustment that has been under way in New England ever since World War I.

New England's problem has been primarily industrial in character, but industry is not, of course, the be-all and end-all of economic development, and certainly the long-established and technically unsophisticated industries that New England lost would be out of place in the region today. A far more accurate guide to progress is given by the growth of sophisticated industries and by the volume of tertiary or quaternary employment generated within the region.[1] And in this respect, New England, far from feeling shame at the closure of old textile mills, can point with pride to its ability to adapt to changing economic circumstances. Furthermore, its experiences may serve as a useful model for other parts of the economic core region, not only the older Atlantic Seaboard sector but also the newer midwestern sector, both of which are now facing similar problems of relative decline, while the greater rates of economic growth are today to be found in the West and the South.

Let us pause for a moment to restate this proposition. We have become familiar, not only in North America, but also in all advanced, industrial communities, with the way in which land use and employment structure within a city change as economic activity increases. Factories that were the core of the original settlement are displaced from its center, and their place is taken by commerce and other central functions of the community or its region. The displaced manufacturing may be relocated in the suburbs or in new and specialized communities elsewhere. Within the city, employment shifts from secondary activity (manufacturing) to tertiary and quaternary. All we have to do to appreciate the New England situation is to transfer this concept from a single city to a group of cities; indeed, to a whole region. In southern New England there has taken place a transition in function; that is, a change in the eco-

---

[1] As a diminishing proportion of the work force in developed countries is engaged in primary or secondary activities, it has become desirable to subdivide the remainder, who now represent a majority, and who used to be grouped together as a tertiary category. The problem is to agree the subdivision. Generally speaking, *tertiary* occupations include nontechnical services—food, cleaning, personal—and transport and trade. *Quaternary* carries an implication of office work and data processing. We then have *quinary,* which implies higher degrees of sophistication and decision making. Since the same firm may employ different workers in each of the categories secondary to quinary, this is not a helpful subdivision but it is, for the present, all we have.

nomic relationship between it and other regions of the United States. Within this region, a concentration of tertiary and quaternary functions serving a far wider area than the region itself has occurred. In relation to insurance ser vices, for example, or to education and research, New England possesses resources of national importance.

In short, success or failure for New England should not be judged solely by the size of its industries or the dilapidation of its farmlands, but rather by the volume of its bank deposits, by its output of science graduates, or by its efforts in research and success in innovation.

How this situation has arisen must now be briefly explained.

Rural New England

European settlement in New England began with the establishment of the Massachusetts Bay Colony in 1620. From the early centers in the coastal lowland the settlers spread westward; but here, even more than in the middle and southern colonies, westward movement was hampered by obstacles, both physical and human. On the one hand, the forested slopes and rock-strewn surface of the interior made progress painfully slow, and on the other, there was the recurrent menace of hostile Indians, who, sponsored by the French in Canada, ravaged the borders of the settled area.

Behind this double barrier of physical limitation and Indian hostility, and, cut off for the first 150 years of its history from the great safety valve of the empty areas to the west, New England experienced serious population pressure. The result was an expansion of settlement, into remote and forested upland margins, where the means of livelihood were meager. With the great new immigrant waves of the early nineteenth century, expansion continued, and indeed was hastened, because of the rise of New England's industrial cities and the resultant demand for food supplies—which, in this period before the coming of the railroads, had of necessity to be provided from local sources.

There has been, since well before the arrival of the first European settlers, a distinction between southern New England (where they landed) and northern New England. In the south, they found an area of mixed woodland, with deciduous tree species, open ground, and marshes. Here the Native American tribes lived within a mixed economy of agriculture,

trapping, fishing, and collecting natural food plants, according to season. Further north, however, the forest grew much more densely; it consisted of softwood species, left little if any space for agriculture, and was mountainous and poor in soils; here the Indian tribes lived by fishing and hunting alone. The conquest of what is now Maine and the north by the immigrants came much later.

The amount of cleared farmland in southern New England reached a peak in the 1850s; in northern New England somewhat later. Agricultural production was varied, and although less grain was being grown (in consequence of the opening of the Erie Canal in 1825), sheep and beef cattle were plentiful in the northwest, while the southeast supplied vegetables and dairy produce to the towns.

In the second half of the century, however, a general agricultural decline set in. After 1850, there was little increase in the rural population, and as the century wore on, a reverse flow developed—from poor farms to expanding industries, from the isolation of the uplands to the fuller life of the towns. The high-water mark had been reached by the tide of settlement. Far to the west, in any case, the Pre-emption and Homestead acts were making it a simple matter for the would-be farmer to acquire quarter sections of land that made New England's rocky margins look like a wilderness. Meanwhile, the westward-spreading railroads brought even closer to the eastern cities the produce of this new agricultural West.

These changes had two effects. First, they brought about a gradual retreat of the farmers from the margins of settlement. Farms were abandoned and scrub crept back to cover the cleared pastures. Essentially, farmers abandoned the remoter and higher regions to concentrate on the better soils of the valley terraces and drumlin slopes and on the more accessible farmlands. It was a retreat from lands that would never have been farmed if the lands to the west had been opened earlier, and it was a move to lands that offered the New England farmers at least the advantages of fertility and accessibility with which to meet western competition. The process of abandonment was not continuous; there were fluctuations, but in general, it began earliest and went furthest in the vicinity of the cities, while in the north of the region the story of advance and retreat unrolled with a time lag of twenty to thirty years behind the south. There are areas in New England that were once as much as 70 percent cleared and farmed; today, there is no break in the forest, and the only evidence of the former occupation of the soil is found when, among trees, one stumbles into the cellar hole of a vanished farmhouse. The figures in Table 12-1 tell a part of this story.

In competition with better-endowed areas such as the Midwest, New England farming faces two main physical limitations: (1) Over much of the upland interior, the growing season has a length of only 90 to 100 days, as against 160 to 200 days on the coast and 130 to 150 in the Lake Champlain area. One particular aspect of this handicap is that the season in the interior is too short for corn to be grown for feed. (2) Much of the area is hilly or mountainous and the soils are commonly thin and stony and generally heavily leached. Farm abandonment often began where glacial action had left the poorest soil cover.

New England seems to have reached its agricultural nadir after World War I. Since then, there has been slow adaptation in agriculture and land use, to meet the changing national patterns of population, markets, and economic competition, just as New England industries have had to adapt to regional changes in their competitive arena.

In agriculture, we can identify several distinct adaptive choices, although each tends to overlap the others. At one extreme, and most commonly practiced during the late nineteenth century, was complete abandonment of farms. Many families simply left their farms to the tax collector if they could not sell them, and sought either employment in the rapidly growing New England factory cities or new farms on better lands in the expanding Midwest. Such abandonment resulted in the decay of farm buildings and the reversion of cleared fields to sec-

ond growth or scrub forest. Even today, flying out of Boston over the southern New England countryside, one has the curious impression not that the cities are spreading across the empty landscape, but rather that the forest is creeping stealthily in to extinguish the city, a sort of silent invader who has penetrated to the very gates of Boston.

At the opposite extreme, occurring mainly during the mid-twentieth century, but still proceeding, has been a rationalization of agriculture through selection of specialized enterprises that are adaptable to changing conditions. The specialties include enterprises particularly favored by location near the markets of Megalopolis, or that are least vulnerable to the physical limitations. In the first category, involving the production of perishable or bulky commodities, dairy products are by far the most important, but poultry, some fruits and vegetables, and greenhouse and nursery products also belong to this category. Dairying has been selective not only as a specialty, but also in its use of the best lowland areas. Vermont is foremost in dairying,[2] with a concentration in the Champlain Lowland, an extension southward along the Vermont Valley past Rutland, and into valleys leading to the Connecticut River and to Lake Memphremagog on the border with Québec. New England dairy farms are modern and efficient (see Table 12-2). Because of physical limitations, they emphasize as feed crops hay, pasture, and some corn for silage; most of the necessary grain concentrates are shipped in from the Midwest.

[2]See H. A. Meeks, *Vermont's Land and Resources,* (Shelburne, Vt., 1986).

Table 12-1. New England Land Use, 1880–1987

	Percent of New England Land in Farms	Land in Farms (thousands of acres)	Land Harvested (thousands of acres)
1880	52.3	20,725	5,053
1910	51.8	20,566	4,790
1930	38.2	15,142	3,890
1945	34.5	13,948	3,800
1954	27.6	11,121	2,493
1964	19.5	7,745	1,929
1970	13.9	5,559	1,607
1974	11.9	4,800	1,451
1978	11.9	4,800	1,412
1983	11.9	4,800	1,371
1987	9.3	4,249	1,372

Table 12-2. Franklin County, Vermont: Agricultural Statistics, 1987

Population (1980)	35,000
Percentage change, 1970–80	+15.0
Area of county (ac/ha)	422,000/171,000
Percentage in farms	50.8
Percentage in cropland harvested	18.5
Value of agricultural products sold	$78 million
Percentage from livestock products	98.0
Number of farms	786
Average size (ac/ha)	273/110
Average value of land, buildings, & equipment, per farm	$265,000
Average value of agricultural products sold, per farm	$99,900
Percentage of operators whose principal occupation is farming	76.3

Early fall farm landscape in the Vermont dairy belt. *J. H. Paterson*

Modern poultry and egg production is a form of "house agriculture" or "factory farming"; the large-scale enterprises are located in relation to processing plants or urban markets rather than quality of land. Although poultry is important throughout New England, the greatest concentration is in south-central Maine, a state in which poultry production accounts for about one-half the total agricultural income, or twice that from potatoes. Greenhouse and nursery products, which account for a large part of total agricultural production in the urban Northeast (one-third of the total value in Rhode Island and about one-fifth in Massachusetts and Connecticut), are usually grown on city margins. Unlike dairying, apple and berry growing in New England is usually a specialization of individual farms rather than whole districts, the grower nowadays often inviting nearby city dwellers to come and "pick your own." There is some concentration of orchards in southeastern New Hampshire and eastern Maine.

Three other crops fit into the second category of specializations, those that are less vulnerable to the physical handicaps faced by agriculture within the region. Most important of these are potatoes. Like apples, poultry, dairy, and many other farm products, potatoes were produced on most farms a century ago—mainly for use by the farmer's family, but with some surplus for sale. However, specialization by regions or by individual operators has long since erased that nostalgic picture of small-scale commercial, diversified farming. By the mid-twentieth century, potatoes had come to be produced only in a few specialized regions. They included one on eastern Long Island, now almost obliterated by urban and suburban expansion, and one in the Aroostook Valley of Maine, which continues in the St. John Valley along the Maine–New Brunswick border. Potatoes are relatively well suited to the short growing season and the mildly acidic, glacial loam soils. The highly specialized and mechanized Aroostook farmers, with their half-underground storage and their cooperative packing and shipping plants, were able to adapt not only to the physical conditions, but also to a contraction of the national market as the role of potatoes in the American diet declined. But in the latest stage of the constantly changing market situation, the highly capitalized and competitive processing and packaging of frozen French fries and "hash browns," which are increasing the popularity and raising the consumption of potatoes once again, have made it difficult for the Aroostook area to compete with producers further south, and the farmers are facing hard times.

Two types of berries, both members of the acid-soil-tolerant heath family, are indigenous to the region. Cranberries are grown in bogs near the site of the first European colony at Plymouth. The soils and water levels of the glacial bogs are modified and carefully managed. Native blueberries have long been harvested for home use and commercial sale along the northern part of Maine's "Down East" coast. In recent decades, however, the plant has been markedly improved or domesticated and most of the greatly expanded market is now supplied by domestic plantings in Michigan, New Jersey, and Washington—one more case of changing geography with changing technology.

Between the two extremes of farm abandonment and specialized, large-scale commercial farming, there is a third series of adaptations to agriculture and land use. It involves the use of land for residence, recreation, and aesthetic purposes, especially by people from the cities of Megalopolis. There are numerous variations. The first was the development of elite resorts, often established as spas or health resorts based on mineral waters, such as Saratoga Springs in the Adirondacks. Later, with better cars and roads and more leisure time, larger numbers of people sought the quiet of the countryside. Many of the more affluent purchased declining or abandoned farms, and renovated the often handsome old farmhouses and buildings as summer homes.[3] Some became full-time residents and commuted to work in the cities. Other people camp or rent rooms and cottages, and many farms and associated woodlands have been converted to summer youth camps. By the mid-twentieth century, skiing and winter sports were attracting winter visitors as well. The thousands of new residents and vacationers need services and thus provide an economic base for the further increase of

population. Much of rural New England has recorded remarkable population gains since 1960. One significant variant type of the new breed of users of rural land are the part-time farmers. Theirs is a kind of compromise between the large-scale, specialized commercial farms of today and the small, diversified holdings of the past. In size and diversity, they are more like the latter; however, the farming operation is often only a sideline, the family deriving its income primarily from other sources.

We can summarize this section on the uses and users of rural land in New England and the Adirondacks by identifying three subregions. The first, including much of northern Maine (the "North Woods") and the heart of the Adirondacks, has always been primarily a wilderness with few farms or settlements. The major change in character and use came with the removal of the fine timber during and after the middle of the nineteenth century. The second and third regions have gone through the sequence of events described; widespread agricultural settlement up to about 1840, then agricultural decline and finally, in the mid-twentieth century, specialized agriculture in selected areas with the remaining land, much of it once farmed, devoted to recreational, residential, and aesthetic uses. Of these two subregions, one is a part of Megalopolis; cities are dominant; and there is daily commuting to city work by a relatively dense rural population. The other is situated between Megalopolis and the remote subregion; it is an area where rural land uses are dominant, and the commuting is more often seasonal.

It was a New Englander, Henry David Thoreau, who, in the mid-nineteenth century, built himself a cabin at Walden Pond and encouraged others to venture out into the open air with the words "In wildness is the preservation of the world." Thoreau's own adventures were with something less than wildness: Walden Pond is only 16 mi (25 km) from the Massachusetts State House in Boston. But he would be gratified, if he could return today, to discover how many artists, poets, and playwrights have taken his advice and are living in rural New England and how many of the region's work force are employed in Boston, but drive the Interstate home in the evenings to old farmhouses and rural seclusion.

Industrial New England

The Revolutionary War opened an era of great industrial opportunity for the United States. The old re-

---

[3]In New England, the landscape impact of this tendency is curious: one sees mile upon mile of abandoned fields, with brush invading them and every appearance of neglect, and often containing barns in the last stages of disrepair, but with the farmhouses themselves maintained in a state of such splendor as they can never have known during their life as working farmsteads.

Maine: Farmhouse becomes summer residence. Renovation of New England farmhouses began as early as the 1920s. This towered residence in Damariscotta, a classic farmhouse of a century ago, has been transformed into a spacious "cottage," as owners of such houses affectionately call them. *Joyce Berry*

strictions on native manufacturers that had been a feature of the colonial period (and indeed a contributing cause of the Revolution) were gone, and the United States was estranged from Great Britain, its principal source of manufactured goods. The Revolution, too, was followed by the beginnings of a large-scale westward movement, a movement on which the Louisiana Purchase of 1803 set the seal, and as the nation's territory expanded, so too did the demand for the means of conquering the wilderness.

In this era of opportunity, New England emerged as America's first industrial region. This resulted partly from the fact that New England's interests had long been commercial rather than agricultural, which, in turn, can be attributed to the lack of opportunity offered by the region's farming. Partly it was the product of New England's labor situation; not only was there a surplus to be tapped in the poor farm areas, but there was also an immigrant stream that brought to the ports of the Northeast a supply of skilled craftsmen from Europe. Perhaps the most important factor was that capital and talents, gathered by extensive experience in trading and shipping, were invested in the new manufacturing ventures. Many of these began to produce the very essence of the trading–shipping business: both the things traded, such as cloth, and the means of trading, such as the ships themselves. The new factories were located in the port cities, the great centers of trade, and, as demands for power grew, at the site of falls or rapids along streams such as the Merrimack.

From these early beginnings, the industrial history of New England has run remarkably parallel to that of Old England. Both set out to be "workshops" supplying nonindustrial areas; both lost their early lead through the competition of newer rivals and through changes in the source of industrial power; both find themselves today with a wealth of experience in manufacturing, but also with out-of-date industrial equipment that places them at a disadvantage competing with other regions. Both, too, have made strenuous efforts to remedy their situation.

Table 12-3. New England States: Percentage Shares in Nonagricultural Employment, by States, 1985

	Total Workforce (thousands)	Construction	Manufacturing	Transport, Public Utilities	Wholesale & Retail Trade	Finance, Insurance, Real Estate	Services	Government
Maine	459	5.0	23.1	4.1	23.5	4.6	20.7	18.7
New Hampshire	466	6.6	26.4	3.4	24.2	5.4	21.2	12.7
Vermont	224	4.5	22.3	4.0	22.3	4.5	23.7	16.5
Massachusetts	2926	3.7	22.6	4.3	23.2	6.5	26.6	12.9
Rhode Island	426	3.5	28.2	3.3	21.8	5.4	24.2	13.6
Connecticut	1569	4.2	26.2	4.4	22.3	8.3	22.5	12.0
U.S., 50 states	97,614	4.8	19.8	5.4	23.7	6.1	22.5	16.8

Southern New England is still one of the most highly industrialized regions of the continent (Table 12-3). Rhode Island has 28.2 percent of its nonagricultural labor force employed in manufacturing, the same figure as South Carolina and exceeded only by North Carolina with 31 percent. But changes took place that turned most of New England's earlier advantages into disadvantages. The industries that once made it famous were faltering, and their very concentration in this region proved a source of weakness, for as they contracted, whole cities and areas lost the mainstay of their employment.

The leading industry of southern New England up to World War I was textile manufacture; at the end of the war, the textile mills employed nearly 450,000 workers. The second of the traditional industries was leatherworking and shoes. Both of these old industries were rather narrowly localized; the main areas were northeastern Massachusetts, Boston, and the Merrimack Valley in southern New Hampshire, together with a district comprising Providence in Rhode Island and the Massachusetts cities of New Bedford and Fall River.

But from World War I onward, New England's position as a manufacturing region was clearly weakening. Changes had taken place, which left it at a comparative disadvantage after its early start. Among them were:

1. *A change in industrial materials.* New England's industries developed in a period when a large proportion of all manufactured goods were made of wood, cloth, or leather. The first was available locally; the latter two were brought by the region's dominant shipping industry. But industry became increasingly metal-based and more recently came to depend on a whole range of synthetic materials, none of which New England produces itself.

2. *A westward shift in the nation's "center of gravity"*—both in production and markets—which left New England on the periphery. This was particularly damaging to a region that imported not only its food, but also its fuel and raw materials, as well as exporting its manufactures.

3. *A change in sources of industrial power.* Great Britain's supremacy in manufacturing was made possible by her coal, and her industrial advantages dwindled with the coming into use of newer fuels. But New England's asset—waterpower—belonged to a yet earlier phase of industrialization: the phase passed and the advantage to a region that had neither coal nor petroleum was short-lived (although with the coming of hydroelectricity it has, of course, in some measure returned). The outcome was that fuel had to be imported and so became a high-cost item in the region's manufacturing. (As a consequence, the utilization of energy in New England industry is on the average less than half as much per employee as it is for the nation as a whole.)

4. *A change in social conditions.* The trade unions in New England grew strong; the state governments brought into force a body of law designed to protect the workers, and the region became unattractive to plant owners, who hoped to find more amenable workers and less restrictive legislation elsewhere. At the same time, working conditions in the old mills were unattractive, so that there developed a double exodus from the traditional New England industries—an exodus of firms from the region in search of more favorable operating conditions, and an exodus of workers from the industries in search of other employment.

New England had, in fact, two problems superimposed on one another. One was that from 1919 onward it was losing ground generally as an indus-

New England's maritime traditions. A variety of wooden and iron-built craft representing different phases of New England maritime life—fishing, whaling, trading—are gathered together at the Sea-port Museum at Mystic, Connecticut, formerly an important ship-building center in its own right. *Claire White-Peterson Photo, Mystic Seaport Museum, Inc.*

trial region.[4] The other was that its leading industry, textiles, was particularly sensitive to competition from other areas (among which the South was by far the most serious rival) and new technologies—there were newer mills and newer synthetic fibers to cut into New England's business. And the combined impact of these forces was felt most acutely in the areas where the old industries were concentrated.

In 1939, the textile and leather goods industries between them still accounted for over 40 percent of New England's industrial employment. But after World.War II, numbers employed in the textile mills fell off rapidly—to 280,000 in 1947, 76,000 in 1972,

[4]"For about twenty years prior to 1939, manufacturing industry in New England was declining in both relative and absolute terms." R. C. Estall, *New England: A Study in Industrial Adjustment* (New York, 1966), p. 17. It would be impossible to write about changes in New England manufacturing without being in some degree indebted to Dr. Estall and his work, and the present author readily acknowledges the debt.

and 51,000 in 1982. Employment in the leather industry declined more slowly, but the competition from cheaper, imported leather goods took its toll. Furthermore, within the region the industry migrated, so that the losses in the oldest leather manufacturing areas were very serious.

What could New England do in the postwar years to reform its economy and recruit its industrial strength? Clearly, two tasks lay before it. One was to develop other industries to replace the older, faltering ones. Given the circumstances in the region, the most suitable type of new industry would be one that requires large amounts of skilled or at least semi-skilled labor, but little fuel or bulky raw material. This kind of industry already flourished in southwestern New England: the problem was to establish it in the hard-core area of industrial decline, in eastern Massachusetts and Rhode Island.

The other task was to expand alternative employment, in order to reduce the region's dependence on manufacturing. In New England, with its declining agriculture, this could only be done by concentrating on such activities as commerce and research to serve a national market.

How successful has this adaptation of the regional economy been? In spite of the loss of so many jobs in the textiles industry, total industrial employment rose from 1939 to 1963 by 26 percent, and by a further 9 percent between 1963 and 1969. In the past decade, while the older manufacturing regions of the Northeast have, as we have already seen, been losing jobs rapidly, New England has come close to holding its own. For Massachusetts, the principal manufacturing state, the employment figures were 1972: 619,000; 1977: 613,000; 1982: 643,000; 1987: 591,000. Rhode Island had 118,000 employees in manufacturing in 1972, and 112,000 in 1987. The principal replacements for the older industries appeared in New England during World War II, and have taken root there very firmly in the years since 1945. They are the manufacture of transport equipment (which, in this case, mainly means things that fly) and of electrical machinery. New England has plunged into the space age with all the vigor at its command, backed by its long industrial experience and the resources of its unrivaled collection of universities and technical institutes.

These industries are ideally suited to New England's situation. For one thing, they require minimal amounts of raw material relative to the value added by manufacture, so that total costs of assembling materials and of shipping goods are cut down. For another, they demand a combination of high-grade research and skilled assembly, which are the two things the region is best qualified to provide. In particular, the electronics industry provides plenty of opportunity for the employment of female labor with a background training in factory work, such as the old textile industry offered to so many New England working women. Furthermore, it is part of the character of these industries that the presence of one tends to attract another so that, once the process has started, it builds up a kind of momentum. As we saw in Chapter 8, the Boston area now ranks behind Los Angeles and the San Francisco Bay area in terms of employment in high-tech industries.

The new industries have been responsible for an economic turnaround in the 1980s sufficient to earn the title of the Massachusetts Miracle. "In the early 1970s" says Lampe, "it looked as though the economy of the Commonwealth of Massachusetts was on its last legs. . . But by the mid-1980s Massachusetts was healthier than it had been in decades."[5] Unemployment had fallen, and so had taxes, to below national averages. Defense spending had created demand. Skill for skill, Lampe argued, labor was cheaper than elsewhere, and entrepreneurship was aggressive and fast-moving.

Riding the defense budget roller coaster, of course, has its perils, as other areas (such as southern California) have found out. It induces travel sickness, and it would be unfortunate if, having successfully thrown off one form of industrial overdependence, New England were to succumb to another. But to some parts of the region, the new industrial activity has brought salvation, and it is unwise to look a gift-horse in the mouth too closely. The principal location of the new electronics firms has been in the outer suburbs of Boston, and especially along its ring road, Highway 128. From here, there has been during the past fifteen years a spread northward into smaller centers—in some cases into the very towns (indeed, the very premises) the textile industry had abandoned—places like Lowell, Massachusetts, or Nashua, New Hampshire. Meanwhile, the surviving elements of the textile and leather industries have tended to migrate northward into Maine, where labor is cheaper and less hard to find than in the competitive labor market of the Boston hinterland.

In fact, the area north of Boston has undergone striking development in the past twenty years. Industrial revival is only one reason for a sharp increase in population in the Merrimack Valley, southern Maine, and southern and western Vermont. Other contributing factors have been the growth of tourism, especially winter sports, and the expansion of the "second home" habit, already well developed among New Englanders. Thanks to the expansion of the freeway network, the effective commuting range of the Boston metropolis now extends well up into New Hampshire and Maine, and so in the 1970s and 1980s, this area has shown population-growth figures the like of which have not been seen in New England since the late nineteenth century.

But the old textile areas in southeastern New England have not experienced this industrial replacement. For them, the need has been to expand alternative, nonindustrial employment. At this they have been quite successful, but there has been a consid-

[5]D. Lampe, ed., *The Massachusetts Miracle: High Technology and Economic Revitalization* (Cambridge, Mass., 1988). Quotation from p. 1.

Lowell, Massachusetts: An old mill building finds a new use. The 1880 structure was built by the Tremont & Suffolk Company for the manufacture of cotton goods. It now houses offices and a technology center. *Lowell Historic Preservation Commission*

erable out-migration from New England to other regions of the United States. In the years before World War I, this was masked by the flow of immigrants from Europe entering the region, which in many years reached one-quarter million, and sometimes more. But the habit of moving on from New England is a long-standing one: after all, it was New Englanders who led the way westward through New York and into the Great Lakes area, and when in the 1850s, the question arose as to whether the new state of Kansas (admitted to the Union in 1861) should be a slave state or free, it was from New England that the antislavery campaigners drew a swift tide of settlement to tip the balance.

There remain local variations in the character and well-being of the region's economy. Even in Connecticut, at the "near end" of New England, Lewis and Harmon in their geography of the state[6] distinguish between what they call "mainstream" and "ebb" areas. The latter are the northwest and north-

east of the state. The "mainstream" areas are the coast and the Connecticut River valley. It is in these that the principal cities are located: Springfield–Holyoke (530,000); Hartford (768,000); New Haven (530,000); and Bridgeport (450,000), with their very wide range of light industries—mostly in metal goods—and Hartford's nationally known insurance business.

Northeastward across New England, the importance of the older industries tends to increase. In centers like Providence (655,000) and Worcester (437,000), there is a rough balance between older and newer industries. For example, in Providence, the capital of Rhode Island and second only to Boston among New England's cities, textiles are still made, but one-quarter of industrial employment is accounted for by a long-established jewelry industry, and there are important manufactures of hardware and machine tools. Beyond this intermediate zone lie the old one-industry cities: the textile centers like Fall River–New Bedford, Lawrence, and Lowell, or the shoe towns of Lynn and Haverhill. It is here that the problem of adapting to a new age has been great-

[6]T. R. Lewis and J. E. Harmon, *Connecticut: A Geography* (Boulder, Colo., 1986).

est—but here, too, that most credit lies when that adaptation is successfully made.

The core and headquarters of this industrial area is to be found in metropolitan Boston, with its 2.8 million inhabitants. In terms of employment, the largest industrial groups are electrical engineering, other types of machinery, and instruments. Textiles and leather goods, while still part of the industrial structure, no longer play a leading role, but there is a substantial clothing industry, and a long-established printing and publishing trade.

Comparing the new industries of the Boston area with the old, M. P. Conzen and G. K. Lewis remark, "The electronics industry has become a major component of the Greater Boston economy. This time the investment came not from accumulated commercial capital, but from the assembled intellectual capital of institutions like Harvard University and the Massachusetts Institute of Technology."[7] It is interesting to note that on the now-famous Highway 128, which encircles Boston, and which has become the favored location for high-tech firms, the greatest concentration of these firms is on the part of the ring road—its middle-western section—that lies nearest to the two institutions named.

BOSTON

Boston is the undisputed regional capital of New England—for finance and for the media—as well as the capital city of the state of Massachusetts. Its original site—on a peninsula separated at certain states of the tide from the mainland—was even more circumscribed than that of New York, but reclamation among the low morainic hills and along the sandy shores of Massachusetts Bay has been relatively simple. So the city has been able to enlarge its site as it increases its population, and to spread inland and to reclaim the Bay with a minimum of physical difficulty. The early constrictions of its site, however, gave central Boston a pedestrian-scale size, which is attractive among modern American cities: its Common is a scaled-down counterpart of New York's Central Park, and most of its important buildings are within walking distance of the Common. Socially, Boston shares the character of other port cities of the Atlantic coast, with its immigrant quarters and its black ghetto, Roxbury. Its immigrants include per-

haps the most famous of all such groups in the New World—the Boston Irish.

Whereas in New York, with its greater ethnic variety, no single group can be said to have attained dominance, the Irish had done so in Boston by the end of the nineteenth century. Originally concentrated in the areas of dense and low-grade housing of the inner ring, the Irish, as has been the way with other successful immigrant groups, have followed dispersion paths throughout the conurbation, and today no single area is markedly Irish. But they were followed by the Italians, and the latter, as more recent arrivals, are still congregated very strikingly in a Little Italy in the city's North End, on the tip of the old peninsular site.

Important though Boston has always been throughout New England's history, it has usually been overshadowed by New York, and has sometimes seemed aloof from the concerns of the other Atlantic Coast cities. In the rivalry between the Middle Atlantic cities, which we reviewed in the last chapter, for example, Boston took little part. Most of its interests were maritime; its railroad building was confined to its New England hinterland, and although the Taconics are a much narrower obstacle than the Appalachians, in an economic sense they seemed to shut Boston off from the interior. Southern New England tended to ship its goods via New York, and the equalization of freight rates to Europe from all the northern Atlantic ports robbed Boston of the advantage it might have enjoyed through its position on the great circle route to Europe.

All these limitations seemed, until recently, to be epitomized by its port. Despite its splendid harbor and good equipment, in the early 1970s it was handling less than one-quarter of New York's business; what is more, the bulk of the freight movement was simply the coastwise receipt of fuel supplies needed by a fuel-less region.

If the port symbolized the city then, it may perhaps do so in the future, too. For the port of Boston (or Massport, as it is now known) has staged a revival. There are two reasons—good labor relations, giving strike-free operation, and whole-hearted commitment to "containerization" and the facilities it requires. Remarkably, it seems as if the northeastern United Sates has room and business both for a resurgent port of New York (as we saw in the last chapter) and a revitalized Boston.

As a delightful bonus, Boston has not only increased its port traffic, but has succeeded in renovat-

[7]M. P. Conzen and G. K. Lewis, "Boston: A Geographical Portrait," in Adams, *Contemporary Metropolitan America*, vol. 1, p. 67. This study of Boston (pp. 51–138) is warmly to be recommended.

Boston: The harbor from the southeast, with the international airport on the right. This fine and historic harbor was long overshadowed by New York's farther south, but the past few years have seen a remarkable upsurge in its fortunes, thanks to good labor relations, low pilferage rates, and the adoption of new techniques of freight handling. *Massachusetts Department of Commerce and Development*

ing its port buildings. The old warehouses of the waterfront, grim, ugly, and massive, were until the 1970s an apparently immovable, quite useless reminder of the port's earlier functions. The waterfront area was derelict. But if the warehouses were too solid to remove, they were not beyond conversion, and converted they have been, to form a shore area of apartments, shops, restaurants, and marinas that has become a sought-after residential setting, the focus of Boston's life on summer evenings, and a major tourist attraction.

New England, like any other region, has its strengths and its weaknesses. Most of the regional surveys carried out over the past decades have concentrated on the weaknesses, such as the aging industries and the high cost of fuel and power compared with levels faced by competitors elsewhere. But there are strengths, too, in the regional economy. The cost of skilled labor is low and so, too, is its turnover rate. Per capita incomes in southern New En-

gland are high; in Connecticut, the highest in the nation, apart from Alaska with its quite exceptional economy. If electricity is more expensive than in other regions, then New England's labor-intensive industries use less of it per employee than those other regions. If some industrial areas are crowded and lack development space, then this drawback is counterbalanced by the economies of agglomeration—the advantage of being close to suppliers or sources of labor.

A New Englander might, however, argue that much of this is quite incidental in assessing the region's strengths and prospects—that it is better off without its marginal farms and textile mills; that the railroad era, with its fraud and stock watering, was an episode best forgotten, and that the true wealth of a region resides in its people. In this respect, New England can claim to be truly wealthy. Starting back in the days when the New England colonies formed the most prosperous block on the Atlantic Coast, and the

Atlantic Coast was America, this region has developed what has been called "a concentration of talents." Its resources are not primarily natural but human. If the realization of this fact has given New England a somewhat patrician attitude to younger and more obviously productive regions of the nation, this is understandable and so, too, may be a certain resentment against it. But no one can deny the great contribution made to the economic and cultural life of the United States by New England's men and women of business and letters.

13
The Canadian Heartland

Toronto: The lake shore. In the middle of what was formerly a maze of railroad sidings and sheds, there now rises the huge Canadian National Tower and the new 55,000-seat Skydome, with a retractable roof. The main station is to the right, and beyond the railroad lie the buildings of the CBD. *Metropolitan Toronto Convention and Visitors Association*

The economic core of Canada—or alternatively the part of the continental core that lies within Canada—can be divided, like its United States counterpart, into subsections. In Canada there are two such sections. These differ from each other in the period of their growth and in the character of their industry and commerce. They are quite distinct in their location and are separated by a physical borderland—the southward extension of the Canadian Shield to the St. Lawrence and beyond. In Canada, also, although not in the United States, these sections are distinguished by another factor: their residents speak different languages.

The core region lies along the St. Lawrence and the southern Great Lakes, delimited on the south by the international boundary and on the north by the edge of the Shield. It stretches, with the break in the middle we have already noted, from Québec City in the northeast to the tip of the Ontario Peninsula in the southwest and, in consequence, it is sometimes known as the "Québec–Windsor Axis" or "Canada's Main Street."[1] The two subsections can then be identified as the East Axis and the West Axis. The focus of one is the metropolis of Montréal; of the other, Toronto. One is predominantly French-speaking, the other, English.

With no more information than this at our disposal, we can perhaps go on to predict that certain other things are true of these two sections of the core. Under these circumstances, we can anticipate that the two sections will compete with each other much more directly and consciously than do the sections of the United States core—say the Middle Atlantic section and New England. We can predict, too, that such a French–English dichotomy will lead to some du-

plication and so to waste, for to divide an economy—work force, advertising, research, legal structure—into two or more parts is always likely to lead to duplication of effort, and that is waste.

We shall later examine the relationship of the two sections to each other, but first let us consider them separately.

The Eastern Section: The Lower St. Lawrence Valley

This section is small but very distinct. On the one hand, it possesses a physical separateness that leaves it with few connecting links with the remainder of the continent. On the other hand, and more significant, it has the cultural distinctiveness proper to an area in which some 80 percent of the population speak by preference a language different from that of the other 265 million inhabitants of Anglo-America. This is the heart of French Canada.

A map of population distribution shows the extent to which this region is isolated by natural barriers from its surroundings. On the north shore of the St. Lawrence, population density thins rapidly up the slope to the plateau surface of the Laurentian Shield, and there are few communities more than 50 mi (80 km) from the shore; beyond lies the immense emptiness of northern Canada. On the west, the Shield encloses the St. Lawrence and Ottawa valleys, swinging south to cross the former at the Thousand Islands Bridge. Here there is, on the population map, only a slender "connecting link" of settlement to join Lower Canada to Upper Canada, and travelers between Montréal and Toronto are conscious that they have crossed an economic and cultural no-man's-land, even though the Shield presents no relief obstacle. On the south, settlement spreads across the broad plains of the St. Lawrence and Richelieu valleys until it ends rather abruptly along the edge of the

[1]See M. Yeates, *Main Street, Windsor to Quebec City* (Toronto, 1975). See also D. F. Walker, *Canada's Industrial Space-Economy* (Toronto and London, 1980).

Adirondacks and the mountains of northern New England. There, the map reveals the one major link with settlement in other regions, where the Champlain Lowland cuts across the international boundary, and Vermont adjoins southern Québec. Finally, on the east, settlement becomes progressively sparser as the river widens into the Gulf of St. Lawrence, until only a string of fishing and logging villages lines the narrow corridor that joins Laurentian Canada to the Maritime Provinces.

Yet for all its separateness, this region is, without question, part of the Canadian core—and was, in fact, for two centuries the only area with any pretensions to that title. It was the base from which westward penetration was directed; the only ready gateway to and from the northern part of the continent. Today, it contains one-quarter of Canada's population, one of its two largest cities, and its busiest port. Its transport links across the adjacent empty areas are good enough to enable it to handle a major share of Canada's foreign trade. And when the St. Lawrence Seaway was opened in 1959, it acquired a "back door," which reduced its isolation and set it astride an international routeway.

It is 400 mi (640 km) from the Thousand Islands Bridge to the end of continuous settlement on the north shore of the St. Lawrence beyond Québec; some 650 mi (1040 km) from Ottawa to Gaspé. At their broadest, the lowlands stretch for 120 mi (192 km) from northwest to southeast. Only in the junction area between the Richelieu–Champlain Lowlands and the St. Lawrence Valley are there wide stretches of fertile soils. There, however, much of the land is of very good quality, as a result of the deposition of marine sediments in postglacial times, when the lowlands were submerged beneath the gulf known as the Champlain Sea. It is this junction area which forms the economic heartland of Québec, dominated by the spreading agglomeration of Montréal; a fertile plain out of which rise abruptly a series of volcanic "islands," many of their lower slopes planted with apple orchards, but now increasingly being used as sites for new homes.

Climatic conditions in the valley are also distinctive: winters are severe, snowfalls are heavy, and weather changes are frequent. These conditions reflect, as we noted in Chapter 1, the convergence of storm tracks on the Great Lakes–St. Lawrence line. The passage of the depressions is responsible for the variability of conditions and for the considerable winter precipitation, which brings snowfall of 100 in

(2.5 m) per annum or more to many valley stations, even at sea level.

Over most of the valley, precipitation is more than 35 in, evenly distributed throughout the year. The January mean temperature is 14°F (−10°C) at Montréal and 10°F (−12°C) in Québec City, and it falls rapidly, on the slopes above the river, to 0°F (−18°C) in the Laurentide Mountains. The St. Lawrence River and the Seaway have both normally been closed to shipping between December and mid-April, but in the past few years the river itself has been kept open for a limited amount of winter traffic as far upstream as Montréal. The frost-free period in the valley is usually limited to 120 or 130 days (Fig. 13-1). July mean temperatures fall from 70°F (12°C) at Montréal to 58 or 60°F (14–15°C) downriver opposite Anticosti Island. In spite of its long coastline, the Gaspé Peninsula has the wide annual temperature range (50 to 55°F or 28 to 30°C) typical of eastern continental margins in these latitudes—Bangor, Maine, for example, has a range of 48°F (27°C) and Vladivostok in eastern Russia has one of 63°F (35°C).

Freezing winters and heavy snowfall affect life in Québec in a number of ways. One is to restrict winter employment opportunities, thus adding a problem of seasonal unemployment to the difficulties of the region. Another more pleasant impact of the climate is upon urban services, for in the past twenty-five years Montréal has created for its shoppers a whole city underground, where one can wander among stores and services without ever being aware of the weather outside. With the reconstruction of transport terminals and the building of the Métro lines, there is little that the city cannot now provide in this world underground.

SETTLEMENT AND LANDSCAPE

For geographers, the St. Lawrence Valley provides a useful illustration of the maxim that to interpret the landscape of a region, it is necessary to be familiar with its settlement history.[2] The valley was settled by the French in the seventeenth and eighteenth centuries, and although by the end of French rule in 1763 there were only some 65,000 French (whereas the British colonies further south in the continent had more than 1 million inhabitants), they nevertheless created by their presence a landscape whose distinctiveness remains to the present day. Not only the

_____

[2] A good starting point for studying this subject is R. C. Harris and J. Warkentin, *Canada Before Confederation* (Toronto, 1974).

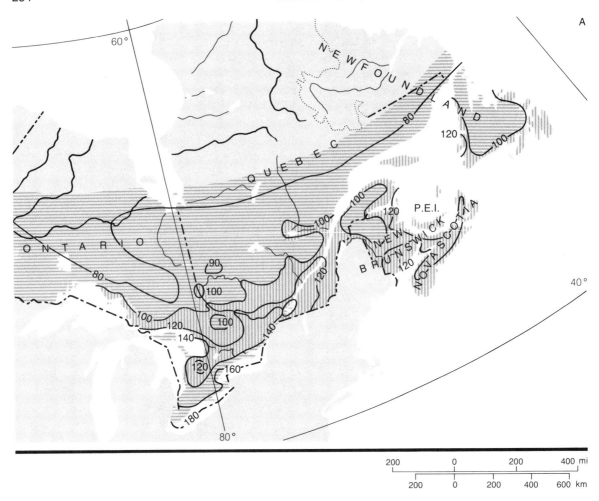

Fig. 13-1. Eastern Canada: (**A**) Frost-free season and (**B**) Growing season. The figures show the number of days in each season. As a commentary on the figures shown on this map, most commercial crops grown in North America are in areas where the frost-free season is, as a minimum: for small grains, 90–100 days; for grain sorghum, 130–140 days; for corn cut ripe, 140+ days; for cotton, 180–200 days. Both maps are taken from the *National Atlas of Canada*, 4th ed. (Ottawa, 1974). © Her Majesty the Queen in Right of Canada with permission of Energy, Mines and Resources Canada. Compare Fig. 18.2.

place-names of the Lower St. Lawrence, but also the rural settlement pattern, reflect the legal and social arrangements of the French colony that later became Lower Canada and, on federation in 1867, the province of Québec.

French settlement in Canada may be dated from 1608, when Québec City was founded by the man whose leadership dominated the whole enterprise—Samuel Champlain. Although Champlain planned to establish the colony on a firm agricultural basis, it was the fur trade that attracted the French, and the fur trade to which their main efforts were devoted. The permanent settlements on the shore of the St.

Lawrence languished, while the French pioneers spread over the interior the impermanent form of occupancy and control that fur trading implied. Officials, missionaries, and freelance fur traders—the coureurs de bois—pushed swiftly inland, becoming embroiled in intertribal Indian wars; adopting Indian modes of life and travel; constantly seeking new fur supplies as they pushed westward across the northern Great Lakes to the Upper Mississippi. In 1670, French sovereignty was proclaimed at Sault Ste. Marie. In 1682, La Salle followed the Mississippi to the sea and claimed a vast Louisiana for the king of France.

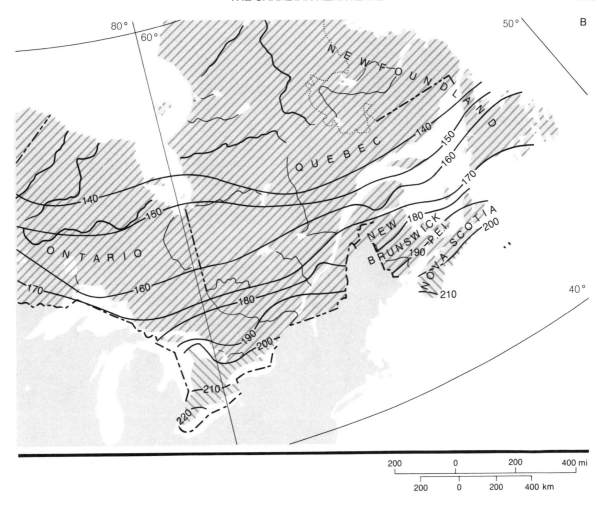

In the meantime, the neglected settlements on the St. Lawrence had achieved by 1660 a population of about 3000. They had spread along the river and were later to spread along its tributary, the Richelieu, in a single line of waterfront settlement. This pattern developed partly because movement by river was simpler than ashore; partly because of the importance of fisheries in the early colonial economy and partly because the river verges offered unforested patches, where the labor of forest clearance might be avoided. Holdings were laid out in long, narrow strips ("long lots"), at right angles to the river frontage, with the homestead close to the water's edge, so

that, as one writer has expressed it, at the end of the French régime, a traveler could have seen almost every house in Canada as he made the canoe trip along the St. Lawrence and Richelieu.

On the inland side, the limit of the holdings was usually only a distant line toward which clearance of the forest slowly progressed. But as the population of French Canada increased, not only were the holdings subdivided into still narrower frontages, but a second line (or *rang*) of settlement was also laid out parallel to that along the waterfront. Here, the process of parceling out and clearing the narrow strips was repeated, the rural road replacing the river as the base

Québec: The legacy of France. Long lots on the north shore of the St. Lawrence near Québec City. *National Air Photo Library of Canada*

line. Thus there developed a pattern in marked contrast to that of the areas further to the west and south, where survey based on mile-square sections produced the familiar gridiron settlement pattern of the American interior. The French pattern has survived in Québec; indeed, in a modified form it has been used in the most recent expansions of settlement in the province—the Abitibi and Temiscaming areas.

At no time during the period of French rule did agriculture develop real strength: it was carried on merely as a support for the fur trade and the local population serving that trade. Periodically it failed even to fulfill this modest task, and famine resulted. This agricultural weakness was almost certainly one of the causes of the French loss of Canada, for France was competing with a rival very differently placed:

"of . . . fundamental importance in the final French withdrawal from St. Lawrence and Cape Breton was the agricultural backwardness of New France as compared with New England. The New England colonies early developed agricultural resources more than sufficient to provision the British staple trades. France could scarcely withstand an opponent upon whom she continuously relied for foodstuffs."[3]

At the end of French rule in Canada in 1763, the settled areas still extended only a short distance back from the shores of the St. Lawrence and Richelieu. The period of British rule that followed, especially in the early nineteenth century, saw an expansion of the settled areas, first into the broad lowland south of

[3]V. C. Fowke, *Canadian Agricultural Policy* (Toronto, 1946), p. 6.

the St. Lawrence and east of the Richelieu, which is known as the Eastern Townships. Here the original settlement was predominantly British, and the place-names were sturdily Anglo-Saxon. Sherbrooke, the urban focus of the region today, was founded in the 1790s by Loyalists moving north from New England in the aftermath of the American Revolution. But the population of the older French areas increased rapidly (by 1830, it had risen to about 400,000), and it overflowed into the Eastern Townships. More, it was encouraged to do so, and we have the paradoxical state of affairs that the British Crown encouraged French Canadians to settle in the English-speaking Eastern Townships in order to keep at bay a threatened wave of U.S. immigrants. Today, the Townships have six times as many French speakers as English, Sherbrooke is 90 percent French-speaking, and such place-names as St. Germain de Grantham and Ste. Anne de Stukely reflect this changing cultural affiliation.

The French Canadians continued to increase in numbers, but owing in part to the agricultural weakness of their home base—the St. Lawrence lowland—they were obliged to seek other outlets for their excess population.

These they found to the north, south, and east. Northward they spread up onto the Laurentian Plateau, carrying their subsistence agriculture with them and pushing forward the farm frontier. In most of these areas north of the St. Lawrence, the agricultural acreage reached a maximum about 1920, whereas the area of all land in farms in the province reached a maximum in 1941 of 18.1 million acres (7.33 mil-

lion ha). Subsequent changes in area are shown in Table 13-1. These decreasing figures are a commentary on the thrust into the north of the Québec pioneers and their subsequent withdrawal as they abandoned the thankless and unrewarding task of farming on the fringe and sought work in cities and industries (from which, as it happens, quite a number of them had originally come).[4]

Eastward, French Canadians increased in numbers all along the south shore of the St. Lawrence and into New Brunswick, although they have never recolonized in any strength the original Acadian settlements in Nova Scotia. Southward, expansion of French Canada was possible only by crossing the border into the United States. But this proved to be no obstacle to movement as pressure mounted in the St. Lawrence Valley. The industrialization of New England was in full swing, and there were plenty of jobs avail-

[4]Already in 1972, it has been estimated that nine out of every ten farms opened by the Department of Colonization in Abitibi, north of Ottawa, had ceased to be operational. Eric Waddell comments, "For almost a century, from 1870 to 1950, the intellectual elite of Quebec promoted a myth of the North, a vision of a promised land where French Canadians were destined to settle, and where spiritual and material regeneration were to be assured through a tactical retreat from Anglo-America. . . . This dream of salvation and regeneration in the North failed to materialize." Eric Waddell, "Cultural Hearth, Continental Diaspora," Chap. 5 in L. D. McCann, ed., *A Geography of Canada: Heartland and Hinterland* (Scarborough, Ont., 1982), p. 145.

The present author owes it to McCann and his colleagues to point out that, although quotations cited here are drawn from the 1982 edition of his book, there is a later, 1987, edition which can be consulted.

Table 13-1. Province of Quebec: Area Under Agriculture, 1951–86

	1951	1961	1971	1976	1981	1986
Number of farms	134,336	95,777	61,257	51,587	48,144	41,448
Area in farms (million ac/ha)	16.8	14.2	10.8	9.9	9.3	8.9
	6.8	5.7	4.4	4.0	3.8	3.6
Area under crops (million ac/ha)	5.8	5.2	4.3	4.6	4.3	4.3
	2.3	2.1	1.7	1.8	1.7	1.7
Farmland but unimproved (million ac/ha)	7.9	6.3	4.3	4.0	3.5	3.7
	3.2	2.6	1.7	1.6	1.4	1.5
Area under wheat (ac/ha)	—	10,600	39,100	78,800	101,100	160,500
		4,300	15,800	31,900	40,900	65,000
Area under oats (ac/ha)	—	1,298,500	695,000	556,600	370,900	254,400
	—	281,400	225,300	150,200	103,000	
		525,700				
Area under hay (ac/ha)	—	3,312,200	2,698,700	2,765,400	2,385,700	2,435,400
	—	1,341,000	1,092,600	1,119,600	965,900	986,000

Note: In each line showing areas, the upper row of figures represents acres, and the lower hectares.

able in its factories, while back in the Valley crop failures were recurrent. Between 1840 and 1900 it is estimated that French Canada lost about 600,000 people to the United States, the great majority of them to New England. In 1900, in fact, there were 573,000 French Canadians within the region, with the largest concentrations in Providence, Worcester, Fall River, New Bedford, and the cities of the Merrimack Valley. The chief draw was the textile industry, in which developing technology had reduced the level of skill required of operatives, and so reduced its attractiveness for the local American workers, who sought better jobs elsewhere. By 1900, in most textile towns, one-third of the operatives were French Canadians.

The special relationship between Québec and New England has been a reciprocal one, which has continued through many vicissitudes. While Québecers were moving south to the mills, the Eastern Townships in particular were attracting New England vacationers and, more recently, there have been many buyers of land and summer homes. The market has risen with the construction of expressways that have brought the cities of the two regions effectively closer, and has fluctuated with the availability of land in the upper New England states, or with restrictions on its use, like rural zoning, on both sides of the border.

The agriculture of modern Québec makes it a part of the Dairy Region, with 57 percent of the crop acreage under hay and a further 9 percent under oats. The cities of the St. Lawrence Valley provide markets for fluid milk, and butter and cheese are manufactured in large quantities; Québec accounts for more than one-third of Canada's butter and cheese output. In addition to this basic farm activity, however, the valley's agriculture includes the production of a number of special crops. The presence of Montréal and its suburbs has encouraged the rise of market gardening in their vicinity. Small fruit crops are numerous in the valley, and Québec produces apples, as well as strawberries and a large part of the country's output of maple syrup and maple products.

As is the case everywhere along the northern fringe of the Dairy Region—in northern New England, for example, or in northern Michigan and Wisconsin—there is a marked falling-off in the intensity of land use and activity on the remoter fringes of Québec's farmlands. Even in the long-settled Eastern Townships, only about one-half of the farmland is improved. Away from the valley markets and the

creameries, agriculture approaches a subsistence level, and farm income is supplemented, to an increasing extent, by fishing or by work in the forest or in industry. More than a quarter of all farms produce less than C$5000 a year in off-farm sales. And a final statistic places all these others in perspective: within the entire Province of Québec, the land in farms represents just 2.8 percent of the total area.

POWER AND INDUSTRY

The St. Lawrence Valley possesses no coal and no petroleum. It does, however, have two resources that are important by any standards: it is the world's leading producer of asbestos, and it controls more than one-third of Canada's installed hydroelectric generating capacity. The asbestos is mined in the Eastern Townships in the neighborhood of Thetford Mines and Asbestos, and the bulk of the production is exported to the United States. The hydroelectric power is produced mainly on the southern edge of the Canadian Shield, which, in the course of its erosional history, has developed fall lines similar to that at the eastern edge of the Appalachian System (see Chapter 1). Such fall lines, or breaks of slope, create excellent sites for the production of power, and conditions approach the ideal on the rivers that descend through chains of lakes from the elevated southeastern corner of the Shield to the St Lawrence (Fig. 13-2). Of these rivers, the early power producers were the Saguenay, flowing out of Lake St. John past the great Shipshaw power stations; the St. Maurice, on which the development around Shawinigan accounts for one-half million horsepower of generating capacity and other works above and below the falls for a further 1.5 million; the Ottawa, with its tributary the Gatineau; and the St. Lawrence itself. These rivers, affording suitable sites within easy reach of the St. Lawrence Valley power users, were exploited first, and a large part of their potential has been realized. Since, however, demand has continued to rise, more remote reserves have been tapped, such as those of the Bersimis and Manicouagan rivers, 200 mi (320 km) below Québec on the empty north shore of the St. Lawrence; this in addition to the considerable bonus presented to the Valley when the St. Lawrence Seaway was built and its by-product, power, was deposited at the very doorstep of the Valley towns.

But the story does not end here. The larger potential and the bigger possibilities lie beyond the watershed—on the rivers that run down to Hudson Bay and out to the Atlantic along the coast of Labrador.

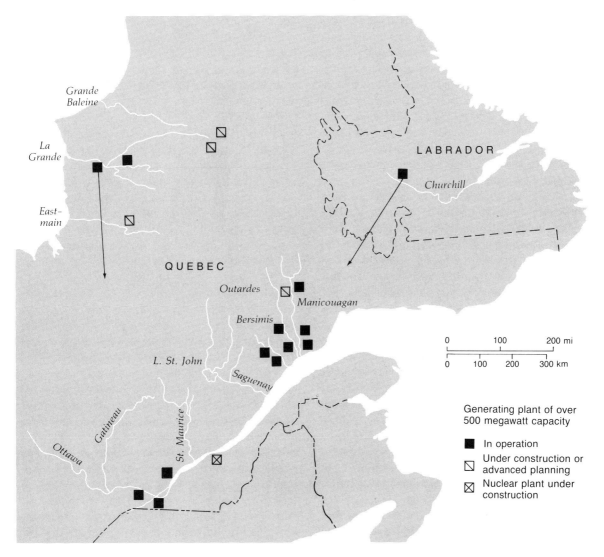

Fig. 13-2. Québec and Labrador: Hydroelectric development. The map shows the principal rivers so far harnessed as sources of power. The most recent of these is the La Grande complex on the eastern side of Hudson Bay. Only the major power plants are shown. Power from La Grande and from Labrador's Churchill Falls is transmitted south over 735 kV lines to metropolitan Canada.

To develop them would mean carrying in workers, equipment, and generators, maintaining them in the wilderness, and eventually transmitting the electricity out again over 600 mi (1000 km) to users in the Valley. It was, and is, a formidable challenge. But it has been met; two giant enterprises, at least, have been successful. One is the Churchill Falls scheme in Labrador. The scheme has created over 5 million kw of generating capacity in Newfoundland–Labrador, for most of which, given the region's lack of development or population, the province has no use. The greater part of the output is therefore acquired by Hydro-Québec. The second scheme, already referred to in Chapter 8, is on the La Grande River on the east side of James Bay. Power has been flowing from the first turbines since 1979. When the scheme is completed, its estimated capacity will be over 13 million kw.

Alongside these two major resources of the Valley, we should place a third: the iron ores of the Québec–Labrador belt. They are not located in the Valley, but they acquire value only when they are brought down to tidewater, at Sept Iles or Port Cartier. In this sense they belong to the region, and the more recent of the mining developments are unquestionably in Québec.[5] The ore bodies form a long chain of deposits running generally from southwest to northeast and north, and were initially opened up in the 1950s, when the town of Schefferville was built to service the first mining. Railways were laid to bring the ore down to tidewater. The Wabush mines followed in 1965, and markets could be reached by shipping ore from the St. Lawrence ports, either westward to the Great Lakes, or eastward and south to the Atlantic Coast steel mills.

The Québec–Labrador Ungava ores are of high quality, but they are geographically remote. In times of falling demand for iron ore (due, as we saw in Chapter 8, to foreign competition and a falloff in Anglo-American steel production; due, also, to an increase in the use of scrap in furnaces), all Canada's iron ore production has been cut back and Schefferville, the original ore town in the wilds, has been closed. Like so many other of the world's mineral developments, this one depends on price and demand for its continuance.

As a location for industry, the St. Lawrence Valley enjoys the advantage conferred by its power supplies, to add to the advantage of its position as the funnel through which goods leave and enter Canada's eastern side. Although this latter advantage is limited in winter by the closure of the Seaway, the Valley exerts sufficient attraction upon industry to employ more than half a million workers in manufacturing—30 percent of Canada's total. (See Table 13-2.)

Much of this industry is textile-based: the textile and yarn mills are located in the small towns east and northeast of Montréal, and the clothing industry is based (and here there is a parallel with New York) in the metropolis. This is one of a number of light consumer goods and luxury industries commonly found, together with food products, in a major regional center. Montréal also refines petroleum, which it receives by sea and by pipeline, either from the Canadian western fields or from Portland in Maine. Engineering industries include a small share in Canada's automobile, aircraft, and electrical goods man-

Table 13-2. Province of Québec: Industrial Employment by Category of Industry, Census of 1986

Clothing	63,539
Food and kindred products	46,301
Paper and allied products	43,259
Transportation equipment	37,636
Electrical and electronic products	37,257
Wood industries	34,250
Fabricated metal products	35,943
Printing and publishing	32,168
Primary metal industries	27,020
Chemicals and products	25,299
Furniture and fixtures	20,446
Textile products	16,009
Machinery industries	14,927
Primary textiles	13,976
Total, all industries in the province	520,459

ufacturing. All in all, Montréal accounts for over 15 percent of Canada's industrial output.

But the Valley's particular attraction is the availability of electric power, which in Québec draws two outstanding power users: the manufacture of woodpulp and paper and the smelting of metals, especially aluminum. Some 80 percent of Québec's power consumption is accounted for by these two industries.

The Canadian Shield is not merely a rich source of hydroelectricity; it is also a great forest area, and these two resources combine to produce the pulp and paper industry. Although this industry represents only a part of the total forest products output of Québec, the whole of which gives full- or part-time employment to many thousands of workers, nevertheless it consumes about two-thirds of the annual cut from the forests. The industry is located in close proximity to the power sources, along the Saguenay, the St. Maurice, and the Ottawa rivers, and large mills are situated in Trois Rivières, Québec City, and Ottawa–Hull. The province of Québec accounts for nearly 40 percent of Canada's output of pulp and paper. These form the nation's principal exports, the item of greatest value being the sale of newsprint to the United States. Since the demand for this is likely to increase rather than to decline, it seems probable that Québec will experience a gradual northward shift of the forestry frontier away from the St. Lawrence and into the vast areas of timber reserves that cover the Shield.

Aluminum smelting, another power-hungry in-

[5]The writer uses what may appear a curious phrase because the original strikes were made in an area that has been in dispute since 1927 between Québec and Newfoundland. In 1927, the Privy Council in London fixed the border between Québec and what was then the separate colony of Newfoundland, but this boundary was never recognized by the Province of Québec.

dustry, is carried on at three main locations, Arvida on the Saguenay, Shawinigan on the St. Maurice, and at the mouth of the Bersimis. The fact that Canada itself produces no bauxite, so that all the raw materials for the smelters must be imported, and the additional fact that a large part of the smelter output is exported again, is evidence of the attraction of the power factor to this industry. The same attraction will doubtless continue to draw new industries, such as chemical manufactures, to this area at the edge of the Shield, as more power becomes available.

The import of bauxite and the export of iron ore which we have been considering bring considerable traffic to the ports of the St. Lawrence north shore. Apart from Québec City, which is a general cargo port like Montréal, but on a smaller scale, there are (in rough order of traffic handled) Sept Iles, Port Cartier, and Baie Comeau. Half a century ago, some of these ports were nothing but Hudson's Bay Company trading posts in the wilderness: they have grown with mineral development and, in the way such places have, they may die with its decline.

In the industrial structure we have just described, there are two serious sources of weakness. One is a general absence of modern, innovative industries— the kind of "propulsive" industry that brings others into being. The other is the heavy reliance on textiles and clothing. Even allowing for the fact that Canada's climate makes the Canadian market for clothing rather specialized, textiles and apparel are just the reverse of innovative or propulsive industries; they are slow developers, technically conservative, and open to intense competition from countries with much lower wage levels.

It is evident that, as Montréal goes, so goes Québec. When the St. Lawrence Seaway opened in 1959, it was apparent that changes would take place, not only in the Great Lakes area, which was made accessible to oceangoing traffic, but also in Montréal, which had, until then, been the terminus for seagoing vessels, and this change was viewed in the city with some concern. The last time that the effective head of St. Lawrence navigation was moved upstream was in 1870, when the channel was deepened to Montréal. As a result Québec City, the older settlement and political capital, found itself at a disadvantage; with little productive hinterland in its own right and more remote than Montréal from the source areas of St. Lawrence trade, its industrial and commercial functions tended to languish, and, although its communications and its harbor are good and are used in winter, it is as a cultural and political center that the city has developed.

The question naturally arose whether the transfer of the head of navigation to Chicago and Duluth would diminish the importance of the Montréal area in the same way. The city carried on a considerable transshipment business, with all the opportunities for local manufacture that transshipment brings. To a limited extent, it remains in this business, since the St. Lawrence channel has a depth of 35 ft (10.6 m), whereas the Seaway is constructed as a 27-ft (8.21-m) channel and, as we have already seen, growth of traffic on the Seaway has been hampered by this fact. To some extent, Montréal exchanged its role of terminus for that of gateway—a "tollgate" function, as it has been described—but it was only a partial exchange and proved no particular hardship. In any case, the city's economic base had broadened, its advantages of position were considerable, and its land transport links made it the largest center of communications in Canada. It is a natural focus of routes— which Québec City never was—running northwestward to Ottawa (718,000), the federal capital, and southward to the Hudson and New York, as well as east and west along the St. Lawrence.

Rather than fear for Montréal's prosperity on this account, it might be considered more legitimate to complain that it has grown too big and powerful in relation to the remainder of French Canada. It has become a commonplace to contrast the metropolis with the "desert" that makes up the rest of Québec Province. Montréal, with 3.1 million of the province's 6.9 million population, earns two-thirds of all personal income within the province, and it tends to drain its hinterland of productive activities and skilled manpower. Yet it can, with equal justice, be argued that without Montréal, French Canada would be little more than a cultural curiosity—a backward, rural area in which were preserved the relics of a culture and community of a past era. That French Canada is a force to be reckoned with can be largely attributed to the fact that it has at its core a city of world standing, offering a wide range of employment opportunities and services and equipped commercially to hold its own in the competitive economic conditions of North America today.

Before we end this, we shall have to assess further Montréal's position, judging it this time not so much in continental terms as in its rivalry with its neighbor, Toronto. For the moment, we remain in the province of Québec to consider one other theme: the course it

Montréal from Mt. Royal. A view over the city center southward, showing the St. Lawrence River and, immediately beyond it, the line of the St. Lawrence Seaway (see Chapter 9). Nearest the camera—the buildings of McGill University. *Ministère du Tourisme de Québec*

set in the 1970s toward separatism—toward the severance of many of its links with the federation and, in the shorter term, the requirement that the French language be used in business and factories. English language was to be curtailed—a particularly critical issue in Montréal, where most of the province's 20 percent of non-French speakers are concentrated.

This tide of separatist feeling and political pressure has ebbed and flooded over many years, several times apparently flowing to the point of final separation, only to retreat again. At present, it is on the increase, and the future is all uncertainty. We can, however, legitimately ask two questions: What chance would an independent Québec have of economic survival? and What are the short-term economic or locational effects of the sort of pro-separatist measures just described?

The economy of the present province of Québec is far from buoyant. The expansion of employment has been painfully slow—slower even than in the Maritimes. The agricultural base, as we have seen, is shrinking, although forest and power resources are huge. Perhaps, Québec could live on exports of power and pulp, but Montréal's industries would lose their national markets. Tourism and the traffic on the international Seaway would bolster the economy, but the traffic would be passing through rather than originating or terminating. These do not sound like the makings of an economy that could hold its own, at least by North American standards of living.

In the short term, measures like compelling firms to use French in their transactions have one direct effect—that of frightening them off. It is not, after all, as if the French-speaking market represents the major share of the Canadian market; on the contrary, it is less than one-third. We can already observe the effect of this type of language law in use—in Belgium,[6] where at least the Flemish-speaking politicans who carried through the law represented a majority of the total population. But the Belgian example is not an encouraging one; it is not by this means that new jobs are created. All that happens is that firms that are footloose move away, as finance houses and insurance companies—whose business does not involve any material ties—have already moved from Montréal to Toronto. It is difficult to see the economic profits of separation; easy to count up the losses.

The Western Section: Southern Ontario

SETTLEMENT AND LAND USE

The second section of Canada's Core began to develop much later than the first, but has since grown, in many respects, to surpass it. It was only after the American colonies won their independence that settlement in this section began in earnest and, by that time, the French had been on the Lower St. Lawrence for 170 years. The peopling of Upper Canada—or what today is Southern Ontario—and the organization of its government, came about partly through fear of the victorious young republic to the south, flexing its muscles and talking of further conquests, and partly through the arrival north of the Great Lakes of the so-called Loyalist exiles, who preferred to remain British rather than become American, and were prepared to move northward to do so.

By 1791 Upper Canada had a form of government of its own, distinct from that of Lower Canada: amalgamation of the two occurred only in 1848. As the nineteenth century opened, immigration from the poverty regions of Scotland and Ireland, together with the organizing capacity of colonizers like Sir Alexander Galt, let to a rapid increase in numbers of new arrivals, so that by mid-century the population of Upper Canada equaled that of Lower Canada. But this population was almost wholly English-speaking, and the framework of its laws and customs differed sharply from that of French-speaking Lower Canada.

So, too, did its settlement pattern. The distinctive French landscape of "long lots" gave way to a series of gridirons (see Chapter 5), and the wider expanse of flat and fertile land available around and between the Great Lakes drew settlement away from the shorelines and riverbanks. With the switch from linear to areal development, Upper Canada was able to support a hierarchy of crossroads settlements, villages, and small towns comparable to that which, at the same period, was growing up south of the border in the American Midwest (see Chapter 14). Although in due course Toronto was to outrank all the other cities of Ontario, it had plenty of rivals in the early days, and that was never the case with Montréal. Upper Canada provided the young nation with what it had previously lacked—a firm agricultural base. Later on, it was to provide it, also, with the greatest share of its industrial strength and its outstanding commercial center.

Southern Ontario and the southern Prairie Provinces are the two major agricultural areas of Canada. The Prairie region is much the larger, but farming in the Ontario Peninsula is more diversified, intensive, and productive. Ontario has only 9 percent of Canada's land in farms, and 12 percent of its area in crops; yet it contains, almost entirely within the peninsula, about 26 percent of the nation's farm units and accounts for nearly 30 percent of Canada's cash receipts from farming. Over two-thirds of the farms are in the size range 70 to 560 acres (28 to 227 ha). Almost all are family farms, owned fully or in part by the operator (see Table 13-3).

Southern Ontario's location in the Canadian heartland contributes to the high quality of its farming. Unlike the Prairie farmers, those in this region

---

[6]Jane Jacobs, in her thought-provoking *The Question of Separatism* (London, 1981) argues that separations can be carried out, citing as her example the successful separation of Norway from Sweden in 1905. As Jane Jacobs makes clear, the Swedish–Norwegian case depended on an almost superhuman degree of self-restraint and tolerance on both sides, and the separation took place, in any case, a long time ago, in a period before the media attained their present dominance. To compare the differences in temperament between two Scandinavian peoples with those between French-speaking and English-speaking Canadians is to strain the credulity of those acquainted with the groups concerned.

The present author prefers the Belgian example, which he cites in the text above, even though this has led, to date, only to a partial separation of two peoples. In the case of Belgium, a government dominated by the Flemish-speaking majority secured in 1970 passage of a law whereby every order, instruction, or invoice issued in Flemish-speaking areas must be in Flemish, and in French-speaking areas in French, regardless of the language of the recipient. This was followed in 1980 by the effective division of the nation into three, with a Flemish Region, a French Region, and an officially bilingual Brussels Region around the national capital. Since the passage of these acts, divergences between the regions have steadily increased.

For further commentary, see S. Crean and M. Rioux, *Two Nations: An Essay on the Culture and Politics of Canada and Quebec* (Toronto, 1983).

Table 13-3. Ontario Agricultural Statistics

	1951	1961	1971	1981	1986
Number of farms	149,920	121,333	94,722	82,448	72,713
Area in farms (million ac/ha)	20.9 / 8.45	18.6 / 7.52	15.9 / 6.46	14.9 / 6.03	13.9 / 5.65
Average size (ac/ha)	138 / 56	153 / 62	168 / 68	181 / 73	193 / 78
Area of improved farmland (million ac/ha)	12.7 / 5.14	12.0 / 4.87	10.9 / 4.40	11.2 / 4.53	10.7 / 4.33

Note: In each line showing areas, the upper row of figures represents acres, and the lower row hectares.

market most of their produce within the region. Furthermore, the cities that provide those markets encourage diversity in farm production and, at the same time, supply most of the goods and services needed by the farmers; that is, agriculturally the region is largely self-contained.

Southern Ontario and the Lower St. Lawrence Valley are, however, also parts of the great North American dairy region (Fig. 7-3). Land resources and their agricultural use have much in common with those across the border. For that reason, the patterns of agriculture in Ontario will be further discussed in conjunction with those in southern Michigan in the next chapter.

MANUFACTURING IN SOUTHERN ONTARIO

The industries of the Lower St. Lawrence Valley depend, for the most part, on the region's resources and the metropolitan role of Montréal. Those of southern Ontario depend, in part, on the resources of the province, its forests, and its farms, but even more on a single line on the map—the international boundary with the United States.

Let us consider for a moment the situation that would exist if the line had *not* been drawn in the late eighteenth century, and the two countries of northern North America had become one. Then, along the Great Lakes waterway there would have been little if any advantage of location on one shore or the other. With all points equally accessible to shipping, some industries might have been set up on the southern shores and others on the northern. Because the coalfields are on the south side, the heavy industries would probably have been sited there: the northern side might perhaps have attracted food processing and lumbering. What is clear is that there would have been no real duplication: one industry of each type could have supplied both shores.

But duplication is exactly what the drawing of the boundary line in due course produced. There was no economic compulsion to make steel in Hamilton and The Soo,[7] on the north shores of the lakes, as well as in Cleveland and Buffalo, to the south. But the line divided the producers and consumers of this naturally unified area: to the economic dimension it added a national one, which brought with it considerations of policy, tariff barriers, and protectionism by the weaker against the stronger. The Canadian shores of the Great Lakes developed industry by processes not so much natural as political. Canadians established industries in order not to be dependent on U.S. sources, and U.S. firms established Canadian branch plants in order to gain a foothold behind the tariff wall the Canadians had erected.

We have already raised (Chapter 8) the question as to whether such an arrangement has or has not been beneficial to Canada: the question itself is answerable only in terms of Canada's national aspirations. What is clear is that, on the whole, southern Ontario has benefitted greatly by the arrangement because it was so placed as to obtain most of these "duplicate" industries. So long as demand for their product was strong, the resultant investment and employment contributed to the well-being of southern Ontario. What happens when demand slackens, the workers in the "duplicate" plants then discover all too dramatically.

Be that as it may, Ontario has built up its industries until they employ 40 percent of the Canadian work force in manufacturing. The earliest commerce of the

[7]Sault Ste. Marie, the original French name for the Falls of St. Mary between Lakes Superior and Huron, is also the name of twin towns, one on the Canadian side and one on the American side, of St. Mary's River. The falls, the river, and the two towns are all known throughout North America simply as The Soo. The Canadian Soo, the larger of the twin towns, make steel.

area was an export trade in wheat and timber, both of which were shipped out via Montréal and Québec City. But with the coming of the railways, small manufacturing towns grew up along the mainly east–west routes across the peninsula of southern Ontario, while the ports of Lake Ontario and Georgian Bay developed the processing of goods passing through them. By 1880, southern Ontario had outstripped Québec in industrial employment and, although Toronto easily dominated the province's manufacturing, there was a wide scatter of smaller industrial centers, many of them profiting from Canada's industrial protectionism in that American manufacturers from the Midwest chose them as sites for branch plants behind the tariff wall. These industries were largely fueled by Appalachian coal, at least up until the turn of the century when hydroelectricity first became available, in the west from the Niagara Falls, and in the east from the St. Maurice River.

Local specializations were apparent from the start. Sarnia, a port city and close to the first Canadian oil field at Petrolia, developed as Canada's leading oil refining and petrochemical complex. Windsor, across the river from Detroit, acquired its automobile industry almost by contagion, although it is now rivaled by Oshawa, near Toronto. Hamilton is Canada's leading steelmaker; London and Kitchener, Brantford and Guelph, produce machinery, rubber goods, and household equipment as they earlier produced farm machinery. Toronto itself participates in most of Ontario's manufacturing: it possesses one-quarter of Canada's entire industrial labor force. And in recent decades, Ontario, in general, and Toronto, in particular, have benefited from another source of industrial strength—the skills and the commitment of immigrants to Canada, one out of every two of whom make for the province.

Because of the border, both the industry and the urbanization of southern Ontario are more or less mirror images of industry and urbanization in northern Ohio, eastern Michigan, or Upper New York State (Table 13-4). And Toronto, the focus of Ontario's urbanization, belongs clearly to the international set of cities we shall be considering in the next chapter: the Cities of the Great Lakes. But like Chicago in the United States, although it belongs to that set it also transcends it. It has a Canadian context and dimensions the other members of the set do not have.

Like them, Toronto began with a fort, an anchorage, and a trading post. Like them, it grew with the construction of canals and railroads—grew, as every

Table 13-4. Ontario: Industrial Employment by Category of Industry, Census of 1986

Transportation equipment	158,781
Fabricated metal products	94,119
Electrical and electronic products	91,708
Food and kindred products	75,304
Printing and publishing	65,446
Primary metal industries	62,230
Chemicals and chemical products	52,411
Machinery industries	49,473
Paper and allied products	42,316
Plastic products	25,772
Textile products	14,893
Rubber products	13,588
Primary textile industries	10,765
Total, provincial workers in industry	956,400

visitor noted, into one of the most British cities in the world, its population undiluted by French-Canadian or other European elements; a solid city, perhaps rather virtuous and perhaps a little dull. Then followed the immigration of the years after World War II, a massive influx (as we saw in Chapter 2) of non-British, and the conversion of Toronto into a cosmopolitan city, whether virtuous or not, but certainly no longer dull. Its rise as a financial center altered its skyline more even than its industrialization: its port expanded with the opening of the Seaway, and by 1991, its 3.9 million inhabitants made it the largest census metropolitan area in Canada (Table 13-5).

Forty miles (64 km) west of Toronto is Hamilton (599,000), the Canadian counterpart of the U.S. steel cities on the southern lake shores. But the western end of Lake Ontario is, in any case, one of the most favored locations in North America for industrial development. Traffic through the Great Lakes converges upon the Niagara River and its bypass, the Welland Canal. Eastward runs the route via the Mohawk Valley to the Hudson and New York. The northern end of the Appalachian coalfield lies less than 100 miles (160 km) to the south, and in the heart of the region is to be found its prime resource, the power of the Niagara Falls, exploited jointly by Canada and the United States. It comes as no surpise, therefore, to find that on the Canadian side of the Niagara frontier there has developed an urban–industrial shoreline that is almost continuous all the way from Niagara to the eastern side of Toronto. In addition to Hamilton and Toronto, this zone contains such other manufacturing cities as St. Catharines and Oshawa.

Table 13-5. Canada: Population of Census Metropolitan Areas, Censuses of 1986 and 1991

Census Metropolitan Area	1986 Population	1991 Population	Absolute Change	Percent Change
Toronto	3,431,981 A	3,893,046	461,065	13.4
Montréal	2,921,357	3,127,242	205,885	7.0
Vancouver	1,380,729	1,602,502	221,773	16.1
Ottawa–Hull	819,263	920,857	101,594	12.4
Edmonton	774,026 A	839,924	65,898	8.5
Calgary	671,453 A	754,033	82,580	12.3
Winnipeg	625,304	652,354	27,050	4.3
Québec	603,267	645,550	42,283	7.0
Hamilton	557,029	599,760	42,731	7.7
London	342,302	381,522	39,220	11.5
St. Catharines–Niagara	343,258	364,552	21,294	6.2
Kitchener	311,195	356,421	45,226	14.5
Halifax	295,922 A	320,501	24,579	8.3
Victoria	255,225 A	287,897	32,672	12.8
Windsor	253,988	262,075	8,087	3.2
Oshawa	203,543	240,104	36,561	18.0
Saskatoon	200,665	210,023	9,358	4.7
Regina	186,521	191,692	5,171	2.8
St. John's	161,901	171,859	9,958	6.2
Chicoutimi–Jonquière	158,468	160,928	2,460	1.6
Sudbury	148,877	157,613	8,736	5.9
Sherbrooke	129,960	139,194	9,234	7.1
Trois-Rivières	128,888	136,303	7,415	5.8
Saint John	121,265	124,981	3,716	3.1
Thunder Bay	122,217	124,427	2,210	1.8

A = Adjusted figure due to boundary change.

One Core or Two?

The western section of the Core differs from the eastern in a number of ways, as we have seen. Ontario has been gaining on Québec ever since the opening of the Canadian West turned the nation's attention away from its ties with Europe: from then on, it became a question of which link, the western or the transatlantic, was more important to the manufacturers and businessmen of the Core. Ontario and Toronto are, in any case, nearer to the center of the continent and its markets than Québec and Montréal.

In terms of population growth and manufacturing, the western half of the axis has a clear lead over the eastern. Toronto's rate of population growth placed it ahead of Montréal for the first time in the census of 1981. With regard to its control of Canadian manufacturing, service industries and resource-based development Toronto is far ahead: it has always specialized in the promotion and financing of mining, and its grip on this sector of the economy shows up

particularly clearly. In the financial sector, Semple and Smith found the balance of power more evenly divided between the two cities: the total assets of Canadian banks headquartered in each were roughly equal, and the assets of Montréal's trust and holding companies exceeded those of Toronto. On the other hand, the life insurance companies were virtually all concentrated in Toronto and so, too, were the leading real estate companies. Overall, in the late 1970s, the financial strength of Toronto exceeded that of Montréal by a ratio of 3:2, but a much larger proportion of Toronto's financial business than of Montréal's was in foreign hands.

In summary, the balance between east and west has been gradually shifting for a century or more. This has most recently been the case with financial–commercial functions, which do not depend on specific work sites, as do factories and farms. Their locational habit is simply to cluster together as closely as possible, as they do in New York City or London, and as they have done in Montréal, since Confederation or before. But if once a move starts—perhaps by the passing of some new law or financial regulation—and if on this account the business climate becomes less favorable to commerce, then a few major enterprises may transfer to another center, and in so doing set a fashion others follow. That is what has happened between Montréal and Toronto: their financial roles have been reversed since World War II. But the lead would probably be longer than it is were it not for the cultural and political forces that keep alive the rivalry of French and English, Québec and Ontario. For just as there is a duplication of industry between Canada and the United States, since they are two separate nations, so there is a duplication of economies between two separate cultural groups, French-speaking and English-speaking, since French

Canada refuses to be eclipsed by or integrated into the main economy.

In view of this rivalry, it is a wonder the two sections function together at all, but they evidently do. Between them they contain one-half of Canada's population, and nearly three-quarters of its industry. As Donald Kerr has written, "although Ontario and Québec are sharply divided on cultural and linguistic grounds, they continue to function as a single economic region. . . . At present, the daily flow of freight along the Heartland's transportation facilities exceeds the inter-provincial trade of all other regions in Canada. . . . Each [Ontario and Québec] is the best customer of the other."[8]

[8] D. Kerr, "The Emergence of the Industrial Heartland," Chap. 3 in McCann, *A Geography of Canada,* p. 97.

14
The Interior

The Corn Belt: A fall landscape in Hamilton County, Nebraska. Corn and soybean crops occupy most of the fields. The level surface and the absence of trees, except planted around farmsteads, are characteristics of the landscape. *U.S. Department of Agriculture, Soil Conservation Service. Photo by Gene Alexander*

We come, in this chapter, to consider the great central region of North America we shall call the Interior. In doing so, however, we must begin by recognizing the problem we confront if we are to justify the use of this regional division. It is that of reconciling two conflicting regional concepts, or—since that is, in fact, impossible—of effecting a compromise between them. One of these concepts is economic: it is that of the core region, whose subdivisions we have been examining in the last three chapters, and whose fourth section we have now to consider. The other is cultural, and involves one of the most commonly recognized realities in American life: the Midwest.

In economic terms, there is an area that stretches north of the Ohio River from the Appalachians to a line west of Chicago, an area that forms the fourth and longest section of the economic core. It contains a number of control points like Chicago, Detroit, and Pittsburgh, where major corporations have their headquarters, as well as a large share of the nation's mining and industry. Unlike the other three sections of the Core, however, this one also plays a key role in the agriculture of the continent. Industry, commerce, and urban growth have entered the region from the east, without reducing the importance of its agriculture. Then at its western edge is what we may call the "hinge"—where the core ends and the western periphery begins, and the character and functions of the one overlap the other.

The terms *core*, *periphery*, and *hinge* mean little to the average citizen of middle North America, at least in the sense that we have used them in this book. By contrast, our other, cultural concept—Midwest—is known to, and used by, virtually every American. There is universal agreement that a Midwest exists; that it has a character all its own; that it has some kind of cohesion. The only problem is that of defining it. It is not precisely coincident with the westernmost section of the Economic Core, or even the Core-

plus-Hinge. To delimit it by means of narrow boundaries on any basis is impossible. As the British writer Graham Hutton pointed out, the Midwest includes the states of Ohio, Michigan, Indiana, Illinois, Wisconsin, Minnesota, Iowa, and Missouri, but in the cultural sense it can be said to include much of the area of heavy industry in Pennsylvania and West Virginia. "In other words, the real Midwest, the Midwest of the midwesterners, is the core composed of most of the area of these eight states; but beyond that core you will still find a Midwest, thinning out into something else the further you go from the center."[1]

That was written nearly half a century ago and it seems that, in the interval, things have changed; that the Midwest, in general perception, has moved further west. In 1960, J. W. Brownell's postal survey[2] revealed that there were people in rural areas of North Dakota who considered themselves Midwesterners, an astonishing proposition for anyone from Indiana or Illinois to accept, to whom the Dakotas are little less remote than the moon. Subsequently, the search for the Midwest was taken up by J. R. Shortridge, in a long series of papers that culminated in a book, *The Middle West: Its Meaning in American Culture*.[3] He confirmed that the region as perceived had gone west; that Ohio and even Indiana were regarded as imposters when it came to naming midwestern states (but what regional title can those two states now be given?), and that the Dakotas were *in*.

What underlies this midwestern sectionalism? To

[1] G. Hutton, *Midwest at Noon* (London, 1946), p. 4.

[2] J. W. Brownell, "The Cultural Midwest," *Journal of Geography*, vol. 59 (1960), 81–85.

[3] J. R. Shortridge, *The Middle West: Its Meaning in American Culture* (Lawrence, Kans., 1989). A possible explanation of this phenomenon, suggests Shortridge, is that "Midwest" has always had rural/pastoral connotations and, as the eastern part of what we always thought of as the Midwest has become increasingly industrialized, so it has forfeited its right to the regional title.

explain it, we might refer to such economic factors as the firm midwestern balance between industry and agriculture; to the farmers' freedom from the marketing problems of East and West; to the wide range of midwestern manufactures, all making for a high degree of regional self-sufficiency. We should necessarily take account, too, of the historical factor—the uniformity of the conditions under which, in the first half of the nineteenth century, this vast tract was rapidly occupied. Much of the explanation lies outside the scope of the present volume: suffice it to say that there results from these factors a marked degree of economic and cultural individuality within the region, and this we are reluctant to disturb by dividing it, as we might otherwise legitimately do, into parts.

The fact is that there is an agricultural pattern to the Interior and an industrial pattern, but the two are quite different and are best considered apart from one another. Furthermore, the urban–commercial pattern, although naturally owing much to both of the other two, is distorted by the diagonal slash of the Great Lakes across the region, which profoundly modifies the distribution of population and cities and the movement of goods. We shall therefore consider each of these three patterns in turn.

First, however, we must briefly review the processes by which the Interior was penetrated by European peoples. The earliest movement into the area was by the French; by their fur traders, their missionaries, and that roving band of unlicensed trappers, the coureurs de bois. They suffered severely, the missionaries in particular, at the hands of the Indians south of the Great Lakes, and built forts at strategic points—Niagara, Detroit, the Straits of Mackinac—to afford themselves some protection and to counter an eventual challenge by the English. The French empire, for all the remarkable achievements of a handful of explorers, was a fragile affair; the fur trade was the very antithesis of a sedentary, colonizing activity, and in only a few areas west or south of the St. Lawrence line did French efforts extend to colonization by settlers. The major effort was devoted to the lower Mississippi, and the creation of plantations upstream from New Orleans. But a smaller zone of settlement developed in what is now Indiana, along the Wabash River while, on the middle Mississippi, St. Louis became a focus for French activity over the hundreds of still-empty miles between north and south.

The belated arrival on the scene of British forces and settlers in the mid-eighteenth century was the

signal for a series of wars against Frenchmen and Indian tribes in turn. Soon the insubstantial hold of the French upon this vast Interior was revealed: they were defeated in the Seven Years' War of 1756–63, and sold out their remaining interests in North America by the Louisiana Purchase of 1803. With the Indians reduced to a kind of docility by either treaty, or removal, or both, settlers from the East were free to occupy the West and, under the generous land policies we reviewed in Chapter 5, they began to do so.

There were two relatively straightforward routes into the Interior for these early settlers—by the Great Lakes and by the Ohio River.[4] Both, however, suffered from the drawback that the pioneers had first to *reach* the water route—the Great Lakes at or near Buffalo, and the Ohio at its forks, at what was soon to be known as Pittsburgh. But in the early days either of these obstacle courses was probably preferable to the third route, the overland road that Daniel Boone had reconnoitred, from the Cumberland Gap in what is now Tennessee through to the Bluegrass country of Kentucky, and on to the lower Ohio.

The northern, or Great Lakes, route was transformed, in 1825, by the opening of the Erie Canal, which provided an all-water route to the Interior. But it was a route that was blocked by ice in winter, and that led northwest rather than west; that is, into lumbermen's country rather than farmers'. Its usefulness increased only after (1) canals were dug from the Lake Erie shore over to the tributaries of the Ohio, and (2) settlement reached out to the west of Lake Michigan, an area to which the lakes route did indeed provide a way. The Ohio River route had its drawbacks, too; the problem of reaching Pittsburgh overland, and the perils and falls of the river itself. But its advantages were considerable: it led where most of the migrants wanted to go; its current took them there, and it was a route on which every man could build his own raft, tie up ashore at night, and use the raft timbers to build a cabin on arrival at his chosen stopping place.

By these main routes the Interior was opened up before the coming of the railroads. The new arrivals came from all parts of the East Coast, as well as a small proportion of them directly from Europe. In 1850, for example, when the population of Ohio was just under 2 million, almost 40 percent of them had

[4]Which route emigrants chose might well depend on their point of origin. See J. C. Hudson, "North American Origins of Middlewestern Frontier Populations," *Annals,* Assoc. Amer. Geog., vol. 78 (1988), 395–413.

been born outside the state, 10 percent of them in Europe (mainly Germany), and New York and Virginia had contributed almost exactly equal numbers to the westward flow into the state.

As communications improved, so the migrants' moves became less dramatic: families would migrate from New England to western New York State—a great gathering ground for New Englanders, particularly religious splinter groups—and then move on by easy stages into Ohio and Indiana. Further south, families like Abraham Lincoln's were making similar moves from Kentucky into southern Indiana, Illinois, and Missouri. By the time the railroads reached the Mississippi in the late 1840s, the westward movement was under way on a broad front. By then, too, the gridiron survey was in force everywhere north of the Ohio, and that regularity of form and spacing, which was to dominate settlement all the way across the Interior and beyond, was making itself felt.

The agricultural geography which marks the Interior today, and which we are shortly to consider, did not at once emerge.[5] In the almost breathless haste of this wave of frontier settlement, farmers used the techniques and the stock they had brought with them from the East, and only slowly adapted their practices to the new environment, and to the market conditions that urban growth and an ever-denser rail network offered them. It was by a process of trial and error that today's patterns and practices emerged.

The Agricultural Interior

The farmers of the Interior are generally favored by natural conditions. The region possesses vast, smooth plains, which make cultivation easy, and wide areas of remarkably fertile soils developed on glacial drift. It has an adequate rainfall (30–40 in [750–1000 mm]), of which one-third to one-half falls in the months May to August inclusive, and hot summer weather to offset the short, frost-free period of a continental interior. If the city dweller finds cause for complaint in the stifling summer heat, the icy winds of winter, and the all-too-brief "in-between" spring weather, these things are quite acceptable to farmers, whose methods are adapted to the climate, who delight to "hear the corn growing" during hot summer nights when the cities are sleepless, and whose chief fear is the occasional thunderstorm, hailstorm,

or tornado that may spring up suddenly on a hot afternoon and lash down with tropical violence on the homestead and crops. This hazard apart, they have little cause for complaint in the natural conditions.

The effect of this favorable combination of circumstances is to produce an agriculture of intensive land use, with high land values, a high percentage of the farm area under crops, and the conversion of much of the output of the fields into livestock products. Because of the historical circumstances under which the area was settled, all this is typically associated with family farms, where hired labor is at a minimum and the average farm size, until well after World War II, was still close to 160 acres (64.8 ha)—the historic quarter section that, as we saw in Chapter 5, dominated land legislation throughout the nineteenth century.

But if these are the general characteristics of Interior farming—high land values, intensive use of capital and technology, a dense livestock population, and high productivity—it must at once be added that over so vast an area all of these circumstances do not apply everywhere. Natural conditons limit and modify their application in various parts of the region, creating the local differences that will be described later. But the area where they are most fully developed, where most of them apply most of the time, where lies the heart of this great agricultural region, is the Corn Belt.

THE CORN BELT

The Corn Belt is one of the best-known entities in American geography, its fame enhanced by Russell Smith's classic phrase, "The Corn Belt is a gift of the gods."[6] Yet to define it is almost as difficult as to set limits to the Midwest itself.[7] It is an area where corn and soybeans dominate crop acreage and where these crops, plus hogs and beef cattle, are the major sources of farm income. This emphasis gives way from the Belt's center to wheat and sorghum in the southwest, to cattle ranching in the Nebraska Sand Hills and Missouri River roughlands, to spring wheat and barley toward the northwest, to dairying in the

---

[5]See A. G. Bogue, *From Prairie to Corn Belt* (Chicago, 1963); or J. E. Spencer and R. J. Horvath, "How Does an Agricultural Region Originate?" *Annals,* Assoc. Amer. Geog., vol. 53 (1963), 74–92.

[6]J. R. Smith and O. Phillips, *North America* (New York, 1942), p. 360.

[7]In 1968, W. E. Akin in *The North Central United States* (New York), took as a definition of intensity of Corn Belt land use those areas that produced more than 12,000 bushels of corn per square mile (4600 bu per sq km). These areas were to be found at that date in most of northeastern Illinois, northern and eastern Iowa, and south-central Minnesota.

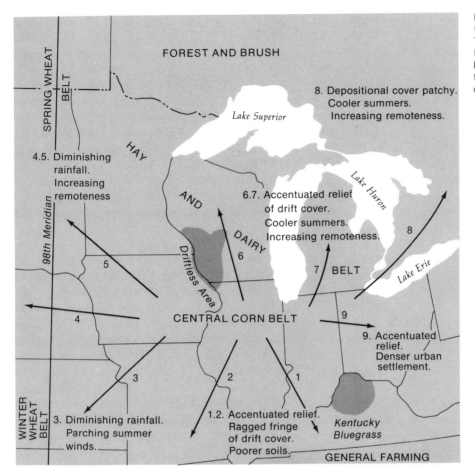

north and northeast, and to various crop and livestock specialties to the south (Fig. 14-1). But, more important, it is the unusual intensity and productivity of land use that diminish with distance from this remarkable area.

The existence and character of the Corn Belt are to be explained, in the first instance, by a fact of physical geography: in the heart of the Interior, south of the Great Lakes, is a belt where conditions for farming approach the ideal. A mantle of drift, the product of continental glaciation, has smoothed over an already gentle relief. Large areas along the Mississippi and Missouri rivers are covered with loess. Under the annual growth and decay of tall-grass prairie in the central and western parts of the region, the glacial mantle and loess were converted into deep, fertile black prairie soils (mollisols). Once the forests that covered the eastern part of these plains were cleared, and once the numerous swamps were drained, there was, except for stream valleys, no major obstacle in

the way of the plow for hundreds of miles. Away from this fertile core area, on the other hand, whether north into the lake and drumlin country of Wisconsin, south into the Ohio Valley and beyond, or east into the Appalachian foothills, the terrain becomes more broken and the soils more patchy.

For the second stage of our explanation, we must enter the field of agricultural economics. As we saw in Chapter 7, the combination of gentle relief, rich soils, and warm, humid summers gives to the Corn Belt a comparative advantage and, hence, first choice, within mid-latitude limits, as to which agricultural products it will produce. We also saw that wheat vied for first place for a time after settlement. But the winning combination, during most of the first century of occupancy, was one of hogs and beef cattle fed with corn. Livestock were at first driven laboriously across the Appalachians from settlements in the Ohio Valley to markets along the Atlantic Seaboard. Later, as the Midwest was settled, Cincinnati became the first major center of hog and beef slaughter, soon to be surpassed by Chicago and joined by St. Louis, St. Paul, St. Joseph, and Omaha. As the size and affluence of America's population increased, so did the preference of the people for corn-fed, fat-marbled beef and pork. And, logically, so also did the application of science to the business of breeding improved corn, hogs, and beef cattle to fit both the environment and consumer tastes.

The production of corn, beef cattle, and hogs is, however, by no means limited to the Corn Belt. Corn growing extends northward into the dairy region, although there, as the growing season becomes shorter and cooler northwest, the whole corn plant is, on an increasing proportion of fields, cut while still green, chopped, and stored in silos for winter feed. In New England, almost all field corn (as distinct from sweet corn for human consumption) is chopped for silage. But it is also the case that, since 1950, the development of strains of hybrid corn adapted to a shorter growing season has made possible the extension of the true Corn Belt along its northern fringe, into what was formerly the southern edge of the dairy region.

Corn growing also extends east and south from the Corn Belt. Corn ranks among the three leading crops, by acreage, in most states east of the Mississippi and south of New England. In the same way, cattle and hogs are leading livestock products, just as they are in the Corn Belt. But what creates the contrast between the Belt and its margins is the intensity and productivity of the land use. In the central Corn Belt, both are high, whereas toward these southern and eastern fringes, broken relief and less fertile soils result in lower levels of intensity and output.

West and southwest from the Corn Belt, production of grain corn decreases most abruptly. In Kansas, decreasing rainfall and the hot, dry winds of summer force the replacement of corn by winter wheat and grain sorghum, whereas in Nebraska, the Corn Belt ends abruptly at the edge of the Sand Hills, where beef cattle are raised on huge ranches. Many of the calves and yearling animals are shipped into the Corn Belt for feeding to the slaughter stage. Irrigation in the Platte River Valley, and more recently, irrigation by deep pumping and radial sprinklers on the loess plains of southeastern Nebraska, extend a finger of the Corn Belt westward (Fig. 7-3).

Evolution of Corn Belt Farming

Even so superior a farming region as the Corn Belt is not uniform throughout. During the first century of

settlement, there developed two major and several minor areas of "cash-grain" farming—so called because farmers sold their corn and oats (the second most important feed grain of the Corn Belt), rather than feeding them to their livestock. Some of the cash-grain found its way to livestock feeders in other parts of the Corn Belt, some was processed into such food products as breakfast cereal or cooking oil, or into starch, and some entered export trade. Geographers have attempted to explain the cash-grain specialization as being related to transport and market facilities, to terrain and soils, or to such historical factors as late settlement and larger-than-usual land holdings. It is logical that farmers tend to use their best land for crops (earlier, for the rotation of corn, oats, and hay; since about 1950, for corn and soybeans) and to use wet or hilly land near a stream for pasture. If a farm, or an area for that matter, has largely superior land, the tendency is to use it for growing corn, not for raising livestock. On the other hand, poorer land is more often used as pasture. In southern Iowa and northern Missouri, for example, raising beef cattle predominates; these regions grow so little corn that they have been excluded from the Corn Belt as outlined in Figure 14-1.

We can now summarize the variations on the corn–oats–hay–hogs–beef cattle theme of Corn Belt farming (omitting soybeans for the moment). Corn Belt farmers fall into four categories:

1. The "all-around" farmer, who grows all the crops so far mentioned, raises his own animals, and feeds his animals with his own crops.
2. The cash-grain farmer, who specializes in growing corn and other crops for sale, but who raises little or no livestock.
3. The hog or beef "raiser," who raises pigs or cattle up to the weights (approximately 40 lb [18 kg] for hogs and 700 lb [315 kg] for cattle) at which they can be sold to the feeder.
4. The "feeder," who then feeds the animals up to market weights—approximately 220 lb (100 kg) and 1100 lb (500 kg) for hogs and cattle, respectively. The feeder may bring in animals, not only from raisers within the Corn Belt, but also from cattle ranchers in the west; more recently, also, from southern raisers.

Some areas specialize in hogs—for example, eastern Iowa and Illinois west of the Illinois River. Some specialize in beef feeding—for example, the Missouri River valley of Nebraska–Iowa, and northern Illinois

Soybeans, the "miracle crop" of American agriculture. *American Soybean Association*

along the line of the early routes by which Iowa beef cattle reached Chicago. But the genesis of these regional specializations is complex. Some of them are the product of economies of scale on individual farms; others of external economies, such as the provision of specialized services offering advice, financing, marketing, transport, or supplies.

Much of the account so far describes Corn Belt farming until about 1950. But since then two changes have taken place. The first has been the rise of soybeans to become the second crop of the Corn Belt, and even to challenge the supremacy of corn.[8] This east Asian crop was introduced as a forage crop early in the twentieth century, but had little impact outside the southern Atlantic Coastal Plain. By 1940, however, its potential as an oilseed crop had been recognized in the Corn Belt. Since then, its expansion has

been phenomenal, as the United States began to supply not only domestic and European markets, but also those of Japan and the Far East, the very area where the crop originated.

The new crop fitted well into the Corn Belt system; the same basic farm machinery that was used for corn could be employed for soybeans. Not only have soybeans become a major crop throughout the Corn Belt, where the expansion was at the expense of hay and especially oats; it has become the first crop, by acreage and value, throughout the Mississippi Valley and important elsewhere in the South, far surpassing cotton. By 1987, the relationship of corn, soybeans, and their combined acreage, as a percentage of total field crop acreage, for core Corn Belt states was: Iowa, 49:38:87; Illinois, 44:42:87; and Indiana, 44:39:83.

Soybeans are crushed for oil in processing plants scattered through the producing areas, the huge one at Decatur, Illinois, being the oldest and largest. The oil is used in edible oils and margarine. The meal left after crushing is high in protein and is a major ingre-

[8]Within the Corn Belt, principal areas of soybean concentration are at present to be found in north-central Iowa, southwestern Minnesota, east-central Illinois, and western Ohio–east-central Indiana.

dient in manufactured livestock feeds—thus soybeans, in part, come back to "feeder" farms as well as dairy farms. But the continued expansion of soybean acreage, even during times when other crops face problems of surplus, has been fostered by the export market—more than one-half of the American crop is exported.

The second major change has been a shift in the ratio of production in the Corn Belt between cash crops and feed crops. The cash crops are primarily corn and soybeans sold off the farms. The feed crops are corn, oats, and hay fed to livestock on the farms where they are grown. An increasing proportion of the Belt's output is being sold as cash crops. The reasons for this trend seem to be (1) that yields, especially of corn, have been rising; (2) that overseas demand has increased rapidly, both in developing countries with high rates of population growth and in developed countries like Japan and the former Soviet republics, where supplies of animal feeds are required to support meat production; and (3) that large-scale cattle feeding in the central and southern Great Plains has expanded greatly, in direct competition with the Corn Belt.

The effect of the two changes we have noted has been to transform the Corn Belt's corn–oats–hay–hogs–beef cattle combination into one of corn–soybeans–hogs–beef cattle. There have been some minor changes as well. Production of oats has declined markedly, and that of hay to a lesser extent: by area, hay ranks third in Iowa, and fourth in Illinois and Indiana; oats rank fourth in Iowa, fifth in Illinois, and sixth in Indiana. Wheat, which for a short time following settlement was the region's premier crop, has made a modest comeback in the last decade, to rank third in area in Illinois and Indiana, although it is far behind corn and soybeans.

The Corn Belt Farms and Landscape

The characteristics that distinguish Corn Belt farming may be summarized as follows: (1) Farms are, overwhelmingly, family farms, but large in scale. Average size is 240 to 300 acres (97 to 121 ha), although many farms are twice as large. Hamilton County in central Iowa represents the Corn Belt in Table 14-1. The average value of land and buildings on its farms is $492,000; average value of products is $130,500. (2) Inputs of capital and technology are intensive. For example, the average value of machinery on Hamilton Country farms is $60,000. Farmers there and throughout the Corn Belt are avid students of crop

Table 14-1. Hamilton County, Iowa: Agricultural Statistics, 1987

Population (1980)	18,000
Percentage change, 1970–80	−2.8
Area of county (ac/ha)	368,000/149,000
Percentage in farms	94.4
Percentage in cropland harvested	88.4
Value of agricultural products sold	$130.5 million
Percentage from livestock products	56.9
Number of farms	1026
Average size (ac/ha)	339/137
Average value, land, buildings, & equipment, per farm	$492,000
Average value of agricultural products sold, per farm	$130,473
Percentage of operators whose principal occupation is farming	74.0

and livestock improvements and market news. (3) Productivity is high. High capitalization and the application of technology to superior land make the Corn Belt uniquely productive and prosperous. (4) The farming and the nonfarming economy are interdependent. The close relationship between farms and the urban-centered economy was developed especially in the Corn Belt, and functions through a hierarchy of central places, from farmstead to small trade center to medium-sized and large cities to portside metropolis. (5) Farms and regions are specialized. Although the Corn Belt is one of the most uniform large agricultural areas anywhere, there are, as we have described, significant local specializations.

The Corn Belt landscape is one of flat to gently undulating terrain divided into very large rectangular fields, one distinguished from the next only by the change from rows of corn to rows of soybeans and back again—mile upon mile. Except in livestock-raising areas, which have fenced pastures, the earlier fences have been removed to make way for larger fields and machines. Even in livestock-feeding areas, fencing is confined to small feeding yards. The only real interruptions are the roads that outline each square-mile section or half section of the original land (see Chapter 5). In western Ohio and Indiana, small woodlots, relics of a former forest cover, dot the landscape; further west virtually the only trees interrupting the fields are those clustered at farmsteads, usually along their windward sides. Farmsteads are substantial collections of buildings and fenced yards, especially in livestock-feeding areas where feeds must be stored and animals protected in winter. In cash-grain areas, where land values are extremely

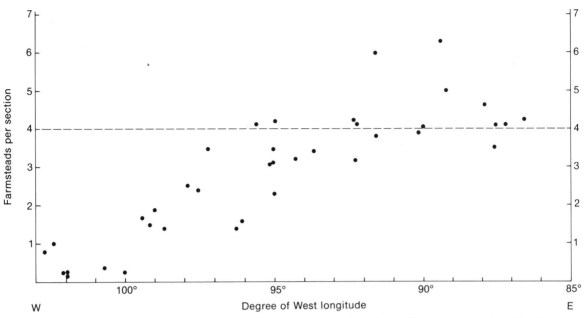

Fig. 14-2. Farm settlement in the Interior: Diagram of a map exercise on U.S. Geological Survey 1:24,000 topographic maps along a line from western Indiana to eastern Colorado. On each map (approx. 48 sq mi or 124 sq km) farmsteads per square mile were counted. There is, of course, no assurance that all the farmsteads are today (1) occupied and (2) farmed as separate units: the reverse is likely to be the case. Nevertheless, the farmsteads exist and the diagram emphasizes the pervasive influence in the Interior of U.S. land legislation based on the quarter section of 160 acres, giving four farmsteads per square mile (2.6 sq km). This density persists far out to the west, where, however, increasing aridity thins the farmsteads.

high, the farmsteads are surprisingly modest—the residence often houses a tenant farmer; livestock and related structures are absent. Because farms are large, farmsteads average only two or three per square mile.[9]

TRANSITION TO THE WEST

The pre-eminence of the Corn Belt as an agricultural region, and of corn as its major crop, is based on nearly ideal conditions of terrain, soil, precipitation, and summer temperatures. Of these, it is precipitation, specifically the decreasing amount, reliability, and effectiveness of it, that requires a change in land use westward.

To the northwest and southwest, the change is gradual. To the southwest, the corn–soybeans dominance in cropping gives way to winter wheat and sorghum. To the northwest, corn gives way even more gradually, to a greater diversity of crops in which spring wheat becomes the most important among hay, oats, barley, and, very recently, sunflow-

ers. Cattle maintain their importance in both directions, especially to the southwest; hogs decline more sharply.

Family farms continue to be the rule, although they increase steadily in acreage away from the Corn Belt. The intensity of land use also declines only gradually (Fig. 14-2). Crops are planted every year and depend on natural rainfall, as in the Corn Belt. In fact, there is more supplemental irrigation in the southwestern corner of the Corn Belt (southeastern Nebraska) than in either of the wheat regions until one reaches the southwestern margins of the winter wheat region. In other words, the two wheat regions, as shown on Figure 7-3, are an integral part of the great Cropland Triangle of North America whose approximate apexes are at Edmonton, Alberta, and Lubbock, Texas, and in central Ohio. The really abrupt changes in the character of farming occur not at the western boundaries of the Corn Belt, but rather within the western parts of the wheat regions, where regular rain-fed cropping gives way to summer fallowing and other techniques necessary to farm under increasingly arid conditions.

It is directly to the west, between the two wheat regions, that the Corn Belt ends abruptly. The

[9]For further comment on the modern Corn Belt, see J. F. Hart, "Change in the Corn Belt," *Geographical Review*, vol. 76 (1986), 51–72.

abruptness is forced by terrain and soils rather than by a sharply different climate. Central Nebraska north of the Platte River is occupied by the Sand Hills, a maze of grass-covered dunes formed during the glacial epoch. Fortunately, settlers learned early that this prairie grass cover could not be plowed or the hills would become active, blowing dunes again. The grass cover, moderately high rainfall, and ephemeral lakes and marshes in the interdunal basins support prosperous beef-cattle ranching on generally huge holdings.

Further north, across the Niobrara and west of the Missouri River, the relatively high-lying strata of the Missouri Plateau have been severely dissected by east-flowing streams, such as the White and Cheyenne. Areas flat enough for cropping are relatively small and scattered and are used for wheat or for hay to help support the predominant beef-cattle ranching, which is adapted to the rough terrain.

TRANSITION TO THE NORTHEAST—
THE DAIRY REGION

The third leg of the great North American Cropland Triangle is the Dairy Region (Fig. 7-3). It stretches from central Minnesota to the middle St. Lawrence Valley and Vermont. The Corn Belt–Dairy Region transition coincides, in part, with the transition from older to more recent glaciation—the Wisconsin stage—and, more closely, to the transition from natural prairie to forest. It is marked by accentuated relief, less deep and fertile soils, more areas of swamp and marsh, and a decrease in the length of growing season. The Dairy Region is, in other words, at a comparative disadvantage, in relation to the Corn Belt, for growing corn and soybeans, or for intensive cropping.

Because the Dairy Region is somewhat marginal, compared to the Corn Belt, the evolution of agriculture has been more difficult and the result more diversified. As described in Chapter 7, both regions experienced a wheat boom, but the Dairy Region had less choice for a suitable crop replacement. The dairy specialization evolved (1) as an adaptation to the land resources—silage corn, oats, and hay for dairy cattle could be grown in the relatively short season, and rough or wetland could be used as pasture—and (2) as a response to a growing market for dairy products in the burgeoning cities, especially those along the southern shores of the Great Lakes.

Specialization began modestly, with small surpluses of farm butter or cheese being marketed in nearby trade centers. Small crossroad cheese factories began to appear in New York after 1850, and in Wisconsin and Ontario a decade later. Eventually, southern Wisconsin had a network of such factories and almost any farmer could reach one with cans of fresh milk, by a journey of two or three miles. With the coming of motor vehicles, the state was the first in the country to pave almost all its country roads— a concession to the need for increasing quantities of milk to arrive from farms, fresh daily. Since 1950, most crossroad cheese factories, like country schools, have been converted to residences or other uses; cheese is produced in large-scale, milk-processing plants in towns and cities.

The Dairy Region has two cores of specialization and production—one in southern Wisconsin and the other in New York and the adjacent St. Lawrence– Lake Champlain Lowland (forming a donut-ring around the Adirondacks)—and an outlier in southeastern Pennsylvania (Fig. 7-3). The eastern core specializes in the production of fluid milk for Montréal and the Atlantic coastal cities, although much cheese is produced in the St. Lawrence Valley.

The Wisconsin dairy area may be viewed as a thicker "donut-ring" around a small "hole"—the Central Wisconsin Sand Plain. The southern and eastern parts of the ring constitute the major milkshed for the Chicago–Milwaukee urban area. The northern and western parts of the ring are major producers of cheese; Wisconsin's production is almost two-fifths of the national total. Much milk in western Wisconsin and Minnesota is manufactured into butter, each state producing nearly one-fourth of the national total.

Dairy farms in the region are overwhelmingly family farms. As in the Corn Belt and elsewhere, the trend has been toward fewer, bigger, and more highly capitalized farms. In Wisconsin, there were 165,000 farms averaging 142 acres (57 ha) in 1952, and 75,000 averaging 220 acres (89 ha) in 1987. The average size of milking herds increased; the average production per cow increased by 22 percent between 1952 and 1981.

Note in Tables 14-2 and 12-2 that in Dodge County, Wisconsin, and Franklin County, Vermont, which are typical of the two core areas, average farm capitalization is near $260,000; average value of annual production is about one-half that, being higher in Wisconsin than in Vermont. Most farms in the dairy area of the Midwest produce their own feed (primarily alfalfa, corn for grain and silage, and oats)

Table 14-2. Dodge County, Wisconsin: Agricultural Statistics, 1987

Population (1980)	75,000
Percentage change, 1970–80	+8.8
Area of county (ac/ha)	568,000/230,000
Percentage in farms	76.9
Percentage in cropland harvested	55.0
Value of agricultural produce sold	$181 million
Percentage from livestock products	84.7
Number of farms	2151
Average size (ac/ha)	203/82
Average value, land, buildings, & equipment, per farm	$263,000
Average value of agricultural products sold, per farm	$84,216
Percentage of operators whose principal occupation is farming	76.1

except for protein supplements (such as soybean meal) and vitamins. There has been a decline in the use of pasture except in more northerly or rougher areas—most dairy farmers prefer to bring the feed to the animals. Many sell surplus corn. In the eastern dairy core, where grain corn does not mature well, farmland is devoted mainly to hay and pasture, with some corn for silage; grain is shipped in from the Corn Belt.

Sale of breeding stock, surplus calves, and older cows culled from milking herds is an important source of income, especially in Wisconsin, on farms with highly developed herds. Already by the early 1980s, prize breeding cows were changing hands for a million dollars or more.[10]

Most large dairy farms have dropped their subsidiary enterprises such as poultry or hogs. Areal specialization, however, is not complete. Cash corn and soybeans dominate on such outlier islands of prairie soil as the Arlington (north of Madison) and Janesville prairies, a reflection of the northward expansion of the Corn Belt crop pattern. In eastern Wisconsin, sweet corn, green peas, snap beans, and beets for processing are grown, often under contract to local canning factories; Wisconsin leads the nation in all four vegetables.

[10]The value of prized animals has multiplied with the development of genetic engineering. Artificial insemination, now widely used, permits a single prize bull to sire thousands of offspring. With embryo transplant, developed during the last decade, an artificially inseminated cow can produce embryos that are then transplanted into host cows, which bear the calves. Combined with cloning or embryo splitting, it makes it possible for upward of thirty calves to be produced from a single cow in a year instead of the usual one.

Because of the need to house animals and to store feeds and machinery, dairy farmsteads, especially in Wisconsin, are impressively large and elaborate. The evolution in technology and scale can be "read" from the farmsteads—for example, huge barns with high lofts for loose hay replaced by newer, low buildings designed for machine handling of hay bales; small, old silos dwarfed by huge, new self-unloading ones of concrete or steel; conveyors for automatic feeding or manure handling; milkhouses with stainless steel cooling tanks to which milk is piped directly from the cow and from which it is pumped into stainless steel tank trucks for delivery to processors. The well-kept farmsteads, dispersed on relatively small holdings among often irregular or contour strip fields interrupted by lakes and wooded hills, make for a beautiful landscape.

PENINSULAR ONTARIO AND SOUTHERN MICHIGAN

In the central part of the Dairy Region, dairying generally maintains first rank, but diversification is much greater. Dairy products, not including dairy animals, account for less than 25 percent of all sales of farm products in Michigan, as compared to nearly 60 percent in Wisconsin. The Ontario Peninsula is even more diversified.

The lake-bordered peninsulas of southern Ontario and Michigan are structurally a continuation of the Central Lowland. In Ontario, the Paleozoic formations of the Lowland meet the Shield along a line from Georgian Bay to the Thousand Islands in the St. Lawrence. Most of the surface features in both peninsulas, however, are of glacial origin. During the Wisconsin glaciation, a lobe of ice pressed southward along each of the three present basins of Lakes Michigan, Huron, and Ontario–Erie, the latter two almost encircling southern Ontario. When the ice retreated, it left sandy, hilly morainic ridges roughly parallel to the lake shores, with gentle till plains in the center of the peninsula and flat, lacustrine plains beside Lakes Huron and Erie. The gray-brown forest soils (alfisols) of the till and lacustrine plains are moderately fertile.[11]

Climatic conditions also favor agriculture. The Great Lakes have a marked modifying influence on temperatures, especially in areas near the eastern or

[11]Some 53 percent of southern Ontario is covered by soils in capability classes 1, 2, and 3; that is, by soils suitable for most field crops. For most of peninsular Ontario southwest of a line from Toronto to Goderich, the figure is in excess of 80 percent.

The Niagara Escarpment, Ontario: The Fruit Belt. *Ontario Ministry of Agriculture and Food*

generally leeward lake shores (Fig. 13-1). Winters are warmer, the seasonal temperature range is less, and the growing season is longer than in areas further east or, especially, west. The length of the growing season (number of days with an average temperature over 42°F [5.6°C]) exceeds 220 days in extreme southwest Ontario, a figure reached elsewhere in Canada only in the vicinity of Vancouver Island.

Peninsular Ontario has become Canada's major area of relatively intensive, productive agriculture. It was settled, first on eastern margins, by American Loyalists escaping the Revolutionary War, but mainly after 1800, by British and some German immigrants. By 1850, the better lands south of the Shield had been occupied and, until the western Prairie lands became available half a century later, settlement and land use intensified. Development of southern Ontario as Canada's urbanized, economic heartland assured area farmers of the best market location in the country, enhanced even more by the international boundary that cut them off from competitors in the United States.

The nearby markets encouraged specialization in dairying and also many other products such as beef, pork, poultry, feed crops, soybeans, fruits, vegetables, and tobacco. Dairying, which supplies fluid milk to the large cities, is important throughout the Peninsula, but is most concentrated in the central area midway between the southern tip of Lake Huron and Lakes Ontario and Erie. Butter manufacture is important to the north and west, cheese further east near the Québec border. Hay, corn for silage, and oats are important feed crops. They are used to feed hogs, especially in the area of dairy concentration, and beef cattle, poultry, and some sheep throughout the Peninsula. Grain corn and soybeans are increasingly important crops in the southwest— the tip of the Peninsula may be considered a part of the Corn Belt. The lacustrine clay soils there are also well suited to vegetable production.

Growing of fruits, including grapes, apples, peaches, cherries and small fruits, is concentrated in the Niagara Peninsula, especially between the Niagara Escarpment and the Lake Ontario shore. The orchard and vineyard district continues eastward into New York and, to a lesser extent, westward along the southern shore of Lake Erie. Specialized fruit-growing districts are also scattered along the eastern shore of Lake Michigan, with cherries most important in the north and a diversity of fruits in the south. Climatic modification is important, especially the effects of the slow-warming lake waters in spring,

The Niagara Escarpment: Hand-picking grapes. *Ontario Department of Agriculture*

which help to retard bud and blossom development until the danger from cold air masses is lessened. The Niagara fruit belt is especially strategic to central Canada—there is no other diversified fruit-growing area in the country closer than the Okanagan Valley of British Columbia. But the limited area suitable for orchards and vineyards is being seriously overrun by the urban expansion of Hamilton and St. Catharines (see Fig. 14-3). Tobacco growing has become an important specialty on the sandy soils of a glacial-age delta that bulges into Lake Erie and ends in Long Point. The flue-cured product supplies most of the Canadian market and yields a surplus for export.

In summary, the transition from Corn Belt to Dairy Region is both gradual and subtle—the differences are more in intensity of total land use than in intensity of use on individual farms; more in end products than in crops and livestock (milk versus corn–soybeans–meat); more in breeds than in kinds of animals (Holstein-Friesian versus Hereford and Aberdeen Angus); more in the variety than in the geometry of the landscape (the woods, hills, and lakes of the Dairy Region offer much more variety—even the rectangular road pattern there is frequently modified in order to avoid hills, lakes, or swamps).

Here in the Dairy Region, farmers confront the same fundamental problem as do their counterparts in the Corn Belt or Wheat Belt. It is the problem of surplus production and could be said to arise from success—from efficiency and productivity. The principle of supply and demand suggests that increase in demand will bring an increase in price and thus encourage an increase in production; when the demand is satisfied, prices will fall and producers will curtail their output. In the business of agriculture, composed of a multitude of small, competing firms (farms), producers seem to follow the principle on the upswing, but often not on the downside. In dairying, for example, when the price of milk declines, farmers tend to increase production in order to maintain the same level of income.

The U.S. Congress has established a minimum support price for milk. When the market price falls below this level, the Commodity Credit Corporation

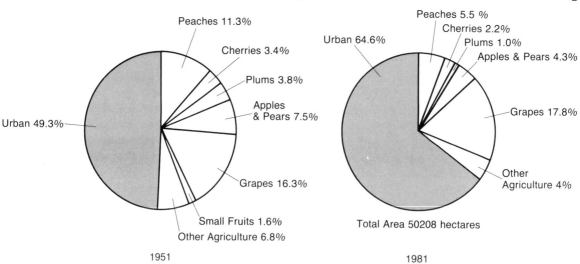

A

Peaches 11.3%

Cherries 3.4%

Plums 3.8%

Apples
& Pears 7.5%

Urban 49.3%

Grapes 16.3%

Small Fruits 1.6%

Other Agriculture 6.8%

1951

B

Peaches 5.5 %

Cherries 2.2%

Plums 1.0%

Apples & Pears 4.3%

Urban 64.6%

Grapes 17.8%

Other
Agriculture 4%

Total Area 50208 hectares

1981

Fig. 14-3. The Niagara Fruit Belt: (**A**) 1951 and (**B**) 1981. The diagrams show how land use has changed through urban encroachment along the western shore of Lake Ontario, and also indicate which types of production have suffered most. For a fuller commentary on these changes, see R. R. Krueger, "Urbanization of the Niagara Fruit Belt," *The Canadian Geographer*, vol. 22 (1978), 179–94. (Diagrams courtesy of Dr. A. J. Strachan.)

(CCC) buys and stores surplus butter, cheese, or powdered milk. But with huge surpluses building up, the government has recently been forced to reduce the support level in an effort to curtail production. Dairy farmers have objected strenuously to this change, arguing that in order to maintain their incomes they will have to increase production, not reduce it. The program of the government of Ontario is more logical and, in its working, more satisfactory in that the price supports are available only if production is curtailed.

TRANSITION TO THE NORTHLANDS

The northern margins of the Dairy Region, and hence of the Agricultural Interior, are abrupt in Canada and only a little less so in the United States. In Canada, the boundary is marked by the edge of the Canadian Shield with its metamorphosed crystalline rocks, thin and infertile soils (or no soil at all), and a short growing season, all unfavorable for agriculture. In the United States, the Laurentian rocks are more widely covered by glacial drift, in the northern half of the lower peninsula of Michigan deeply so, but the podzolic (spodosol) soils are generally acidic and infertile. Only in north-central Wisconsin do moderately fertile soils of the Dairy Region overlie the southern margin of the Shield.

These soils, however, have seen a varied use in the century and a quarter of their occupance by the white man. There was a first phase represented by the lumber boom in the North Woods. The opening of the dry, treeless West in the 1870s, and 1880s led to a great demand for timber, a demand that the Great Lakes area could meet. Northward the lumbermen led the way, clearing ground for the farmers who would follow to grow their supplies. Disregarding or burning all but the particular timber they sought, they cut their way through the area between 1875 and 1905, and then, as abruptly, left for the mountains of the West. The farmers who had settled the North Woods (like their neighbors who had followed the copper miners into Upper Michigan a few years earlier) found themselves virtually marooned in a wilderness without local markets; the state governments found themselves left with the cutover land (Fig. 14-4).

As a result, the northern margin of the Dairy Belt advanced into the northern Great Lakes area, and then retreated again. Once the stimulus of the lumbermen's demand for supplies was removed, farming on the poor soils and the cutover generally proved uneconomical. Consequently, there has been a steady out-migration from the area, and the few farmers who remain supplement their incomes with work off the farm.

Today, by far the largest money-maker here is

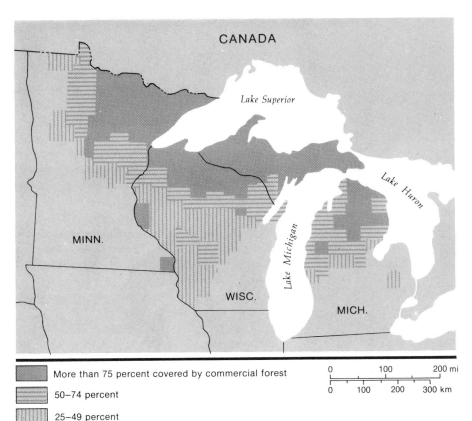

Fig. 14-4. Forest areas of the Northern Great Lakes Region: Viewed from the other aspect of the land-use pattern, this is a map of the northern margin of the Dairy Region.

More than 75 percent covered by commercial forest

50–74 percent

25–49 percent

tourism or recreation. It was first stimulated in the 1880s by the region's railroads, which built resort hotels. Today's visitors are more likely to own or rent a lakeside cabin, and to spend much of the summer there. In winter, there is a small winter-sports development and a hectic period in November, when everyone who can hold a gun and pay for a license (or so it seems) goes north to shoot deer. The old farmer's supply centers and lumber towns have found a new life as resorts.

TRANSITION TO THE EAST AND SOUTH

The eastern limit of the Corn Belt is generally taken to lie in western Ohio. South of Lake Erie there lies the dead-level plain now occupied by the Maumee River, but formerly part of the bed of a larger Great Lakes system. This plain is today splendidly farmed by a population predominantly German in origin, and it may be taken as a convenient terminal point. Eastward from here the land rises gradually to the plateaus of the Appalachians; rivers are deeply incised and the local relief becomes more pronounced. To the east, too, there lies the heavy industry region of Ohio and western Pennsylvania, whose urban

markets exercise a powerful influence on farm production. This eastern margin of the Interior is therefore characterized, on the one hand, by a decrease in the amount of land in crops but, on the other, by a close adaptation to local market conditions.

The resulting pattern shows a marked similarity to that which has developed under comparable conditons on the Middle Atlantic coast, with dairying widespread over the uplands, and intensive concentrations of truck farming at strategic supply points near the cities. As in southern Ontario, the only significant variation of this pattern occurs along the shores of Lakes Ontario and Erie, where a fruit- and vegetable-farming area extends the full length of the lakeside, famous both for the orchards of the Finger Lakes in western New York State and for the greenhouses around Cleveland.

Southward, toward the Ohio River, the factors of change are mainly physical. The drift cover to which the Corn Belt owes so much of its prosperity becomes fragmentary, and dissected more and more deeply by the south-flowing tributaries of the Ohio. As the surface becomes rougher and the soil cover patchy, the intensity of farming which characterizes the central

Kentucky: In the Bluegrass country. *Kentucky Department of Travel Development*

Interior is lost. To find it again one must cross the Ohio into Kentucky where, on the limestone soils of the Bluegrass country, the manicured farms, and the equally cosseted horses that live on them, generate a landscape of almost unbelievable well-being.

Further west, the near-ideal physical conditions of the Corn Belt give way to less fertile claypan soils in southern Illinois and northern Missouri. The Corn Belt barely touches northern Missouri and, going south across the state, pasture farming and beef raising become more general, as the approach to the hill lands of the Ozarks produces a more broken terrain (see Fig. 14-5).

The Industrial Interior

The continental Interior contains a large proportion of all North America's industry. It is, in fact, the combination of intensive agriculture and widespread in-

dustrialization that gives the region its character and that underlies its prosperity. The eastern half of the Interior is also the western half of what has been known as the Manufacturing Belt. Even beyond the limits of this Belt, although the industrial centers are more widely dispersed, the manufactures they support form no less vital a part of the midwestern economy.

In the light of what has been said in the last section about farming in the Interior, we can visualize the agricultural pattern of the region as represented by a kind of trend surface, domelike in shape, whose highest points (whether represented by indices of farm income or of input intensity) are to be found in the central Corn Belt. We can now attempt the same sort of characterization of the Industrial Interior. This time, the type of surface is quite different. It consists of a fairly even gradient, trending downward from east to west; that is, from more industrialized to less

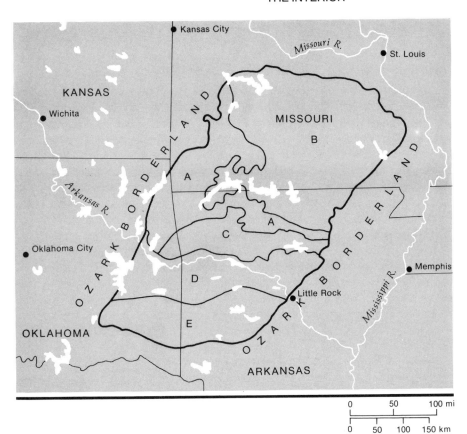

Fig. 14-5. Physical subdivisions of the Ozark region: (**A**) Springfield Plateau, (**B**) Salem Plateau, (**C**) Boston Mountains, (**D**) Arkansas Valley, (**E**) Ouachita Mountains. Lakes and reservoirs are also shown. These represent an important part of the resource base for a naturally poor region, since they make it more attractive to tourists.

industrialized areas. At a certain point going westward, the trend surface reaches zero level, but west of this point it has outliers—islands of industrial development detached from the main body. This industrial trend surface, or rather, the area it covers, is or was the Manufacturing Belt. The stability suggested by the name Manufacturing Belt is, however, belied by its history. What has happened is that, in North America, industry, like population, has tended to move from east to west—but in several waves rather than in a single motion. In other words, our trend surface consists of a number of layers.

Why is this? Industry first developed on a large scale in New England and around the port cities of the East. From there it spread to the coalfields of Pennsylvania and Ohio. Subsequently, it developed in areas that were accessible to the movement of coal, either on the Atlantic Seaboard or along the water routes of the Interior—the Ohio River and the Great Lakes shores. And it kept on spreading, even though it was moving even further away from its coal supplies. It continued to spread in this way (1) because population—that is, markets—continued to build up

in the West and Southwest; (2) because agricultural production was increasing in the West and, eventually, diminishing in the East; and (3) because beyond a certain point in space and time—roughly speaking, the Mississippi River after World War I—industry was being attracted toward new energy sources, which were in the process of replacing coal—oil and natural gas in the Southwest.

This notion of a spread of manufacturing across the Interior enables us to classify American industries on the basis of their mobility. We can distinguish four different situations.

1. *Industries that have never moved from their original locations.* For them there has been no "spread." Anchored by their dependence on skilled labor or by family control, they remain to this day, in the East. An excellent example is provided by the manufacture of small arms. It has often been said that the West was won by the Colt revolver, but the Colt itself was manufactured not in the West, but rather in New England, along with the famous Springfield rifle.

2. *One industry, at least, moved partway and then stopped.* This was the steel industry. As we have al-

Leisure and land use: Lake of the Ozarks, Missouri. *Grant Heilman Photography, Inc. Photo by Grant Heilman*

ready seen in Chapter 8, the steelmakers' base in the great period of expansion and frontier advance after the Civil War was Pittsburgh. From there the industry spread, in the period 1880–1901, to the cities of the Great Lakes. But there it stopped, halted by the application of Pittsburgh Plus (see p. 160). Between 1901 (the founding date of the United States Steel Corporation) and World War II, there was almost no westward expansion for steel. The steelmakers' investment in their eastern mills was too large for them to abandon; instead, they chose to impose on the industry an artificial standstill. Since 1941, however, westward spread has resumed.

3. *Some industries have moved west and are still moving.* Among these, the major examples are to be found among the agriculture-based industries. When the Agricultural Interior was first opened up, the farmer's base, both for equipment supply and for processing of farm produce, was in the East. The long west–east haul for grain or livestock and the return flow of farm supplies were basic—and costly—elements in the North American economy. Over the years, however, this group of industries has followed the farmers westward, more than keeping pace with the agricultural frontier, and so meeting the flow of farm produce ever closer to its source. In the 1850s,

the meat-packing industry, as we have already noted in Chapter 8, was centered in Cincinnati, a city to which it was drawn because of good communications, local salt supplies, and availability of banking and financing services.[12] In the Civil War years, however, Chicago (by then the focus of an increasing number of railway lines) replaced Cincinnati in importance. Then in the 1880s, and more definitely after World War I, Chicago began to lose ground relative to cities still further west, such as Omaha and Kansas City. Finally, the years since 1950 have seen the growth of a truly western meat-packing industry especially in the high plains of Texas, the Colorado Piedmont, and the irrigated valleys of Arizona and southern California.

4. *Some industries in the Interior today have no history of location in the East,* for the simple reason that they are late arrivals on the technical scene and first came into being in their present Interior location. The most obvious examples are in the motor vehicle, petrochemical, and pharmaceutical fields, but branches of electrical engineering and electronics may also be mentioned. Such industries are to be found in the Interior because of market opportunity or ease of access to raw materials, or because they were developed by the people of the region, using its resources.

We can express this pattern of manufacturing in another way. At any given point in the eastern Interior, an observer gifted with long life and infinite patience could have watched successive groups of industries moving past from east to west in the century between 1850 and 1950. First would come the supply and processing industries of the agricultural frontier: milling and meat-packing, and the supply of farm tools and a few simple consumer goods to the frontier populations. Then would come a second wave of more sophisticated manufactures, catering to the needs of a population now grown in numbers and more financially secure: these could only spread west as and when the market could sustain them. Next there would spread before our observer the capital-equipment industries, supplying the means of transport and the machinery needed by plants already established—machinery that, in the past, had been shipped into the region from the East. And finally the observer would note the growth, in situ, of a new brand of industry, one unrelated to the specific needs

or resources of the region: a class of national producers who had simply selected the Interior as a favorable location from which to meet the sophisticated requirements of today's consumers for all kinds of aids to modern living.

Supposing that our observer had been watching from Chicago, he or she could have followed this sequence of events and their impact on the growth of the city's manufacturing. In the first two decades after Chicago was founded in 1833, it built mills and breweries and packed meat under contract to the U.S. Army. No major iron-using industry was yet present, nor did such industries develop until after 1870. But the period from 1850 to 1870 saw the emergence of the meat-packing industry as the giant of the city, with the Union Stockyards (opened in 1865) and the spread of refrigeration beginning to give shape to the industry. Second to meat-packing in this period was the production of men's clothing for the farmers along the frontier.

Between 1870 and 1900, Chicago passed through a period of industrial expansion and consolidation. Meat-packing, clothing, and furniture making emerged as the "big three" in this period, the last of these a by-product of the Great Lakes lumber boom of the 1880s. By 1900 Chicago had become the world's largest producer of upholstered furniture. Only during this period did the city's industry acquire a basis of steel making: by the end of this period, iron and steel ranked fourth among Chicago's manufactures.

Between 1900 and World War I, iron and steel moved into the lead, the new mills rising south of the city around Calumet and Gary. The manufacture of clothing, although the second industry of Chicago, was declining and so, too, as we have already seen, was meat-packing. But the city was now building machinery and equipping a good share of the nation's railroad system, and after 1914, it moved into the position of national manufacturing center it has occupied ever since. By 1980 its leading industries, measured in terms of gross sales of manufactured goods, were primary metals, food products, fabricated metal products, and electrical machinery and equipment.

In the 1950s, with new industrial areas developing elsewhere in the continent, it seemed as if the Interior was reaching its industrial saturation point under existing conditions. In these circumstances, the opening of the St. Lawrence Seaway helped to trigger another phase of industrial development, by bringing

---

[12]Cincinnati became known as "Porkopolis." About the year 1848 a traveler recorded his opinion that the city was "the most *hoggish* place in the whole world."

the Great Lakes, their ports, and their factories into direct touch with the outside world.

This last stimulus to new industrialization in the Interior is now over thirty years in the past. In those thirty years, the position of the Interior, like other parts of the economic core, has not improved. Many of the industries on which the region's leadership was based, such as steel or automobiles, are now slow-growing and burdened with old plants and old equipment. Innovative new industries, such as computer manufacture, have risen elsewhere. The general drift of population toward the Sun Belt, described in Chapter 2, has weakened the position of many of the consumer-goods industries that catered to the needs of a Snow Belt population. In fact, considering the range of unfavorable circumstances with which the Industrial Interior has had to contend, it says much for the region's advantages of location and infrastructure that the loss of jobs in manufacturing has not been greater (see Table 14-3).

Diverse as the industries of the Interior certainly are, they do show regional specializations—the heavy industry group in eastern Ohio and western Pennsylvania, for example, and the agriculture-related industries in the western half of the Interior. Thanks to their advantages of position, the Great Lakes ports—particularly Chicago—generally possess a much wider range of manufacturing than any other of the cities of the Interior.

THE HEAVY INDUSTRY AREA OF PENNSYLVANIA AND OHIO

The industrial area that extends along the valleys of the upper Ohio and its tributaries owes much of its growth to the existence of the Appalachian coalfield,

but some also to the personal influence of past generations of industrialists. It is to be found in an area ill suited to the purposes of heavy industry—narrow, winding valleys that cut into the northern part of the Appalachian Plateaus. Although the combination of horizontal coal seam and steep valley wall aids the miner, and the river system serves both for transport and for industrial water supply, these advantages are seriously offset by industrial congestion in the smoke-filled valleys, where there is neither room for expansion nor attraction, in an era of "clean" industry, for the notoriously mobile American worker to settle.

So great, however, was the initial advantage of proximity to the great Appalachian coalfield that this area is not only one of the most important steel-producing regions in the world, but it is also the center of North America's manufacture of glass and clay products. Within the area localities specialized in one or another of these types of industry; for example, Youngstown in primary metal production and Akron in rubber goods, and the valley of the Kanawha in West Virginia has been called "The Ruhr of the United States chemical industry."

At the heart of this great industrial concentration stands the city of *Pittsburgh* (2.2 million). As recently as 1970, one-half of Pittsburgh's industrial workers were engaged in primary metal production, and a further one-quarter produced machinery and steel goods of various kinds. As a steel city, its main natural advantage was that of location at the junction of the Allegheny and Monongahela rivers—at the hub of the upper Ohio routeways and close to the Appalachian coking coals. Yet it was not location alone that made Pittsburgh what it is. To explain its ascen-

Table 14-3. The Central Interior: Percentage Shares in Nonagricultural Employment, by States, 1985

	Total Work Force (thousands)	Construction	Manufacturing	Transport, Public Utilities	Wholesale & Retail Trade	Finance, Insurance, Real Estate	Services	Government
Ohio	4379	3.5	25.6	4.6	23.6	5.0	21.8	15.2
Michigan	3505	2.8	28.1	4.0	22.3	4.6	21.2	16.5
Indiana	2177	4.1	28.0	5.0	23.4	4.9	18.8	15.3
Illinois	4767	3.8	20.6	5.8	24.5	7.1	23.1	14.5
Wisconsin	1977	3.2	26.0	4.7	23.4	5.3	21.0	16.3
Iowa	1075	3.3	19.1	4.7	25.8	5.9	21.5	19.4
Missouri	2097	4.5	20.4	6.8	24.3	5.8	22.1	15.8
Minnesota	1866	3.8	20.1	5.3	25.0	5.9	23.2	16.2
U.S., 50 states	97,614	4.8	19.8	5.4	23.7	6.1	22.5	16.8

dancy, we must recall a historical coincidence: in the years after the Civil War of 1861–65, three things were happening simultaneously. The first was that the demand for steel, especially steel for railroads, was rising as the West was opened to settlement. The second was that the Bessemer process for making steel was introduced, with the result that steel could be produced both more cheaply and more quickly than ever before. The third element was the appearance in Pittsburgh of Andrew Carnegie, an industrial wizard who, by his bold application of the Bessemer process and by his ability to appreciate the economic realities of steelmaking, came rapidly to dominate the industry. Securing an alliance with Henry Frick, whose comparable talents had given him control of the coke-making phase of the industry (then centered at Connellsville, 50 mi [80 km] southeast of Pittsburgh), Carnegie made Pittsburgh and steel into synonyms before selling out in 1901 to the newly formed United States Steel Corporation. Thus, by the conjunction of location, timing, and personal initiative, Pittsburgh became the center of the steel industry, and its steel men imposed on the industry as a mark of their hegemony the "Pittsburgh Plus" arrangement (see Chapter 8) that lasted until 1924 and itself assured the continuance of the regime.

Since the end of World War II, however, Pittsburgh has both changed in appearance and done its best to change its public image. The most striking evidence of these changes has been the transformation of the area within the junction of the Allegheny and Monongahela rivers (the point at which the original Fort Duquesne and the later Fort Pitt stood) from a slum to an open space fringed with impressive new structures, the whole forming a "Golden Triangle." Several steel mills and railroad yards close to the city center have been removed, leaving the principal steel mills of the present city upstream on the Monongahela. Pittsburgh is no less involved in steelmaking than formerly, but today the city is primarily involved in the control and administration of the industry rather than in the output of its products. The headquarters of U.S. Steel appropriately dominate the Golden Triangle, but there are a number of other major American companies with headquarters here—firms like the Aluminum Company of America (ALCOA) and Heinz food products. After New York, Chicago, and perhaps Los Angeles, Pittsburgh is the most important corporate headquarters in the United States.

Accordingly, the role of manufacturing in the city's employment base has been greatly reduced, as it has in so many of the continent's larger and more mature cities. At the same time, it is Pittsburgh's claim that, for every job lost in the iron and steel industry, at least one has been created in other, modern industries, including the high-tech sector.

But changing the city's image has not been easy. Deeply entrenched valleys and steep slopes make urban development patchy, crosstown movement difficult, and air pollution hard to dispel.[13] The population of the metropolitan area has been falling ever since 1960: so, ever since its 1973 peak, has U.S. steel production. What has been done in the Golden Triangle needs to be done in much of the remainder of the city, and the thirty-year campaign to reduce smoke and air pollution needs to be accompanied by a continuing campaign to bring in new industries. But both these tasks are herculean in their extent. Pittsburgh was too successful for too long at making steel easily to switch over to other activities.

From Pittsburgh, the ribbons of industrial development stretch out along the valley floors of the Ohio system, south and east into the coalfield and north and west toward other manufacturing centers. *Youngstown* (493,000) and *Wheeling–Steubenville* are the principal steel cities, and a number of smaller centers, such as East Liverpool, produce clayware or glassware, much of it for use in the industries of the area. Further south, in West Virginia, the output of the Kanawha Valley ranges from heavy chemicals, such as ammonia and caustic soda, to synthetics for the plastic and textile industries, and includes also steel alloys, glass, and synthetic rubber. To the northwest between Pittsburgh and Lake Erie are *Akron* (223,000) and *Canton* (394,000), with their satellites. Canton is a steel-goods city, and Akron long claimed the title "rubber capital of the world."

Like New England and like Old England, the industrial area of the upper Ohio is suffering today the disadvantages of its early pre-eminence. If there re-

---

[13]"Nearly a seventh of the [Pittsburgh metropolitan] region's area is taken up by hills over twenty-five degrees in slope. . . . Though the numerous hills in the city of Pittsburgh form serious barriers to transportation, the many steep wooded slopes provide a most welcome contrast to the gray tones of the built-up area. . . . Over half of the land area in the four [metropolitan area] counties is either vacant or unusable, largely because of the irregular terrain." P. H. Vernon and O. Schmidt, "Metropolitan Pittsburgh: Old Trends and New Directions," in John S. Adams, ed., *Contemporary Metropolitan America* (Cambridge, Mass., 1976), vol. 3, pp. 1–60. Quotations from pp. 9 and 11.

Pittsburgh: A mill district in 1941. Note the "staircase" street on the steep valley side and the industrial smog in the background around the mills. *Library of Congress*

ally is a Rust Belt in the United States, this is one part of it. Inevitably, its relative importance waned with the change to newer sources of power; with the lack of space for expansion; with the competition of new and more efficient producers further west who have profited by its experience; with the abolition of "Pittsburgh Plus." Changes in industrial techniques have killed some of its activities, such as the coke making at Connellsville, and structural unemployment is a recurrent problem. But in spite of these difficulties, it remains one of the great industrial concentrations of the continent. The attachment of its industries to coal and clay, the wealth of labor skill at its command, and the enormous investment that has gone into its heavy industries are all factors that are resistant to change and that assure the area of a continuing importance.

The Great Lakes and Their Cities

The basin of the Great Lakes contains almost one-fifth of the population of the United States, and one-half that of Canada. The six great cities that lie on the lake shores—Chicago, Milwaukee, Detroit, Cleveland, Buffalo, and Toronto—themselves contain 8 percent of all Anglo-Americans. Both cities and people owe their presence to the movement and manufacture of goods either using the Lakes' water route, or transported along their low-lying shores.

But we can be a little more precise than this about both people and industry. First, we should notice that this great concentration is largely to be found on the shores of the *southern* Great Lakes—south of Milwaukee on Lake Michigan, and south and east of Sarnia at the foot of Lake Huron. Much of the shore-

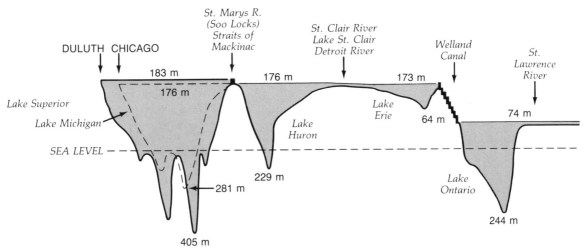

Vertical exag., × 2000
Lake surface elevations given above sea level,
maximum depths, below surface level

Fig. 14-6. The Great Lakes: Longitudinal profile from Duluth to the St. Lawrence River. The heights of the water surfaces and the depths of the lakes are shown in meters. (Reproduced from *The* *Great Lakes: An Environmental Atlas and Resource Book* (1987), published by the U.S. Environmental Protection Agency and Environment Canada.)

line of the upper lakes is occupied, if at all, only by summer cabins or camps; it is a resort and, in part, a wilderness area. Secondly, although the Great Lakes constitute the largest continuous freshwater body in the world, their drainage basin is small out of all proportion to their size. None of the major rivers of the Interior drain into them; on the contrary, rivers like the Wisconsin and the Illinois rise within a few miles of their shores and flow away from them. At the basin's narrowest, at Chicago, the watershed is almost within sight of the shore of Lake Michigan. In other words, as a result of the glacial action we reviewed in Chapter 1, nearly all the water in the Great Lakes region is *in* the lakes, not in rivers draining into them (see Fig. 14-6).

These two circumstances between them ensure that, along the thickly populated and heavily industrialized southern shores of the lakes, pollution has become a serious problem and so, paradoxically, has water supply. The early settlers were content both to draw their water and to tip their wastes in the same lake. To them, it would have seemed incredible that, in the presence of such plenty, their descendants would have been searching for clean water. But the results of 7–10 million people following their practices along the southern Lake Michigan shore would be horrific—indeed, are horrific—since to the problem of waste and sewage, even if solved, must be added the discharge of wastewater from steel and paper mills, and the washings of chemicals from Corn Belt fertilizers.[14]

Pollution had its effect not only on water supply, but also on recreation and on the previously extensive Great Lakes fisheries. So rapid was the deterioration of water quality that action was forced upon the governments concerned. As we saw in Chapter 5, the U.S.–Canadian agreements of 1972 and 1974 marked a turning point, and since then there has been a welcome improvement in the situation.[15] The cost has been high, in constructing a new waste-disposal plant and in reducing industrial discharges, but there were no serious alternatives.

Over the Great Lakes route moves a fleet of cargo ships which, during the nine-month season, have in some years (1978 and 1979, for example) carried over 200 million tons of freight. The volume of lake cargoes in the 1990 season (ending January 31, 1991) was well below 200 million tons, however. All through the twentieth century, the chief item in the tonnage total was iron ore, moving from the Mesabi, Gogebic, and Marquette ranges, and from Steep Rock, Ontario, down the lakes to the steel mills.

[14] An interesting source on the Lakes is *The Great Lakes: An Environmental Atlas and Resource Book*, published by the U.S. Environmental Protection Agency and Environment Canada (Chicago and Toronto, 1987).

[15] See N. M. Burns, *Erie: The Lake That Survived* (Totowa, N.J., 1985).

Thunder Bay and the Canadian grain trade: The former port cities of Fort William and Port Arthur on Lake Superior have been known collectively since 1970 as Thunder Bay. The port serves as the eastern outlet for prairie grain, which is shipped from here down the Great Lakes. It claims to be the largest grain-handling port in the world. The elevators shown have a capacity of over 2 million tons, and the maximum annual throughput to date has been 17.7 million tons. *Thunder Bay Harbour Commission, Port of Thunder Bay*

Moving in the opposite direction was coal from the Appalachian and Interior fields. The third bulk cargo by volume has been cereals—Canadian grain loaded at Thunder Bay and U.S. grain from Duluth–Superior at Lakehead (see Fig. 14-7).

But there have been some recent changes. For one thing, with the falloff in steel production, less ore was needed—in 1990, 50.8 million tons moved downlake from Superior ports, and 10.6 million tons moved west from eastern Canada. For another, coal now moves *both* ways on the lakes—from the Appalachian field via Lake Erie ports (Toledo, Sandusky, Ashtabula, and Conneaut) to a 1990 total of 22 million tons, and from the newer, western fields via Lake Superior ports (15 million tons in 1990). These cargoes are apt to meet at The Soo locks be-

tween Lake Superior and Lake Huron; in the 1990 season the locks handled the following traffic:

Commodity	Volume of Freight (millions of tons)
Grains	12.8
Coal	15.9
Potash	1.35
Iron ore	49.6
Other goods	8.25
Total	87.9

This total is still impressive, when we realize that its transport required nearly 4,000 passages of the locks during the navigation season. The fastest lake

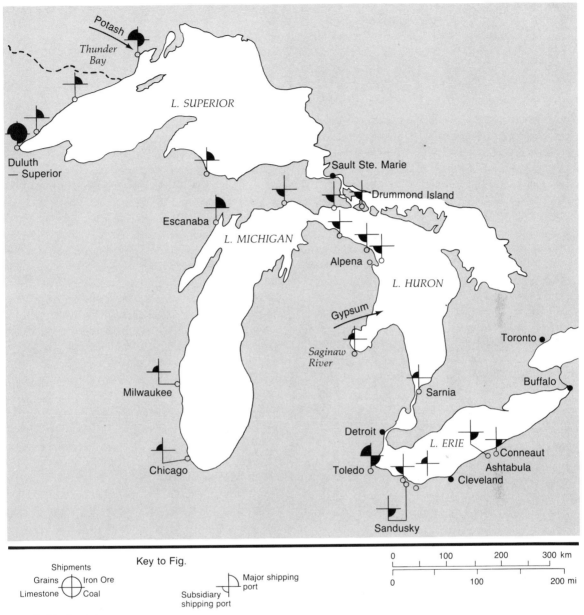

Fig. 14-7. The Great Lakes: Shipping ports for major bulk commodities. The commodities are identified and the ports are classified as either major or subsidiary shipping points. (Source: *Annual Report of the Lake Carriers' Association, 1986.*)

carriers made forty-five to forty-seven trips during the navigation period.[16]

Before we consider each of the cities of the Great Lakes in turn, it is of interest to notice that they have a number of features in common. That this should be

[16]All figures from the *Annual Report of the Lake Carriers' Association, 1991*, Cleveland, Ohio.

true is not surprising when we recall that they share a common situation on a lakeshore and a common history as well—brief as urban histories go, but representing a shared response to the technical stimuli and transport needs of their era.

The series of maps and explanatory comments that make up Figure 14-8 deal in simplified form with this common history. It is a history that had its effective

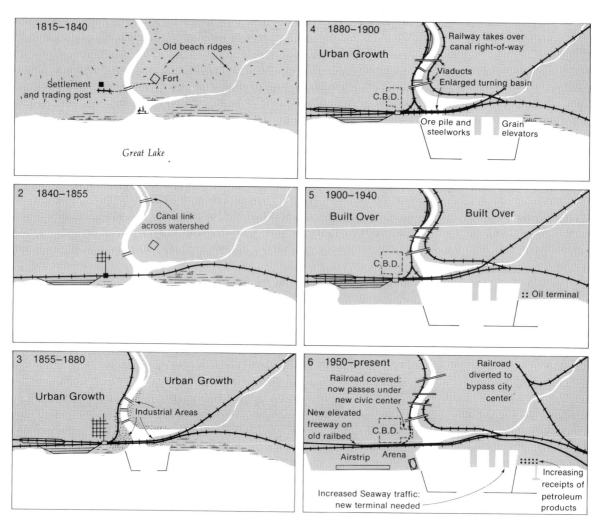

Fig. 14-8. CLE-RON-KEE-O, the model of a Great Lakes city. Its story in maps:

1. *1815–40.* The natural setting: a marsh-fronted lakeshore at the mouth of a creek entrenched in old beach ridges. Anchorage at creek mouth; fort, trading post, and bridge.

2. *1840–55.* A canal parallels the river, replacing the old portage route over the watershed. Earliest railroad follows the lakeshore, with yards on reclaimed shore fill and drawbridge over the river mouth, which is progressively widened for larger vessels. Fort disused; may be resurrected in tourist-orientated twentieth century.

3. *1855–80.* Railroad development; more yards and connections inland. Soo Canal (opened 1855) is one cause of increasing lake traffic which, with the trend to larger vessels, leads to construction of the first harbor works. Industry along the riverside and in the valley bottom. Commercial area growing toward the railroad station.

4. *1880–1900.* Steel industry arrives, with consequent increase in bulk cargoes handled and enlargement of the harbor. Ore stockpiles for winter closure of the Lakes. Grain traffic from lakehead is also increasing; elevators and flour mills at transshipment point. The growing city is now effectively cut off from the lakefront by railroad, industrial belt. Large vessels impeded upriver by low-level bridges; the city constructs the first viaducts over the valley. Canal falls into disuse.

5. *1900–40.* New traffic in petroleum products catered for by harbor extension. Reclamation beyond railroad belt for parks or further industrial sites. The CBD kept from necessary enlargement by the river bluffs, railroad belt, and industrial areas. Old low-level bridges removed to give shipping access to docks upriver; increasing road traffic requires new viaducts. Important decisions about future amenity are called for, but shelved because of the depression of the 1930s. Several fine new public buildings from the New Deal's public works program, but no geographical changes in city's layout.

6. *1960–present.* Declining importance of railroads leads to the removal of many tracks in valuable central areas; yards are resited outside the city. Dwindling passenger traffic closes the main railroad station; the site is redeveloped. Main railroad route now bypasses the city center, and the elimination of several tracks makes possible an expansion of the CBD; the city regains its lakeshore after being cut off from it for a century. But plans for replacing an old railroad by a new freeway lead to fears of a new barrier between city and lake, with heated debate in the city council. More infill of the lakeshore, earmarked for recreational space, marina, etc.; no more industry permitted. But pollution reduces the attractiveness of the lake for recreation. Opening of the St. Lawrence Seaway (1959) gives the city a new role as a foreign trade port.

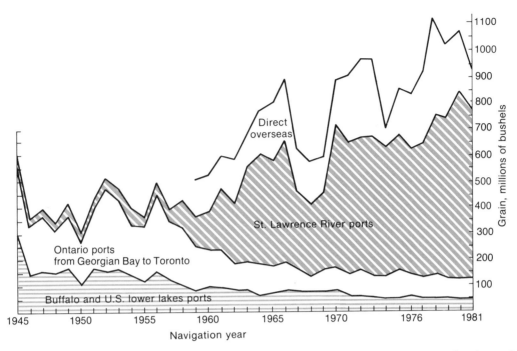

Fig. 14-9. The Great Lakes: The grain trade, 1944–81. Grain moves from western lake ports like Thunder Bay and Duluth: what the graph charts is the change over this period of the receiving points in the east. In particular, the opening of the St. Lawrence Seaway in 1959 led to radical changes in the trade, first by making direct overseas shipment possible in oceangoing vessels and, second, by eliminating the need to off-load grain at Buffalo or Georgian Bay ports, for onward shipment by rail. In these changes, Buffalo, as the graph suggests, has been the main loser.

beginnings in 1825, when the opening of the Erie Canal from the Hudson River brought the first major flow of traffic to the Great Lakes. Pre-existing settlements—forts and mission posts—or simple creek mouths became the sites of cities: portages and canals linked these anchorages with other river systems. In the following decades, although each of these cities felt the effect of local conditions upon its growth as an individual, all of them shared the stimulus of major developments: the opening of the Soo Canal in 1855; the spread of the steel industry between 1880 and 1900; the exploitation of the Mesabi ores after 1890; the opening of the enlarged Welland Canal in 1931; the changes brought about by the St. Lawrence Seaway after 1959; the burden of recession in the early 1980s.

In view of their situation, the most formative events in the lives of these cities were those that brought changes in the sphere of transport, by water or by rail. All of them had to provide space for port installations, railroad yards, and stockpiles (where the winter's supply of ore or coal could be dumped). All of them were involved in bulk traffic. All of them provided the necessary space, in part, by reclamation

along the lakeshore, and each of them at some time in its history has found itself cut off from the lake by the spread of industry or the sprawl of railroad lines; so that it has had to fight, consciously and not always successfully, to establish the principle of the lake as amenity and not merely as lifeline. These cities arose in an era of railroad expansion, steelmaking, and industrial laissez-faire; an era when it must have appeared self-evident that the industries and transport lines serving them were the most important features in the life of the city and therefore entitled to priority in the competition for space. They have now to adjust to life in an era of road building, smoke abatement, pollution controls, and urban planning. It is these changes and the problems they raise that are sketched in Figure 14-9.

Most easterly of the Great Lakes cities is *Toronto*. We have already considered Toronto in its Canadian context in Chapter 13, together with its steelmaking neighbor, *Hamilton*. Here we need only add that Toronto adheres closely to the model of the Great Lakes city we set out in Figure 14-9. Its lakeshore railroad belt—or barrier—was as formidable as that of any of these cities. The barrier effect was then reinforced by the construction of an elevated highway parallel to the railroad and, beyond the barrier, the port installations spread ever further on reclaimed land. Only the steel mills were missing from the model; they were in Hamilton, further west.

This being the case, it is gratifying to be able to report that, in the past decade, Toronto has rediscovered the lake. In the very heart of the railroad belt, Canadian National Railways have erected the world's tallest structure; beyond the belt, a new amenity area has been developed. The city has leapfrogged over its own transportation fetters and can now breathe lake air again. The port is as busy as ever but the main growth of non–port-related industry and of services is inland, along the line of Highway 401, which rings Toronto on the north.

Where the western end of Lake Ontario and the eastern end of Lake Erie are linked to one another by the Niagara River is a strategically located industrial area dominated by *Buffalo* (969,000). Goods brought east by the water route through the lakes can be transferred there to the land routes that lead to the Atlantic coast, and at the point of transfer there have sprung up a wide variety of industries. Lying as it does on the iron ore route, Buffalo was for long a steel city: a steel plant from Scranton, in Pennsylvania (see Chapter 11), was transferred there, but eventually closed. It is still an important manufacturing city; an outstanding industrial feature has been that it was the continent's largest flour-milling center. An arrangement that equalized the railway freight rates on grain and on flour made it equally suitable to mill the grain anywhere between the farmer and the baker, and thanks to its location, Buffalo established a dominant position in the industry. However, the opening of the Seaway was a threat to Buffalo, perhaps more than to any other city, for the grain it milled formerly left the water route at this point and traveled by rail to the Atlantic coast. Now the unique advantage of its position has been lost.

The power potential of the Niagara Falls has given Buffalo's manufacturing another facet. It has attracted to the banks of the Niagara River many industries that require access to large supplies of power—such as chemical manufacture and aluminum. Like the Kanawha in West Virginia and the St. Clair between Lake Huron and Windsor, the short Niagara River flows through a veritable "Chemical Valley."

The principal business of the ports situated along the southern shore of Lake Erie is, as we have already seen, the handling of iron ore and coal moving to and from Pittsburgh and West Virginia, and their manufactures are related to this advantageous position astride the industrial artery. In detail, the sites of these ports have been decided by harbor possibilities and by the existence of valley routes connecting with the interior. This route factor—the linking of the Cuyahoga River to the Muskingum and thus to the Ohio by a canal in 1834—gave *Cleveland* the initial advantage that has enabled it to become the great city of the Erie shore (1990 CMSA population: 2.76 million).

The natural advantage, however, was short-lived. The mouth of the Cuyahoga is narrow and winding, the harbor works were for long neglected, and, as the Great Lakes freighters grew larger, the problem of entering the port became more serious (and more expensive) and business was lost to neighboring ports. To some extent, therefore, the development of Lorain and of Conneaut may be regarded as an overflow from Cleveland made necessary by the limitations of the latter's site. The port's principal unloading and storage area for iron ore and bulky materials has had to be established on the lake shore, since access up the Cuyahoga is so difficult and so time-consuming.

This is not Cleveland's only drawback as a site for a great industrial city. The Erie shore is composed of soft beach materials from earlier lake levels, and these have been deeply cut by streams flowing into the lake. The Cuyahoga itself bisects the city, its steep-sided valley crossed by only a handful of viaducts (onto which all east–west traffic must be funneled), and the floor of the valley filled with industry and railroad tracks. Not only this, but the Cuyahoga squeezes the CBD onto a narrow spur between the river and the lake shore, and makes expansion in any direction except eastward virtually impossible.

Cleveland was the home base of John D. Rockefeller, who, in the 1860s, set out from there to capture North America's oil industry, and who made the city for a short period the refining center of the continent. Since that time, Cleveland has developed a great diversity of industries, mainly heavy: steelmaking, machinery, paints, chemicals, parts of the automobile industry—it has one of the most diverse manufacturing bases of any American city. At the same time, there is about that list of industries an air of past greatness; it lacks modern developments to replace old, declining types of manufacturing. Like Pittsburgh, it has switched from production to control and services: it is an important corporate headquar-

Cleveland: Collision Bend. The lower section of the Cuyahoga River winds sharply, and with the growth in size of vessels serving the steel mills and factories upstream, navigation has become increasingly difficult. In spite of action being taken to cut away the inside of the bend, this particular corner has proved a serious hazard. The photograph is taken from the Terminal Tower; that is, virtually from the center of the city. *J. H. Paterson*

ters, like its neighbor Akron which, as we have already seen, not only employs one-third of its industrial workers in the rubber industry, but plays host to the headquarters of most of America's rubber companies. Cleveland is the base of no less than seven of the fifteen or so shipping lines that own the Great Lakes fleet.

No city on the Erie shore, however, can compare in location with *Detroit* and *Windsor* (250,000), at the western entrance to the lake. Here, where the Detroit River forms a passage half a mile wide between Lake St. Clair and Lake Erie, is an unrivaled position from which to tap the flow of lake traffic. Yet Detroit took only a small share in the traffic of the Great Lakes, and made little attempt to develop its port (perhaps because of the narrowness of the passage at this point on the Detroit River). It required other influences to set it on its way to becoming the great city of 4.4 million inhabitants it is today—the genius of Henry Ford and the growth of a nationwide road network.

Detroit is the fifth largest industrial city in the United States, after New York, Chicago, Los Angeles, and Philadelphia. Like other southern Michigan cities within its orbit—Flint, Lansing, and Pontiac—it depends very heavily on the motor vehicle industry, which is spread all over this region as far west as Kalamazoo. Some of these smaller cities are virtually one-industry towns, and, in Detroit itself, the welfare of automobile manufacturing is all-important: too important for so large a city.

The automobile industry passed through difficult times in the 1980s. In 1982, car sales in the United States fell below 8 million for the only year in the decade; in 1985, sales were up to over 11 million but, of these, almost 3 million were imported. Several hundred thousand workers were laid off in the early 1980s by the industry and its suppliers and, when demand increased again, only a fraction of that number were taken back. It is true that some relief may be given to Michigan and its close neighbors by foreign car manufacturers setting up plants in the United

States, but for a proud and successful industry like automobile engineering to be dependent on foreign competitors for work is not likely to make Detroit happy.

"Detroit is a workingman's town, dominated by the automobile industry" declare Sinclair and Thompson.[17] That being the case, there is no surprise to read on and find that "Detroit is one of the nation's leading black cities, and much of the social change of the last two decades is the result of the rapid increase in the size of the black community." This influx (Table 14-4) has created severe problems for the urban area, and, despite the most strenuous efforts to revitalize a city center isolated from its surroundings by expressways, substandard housing, and empty lots, Detroit has had to contend with what has probably been the worst case of central-city abandonment in the whole United States. Unlike all the other Great Lakes cities, it cannot expand its shoreline by reclamation—the Detroit River is too narrow. What is done must be done onshore, and what has been done is to provide great conference facilities and splendid cultural institutions in equally splendid isolation in this "workingman's town." Housing renewal is more difficult. To the problems of providing acceptable residential areas for in-migrants from the South or from Appalachia, we must then add those of a city population falling by 28 percent in twenty years, and a higher than national unemployment rate. The combination is one to challenge the stoutest heart.[18]

There remains for consideration the greatest of all the urban areas of the Great Lakes Region: that which sprawls around the southern shores of Lake Michigan and contains the cities of Chicago and Milwaukee. From Gary in Indiana to the northern edge of Milwaukee is over 130 mi (210 km), and, although the built-up area is not continuous for the whole of this distance, the traveler along the lakefront might certainly be forgiven by anyone but an enthusiastic Chamber of Commerce for thinking that Milwaukee is merely one more northern suburb of Chicago, instead of a city of 1.4 million inhabitants lying 88 mi (140 km) from the center of Chicago.

Milwaukee possesses that mixture of heavy and

Table 14-4. City of Detroit: Population and Percentage Black, 1910–1990

	Population (thousands)	Percentage Black
1910	466	1.2
1920	994	4.1
1930	1569	7.6
1940	1623	9.2
1950	1850	16.2
1960	1670	28.9
1970	1511	43.7
1980	1203	63.1
1990	1028	75.7

light industry which we have seen to be characteristic of the Great Lakes port cities. A large element in its population is of German origin and this led to its being, for many years, one of the three or four leading brewing centers of the United States, although that particular glory has now departed. It has a splendid harbor and a hinterland that includes the most prosperous part of the western Dairy Belt. As a city, however, it suffers the same disadvantage as Cleveland: the Milwaukee River cuts through its center in a steep-sided valley entirely filled with railroad tracks, one of the solidest barriers to north–south movement that nature and man could have combined to design. Its CBD has been revitalized, with pedestrian covered ways to counter the climate of the lakeside winter, and, for a city of its size, it has an outstanding cultural equipment. It has a soundly based economy, and, even if it is overshadowed statistically by its great neighbor, Chicago, it possesses facilities similar to those of the metropolis and the advantages that go with smaller size.

CHICAGO

Simply to list Chicago among the cities of the Great Lakes would be as misleading as to treat it among the cities of the Interior, for it belongs to both and yet transcends both.[19] It lies on the great Lakes waterway, but is also the focus of the rail routes of the continent. It functions not only as the "big city" of the Corn Belt, but also as the capital of the Midwest and the headquarters of the Agricultural Interior, while

[17]R. Sinclair and B. Thompson, "Detroit," in Adams, *Contemporary Metropolitan America*, vol. 3, 285–354. Quotation from p. 289.

[18]In this connection, it is recommended that the reader consults W. Bunge, *Fitzgerald: Geography of a Revolution* (Cambridge, Mass., 1971).

[19]There is far more geographical literature on Chicago than on any other American city, thanks in large part to the output of both geographers and literature of the city's universities, and especially the University of Chicago's excellent series of *Research Papers in Geography*. In Adams, *Contemporary Metropolitan America*, the article on Chicago appears in vol. 3, pp. 181–283.

Detroit: The waterfront. Seen across the narrows of the Detroit River are the buildings of the CBD, serving as a counterbalance to the urban decay which has long afflicted the city's inner areas.

for many enterprises whose business is nationwide, it provides a more central location than does New York.

It is easy now, with the wisdom of hindsight, to point out how the southward projection of Lake Michigan and the low watershed between the Great Lakes and the Mississippi system gave an inevitable importance to the settlement at the lake head. But early visitors to Chicago were unanimous in condemning the site as swampy, liable to flood, and unfit for habitation.[20] The city's importance grew, in fact, by stages. First it was the head of navigation for the

Great Lakes emigration route to the West. In 1848, the Illinois and Michigan Canal was cut to link Lake Michigan with the Mississippi. Then in 1852, Chicago was linked by railroad with New York. The coming of the railroad marked its real beginnings, and when, during the Civil War, the choice of an eastern terminus for the new transcontinental line went to Chicago by default of its southern competitors, its future was assured. In the 1870s, it was the development of the stockyards and of the clothing and furniture industries that marked its growth; in the first decade of the twentieth century it was the rise of the steel industry on the southern lakeshore. Today the metropolitan area has 6.1 million inhabitants. The consolidated metropolitan area (CMSA), which includes a number of outlying cities—Kenosha with its car manufacturing and Gary, the steel town—has a population of 8.1 million. With these additions, Chicago becomes comfortably the second-largest industrial center on the continent. In terms of area, it is one of the largest urban concentrations in

[20]So marshy was the site of Chicago that, in the late 1850s, the citizens decided to raise the whole city, building by building, to get their feet out of the mud. The scheme took years to complete, and naturally left behind many rooms and passages at the old street level, now underground. These quickly came to house the criminal element and their activities; whence, we are told, the first use of the term "underworld" in its modern sense. Those readers who do not believe a word of this are directed to J. Robert Nash, *Makers and Breakers of Chicago* (Chicago, 1985), p. 57.

the world, with miles of suburbs spreading unchecked over the featureless Illinois plains. By contrast, its central business district is now far too small for it, jammed between the Chicago River, the lake, and the largely disused tracks of half the railroad systems of the United States (see Fig. 14-10).

Chicago's heavy industry and oil refineries are concentrated close to the lakeshore on the south side of the city, where the steel town of Gary was created by the U.S. Steel Corporation and named after its first president. The Union Stockyards also used to be on the south side, 5 mi (8 km) from the city center and, in the days when Chicago was "hog butcher for the world" (the phrase was Carl Sandburg's), the meatpacking plants were grouped around the stockyards, but the last of them closed in 1959. For the rest, industries tend to cluster along the railroads radiating from the city, and especially along the "belt" (or ring) lines, which play a part of particular importance in a city where so many separate railroad companies operated. This pattern calls attention to a general tendency for industry to move out from the overcrowded central districts (where the older industries, such as clothing manufacture, were situated) to the suburbs, leaving the central area—Chicago's famous "Loop"—to be cleared for much-needed road improvements or occupied by commerce. Many industrial districts or parks have been laid out, such as the early Clearing and Central Districts, but with the newer areas located up to 40 mi (64 km) from the Loop, and a great cluster of parks for light industry around O'Hare Airport, northwest of the city.

We reviewed earlier in this chapter the sequence of development of Chicago's industries since 1850. Through the last few decades, the major employers have been the steel industry, machinery manufacture, and communications equipment. But change is continually taking place and, at a recent count, Chicago stood not far behind New York in the high-tech field. The truth is that Chicago, like New York, manufactures virtually everything.

In any case, the city's manufacturing nowadays counts for less than its role as metropolis—its commodity exchange, its office space (which includes the Sears Tower still, after nearly twenty years, clinging to its record as the world's tallest office building), its convention business, its cluster of universities, and its research establishments among their suburban lawns and trees.

Chicago, like New York, suffers from an unusual number of obstacles to movement of traffic; ironically enough, this results from its importance as a traffic center. On the flat lakeshore, almost all the railroads run at ground level and cross each other and the city streets by means of level crossings. Of the few railroads that do not, the main one is the loop of elevated electric railroad which gives the central business district its name. This, by contrast, runs above the street, 20 ft (6 m) up on steel trestles, and the streets beneath are dark, obstructed, and clangorous. Even worse, through the heart of the city runs the canalized Chicago River, traversed by a series of drawbridges. Each time these are opened—and fortunately for the motorists this is not often, since a canal cutoff has been built to enter Lake Michigan south of the city—the traffic of the business district comes to a stop.

But the main problem, which is now also the main opportunity for the city in the twenty-first century, has for years been caused by a kind of iron ring that has enclosed Chicago's CBD. As Figure 14-11 shows, the business district has been confined between the lake, the river, and broad bands of railroad land, serving no less than five stations on the south side. In the days of passenger rail travel, this was the railroading heart of the continent: it was axiomatic that, between any two points, east and west, one always had to change trains at Chicago.

With the virtual disappearance of the passenger train, however, these facilities have become redundant. Two of the stations have disappeared, and a third barely survives. The remaining two can adequately handle today's traffic, and so there is left a huge, empty space for redevelopment, abutting directly onto the southern edge of the CBD. Over the past decade, the process of filling this space has been under way. It is a mixed, largely residential, development; the city has so planned it. Up to the present the CBD, the most obvious candidate for expansion, has not grown to take advantage of this apparently heaven-sent breathing space. Indeed, perversely as it must appear to the outsider, the CBD has in recent years been spreading not south but north, across the obstacle presented by the Chicago River, while its southern end has atrophied. It would be tragic if so unique an urban opportunity were to be squandered.

While rail passenger traffic has been declining, and new harbor works and channels have been constructed to take account of developments in the sphere of water transport, Chicago has also become the busiest air traffic center in North America. Whereas, at New York, air movements are handled

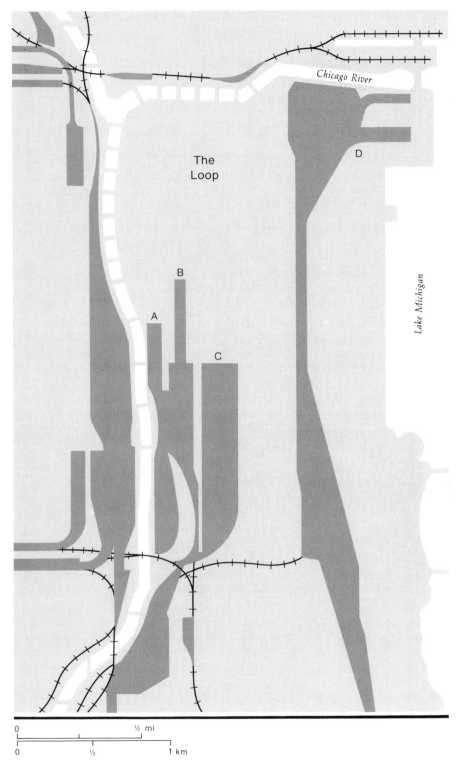

Fig. 14-10. Chicago: The city center at the peak of the railroad era, the late 1920s. The map indicates the extent to which the railroads formed a noose around the center of Chicago, hampering expansion and distorting urban growth patterns. Nevertheless, at this period, the importance of the railroad was such that the arrangement appears unreasonable only in retrospect. By 1985, the areas marked A, C, and D had been cleared of tracks; at B, the entry into once-proud La Salle Street Station was reduced to a pair of tracks. Redevelopment has since seen much of areas C and D covered by a largely residential growth, that at D containing a number of luxury apartment towers.

Chicago River

The Loop

Lake Michigan

A

B

C

D

0   ½ mi

0   ½   1 km

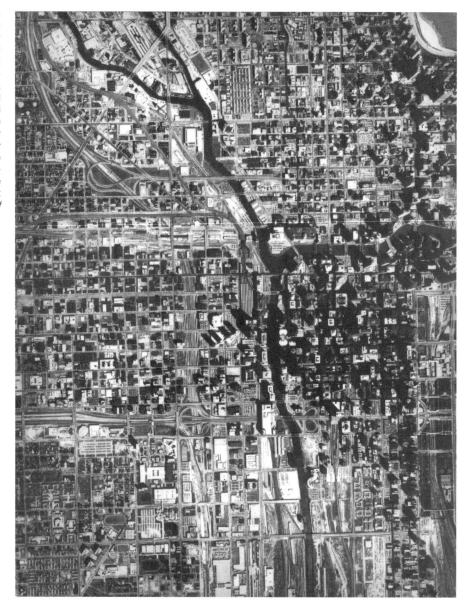

Chicago: The central districts from the air, 1990. The two branches of the Chicago River can been seen merging and flowing into Lake Michigan at the right-hand edge of the picture, where the lakeside park is also visible. The belt of land formerly occupied by railroad terminals (described in the text) is at the lower right; by 1990 much of it was already developed. Also visible at the lower left center is the freeway intersection between the north–south and east–west arteries. *Illinois Department of Transporation*

by three airports, the bulk of these at Chicago take place through a single airport—O'Hare, on the northwestern outskirts of the city. The problem of congestion has become a serious one for, as in the heyday of railroad construction, Chicago's central position within the continent and its densely populated hinterland make it the obvious meeting point for routes, both internal and international, and the buildup of traffic has reached a point where drastic measures may shortly become necessary.

Chicago has been known since its early years for its ethnic communities—Poles and Russians, Germans and Italians, most of them still showing concentrations in distinct areas of the inner suburbs. Steelmaking attracted, to Gary as to Pennsylvania and Ohio, a strong east European element in the work force. Since World War II, however, the ethnic group that has come to dominate all others is the black population which, in 1990, accounted for 1.3 million within the metropolitan area.

Chicago's black city-within-a-city is on the south side, beginning immediately south and west of the CBD and extending for 15–20 mi (24–32 km) south, past Lake Calumet. Its main axes are close to the lakeshore, just behind the lakefront parks and apartment blocks. As the black population has increased, so black Chicago has grown larger, the degree of segregation between black and white becoming ever more pronounced. Meanwhile, another statistic suggests that there may be more social problems to be solved in the future: in 1980, Chicago also had a Hispanic population approaching 10 percent. There have been one or two heartening examples of urban renovation, especially in the old white ethnic neighborhoods north of the Chicago River. But on the south and west sides of the city it is the scale of the social problems that daunts the mind. With 6 million Chicagoans, to accept, absorb, and, if necessary, relocate the 1.3 million members of a single ethnic group is a task whose dimensions would be overwhelming.

The Inland Cities of the Interior

Away from the lakeshores and across the Interior are to be found hundreds of cities and lesser centers whose similarity of appearance, regularity of spacing, and graduations of size are so striking as positively to invite the geographer to seek a general model of urban growth within the region. In studying agricultural regions and their servicing, the two basic models with which geographers have worked are Von Thünen's for land use and Christaller's, as modified by Lösch, for service centers.[21] The first of these embodies the concept of rings of diminishing land-use intensity concentric around a central market, the second, a hierarchy of centers with overlapping hinterlands, within which various grades of service are offered. It was this second concept to which we referred in Chapter 3.

In studying the agricultural pattern of the Interior, we have already considered something that bears a certain resemblance to the Von Thünen rings—a central zone of high-intensity agriculture surrounded by areas where intensity falls away from the center.

We even have near, if not at, the middle point of the central zone a great urban market—Chicago. And we can now go on to recognize that in the settlement pattern of the Interior we have a very fair approximation to the Christaller model.

The problem with most of the geographer's models is that they involve three assumptions: physical uniformity of the surface, initially uniform distribution of population, and freedom from distortion by external forces. Nowhere in the real world, of course, do these conditions obtain. But in the North American Interior, and especially in the western Interior, they come much closer to fulfillment than they do in most other regions.

The condition of physical uniformity approaches fulfillment in the circumstance that in the Interior there are few barriers to movement, few wide variations in agricultural potential, and few obstacles to maximum utilization of the surface. Such variations as exist are, on the whole, regular around the central core, and physical potential for urban growth is everywhere virtually unrestricted.

The condition of an evenly dispersed population in the initial situation was secured, to a remarkable degree, by the method of survey and settlement adopted across the Interior—by the uniform application of the rectangular survey grid and the land laws based on sections and quarter sections (see Chapter 5). More regularly, perhaps, than anywhere else on earth, the landscape of settlement in the Midwest was laid out by law—the farm roads repeating themselves at every mile, north and south, east and west, delimiting the sections, and the farmhouses beside the roads, four to a section, from one horizon to the other. This regular pattern of settlement was imposed on a region with no relics from previous technologies and no remnant cultural landscapes to adapt or overlay; the region was settled, moreover, in a few short decades between 1800 and 1850; that is, in a single technological era and by two generations of settlers at most. For all these reasons, the Interior provided, by the end of the settlement period, a pattern of uniform occupance such as could hardly be matched anywhere else in the world.

The third assumption of the model—freedom from distortion by external factors—was certainly not fulfilled in the eastern Interior, where industry became widespread, but it comes close to fulfillment in the virtually unindustrialized western part of the region. There the relationship between the service center and its consumers of services is very simple:

[21]The basic references translated into English are P. Hall, ed., *Von Thünen's Isolated State* (Oxford, 1966 [originally published in part in German, Hamburg, 1826]); W. Christaller, *Central Places in Southern Germany*, trans. C. W. Baskin (Englewood Cliffs, N.J., 1966 [originally published in German, 1933]); and A. Lösch, *The Economics of Location* (New Haven, Conn., 1954). There is, however, little point in the general reader going back to these original studies: they are summarized and discussed in most of the numerous books on spatial analysis and spatial model building now available.

the consumers are farmers, and no other major activity breaks the pattern; neither are there any large metropolitan areas to generate such activities. Under all these circumstances, we may well expect to find a pattern of service centers developing in a hierarchy of sizes, and very close to that defined by the model.

That a hierarchy of settlements exists is empirically obvious. At the top in splendid isolation is Chicago: below it are the regional metropolitan poles and below them the regional supply centers; seven or eight ranks in all, reaching down to the hamlet or the corner store. In a few areas, even the spacing of these service centers approaches the model distribution. That it generally does not can be explained by two or three factors. One is that it is distorted by the drainage pattern. On the largely featureless surface of the Interior and during the particular era of settlement, the only major site-factor likely to confer a decisive locational advantage on any of the hundreds of towns founded, in that optimistic age, to be the capital of the New West, was a riverbank situation in a region where the rivers formed the main lines of communication. It is no coincidence at all that every one of the Interior's cities of the second rank (which we shall later identify) began its career as a river port or portage point.

The second reason for distortion was that settlements in America were often founded under highly competitive conditions, their location in many cases based not on economic considerations, but rather on political or even personal choice. These settlements fought each other, sometimes literally, for survival; they bribed railroad surveyors and indulged in land frauds to enlarge themselves and weaken their rivals. What we have on the map, therefore, is far from being simply the product of economic principles, such as least-cost location.

The third distorting factor was the tendency, over so large an area, for settlements to develop in lines along the transport routes instead of in an even distribution over the area they were servicing. The importance of the river and the railroad was so great that it easily overrode the diseconomy of a service point located on the periphery of its hinterland rather than in the center.

The place of a settlement in the hierarchy depends on its functions rather than on its population. On the whole, within each rank of settlements in the Interior, the more easterly members have larger populations than those further west; that is, service functions performed in Illinois by a city of 20,000 to 30,000 may be handled by one of only 15,000 in the eastern Dakotas, or local, everyday services may be provided by towns of 2500 in the east and 1200 in the west. Then again, the service centers are much more sparsely distributed in the west than in the east of the Interior. This brings us to the important question of how far apart we might expect service centers of a particular rank to be; what in fact are the dimensions of the Christaller model?

The answer is found to involve sequences of explanation that spread out from urban spacing to touch causes that are physical and legal as well as economic; they derive from land-disposal policy as well as farming system, and from original perceptions of the environment as well as contemporary transport facilities. In order to save extended discussion in the present chapter, some of the factors involved in the answer are presented in the form of a diagram (see Fig. 14-11). It must be stressed that the model applies only to a wholly agricultural region, such as the western Interior, where, as we have seen, there is little to complicate the basic relationship between service and those served.

The number of viable service centers must be a function of the aggregate demand for services per unit of area. What that demand sum may be depends on all the factors shown on the circular diagram. Once the demand exists, then the aspiring centers compete to supply it, on the basis of what we should think of as a gravity model: somewhere along a line between any two centers, there is a point where the cost or advantage to a customer seeking services is equal as between the two places. The more services one offers, the greater the "pull" of gravity toward that center, but always within the limits set by the total demand of the population concerned.

There is probably not a single settlement, in the Interior or anywhere else, that would not like to increase its "pull." Usually, it can only do so by taking business from a neighbor, and perhaps condemning the neighbor to extinction—which is precisely what has been going on in the Interior, through the lives and deaths of scores of central places, for the past century and a half.

We can now move on to discuss some of the members of the urban hierarchy in more detail (Fig. 14-12). The cities of the second rank each dominate a portion of the Interior, and their populations range in most cases from a million upward. All of them, as we have noted already, began life as river ports and it was probably their strategic location on the river sys-

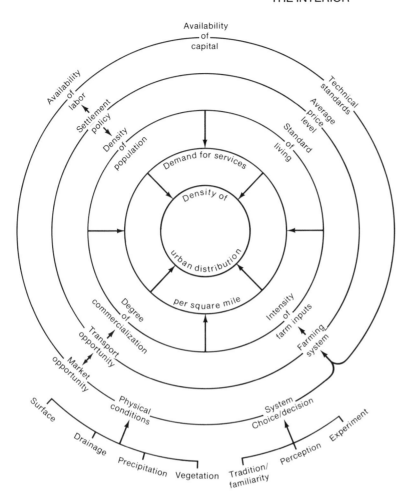

Availability of capital

Technical standards

Availability of labor

Average price level

Settlement policy

Standard of living

Density of population

Demand for services

Density of

urban distribution

per square mile

Intensity of farm inputs

Degree of commercialization

Farming system

Transport opportunity

Market opportunity

System Choice/decision

Physical conditions

Surface

Drainage

Experiment

Perception

Precipitation

Vegetation

Tradition/ familiarity

Fig. 14-11. The Interior and its service centers: Their variable spacing. Service centers within the region are unevenly distributed. If we inquire why this is so, a great many factors go into the answer, and this diagram is intended to present some of them. The diagram should be read from the center circle outward.

tem that enabled them to outgrow their lesser rivals. Into this category fall St. Louis (2.4 million), Cincinnati (1.4 milion), Minneapolis–St. Paul (2.5 million), Kansas City (1.6 million), and probably Indianapolis (1.2 million) and Louisville (1 million). In Canada, Winnipeg fulfills much the same role as these cities, and possesses a similar history.

It was the importance of the Ohio–Mississippi route as the main road to the near west that led to the settlement and early growth of Cincinnati, Louisville, and St. Louis. The falls in the Ohio River at Louisville interrupted river transport; the position of St. Louis at the junction of the Mississippi and Missouri made it the natural fitting-out point for movement westward along the Missouri; and subsequently Kansas City and Independence derived a similar advantage from their situation at the places where the overland trails left the hazardous water route and struck west for Santa Fe and Oregon. Further north the Falls

of St. Anthony marked the head of Mississippi navigation and provided a focus for the growth of Minneapolis. Later came the railroads to lend momentum to the rise of these cities as the natural foci of the routes of the Interior.

Cincinnati has greatly altered since its "Porkopolis" days, when it was a meat-packing and flour-milling town and one of North America's principal ports. The agricultural industries, as we have seen, moved west, but the city's proximity to the heavy industry areas of the Appalachian coalfield provided the basis for a new industrial career, and today Cincinnati has a well-balanced industrial structure and a worldwide reputation for its machine tools, which find a wide market in the adjacent Manufacturing Belt. *Louisville,* on the other hand, has retained the two industries for which it has long been known—tobacco processing and furniture making—as well as acquiring new industry in the form of electrical ap-

Business Centers
● First Rank
● Second Rank
● Third Rank
● Fourth Rank
· Fifth Rank

Fig. 14-12. Business centers of the Midwest. This classification, from the *National Atlas of the United States*, is not used in this chapter because different criteria are employed. In the *Atlas*, precise criteria consist of such indices as volume of retail sales and newspaper coverage. In this chapter, the criteria are broader and, in part, subjective, but in both cases, the idea of a hierarchy of centers is strongly developed, even if precise rankings are open to discussion.

pliance manufacture, a very necessary addition since tobacco manufacturing has shown little growth in recent years in the United States.

The modern city of *St. Louis* is Chicago's only close rival for the position of economic focus of the United States Interior. It is a much older settlement—it was founded by the French as a fur trading post nearly a century before Chicago was incorporated—and with its commanding position on the river system, it seemed destined to be the focus of the Interior. But its commitment to the river and its traffic, and so to the South, told against it when the railroads reached the Mississippi. The railroads focused on Chicago as the riverboats had converged on St. Louis, and the railroads survived, whereas the riverboats did not. St. Louis was left with a decaying waterfront and nothing to use it for. Only in the 1960s was a program of urban renewal started, which swept away the old commercial focus of the port and replaced it by a belt of parkland.

St. Louis is, nevertheless, a great transport center and industrial city. Much of the manufacturing is carried on in East St. Louis, on the Illinois bank of the Mississippi, or in the lower Missouri Valley, where it winds through the northern edge of the city. It builds aircraft and automobiles, brews beer, and makes clothing and hardware. It is the market headquarters for much of the produce of the southern Interior. The combination of excellent transport facilities and numerous local manufactures has given the city an important trade area that stretches not only across the southern Interior but well into the middle South.

It is significant that several of these cities of the second rank in the Interior lie near the western and southwestern edges of the region, adding to their functions within it important connections with areas outside it. These are the cities of the "hinge" to which we earlier made reference—the linking zone between core and periphery. Perhaps the most obvious examples of major cities playing this role within the Interior are Minneapolis–St. Paul and Kansas City: further south, we should need also to include Dallas–Fort Worth.

On any showing—newspaper coverage, agricultural marketing, or mail-order business—the hinterlands of these cities stretch far out into the Great Plains. They serve, in fact, both the Agricultural Interior and the Great Plains, both the intensive-farming areas of the region we are considering and the drier lands of less intensive use that lie further west. If, for purposes of regional subdivision, we distinguish between the Interior and the Great Plains, then these cities should, strictly speaking, be treated under both headings, for although they are located in one region, they play a part in the life of both.

Let us also notice that, although the regions which these hinge cities serve are largely agricultural, the cities themselves are very far from being mere overgrown farmers' markets. They have splendid cultural equipment—music, theater, galleries, museums—and some of America's most elegant suburbs, whether on the many lakeshores of Minneapolis or along the parkways of Kansas City. It is true that most of them have a railroad belt and huge grain elevators, and they may lack the sophistication of Boston or Philadelphia, but they are self-consciously urban, and proud of their regional dominance.

Kansas City began as one of the settlements on the Missouri River, where the river route to the west turns abruptly north, and where most emigrants, consequently, disembarked to follow the overland

St. Louis, Missouri, from the east bank of the Mississippi. The huge arch is 630 ft (192 m) high and is officially called the Jefferson National Expansion Memorial, but in view of its position here at St. Louis, and of the city's role in the nation's history, it is also very appropriately known as the Gateway to the West. Along the Missis- sippi bank beneath the arch, old riverboat terminals have been cleared away, providing a view of the civic center, with the city hall, court, and federal buildings. *Jefferson National Expansion Memorial/National Park Service*

trails. Its competitors north and south of the river, Liberty and Independence, have never grown great because they lie some way from the Missouri's banks, with the wide floodplain between them, but Kansas City itself grew up on high bluffs directly on the south bank, free from flood danger; indeed, cuttings had to be made through the bluffs to create wagon roads for goods off-loaded from river craft. It grew up as a receiving and processing center for grain and livestock from the Corn Belt and Wheat Belt and for cattle from the ranges. Its share of the nation's agricultural processing industries has steadily increased, but it has also achieved, in recent years, a broader industrial base, which has changed it from a specialized to a general manufacturing city. This has

been partly the result of proximity to the southwestern oil and gas fields, which have provided industrial power, and partly a product of the general westward-moving tide of industrialization, which has brought to the city numerous concerns seeking a central location within the United States.

Today, therefore, Kansas City is an outlier of the Manufacturing Belt and not merely a convenient processing point for farm produce. To use again the terms mentioned earlier in this chapter, the most recent wave of industrialization has crossed the Interior and reached as far west as Kansas City, bringing to it sophisticated modern industries to reinforce the older midwestern "regulars." Yet with Kansas City, as with Chicago and St. Louis, much of the city's importance derives from its roles of market and transport center. This is particularly true of the cities at the margin of the Great Plains where the next major market centers lie hundreds of miles to the west and where, in consequence, the cost of distribution or collection over the sparsely settled rangelands is exceptionally high. Thus Kansas City's wholesalers operate over an area that includes much of the Winter Wheat Belt and the southern Great Plains, and the role of its warehouses and wheat market is no less vital to the region's economy than that of its factories. As a trade center, it has the advantage of excellent rail and road services, and local opinion in Missouri is divided on the question of whether it or St. Louis can claim to rank second to Chicago among the transport centers of the Interior.

Minneapolis and *St. Paul,* the "Twin Cities," whose centers lie some 8 mi (12.8 km) apart and on opposite banks of the Mississippi, fulfill a similar role for the Spring Wheat Belt and the northern Great Plains. With several transcontinental railroad lines running west from Minneapolis, the Cities' tributary area stretches well into Montana and is limited on the north only by the international boundary. Between the sections of the metropolitan area there is a marked "division of labor." St. Paul is a combination of state capital and railway junction, whose merchants deplore a general St. Paul habit of going to Minneapolis to shop. Minneapolis, in turn, having the waterpower of the Falls of St. Anthony at its disposal, developed the early industrial core, participated as a mill town in the Great Lakes lumber boom, and then settled down to a more stable career as one of the continent's flour-milling centers, with important manufactures of machinery in addition.

Indianapolis, the capital of Indiana, has gradually earned its place among these centers of the second rank by its development of industry, services, and cultural functions like education and publishing, over the period since World War II. Its population increase has been steady rather than spectacular, and it is the only one of these cities not on a major waterway. Nevertheless, its central position within the state, its road and rail routes radiating out in all directions over the smooth Indiana plains, and its situation in the eastern Corn Belt all contribute to its central place functions and its prosperity.

The cities of the third rank are, for the most part, centers whose functions tie them closely to their surrounding farmlands: their industries supply rural needs and their commerce may serve a hinterland of thousands of square miles, although always beneath the shadow of the second-rank metropolis. They vary considerably in size: Des Moines (393,000) and Peoria (339,000) are typical, although in the eastern interior the Ohio cities, such as Dayton or Columbus, that fall into this category have their populations and their functions enlarged, to around 1 million in each MSA, by a wider spread of manufacturing employment. Although many of these places are nationally known for a particular product (Peoria and Moline in Illinois, for example, for tractors and farm machinery), their industries are generally related to local needs; food processing and the construction of machinery are usually leaders. Their employment structure is broadly based. Characteristically, industry accounts for 30 to 35 percent of the employed workers, wholesale and retail trade for 20 to 25 percent, and finance, insurance, and real estate for 6 or 7 percent. This latter set of functions is, however, a very important one in these cities and one member of the group, Des Moines, has the highest proportion of its work force in this category of any MSA in the nation, including New York City.

Below the cities of the third rank, the hierarchy continues downward: cities, towns, villages, hamlets. Although the range of sizes is wide, all these settlements are concerned with the same basic set of functions: marketing, maintenance, and supply. The cities provide food-processing plants—meat-packing, flour-milling, and feed preparation—whereas the small centers merely provide storage: grain on its way out and feeds on their way in. The cities manufacture agricultural equipment; the small centers maintain it. It is largely a matter of scale. In the middle ranks, there are also the administrative and transport functions to be fitted in—the county seat with its courthouse and lawyers' offices or the railroad division point with its maintenance crews. For settlements in their size range, these cities and towns of the Interior have a remarkably wide range of industries and services to offer. But this, of course, reflects the region's economy. To supply, as many of these towns do, the farmers within a radius of 20–30 mi (32–48 km) with their day-to-day requirements, when those farmers operate some of the most highly mechanized farms in the world and have one of the world's highest rural standards of living, has clearly required the growth of a rather specialized type of service center, and one whose prosperity depends heavily on a prosperous agriculture.

At a certain point in the lower ranks of the urban hierarchy, a new question arises: that of survival. Despite the stability and prosperity at the top, the bottom of the hierarchy is somewhat precarious. Villages that once offered a range of services may today be purely residential, and many small settlements are losing population.

To understand why this is so, we need to recognize the effects of two trends, already referred to in Chapter 7.

1. *In the Interior, as in virtually all regions of Anglo-America, the farm population is declining in numbers.* Although in this region, at least the land remains in cultivation, the number of farms is decreasing and so, therefore, is the number of customers for such services as groceries, clothing stores, and schools. The service centers of the Interior were established for a much denser farm population than today's and, what is more, for a population whose effective range of movement was limited by its transport to horse-and-buggy distances. Under these circumstances, it was to be expected that service centers would develop every few miles across this agricultural region, but it must now equally be expected that, given a less numerous, but at the same time more mobile population, the pattern of regional servicing will alter. There are today, in fact, too may service centers competing for a limited amount of business. In Nebraska, as an example, it has been calculated that in 1930 there were 230 farms for every incorporated place, but that by 1960, there were only 170 and in 1978, about 135 farms—or, at a generous estimate, about 500 farm-based customers. Against these figures must then be set the calculations that an adequate high school requires a base population of 6000 or more and that a supermarket offering full economies of scale to its customers requires a population of 8000 to support it. With the upward climb of costs and incomes it can be expected that in the smallest settlement fewer and fewer services will be economical. They will close, and if they all close, the "service" center may cease to exist and become simply a collection of residences.

2. *The remaining farmers require for their operations equipment and supplies quite different in scale and type from those provided in earlier days.* The technical quality of today's farming encourages the farmer to rely increasingly on the larger center: the smaller ones are likely to be bypassed, and once again they may fade out. If, in practice, many of them survive it may be because (whether by luck or good judgment) they come to specialize in a single type of service. Interior farmers do not necessarily patronize one center only. They may well buy feed in one, sell stock in a second, buy groceries in a third, and go shopping in a

fourth.[22] From the farmer's point of view, what this habit produces is a kind of dispersed service center, its components scattered 10–20 miles (16–32 km) apart, which is perfectly acceptable to today's farm operator.

What we then have—to revert to our earlier comment on the gravity model—is not a single gravity model covering all the central places in an area for all of a generalized set of services, but a separate gravity model for each separate service and yielding different patterns of movement for individual services.

What, then, is happening to these smallest settlements of the Interior, as many of their service functions decline? There are two answers to this question. One is that, despite the losses, the settlements survive, but purely as residential places. Farmers retire off outlying farms and move into a nearby center. And within reasonable commuting distance of the cities, urbanites move out to take advantage of low-cost housing and village living. The population holds up even if the services do not.

The other tendency is one to which our attention is drawn by J. Fraser Hart, who for many years has been monitoring rural life in the Midwest and saving us from our own false perceptions. He points out that, although the old service center has gone, its services may be replaced by small-scale industry. His comment is worth quoting in full:

Main Street is dead. The stately bank on the corner has become a beer tavern; some stores have been boarded up; others have been converted into private residences; the handful still open for business are struggling. Despite the hand-me-down appearance of Main Street and contrary to widespread popular belief, most small towns in the Midwest have gained population, however slowly and fitfully, during most of the twentieth century. . . . The trading and service functions of small towns have been drifting up the urban hierarchy since World War II, but the processing function has been moving downward from metropolitan centers to smaller places. The small town has been transformed from a central place serving an agricultural hinterland into a minor cog in the nationwide network of manufacturing centers.[23]

The Interior possesses most of the geographical advantages and almost none of the disadvantages of

[22] A well-documented study of this habit, although made in an area marginal to the Interior, was in H. S. Ottoson, *Land and People in the Northern Plains Transition Area* (Lincoln, Nebr., 1968), pp. 252ff.

[23] J. F. Hart, "Small Towns and Manufacturing," *Geographical Review*, vol. 78 (1988), 272–87. The study is concerned primarily with the state of Iowa.

the other major regions of North America. If it is charged with isolationism, then it is the isolationism of economic self-sufficiency; if it is charged with monotonous uniformity, then at least it is uniformity based on decades of past prosperity. To leave the Interior and travel in almost any direction is to travel down the economic gradient and to enter regions whose problems are manifest in the landscape—soil erosion by wind and water, rural overcrowding or abandoned farmsteads on rugged, infertile lands. The Interior has its problems, too, but they are almost all the problems of prosperity; how to dispose of its huge agricultural output; how to control the prices farmers and industrialists are willing to pay for its land; and how to transact, within the confines of its crowded cities, the volume of business its richness creates.

15
The South

Mississippi: Dawn harvest in the cotton fields. Cotton picking is accomplished by a fleet of modern machines. The driver's cab is air-conditioned. *Rhone-Poulenc Ag Company*

The Old South

Between the Potomac and the Gulf Coast (the latter of which, for a variety of reasons, is best considered separately) there lies an area whose regional distinctiveness cannot be denied. It may be variously defined, and its western limit, in particular, is open to question, but no regional analysis could possibly overlook the Old South (Fig. 15-1). This region developed a distinctive plantation economy, maintained a large black population to operate it, fought a war to preserve it, suffered the bitterness of defeat and the chaos of the aftermath, and has since been struggling to regain both its regional self-esteem and its place in the nation.

It is, perhaps, the extent of the area within which there existed before 1860 the old plantation system[1] that gives the clearest single indication of the extent of the Old South, for it embodied both the economic and the social elements that made up the region's character. In the Civil War of 1861–65, the Confederacy drew little support from the upland areas in the Appalachians to which the plantation system could not spread for geographical reasons. Although the state of Virginia fought on the side of the South, the independent upland farmers' sympathies remained with the North, and they seceded to form the state of West Virginia. Kentucky, southern in so many other respects, was not a plantation state and, after wavering for a time, joined the North.

However, the problem of regional definition remains. It can be resolved by suggesting that there were in fact two Souths, the Upper and Lower, which were different in character; one formed the heart of the Confederacy and the other was at best lukewarm to the cause. "The Lower South, which by 1860 encompassed almost the entirety of the Gulf and Atlantic Coastal Plains, was a land of cotton and slavery, a land dominated economically by the plantation type of agriculture. . . . In contrast, the Upper South was primarily the domain of the slaveless yeoman farmer, an area largely devoid of cotton and the other subtropical cash crops."[2]

But define the area how we may—in terms of its former economy, its war memorials, or its black population—the South remains a reality in American life. And embedded almost equally deeply in the consciousness of the twentieth-century American has been a second impression: the Old South is a depressed area. During the difficult years of the 1920s that culminated in the depression of 1929–32, there crystallized what became the familiar concept of the South in the American mind. Like most such mental pictures, it contained an element of exaggeration, but regional statistics would at this time have borne out most of its points.

Southern agriculture was based mainly on cotton, corn, and tobacco. These crops, grown year after year on the same fields, had eaten the heart out of the land and left the soil particularly liable to erosion. Southern farmers, mostly small tenants, farmed hopelessly on in an era of low world agricultural prices and knew the despair of declining yields and gully erosion without having the means or the will to arrest the process. Farm buildings fell into disrepair, and mules did the work for which more fortunate farmers used tractors. The poverty of the white farmers was only exceeded by that of the blacks, most of

---

[1] The wording of the sentence is important. The Old South never consisted of wall-to-wall plantations. But there were areas within which plantations existed, and other areas within which they were unknown. This is the distinction made here.

[2] T. G. Jordan, "The Imprint of the Upper and Lower South on Mid-Nineteenth Century Texas," *Annals,* Assoc. Amer. Geog., vol. 57 (1967), 667.

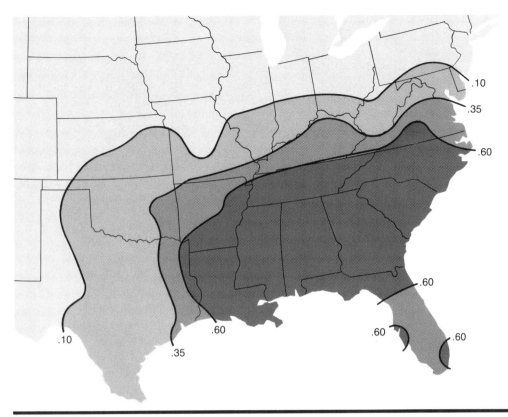

Fig. 15-1. The South: The problem of regional definition. This map, one of dozens which might be constructed in an attempt to decide where, precisely, the South is to be found, is based on a single, simple index: using telephone directories, the investigator has calculated the ratio between the number of times the term "Southern" appears in the directory for selected cities, to the number of occurrences of "American." In what might be regarded as a Southern heartland, this ratio is in excess of 6:10. (Reproduced by permission from J. S. Reed, "The Heart of Dixie: An Essay in Folk Geography," *Social Forces*, vol. 54 [1976], 929.)

whom held land as "sharecroppers,"[3] paying their rent by a fixed proportion of their crop—a crop that might disastrously glut the market one year and be stricken with blight the next.

Southern industry offered little palliative for the region's distress. With an exceptionally high regional birthrate, there was an abundance of cheap labor, and wage levels were far below the national average. New industries were slow in appearing; rather, the region's raw products were shipped north to be processed, and the population remained overwhelmingly rural.

Yet in the mid-nineteenth century, this region could bear comparison with any part of the nation in respect of its wealth and of the leaders it produced in cultural and political fields. Even if we allow for the fact that our concept of the antebellum South is generally romanticized, the contrast with the South of the 1920s is remarkable. This is much more so when we take into account the rich resources of this region, whose natural endowment ranks it high among the regions of North America: long growing season, ample rainfall, extensive areas with gentle slopes and cultivable soils, vast timber supplies, and a variety of minerals that included the unique combination, all in the same area, of iron ore, coking coal, and limestone for steelmaking.

If we attempt to account for the great contrast between the 1850s and the 1920s in the Old South, we should begin by recalling that the prosperity of the earlier years was in some ways only illusory. There was soil erosion in the 1850s and earlier, too, but its extent was concealed by the fact that there was always new land available in the West: abandonment of the old in favor of the new masked the seriousness

---

[3] The U.S. Bureau of the Census defines a sharecropper as a tenant who supplies nothing but his own or his family's labor: all necessary equipment is supplied by the landlord. A tenant who owns any part of the equipment (e.g., a tractor) is a *share-tenant*.

of the problem. The plantation owner and his mobile labor force, the slaves, could move west if necessary—into Alabama, into Mississippi, finally into Texas—and there begin again. Then we must recall that the South's concentration on cotton and tobacco resulted in an unbalanced regional economy, with industry poorly developed and even the production of food crops barely adequate for the population. Again, we must recall that the early southern prosperity was concentrated in one narrow section of the population, so that to many of its people the South's varying fortunes made no difference: they simply remained poor.

But despite all these reservations about the earlier southern prosperity, the fact remains that the Civil War was a great underlying cause of the later southern ills. The South was invaded and occupied, and physical destruction was immense. The southern ports were blockaded, and cotton exports that should have paid for industrial imports never left the quays of New Orleans. Then, in the midst of the war, came Lincoln's emancipation of the slaves. For the South it was, apart from anything else, a staggering economic blow. For generations, southern landowners had been buying slaves as a form of capital investment. To abolish slavery meant the elimination of more than $3500 million of southern capital. When the time for rebuilding came, the loss was acutely felt.

The end of the war, in 1865, brought little relief. Military occupation, enforced liberal reforms, and southern reaction followed each other over the succeeding fifteen years. This was a period, too, when the 3 to 4 million emancipated slaves were trying to adjust to their new positions as citizens and farmers, hindered, on the one hand, by white prejudice (typified by the Ku Klux Klan) and, on the other, by ignorance of farm methods. In practice they were most often forced to turn to their old masters for instruction, and many became sharecroppers.[4]

Beyond the immediate, impoverishing effects of the war and the peace, other factors foreshadowed the future weakness of the southern economy. About 1879 there began an economic revival in the South. Up to that date, much of the war damage remained unrepaired; factories and port installations lay dere-

lict. But after 1879, the North, encouraged by southern propaganda, "discovered" the South as a field of investment. Capital flowed in to repair railroads and factories or to create new industries. An industrial boom resulted but, when it was over, the South was almost as firmly in northern hands financially as it had been politically in 1865. The financial hold (of which the South had by no means ceased to complain even in the 1950s) allowed the North to exploit southern lands and timber resources and has probably had other lasting effects, such as that of maintaining high freight rates between southern factories and the great market areas of the North. Since the railroads were controlled by northern interests, it was possible to exclude southern goods and to enjoy undisturbed occupation of the markets.[5]

Then poverty and lack of education made themselves felt in the all-important sphere of agriculture, where the old mistake of overspecialization was repeated, this time over a wider area that included newly settled lands in the western South. Smallholders wore out their plots by persistent cultivation of the same three row crops—cotton, corn, and tobacco—either because these were the only crops they knew how to grow or, in the case of sharecroppers, because they were necessary to pay the rent. Equipment was as scarce as experience, and only labor was abundant—labor without the capital to make it productive.[6]

These, then, are some of the reasons why, by 1930, the annual income per person in the fifteen southern states was only 45 percent of that in the other thirty-three. During the depression years, after 1929, the South, with its dense farm population and its marked dependence on cash crops, suffered greatly and the differential between the South and the rest of the nation widened. Migration to the towns, which had provided some relief in the 1920s, was halted and even reversed. The passing of the depression left the

[4]"Nevertheless, the Southern planters and small farmers after 1865 could hardly have taken any other road to agricultural reconstruction. Both planters and ex-slaves knew how to grow cotton: it was the *only* thing that many of them knew how to do." W. N. Parker, "The South in the National Economy, 1865–1970," *Southern Econ. Journal*, vol. 46 (1980), 1019–48; quotation p. 1023.

[5]The tone of these statements is intentionally vague, since the validity of the southern charges on this score has been much disputed. It is argued by some historians that this is simply a part of a "conspiracy theory" evolved by southern apologists. We maintain here a careful neutrality.

[6]But, argues Parker in "The South in the National Economy," p. 1045, "In effect, the South's growth path in 1866, the setting of its social controls to give an impoverishing agriculture another seventy-five years of life, could hardly have occurred otherwise except as part of a national economic and social policy which would have redistributed labor and capital within the nation . . . without regard to race, locality, or previous social structure." It is hardly surprising that this did not occur in 1866.

The South: An East Tennessee farm in 1935. This picture, showing a slope that had been unwisely but continuously cultivated under row crops, epitomizes conditions in the Old South in the years of agricultural depression. These were the conditions that the Soil Conservation Service and the TVA, among other government agencies, helped to relieve by spreading knowledge of erosion control and scientific farming. By the end of the 1940s such scenes could still be found in the South, but they were rare; a New South was emerging. *TVA*

South with gigantic problems: a material one of increasing the wealth of its people and a psychological one of throwing off the stigma of the backwardness under which it had labored so long. To solve these problems, certain objectives would have to be realized. In agriculture, these were (1) to apply good management principles to land holdings so as to encourage wider participation in decision making and rewards; (2) to reduce the dependence on cotton, tobacco, and corn through the introduction or improvement of other crops and, especially, livestock; and (3) to encourage innovation and improvement in crop and livestock production and in the use and preservation of land resources.

If southern agriculture was to reach these goals, it was clear that not only would many farmers have to revise their methods, but also that much of the rural population would have to cease farming and find employment elsewhere. Improvements in farming would therefore depend upon the availability of other sources of employment. Only, in fact, by matching agricultural improvement with industrial development could the southern standard of living be raised.

In industry there were also certain clear objectives: (1) The fuller use of local materials, agricultural and mineral. (2) The processing of products within the region, instead of shipping raw materials out. The South would thereby retain within its borders the additional revenue created by turning trees into furniture and cotton into high-quality cloth, instead of passing it to northern workers. (3) The development of its own capital resources to finance local industry. The South might then cease to be a "colony" of the North, and be able to bargain on equal terms with other regions over such matters as freight rates and factory location. (4) The provision of new employment for an unskilled rural population surplus, as the only alternative to mass emigration from the area.

The two sets of objectives for agriculture and in-

The Northeast Railroad Depot in ruins after the siege of Charleston, South Carolina, in the closing months of the Civil War. *Library of Congress*

dustry were (1) to raise levels of production and purchasing power through the fuller development of land and human resources, especially those represented by the rural and the black population and (2) to develop the kind of mutually supporting interdependence between the rural–agricultural and urban–industrial economies that developed much earlier within the economic core region.

From the Old South to the New: Elements of Change

Today the South presents a different picture. In almost every detail, the account just given must be modified in the light of developments since 1933. For the changes that have taken place some of the credit must go to the Roosevelt administration, which, taking office at the low point of the depression in 1933, enacted the New Deal measures that opened the way to recovery. In a more local sphere, much of the credit goes to the Tennessee Valley Authority, established as a part of the Roosevelt program. But no one can deny that, apart from these outside forces for good, there has been a remarkable revival within the South

itself. It has not yet caught up with the rest of the nation, but the gap has narrowed. In 1930, the average per capita income in all sixteen southern states was only 60 percent of the national figure. By 1985, it was 105 percent in Virginia and 99 percent in Georgia, although still only 76 percent in South Carolina, 75 percent in Arkansas, and 66 percent in Mississippi. Clearly, however, progress has been made toward the regional objectives, and we therefore turn now to a description of the South as it is today.

THE PHYSICAL SETTING

The region we are considering is composed mainly of a broad coastal plain and the southern portions of the two mountain systems that extend into it—the Appalachians and the Ozark–Ouachita. The plain is underlain by sedimentary rocks whose strata dip gently away from the mountains and toward the sea. Differential erosion has resulted in a belted coastal plain; that is, one with alternating, elongated belts of upland and lowland, roughly parallel to the coastline.

In the vicinity of the Mississippi River, however, the alignment of the belts changes. Here, between the Appalachians to the east and the Ozark–Ouachita highlands to the west, the coastal plain reaches far northward, to the southern tip of Illinois, in a great embayment. The belts of the coastal plain bend north so as to lie roughly parallel, not now to the coast, but rather to the great river. They form U-shaped crescents around the southern end of the Appalachians. Two of the lowland belts, the Black Belt of Alabama–Mississippi and the Blackland Prairie of Texas, became famous for cotton production on their limestone soils.

The outer half of the coastal plain, a zone up to 50 mi (80 km) wide, is generally flat, low, and swampy, with little agriculture. Inland, elevations rise gradually and drainage improves. Although the coastal plain possesses mainly sandy soils, it is separated from the Appalachian Piedmont by an even sandier belt known as the Carolina Sand Hills, which stretches from central North Carolina to Alabama. In the Piedmont, the soils, developed from metamorphosed crystalline rocks, have a higher clay content.

Most notable for the productivity of its soils, if they are properly drained and protected from flooding, the Mississippi alluvial valley is the axis of the northward embayment. As wide as 125 mi (200 km), and stretching more than 500 mi (800 km) from southeastern Missouri to the Gulf, its level floor is bordered by upland bluffs rising perhaps 200 ft (60 m) above it. Most of the land sufficiently elevated and drained for farming is on the natural levees, which parallel the riverbanks and slope imperceptibly away into the "back swamps." On top of these levees, the U.S. Corps of Engineers has built and now maintains thousands of miles of artificial levees or dikes. Not only agriculture, but also roads, railroads, and towns are concentrated on the natural levees just behind the dikes, creating a pronounced linear pattern of settlement.

At their inner edges, the Piedmont and coastal plain butt against the foothills of the Appalachians, the Ozark–Ouachita highlands, or the escarpment of the southern Great Plains. The Ozarks are hilly rather than mountainous. Their steep slopes and thin, cherty soils are mainly forested, either never cleared (although they have been cut), or reclaimed from abandoned farms. The Ouachitas, like much of the southern Appalachians, consist of parallel ridges and valleys. The only valleys sufficiently open and fertile to support agriculture, however, are the Great Valley of East Tennessee (much interrupted by ridges) and the narrower valley of the Arkansas River separating the Ouachitas from the Boston Mountains. The hill country west of the southern Appalachians is interrupted by significant agricultural areas such as the Bluegrass–Pennyroyal, the Nashville Basin, and the Tennessee River Valley and Sand Mountain plateau in northern Alabama (see Fig. 15-2).

Climatically, the region is well favored from most points of view. Only on its western fringe is it liable to drought; everywhere else the rainfall is over 40 in (1000 mm) per annum, and in the Great Smokies it rises to 80 (2000 mm). Snow seldom falls, and only in the Appalachians does the frost-free period last less than 200 days. On the other hand, much of the rain falls in heavy thundershowers, which increase the danger of erosion, whereas high humidity over most of the area makes for summer lassitude, and cloudiness reduces evaporation and increases leaching of the soil.

Perhaps the outstanding feature of the South's natural endowment is its forest cover. Commercial forest occupies 67 percent of the surface area of Georgia, 66 percent of Alabama, 63 percent of South Carolina, and 54 percent of Mississippi. The Appalachians, the Ouachitas, and much of the coastal plain and Piedmont are forest covered; indeed, from the air it is the forest that dominates the landscape, and agriculture has rather the appearance of being carried on in forest clearings—which, statistically, it is. These forests represent only a remainder and a regrowth of a far greater original cover, but even after the cutting and burning of the past decades, the South still possesses a tremendous asset—nearly 40 percent of the nation's commercial forest land, including the bulk of its hardwood reserves. It produces nearly 60 percent of the United States' pulpwood. Fuel is in good supply within the region. The Appalachian coalfield extends through Kentucky and Tennessee into Alabama, and the western end of the region lies athwart the great mid-continental and Gulf oil and gas fields. Other mineral resources are numerous. Bauxite is mined in the neighborhood of Little Rock, and these deposits account for over 90 percent of the United States' domestic production. Phosphate rock is worked south of Nashville, and copper is mined in the mountains of eastern Tennessee. Finally, any account of the region's resources must include mention of the great amount of electric power available from the rivers of this area, of which the development on the Tennessee is the best-publicized, but by no means the only, example.

THE REVOLUTION IN SOUTHERN AGRICULTURE

Tobacco and the Establishment of the Slave Plantation System

We may note four major revolutions in the agriculture of the South since its earliest beginnings. The first should perhaps be called a beginning rather than a revolution, but its importance can hardly be exaggerated—it set not only the structure of southern agriculture, but also the course of southern history, economy, and society for more than three centuries. This beginning depended upon the adaptation of the slave plantation system to the Indian crop tobacco, and it took place early in the seventeenth century in Virginia and Maryland. It was the start of a course that brought black people and their culture to North America; it led to a North–South industrial–colonial relationship and a devastating Civil War, which strengthened rather than erased that relationship.

Cotton and the Spread of the Slave Plantation System

The second might also not be termed a revolution, but rather a major extension of the patterns that had been established earlier. The production of tobacco spread from the shores of Chesapeake Bay into the Carolina coastal plain and later to the Piedmont and

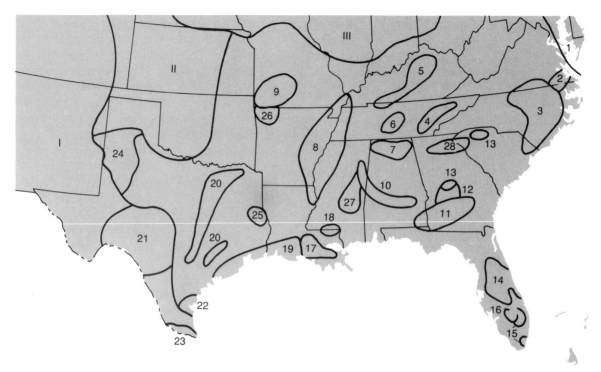

Fig. 15-2. The South: Agricultural areas.
BORDERING REGIONS

I   Livestock ranching
II Winter wheat region
III Corn Belt

SOUTHERN AREAS

1 Chesapeake Bay (tobacco, poultry, vegetables)
2 Norfolk Peanut region
3 Carolina tobacco region (with hogs, corn, soybeans, cotton, peanuts)
4 Valley of East Tennessee (dairy)
5 Kentucky Bluegrass–Pennyroyal tobacco region (with cattle, hogs, horses, wheat, corn)
6 Nashville Basin dairy region (with beef cattle)
7 Tennessee Valley of Alabama (poultry, cattle, hogs, cotton, soybeans, corn)
8 Mississippi alluvial valley (soybeans, cotton, rice, cattle)
9 Springfield Plateau dairy region
10 Alabama Black Belt (cattle, soybeans)
11 Georgia–Alabama coastal plain peanut region (with hogs, pecans, cotton, corn)

12 Georgia pecan area
13 Georgia and South Carolina peach areas
14 Florida citrus region
15 Florida vegetable areas
16 Florida (sugarcane)
17 Louisiana (sugarcane)
18 New Orleans–Baton Rouge (dairy milkshed)
19 Louisiana–Texas coastal plain (rice, cattle)
20 Texas Black Prairies sorghum region (with hogs, cattle, cotton, corn)
21 Edwards Plateau (sheep, goats, cattle grazing)
22 Corpus Christi area (sorghum, cotton)
23 Lower Rio Grande Valley (irrigated sorghum, cotton, vegetables, citrus)
24 Texas High Plains (Llano Estacado) (irrigated cotton, sorghum, wheat; beef cattle feeding)

Poultry concentrations
25 East Texas
26 Northwest Arkansas
27 Eastern Mississippi
28 North Georgia

to smaller areas across the Appalachians. More than half a century after tobacco was first cultivated, cotton growing was still of little importance. The difficulty of extracting the fiber from the seeds made cotton cloth an expensive, luxury product. That was changed by Eli Whitney's invention of the cotton gin in 1791; from a negligible amount in 1790, cotton production rose to 100,000 bales in 1800, and to more than 5 million by 1860. As we said in Chapter 7, cotton, like tobacco, was a labor-intensive crop. And as with tobacco, the slave plantation system became the dominant means of production. The system spread throughout the lowland South, but became especially concentrated in the most fertile, productive areas: the inner coastal plain and Piedmont, the Black Belt and the Tennessee River valley in Alabama, the Mississippi alluvial valley, and the Black-land Prairies of Texas (Fig. 15-3).

The Civil War and the Establishment of the Sharecropper System

At the end of the Civil War, many plantations were devastated; their original owners were bankrupt and were forced to relinquish their land. But the great landholdings stayed intact—ownership was in some cases merely transferred from planter to banker or merchant or war profiteer. Big landowners still had land—what they did not have was labor. The black people, on the other hand, had gained freedom, but little else—no land, no education, not even homes except at the will of the landowners.

The solution to the dilemma was predictable. The former plantation slaves went back to work on the land, raising cotton or tobacco as before, and the multiple-unit plantation or sharecropper system evolved. A plantation that formerly had been worked by slaves under overseers was now divided into 20- to 40-acre (8- to 16-ha) units, each with fields sufficient to grow about ten bales of cotton or 2 to 3 acres (1 ha) of tobacco, the maximum that could be tended by a black family. Landowners continued to furnish the worker family a shack in which to live, a mule, and implements with which to cultivate, and usually half the cost of seed and fertilizer. They even supplied food, clothing, and medicine, all on credit against the hoped-for crop. The sharecropper family supplied the one thing it possessed—labor. At the end of the harvest, the income from the crop was shared between landowner and cropper, the cropper's share often falling short of his debt to the landowner.

Poor white as well as black families entered into the sharecropper arrangement. There were also, of course, many independent small family farms, both white and black, especially in the less fertile, hilly areas. But the plantation sharecropper system dominated the agriculture of the South, and to a considerable degree its economy and society, for a full century. Its presence and its dominance accounted, substantially, for the relatively depressed and retarded status of the southern economy in the 1920s.

As the South moved toward mid-century, its regional agriculture conformed less and less to the models for the nation as a whole which we reviewed in Chapters 7 and 14. The survival of the plantation system militated against the development of family farms. The enlarging scale of farm operations in the rest of the nation found no counterpart here. Capital investment was low, and laborsaving devices were ignored: there was always plenty of labor, but its productivity was low. Urban–industrial markets, so important in, say, the Interior, were few and far between in the South. National trends toward a truly modern, commercial agriculture were passing the South by.

The Mid-Twentieth-Century Revolution in Southern Agriculture

The real revolution, which is still in process, represents a major change. The sharecropper system, overwhelmingly dominant in 1950, has almost completely disappeared, replaced by an agricultural system generally characteristic of present-day America, although traits and influences of the southern past persist.

To assess briefly, and so soon after its occurrence, why and how such a change took place is difficult. We may explore three important sets of factors.

1. *The two world wars and the migration of blacks from the South.* The relative dominance of the southern economy by labor-intensive agriculture depended upon an ample supply of agricultural workers for whom alternative opportunities were nonexistent or limited. During World War I and, especially, World War II, industries expanded rapidly, at first in the cities of the Economic Core, but later in the rapidly growing cities of the West and the South. Beginning during World War I, continuing during the expansion of the auto industry during the 1920s, and burgeoning during and following World War II, the expanding factories and warehouses of the northern and western cities attracted great numbers of new laborers. Many came from the South, at first mostly white people from the failing farms of the Appalachians and hill country, but later both blacks and whites from the Deep South. For many, it was the collapse of cotton prices in the 1930s that served to persuade them to give up what had become a most unequal struggle. But now, for the first time since the importation of slaves more than three centuries earlier, rural black people found alternative opportunity—and landowners found their supply of cheap labor seriously threatened.

Tobacco and cotton are labor-intensive crops, from soil preparation and planting through harvest and marketing. It was generally accepted that harvest, especially, could not be mechanized—a belief that held for cotton until the 1940s, for tobacco as late as the

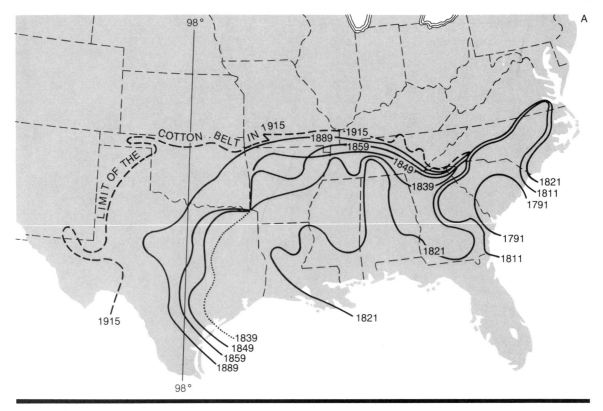

Fig. 15-3. The Old South: The Cotton Kingdom. These two maps give us an impression of the spread of cotton growing across the Old South. Based on the maps accompanying the Census of Agriculture, (**A**) aims to show how the spread of cotton growing was halted northward and, especially, westward, by adverse climatic conditions. Beyond the 98th meridian of west longitude, cotton, like a number of other crops, could generally only be grown under irrigation. (**B**) shows the core areas of cotton production on the eve of the Civil War of 1861–65. They were in Alabama and the Mississippi Valley. Of the states shown here, Texas has today by far the largest cotton production: Mississippi ranks a rather distant second. (Map [A] is from Walter Prescott Webb's classic *The Great Plains* [Boston, 1931], p. 189. Map [B] from S. B. Hilliard, *Atlas of Antebellum Southern Agriculture* [Baton Rouge, 1984], p. 72. Reproduced by permission.)

1970s. But with technological advances in a growing economy mechanization did take place in the case of cotton, first in western Texas and California, to which production had expanded. And mechanization came to the South, first in the form of tractor-drawn cultivators, then cotton harvesters, and finally, in the current decade, tobacco harvesting and handling.

Mechanization of southern agriculture and outward migration of labor, especially of black labor, occurred simultaneously, especially from 1940 to the present; most probably, the cause–effect relationship operated in both directions. Mobilization for World War II accounted for much of the migration, as men were drafted or volunteered for the armed services and as wartime industry boomed; southern blacks, as well as whites, were recruited to work in the shipyards of the West, for example. In any case, the result

is the virtual disappearance of the sharecropper system and its replacement by family, larger-than-family, and manager-operated farms. The latter two types are sometimes called "neo-plantations." In effect, the sharecropper units of a multiple-unit farm or plantation were pieced together again, but in the newer version, the large fields are worked by tractor-pulled equipment or harvesters, rather than by workers and mules. Production of a bale of cotton using mule-power and hand labor once had required more than 150 hours; complete mechanization cut the time to 12 hours and then, as Table 7-2 showed, later on to 5 hours.

Nor did mechanization make itself felt solely by cutting down labor-hours. The greater power of the tractor over the mule enabled farmers to enlarge their range of crops; indeed, to change land surfaces, by practices such as terracing. They could now tackle

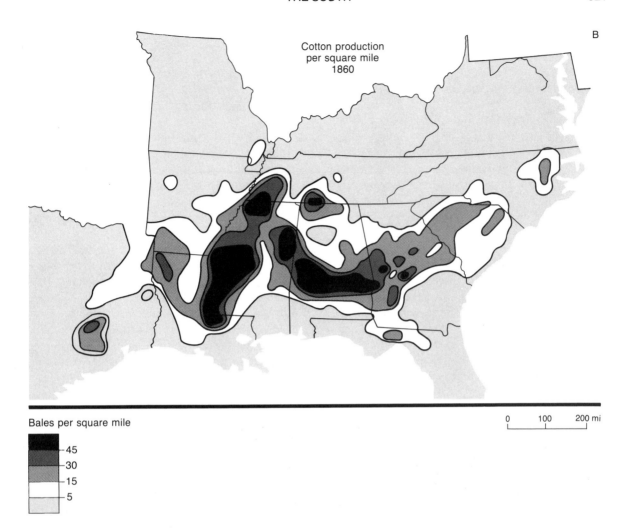

Cotton production
per square mile
1860

B

Bales per square mile

45
30
15
5

0   100   200 mi

heavier soils and steeper slopes (in order, let us hope, to grass them over and check erosion). They could, given the necessary capital, join in a movement for higher yields of their traditional cotton crop, which was being aided by technicians on and off the farm: "Before full mechanization of the cotton crop could be achieved, the combined contributions of engineers, chemists, fertilizer specialists, plant breeders, entomologists, agronomists and other scientists were necessary."[7]

Once again, the landscape as well as the system changed. Worker houses that had been grouped in village-like clusters during the time of slavery, and had then been dispersed onto the sharecropper units for a century, have now mostly been razed. "Other

[7]G. C. Fite, *Cotton Fields No More: Southern Agriculture, 1865–1980,* (Lexington, Ky., 1984). Quotations from pp. 186 and 190.

than the increase in trees, nothing so modified the southern rural landscape in the 1960s and 1970s as the destruction of tens of thousands of sharecropper and tenant houses." New groupings along roadsides house tractor-operators, seasonal laborers, or workers left over from the old system. Some of the large, manager-operated neo-plantations in intensively farmed areas such as the Mississippi alluvial valley have machinery equipment centers reminiscent of the machine tractor stations of Stalinist Russia—they serve surrounding fields that stretch in all directions uninterrupted by farmsteads or houses. Mechanization has tended to increase the scale of operation, squeezing out family farms that are too small to support cotton harvesters or mechanized tobacco equipment. Independent black farmers, especially, have faced difficulties, often in the form of subtle discrimination by those who control credit or even by officials dispensing government aid. The number of farms owned by blacks reached a peak of about 200,000 between 1910 and 1920, but has fallen sharply since 1950. Most black people in the South now live in towns and cities, as virtually all do in the North and West.

The industrialization originally sparked by wartime needs eventually spread to southern cities, where it has continued to grow—in contrast to slower growth or stagnation in the old Economic Core. Urban growth in the South not only provides employment for displaced farm workers; it also encourages agricultural diversification by providing expanded markets for dairy, beef, poultry, and fruit and vegetables.

2. *The Great Depression and attempts at recovery.* The Great Depression in the 1930s accentuated the problems of southern agriculture, but also stimulated change. Government programs, such as those in soil conservation and farm credit, offered ideas and help. Most notable was the TVA scheme, which not only helped to stimulate industry by providing electric power at relatively low cost, but whose programs in soil conservation and crop rotation and diversification also served as models for change and improvement.

During and since the depression years, the federal government has attempted to alleviate the problem of surplus production and consequent low prices by placing controls on production. Various government programs have been applied regularly to cotton, tobacco, and wheat and, at times, to corn, peanuts, rice, and sugarcane, all leading or significant crops in the South. Such programs were based not on output, but rather on acreage. In a system of acreage quotas, each farmer-participant is assigned, for a given crop, a maximum acreage based on previous acreage of the crop. It took little time for even the most backward farmer to perceive that it would pay to reduce acreage, but to raise the yield per acre.

Farmers increased their yields per acre by first using their best land for the restricted acreage. They then increased the number of plants per acre, especially of cotton, and applied fertilizer, insecticides, even supplementary irrigation with an intensity equaled only in the Midwest, where high investment of capital and technology was the rule. So, although the quotas cut back the cotton acreage in the South from 40 million acres (16.2 million ha), in 1928–32 to 14 million (5.67 million ha) in the years 1959–60, the yields doubled during the period and the total crop

declined by only about one-fifth. By 1966, acreage was down to a low of 8.6 million (3.48 million ha), but it had increased again to 9.05 million acres (3.66 million ha) by 1975, and to 11.6 million acres (4.7 million ha) by 1981, five-eighths of which were in Texas. The figure for 1987 was 9.8 million acres (3.97 million ha).

These changes contributed, through competition, to changes in the location of cotton production locally, regionally (within the South), and nationally. The quotas assigned to small family farms were so small that it was not economical to work the quota acreage; as a result, many families dropped out of farming altogether.[8]

Regional changes in cotton farming had already been caused by the boll weevil, a pest that invaded southern Texas from Mexico in 1892 and that, during the three decades that followed, spread eastward along the Gulf Coast and inland throughout the Deep South. Feeding on the seedpods or bolls of the cotton plant, this insect thrives in the humid climate of the Southeast. The cotton farmer suffered until agriculturalists could develop and apply effective pesticides and control methods, including more rotation of different crops.[9]

3. *The civil rights movement.* Before discussing locational change further, we need to recognize a third set of stimuli for change in the South. We have stated that the most basic factor in the southern agricultural and economic revolution has been the new opportunities offered black people, in particular, and poor people, in general. This was provided initially by their discovery of opportunity elsewhere. But most certainly it was aided by the federally enforced civil rights movement which, after the 1950s, began to bring black people into the mainstream of society, offering them, for the first time, an education and the chance to participate in their government. Most black

---

[8]This is the reasoning offered by many of those who have dropped out of cotton producing. However, Prunty and Aiken claimed that the available figures on production costs did not support the idea that any particular acreage quota was too small to be efficient; even a small allotment can be made to pay. (See M. C. Prunty and C. S. Aiken, "The Demise of the Piedmont Cotton Region," *Annals,* Assoc. Amer. Geog., vol. 62 [1972], 291.) It is rather that the farmer prefers to develop larger-scale production of other crops, and so may sell or lease his quota allotment to a farmer who does wish to continue in the cotton business.

[9]In response to the stimulus for diversification, the town of Enterprise in southeastern Alabama erected a monument to the boll weevil with the inscription, "In profound appreciation of the boll weevil, and what it has done to herald prosperity."

people have opted out of agriculture, but that is both cause and effect of the agricultural transformation we are discussing.

SPECIALIZATION AND REGIONS

Unlike the Corn Belt, where cultivation is intensive and continuous from one end to the other, intensive agriculture in the South is restricted to scattered lowland areas having relatively fertile soils. Extensive intervening areas, including many once farmed, are in woodland. Figure 15-2 identifies the more intensively farmed areas and their specializations.

The struggle against the boll weevil, adjustments to the federal system of acreage limitation, and the selective, competitive adoption and application of new technologies, organization, and methods all served to change the location, specialization, and degree of intensity of agriculture.

Foremost has been the demise of cotton as the dominant crop in the South. It has almost disappeared, not only from areas of poor soil or terrain, which have reverted to woodland, but also from former areas of concentrated production, such as the Carolina coastal plain and Piedmont or the Alabama–Mississippi Black Belt. It is still significant in the Georgia–Alabama coastal plain and in the Tennessee Valley of Alabama, but ranks as a major crop only in the Mississippi alluvial valley and areas to the west, especially the Corpus Christi district and the irrigated areas of the Lower Rio Grande Valley and the Texas High Plains (Fig. 15-2). Nearly one-half of the country's cotton acreage is now in Texas, the only state in which cotton is the first-ranking acreage crop. Only in Mississippi and Louisiana does it rank as high as second in acreage; in no state is it the leading source of farm income (see Table 15-1).

King Cotton has given way to King Soybean, which has become the leading crop by acreage in every state south of the Corn Belt and Virginia, and east of Texas and Oklahoma. It is the leading source of income in Alabama, Missouri, and Louisiana and ranks high in many areas. Since World War II, the soybean has found ample domestic and export markets—oil for cooking oil and margarine (at the expense of butter) and meal for livestock feed. Consequently, it has been a popular replacement, where growing conditions permit, for crops such as cotton, which have been subject to troublesome surpluses and acreage limitation.

In the Mississippi alluvial valley, the largest and most intensively farmed and productive agricultural area in the South, cotton remains a major crop, but soybeans have come to occupy more than three times as much area and to provide more income. For example, Bolivar County, Mississippi, a representative county (Table 15-2), has a large percentage of its area in crops, and its farms are large and highly capitalized and mechanized, but there is still a relatively high level of tenancy.

Livestock has been the other major gainer in the revolution of southern agriculture. Dairy and beef

Table 15-1. United States: Area Under Cotton, by States,* 1987

State	1987 Acreage**
Virginia	—
North Carolina	94,186
South Carolina	116,424
Georgia	231,635
Florida	—
Tennessee	411,100
Alabama	346,013
Mississippi	1,028,249
Louisiana	590,257
Missouri	197,598
Arkansas	405,345
Oklahoma	360,299
Texas	4,349,755
New Mexico	79,135
Arizona	381,733
California	1,083,811
U.S. total	9,826,061

*States are listed here in approximate east–west sequence.

**All figures in acres: for hectares, divide by 2.47.

Table 15-2. Bolivar County, Mississippi: Agricultural Statistics, 1987

Population (1980)	46,000
Percentage change, 1970–80	−7.0
Area of county (ac/ha)	590,000/239,000
Percentage in farms	68.3
Percentage in cropland harvested	50.5
Value of agricultural products sold	$79 million
Percentage from livestock products	9.3
Number of farms	478
Average size (ac/ha)	843/341
Average value of land, buildings, & equipment, per farm	$688,000
Average value of agricultural products sold, per farm	$166,000
Percentage of operators whose principal occupation is farming	75.3

The New South: This picture, taken near Oxford, Mississippi, shows high-quality Aberdeen Angus cattle-grazing pastures laid down on former eroded cotton fields, while, in the background, there is a large plywood factory supplied by sustained-yield forestry practiced in part on other old, eroded lands now planted with trees. The South's new industries and new agriculture help to balance one another. *C. W. Olmstead*

cattle, raised primarily on pasture and forage crops,[10] not only fit well into programs of improved land use, erosion control, and diversification; they also find increasing markets among a population that is growing in both size and affluence. The Springfield Plateau, the Nashville Basin, and the Valley of East Tennessee have become important dairy regions. Smaller dairy areas have developed near such major urban markets as Atlanta, New Orleans–Baton Rouge, and Dallas–Fort Worth. Production of beef cattle is widespread. It is especially important in some former areas of cotton dependence, such as the Alabama Black Belt. Hogs are somewhat less widespread, but they are especially important in the Georgia–Alabama coastal plain. The growth in the importance and volume of livestock production has been accompanied by great improvement in breeds and quality of animals and in methods of production. Once-common razorback hogs and scrawny cattle foraging unfenced woodlands have been replaced by well-bred animals feeding in planted pastures, and later sold to feedlots in the Midwest and Texas.

The most spectacular change in livestock production, involving both method and volume, has been in the poultry industry. Since World War II, it has be-

---

[10]About one crop, kudzu, introduced from Asia for forage and erosion control, an outsider is entitled to feel somewhat baffled, at least when he has read J. J. Winberry and D. M. Jones, "Rise and Decline of the 'Miracle Vine,'" *Southeastern Geographer*, vol. 13 (1973), 61–70. Up until about 1935, it appears, kudzu was known, but little used, but when the Soil Conservation Service programs of the 1930s began, it was hailed as the South's answer to erosion. That it cut runoff and soil loss seems clear; that it had pasture value is also undeniable. Less well established are rumors that Southerners rocking in their chairs on the porch woke up to find themselves covered with kudzu vines. The peak of kudzu popularity came in the 1940s. In 1970, the U.S. Department of Agriculture declared it a common weed. It has overrun many areas, and attempts are being made to eradicate it.

come concentrated as house agriculture[11] in huge, vertically integrated, contract-farming systems within small specialized districts. Some farms, or even districts, specialize in hatching chicks, some in producing eggs, some in producing broiler or fryer chickens, and some in raising turkeys. In the broiler chicken industry, which is most concentrated in the South, a large enterprise, often a subsidiary of a feed-manufacturing company, may own a hatchery and/ or a meat-processing plant. The company supplies young chicks to a farmer under contract, who nurtures them following specific directions, and using company-supplied feed often in a company-financed house. At the end of nine weeks, the birds move to the company slaughtering plant, the house is cleaned, and another batch of chicks is moved in.

The system was first developed for egg production in New England, and for broiler production in the Delmarva Peninsula and Shenandoah Valley. Now used from coast to coast (and elsewhere in the world), it is most important in the South. Like other enterprises that depend upon external services, it tends to be highly concentrated in small districts. Only the largest and most important districts are shown on Figure 15-2. How does one explain such scattered concentrations? Perhaps the comparative advantage of some of the districts was not superior agricultural resources, but rather *inferior* ones—failing farms and farmers needed an enterprise (and capital and advice) that did not depend upon a particular soil or climate. Perhaps the advantage consisted of a venturesome entrepreneur who, having established an enterprise, attracted external services that attracted more enterprises. The climate of the South does permit year-round operation with little heating cost—however, air ventilation during hot, humid weather can be a problem.

Unlike cotton, tobacco has maintained its importance in the same specialized areas since early in the settlement period. In the largest such area, the Carolina–Virginia coastal plain and Piedmont, rural population is dense and farms have been small, in keeping with the intensive labor demands of the crop. (See data for Wilson County, North Carolina, Table 15-3.) As indicated previously, it was the last major field crop to be mechanized. The process of

[11]In contrast to *field* agriculture, the chickens or turkeys spend their entire lifetime within *houses,* each house generally containing 5000 to 30,000 birds. A single operator may take care of several houses. In many ways, it is more akin to manufacturing than to farming. It is sometimes called factory farming.

Table 15-3. Wilson County, North Carolina: Agricultural Statistics, 1987

Population (1980)	63,000
Percentage change, 1970–80	+9.8
Area of county (ac/ha)	240,000/97,000
Percentage in farms	60.3
Percentage in cropland harvested	31.1
Value of agricultural products sold	$69 million
Percentage from livestock products	35.0
Number of farms	666
Average size (ac/ha)	217/88
Average value of land, buildings, & equipment, per farm	$356,000
Average value of agricultural products sold, per farm	$104,000
Percentage of operators whose principal occupation is farming	67.6

mechanization has been piecemeal and partial—partial in that early machines were little more than taxis to transport seated or prone workers who still transplanted seedlings or removed plant suckers by hand. Even these devices were not developed until the 1950s and 1960s. Mechanical harvesters have come into use only during the last twenty years.

As with mechanization of cotton farming, the substitution of machines for men and mules has been accompanied by a host of interrelated changes. As employment in agriculture has plummeted, factories have sprouted in the small towns or countryside to utilize excess labor, which has led to a more balanced economy. As with the earlier cotton mechanization and black migration from cotton areas, the cause–effect relationship is not clear, but it probably has operated in both directions. The sharecropper system had largely given way to farms using hired wage labor during the earlier phases of mechanization. Even a sharecropper family with 3 acres (1[+] ha) of tobacco had had to obtain additional workers during periods of peak labor demand. Whereas black people displaced in the earlier mechanization of cotton had little choice but to migrate out of the segregated South, those displaced by tobacco mechanization in the 1970s could find jobs in the new factories. Most remain in the countryside, often in new clusters of roadside homes, and commute to their new employment. Landscape changes also include new metal bulk-curing barns and dilapidated old flue-curing ones; fewer, larger, multi-parcel farms with larger tobacco fields to utilize and justify the large machines; and fewer and fewer mules.

Rice and peanuts are other significant crops that are planted predominantly in the South. Peanut production is most concentrated in the small, specialized area inland from Norfolk, with its sandy soils, and in the much larger and more diversified farming area of the Georgia–Alabama coastal plain, but it is also important in Texas and Oklahoma. The peanuts are primarily for human consumption, but the crop is often associated with raising swine, which feed on surplus nuts and crop waste. Baker County, Georgia, is a representative county (Table 15-4). Rice is most important on the level, diked fields of water-holding soils in eastern Arkansas and on the Louisiana–Texas coastal plain.

Workers and Industry in the South

The revolution in southern agriculture has been accompanied by, and indeed could not have occurred without, comparable change in other parts of the economy. The removal of thousands of sharecroppers and laborers from plantations and farms necessitated either mass emigration from the region or provision of new jobs within it (see Table 15-5).

In fact, these are the two means by which these changes have been brought about. Several million persons have emigrated to the other parts of the country during the three decades prior to 1980, while in the years 1972–80 alone, 3.2 million jobs outside farming were created in the South Atlantic census division; 1 million in the East South Central division, and almost 3 million in West South Central (Table 15-6).

Migration out of the South did not begin in 1940: it had been going on for decades and consisted of two streams separated by their color. Young white people

Table 15-4. Baker County, Georgia: Agricultural Statistics, 1987

Population (1980)	4,000
Percentage change, 1970–80	−1.7
Area of county (ac/ha)	227,000/92,000
Percentage in farms	51.6
Percentage in cropland harvested	18.6
Value of agricultural products sold	$23 million
Percentage from livestock products	19.5
Number of farms	155
Average size (ac/ha)	756/306
Average value of land, buildings, & equipment, per farm	$783,000
Average value of agricultural products sold, per farm	$152,000
Percentage of operators whose principal occupation is farming	69.0

left because of the limited range of opportunities offered by the region in all occupations other than farming, or because of fear of job competition from blacks at the unskilled level. Black people, drawn by news of jobs in northern cities, moved to get away from the region where their ancestors had been slaves and where they themselves were still treated as second-class citizens. We have already noted the lure of Harlem in the early years of the century (see Chapter 11). World War I saw a northward migration on a new scale, as white workers in the North left to join the armed forces, and blacks were brought in to replace them. World War II witnessed a return of these conditions on a larger scale. Throughout the forties, the average annual out-migration of blacks from the Old South and the border states rose to around 160,000. For the fifties and sixties, the figure was only slightly less—145,000 in each decade—al-

Table 15-5. Southern Agriculture: Selected Statistics, 1950–87

	Number of Farms (thousands)				Average Farm Size (acres/hectares)			
	1950	1964	1976	1987	1950	1964	1976	1987
Virginia	151	80	72	45	103/42	149/60	153/62	194/78
North Carolina	288	148	125	59	67/27	97/39	104/42	159/64
South Carolina	139	56	47	20	85/34	144/58	232/94	224/91
Georgia	198	83	73	43	130/53	215/87	233/94	247/100
Kentucky	218	133	124	92	89/36	122/49	129/52	152/61
Tennessee	231	133	124	80	80/32	114/46	121/49	147/59
Alabama	211	92	77	43	99/40	165/67	195/79	211/85
Mississippi	251	109	84	34	82/33	163/66	202/82	315/127
Arkansas	182	80	69	48	103/42	207/84	246/100	298/121

Table 15-6. Southeastern States: Population Composition by Race or Ethnic Group, 1980 (in thousands)

	Total	White	Black	American Indian	Asian*	All Other Races	Spanish Origin
Virginia	5347	4230	1009	9	64	34	80
North Carolina	5882	4458	1319	65	20	21	57
South Carolina	3122	2147	949	6	11	9	33
Georgia	5463	3947	1465	8	23	20	61
Florida	9746	8185	1343	19	55	145	858
Kentucky	3661	3379	259	4	9	9	27
Tennessee	4591	3835	726	5	13	11	34
Alabama	3894	2873	996	8	9	8	33
Mississippi	2521	1615	887	6	7	5	25
Arkansas	2286	1890	374	9	6	6	18
Louisiana	4206	2912	205	169	16	20	99

*This column represents a compounding of separate figures for Chinese, Japanese, Filipino, Asian Indian, Korean, and Vietnamese. All other races, whether Asiatic or not, are included under the heading "all other races."

though by the end of the sixties, the tide was ebbing; racial violence in northern cities was a dissuader, and the crest of the migratory wave had passed. Meanwhile, the growth of the New South had been drawing back to it a small counterflow of white managers and businessmen from outside the region. If we exclude from our calculations Florida, which drew over 1.3 million white residents during the decade, but most of them for leisure rather than for work, then the states from Virginia in the north to Texas in the southwest gained 700,000 whites and lost 1,445,000 nonwhites between 1960 and 1970.

Had the safety valve of interregional migration not existed, the southern situation might have become very serious indeed. It is perhaps interesting to consider what the present state of the South would have been if secession had become permanent and if the transfer of southern labor to northern markets had been blocked by political barriers while the rural population went on increasing. From such circumstances as these, the rural slums of the Caribbean and East Asia came into being.

This story of inter-regional migration must, however, be brought up to date by recalling the facts cited in Chapter 2. Although by 1970 there was a net migration of white Americans into the South, the flow of blacks was still outward, as it had been for decades. But in the years since 1970, the historic northward drift of Southerners, black and white, has been countered by a large in-migration of both (Table 15-7). Old slave states like South Carolina and Georgia have been attracting black newcomers. It has been a momentous reversal.

To all appearances, the in-migration reached a peak in the 1970s and has been falling off since 1980. As Table 15-7 shows, the state of Mississippi, which for years had had a static, or even a declining, population, showed an increase between 1970 and 1980 which was above the national average, and a small net in-migration. For the period 1980–85 the net trend for the state, as for Alabama, was outward once again. While nothing, apparently, will halt the flow of millions into Florida, for other parts of the South the future looks less certain. It would be surprising if so dramatic a change in migration patterns were to begin and end in a single decade, but we must simply wait and see.

Within the region, as we have already noted, new jobs have been created. Here the swing from agricultural to nonagricultural employment (and so from rural to urban surroundings) has been very marked. Some of the new employment has been in industry, but the bulk of it has been in professional, technical, and government service; that is, in sectors in which the South had traditionally been deficient. Forty, or even twenty, years ago, the figures in Table 15-8 would have shown far larger departures from national averages than do the state employment percentages today.

The changeover from farming to other occupations affected the whole population; indeed, it has affected the whole continent. But it was most striking in the case of the blacks. In 1940, 35 percent of the black population of the South was classified as "rural-farm"; 41 percent of all employed male blacks there were farmers or farm laborers, and so were 16

Table 15-7. Southeastern States: Net Migration, and Percentage Population Change from All Causes, 1970–80 and 1980–85 (in thousands)

	Migration Balance, 1970–80	Percentage Change in Population, 1970–80	Migration Balance, 1980–85	Percentage Change in Population, 1980–85
Virginia	239	+14.9	161	+6.7
North Carolina	278	+15.7	188	+6.4
South Carolina	210	+20.5	91	+7.2
Georgia	329	+19.1	273	+9.4
Florida	2519	+43.5	1437	+16.6
Kentucky	131	+13.7	−52	+1.8
Tennessee	297	+16.9	38	+3.7
Alabama	97	+13.1	−4	+3.3
Mississippi	31	+13.7	−22	+3.7
Arkansas	184	+18.9	6	+3.2
Louisiana	100	+15.4	29	+6.5
U.S., 50 states		+11.4		+5.4

Table 15-8. Southeastern States: Percentage Shares in Nonagricultural Employment, by Selected States, 1990

State	Total Work Force (thousands)	Construction	Manufacturing	Transport & Public Utilities	Wholesale & Retail Trade	Finance, Insurance, Real Estate	Services	Government
Alabama	1637	6.0	23.5	5.1	21.7	4.5	19.4	19.9
Florida	5403	6.0	9.7	5.1	27.0	6.8	29.6	15.6
Georgia	2995	4.9	18.7	6.6	24.9	5.4	21.4	17.8
Mississippi	937	3.9	26.2	4.8	21.2	4.1	17.2	21.8
North Carolina	3129	5.3	27.6	4.9	22.9	4.3	19.1	15.8
South Carolina	1549	6.5	24.7	4.3	22.5	4.3	19.1	18.3
Tennessee	2195	4.2	23.8	5.3	23.5	4.6	22.2	16.0

percent of the employed black women. By 1970, the rural-farm element was a mere 2 percent.

So a movement to town has taken place. In the South, this movement has had a distinctive character: it has not contributed so much to the expansion of the large urban centers—there are, in any case, few of them in the South—as to the growth of the small industrial towns. It is not primarily to Atlanta or Birmingham that the workers have moved, but rather to the towns of the Carolina Piedmont and the Tennessee Valley. It is true that the population of Dallas increased phenomenally between 1940 and 1990, but the circumstances creating this situation had their origins (see Chapter 17) largely outside this region.

This type of urban development is to be explained, in part, by the character of southern industrialization. In a region where one of the main attractions to industry is the availability of labor, there is a genuine incentive to locate plants in the rural communities where the labor surplus is to be found. Furthermore, the power resources of the South are mainly electricity and oil (or gas), which allow considerable flexibility in locating plants. It has therefore been unnecessary, and certainly southern opinion has judged it undesirable, to crowd workers into manufacturing cities; instead, the factories are located in small centers. Although there is, of course, an element of risk in linking a town's employment exclusively to one or two plants, it is outweighed in most cases by the advantages of a small city atmosphere and a freedom from the ills of industrial life on the nineteenth-century pattern. For this satisfactory state of affairs, much of the credit must go to the planning commissions of the various states, who have encouraged the policy of rural industrialization to provide work for the rural population. Local authorities also offer a variety of inducements to suitable industries, and business groups have been active in creating local enterprises.

The net result is that, in the interior of the Deep South, only the Carolina Piedmont and metropolitan Atlanta and Birmingham can be termed industrialized areas. Industry is dispersed throughout the whole South, and only if we consider intensity of employment in manufacturing are the small centers, with their heavy dependence on industry, emphasized.

Nevertheless, the southeastern states have secured an impressive share of the total U.S. increase in industrial employment during the past four decades, and it is worthwhile pausing to examine the figures in Table 15-8 more closely, to discover which areas, and which industries, are involved.

In 1947, the southeastern states contained 13.6 percent of the national work force in manufacturing. By 1987, this proportion had risen to 30.8 percent; the South's value added by manufacture matched employment, with 30.3 percent of the national total. During the period since 1947, total U.S. employment in industry had risen by a little over 30 percent, whereas the growth in the South was nearer 120 percent. Furthermore, a number of states had succeeded in shrugging off the national industrial downturn of the early 1980s, which saw total employment fall, and they had shown an employment increase in every year tabulated. North Carolina, Georgia, and Florida had achieved this and they contained, as it happens, the most significant industrial growth areas of the Southeast. North Carolina (Table 15-8) had the highest proportion in the nation of its nonagricultural work force in industrial employment.

Since before the Civil War, the South has been noted for its manufactures of textiles and cigarettes. But these traditional southern industries, although they remain important, are not primarily responsible for the expansion since 1940. That has been produced by the emergence of three other classes of industry. The first consists of branch plants of concerns producing for nationwide markets. As the southern standard of living has risen, so the consumer-goods market in the Southeast has become increasingly attractive to producers, who have established local supply points to meet the growing demand. In this category fall such manufactures as those of agricultural machinery and household equipment. The second class consists of firms that have migrated to the South to secure a more favorable business location, especially lower costs for labor and heating and,

often, lower taxes. The third class consists of industries that have only recently come into existence— the aerospace and petrochemical industries are good examples.

What advantages does the South offer to industry? Electric power, oil, and gas, certainly. A growing market may be another attraction: a region with a rising standard of living. A labor force that is generally cheaper, and often less unionized, than that of the older industrial regions, but that is also more prepared to regard working in industry as a means of gaining status rather than losing it. Under all these circumstances, it is not in the least surprising that the South has been gaining a quite disproportionate share of the new industrial jobs in the United States. Something of a tremor ran through the industrial world recently when aircraft builders decided to construct their newest model in Georgia instead of in southern California (see p. 495).

The industrial development of the South has been varied. Forest products, cotton, tobacco, bauxite, and oil have been the native raw materials: motor-vehicle assembly, clothing manufacture, aluminum smelting, and fertilizer production have been brought into the region along with a score of other industries spilling out of the older industrial areas further north. As we have already seen, the new industries have been located typically in small towns, each of which can then claim (and does claim) to be the national focus of its industry. So Dalton in Georgia is "the nation's tufted-textile capital" (it produces some 70 percent of the tufted carpet made in the United States); Marietta, near Atlanta, was the site for the plant where Lockheed built what, up to the time of writing, is the world's largest plane—the Lockheed C-5 Galaxy— and where the company still employs some 9,000 workers.

Few areas of the South, in fact, resemble the industrial regions further north, even where the actual concentration of manufacturing employment in the South is as high as or even higher than that in the older industrial regions. One area that does, however, bear such a resemblance is the steel region around Birmingham (911,000); developed in the 1880s, it was originally built on the initiative of the Louisville and Nashville Railroad. Here, where coal and iron ore were found within a few miles of each other grew up the "Pittsburgh of the South," its night skies lit by the glare of blast furnaces. There is a well-developed railroad network, and water transport has been made possible by a canal joining Birmingham to the Warrior River, which gives access to the Gulf by the way of the Tombigbee.

The New South in the Nation

Students of American affairs have for so long been accustomed to treating the South as a special case economically that perhaps the best tribute that can be paid to southern progress is to record the fact that it is now possible and legitimate to treat it as part of the nation. It has its problems, but they are common to all the regions of North America—and the South, as it happens, is closer to solving them than are some other regions with richer economies and longer histories of prosperity. The South may or may not have joined the Union in spirit, but as a fact of economic geography, it is today inseparable from the nation. Wright argues that "it is now virtually impossible to find an essentially regional southern identity in economic life."[12] The economic issues in the modern South are not sectional, but national. The economic goals of the nation are sought also by the South, and with a considerable degree of success. To test the truth of these statements, it is only necessary to review the familiar regional issues in American life and resource use today. Among these issues are:

1. *Interregional population movements.* Since the arrival of the carpetbaggers after the Civil War, almost nobody had moved to the South (excepting Florida); it was a place to move away from. In the interregional movements of the last two decades, the whole South has figured as a major receiving area, and some parts of it are beginning to experience the problems of urban pressure.

2. *Industrial competition and adaptation.* Bidding for new industry (which, in turn, means new jobs and new tax revenues) is a way of life in every region of North America. The South has been highly successful in attracting new industry. It is true that its long-established textile industries (that employ some 600,000 workers) have been feeling acutely the pressure of overseas competition and have been forced into a program of modernization and self-improvement, but so have a score of other American industries: regionally, the effect of the shakeup has been beneficial after several decades of easy southern dominance. All through the Piedmont industrial area (North Carolina, South Carolina, and Georgia), tex-

tiles, despite the competition, remain the largest group by value of industrial products.

3. *Agricultural efficiency.* We have already noted the problems of American farmers, nationwide, caused by the fact that, although their efficiency and investment may both increase, it is not easy to increase the size of their farms. The problem is at its most acute in the Midwest. It exists also in the South where, as we have seen, it has been alleviated by the renting of relinquished farmland and crop quotas to form "neo-plantations." In this respect, the South is no different from the rest of the nation.

4. *Development of alternative employment sources.* In a nation where agriculture and industry together employ a mere one-third of the labor force, all regions confront the problem of what to do for alternative employment. In this respect, the South is a little better placed than some other areas, since it has historically been deficient on the tertiary side of employment, and its proportional gains in government employees, legal and educational services, and finance have been greater than those for the nation as a whole in the past two decades.

5. *Government spending.* The federal government's spending and placing of contracts are influential factors in the economy of regions and states—as witness the fierce political competition to secure them. In this competition, the South has held its own in spite of such drawbacks as a comparative lack of research institutes or major universities. On a per capita basis, the South gets a fair share of federal grants and, thanks to a number of key defense and space installations in the region, government spending has been well sustained. Of the federal government's expenditures on defense in 1985, nearly 36 percent of them were made in the three southern census divisions, including Texas: if Texas is excluded, the proportion was still 28 percent.

6. *Poles of growth.* It became apparent in the era of regional planning that one of the most significant elements in regional development is the "self-generating" growth of the metropolitan center. The nation contains only a limited number of such growth poles at any one time; they wax and wane as the balance of regional advantage shifts. But it is clear that in the present period the South possesses two such poles (we exclude Florida and Texas for consideration in the next chapter). One of these is Atlanta, Georgia (2,914,000). The other is not a single city, although its focus might be identified as the twelve-county area that surrounds Charlotte, North Carolina

[12]G. Wright, *Old South, New South* (New York, 1986). Quotation from p. 273.

Atlanta, Georgia: The city center. Atlanta has a spectacular set of center-city buildings, especially hotels, since it has set out to become a major convention center. It is also a transportation hub of national importance. The gold-domed capitol building still asserts itself among its taller and newer neighbors. Photo by *Kevin C. Rose*

(1,181,000), which has come to be known as "Metrolina." Rather, it is the urban region of the Carolina Piedmont, which extends from the small, flourishing cities of Greenville and Spartanburg in South Carolina, to the Winston–Salem–Raleigh urban axis in North Carolina.

These two areas possess all the attributes of growth poles. Atlanta has shown a remarkable power of attracting business over the past two decades, while equally remarkable has been the output of research and planning initiatives for the South as a whole, generated by the Piedmont cities. Apart from these two, there are lesser, but still striking developments within the region that attest its vitality—the growth of Memphis (990,000) as a great agricultural market and processing center, and the rise of the Tennessee

Valley towns like Knoxville, Bristol, Chattanooga, and Huntsville in the land of the TVA.

It is a little more than a century since Sherman took Atlanta and tore up the railroad tracks of which it was then the focus. The city he destroyed has become one of the fastest growing in the country. Not only are its railroads in full working order; but by a curious turn of the wheel of history, it is transport which has underpinned the city's modern growth. Its Hartsfield airport is one of the largest and one of the three or four busiest in the world; in 1990, it handled 48 million passengers. It is a focus of Interstate highways and has a number of large trucking companies. As a result of these excellent communications, it is the eighth-largest wholesale center in the United States: in fact, warehousing and wholesaling account for al-

Research Triangle Park, North Carolina, one of the development centers of the New South and its industrial economy. *North Carolina Department of Commerce*

most as large a share of urban employment as does manufacturing (130,000 employees in 1990, as against 173,000). It has been chosen as a national or regional headquarters by many firms and government departments, and claims that 420 out of the 500 largest industrial corporations in the United States (the "Fortune 500") maintain some kind of operation in the city. It recently attracted the headquarters of the largest parcel carrier in the nation, and has been chosen by all the major telecommunications firms as their southeastern headquarters. It has made strenuous bids for convention business, and has a glittering array of central hotels, surrounded by a kind of desolate no-man's-land of car parks and dereliction into which, no doubt, it hopes that its CBD will one day spread. It also has the 1996 Olympic Games to look forward to and to spur it on.

The other major growth pole, Metrolina on the Piedmont, is a multinucleated area of industrial cities whose principal interests are: (1) tobacco processing, for which Winston–Salem–Raleigh, Durham, and, further north, Richmond, Virginia are noted (in 1980, North Carolina accounted for 43 percent of the total

U.S. crop; adjacent South Carolina and Virginia produced 13 percent); (2) industries based on the Appalachian forests, such as furniture making and the manufacture of paper and cardboard; and (3) textiles. This is the center of the nation's cotton and synthetic textile manufacture. The Carolinas contain 70 percent of the cotton-spinning spindles of the United States and a very large share of the worsted spindles and broad fabric looms as well. The district is also noted for its synthetic fiber mills: those at Asheville and Roanoke are among the largest in the world. Over 70 percent of Metrolina's industrial employees are involved in textiles or textile-related industries.

But poles of growth are not created by tobacco or cotton textile industries; as it happens, those are some of the slowest growers in the industrial seedbed. The more remarkable feature of Metrolina is the way in which it has attracted research and development. Its Research Triangle Park is, it is claimed, the largest such concentration in the United States today, spreading as it does over many miles of wooded Piedmont.

This is particularly important in a southern con-

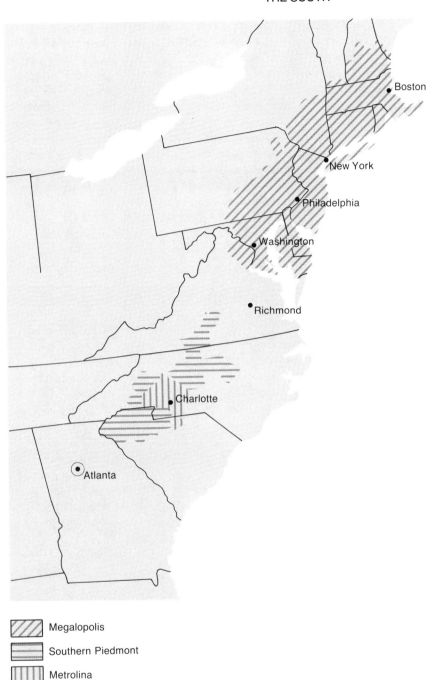

Fig. 15-4. Megalopolis: The present and the future. The map shows the urbanized areas of the eastern United States, and suggests the development of a Megalopolis-type extension into the South, centered upon the "Metrolina" area of the Carolinas. (But see the accompanying text.)

Megalopolis

Southern Piedmont

Metrolina

text. One of the weaknesses of the southern economy in the years since World War II has been that, although it was gaining new jobs in industry, it was not experiencing a parallel diversification of industry. The old textile region, in particular, badly needs non-textile employment. The presence, in its heart, of this center of innovation should do much to create the needed alternatives.

One other feature of these two southern poles of growth must be mentioned—the one illustrated by Figure 15-4. The map, which is taken from the *Atlas of Metrolina*, calls attention, in a pleasantly partisan manner,[13] to the way in which the Piedmont growth areas carry on the line of the North's Megalopolis into the South. This "line" is, in fact, well known to geographers; it is the Fall Line (see Chapter 1), along which a series of settlements sprang up in colonial days, most of them to take advantage of waterpower on the rivers at the break of slope. What has been for years a kind of geographical curiosity may well, the map suggests, be the urban axis of the future—an axis continuous from the Maine side of Boston in the north to Atlanta in the south; a sort of Macro-Megalopolis stretching from the Merrimack to the Chattahoochee.

But we must end on a note of caution; specifically, with an answer to the question: Are there then no problems remaining that are southern rather than national in character? There are, in fact, at least two, and they are linked.

The first is that although the South has joined the nation economically, it has joined it at the lower end. South of Virginia no state, not even Florida or Texas, has an average per capita income as high as the national average. By this criterion, Mississippi comes in fiftieth out of fifty states, Arkansas forty-eighth, and South Carolina forty-sixth. Even if these figures are

adjusted for cost of living, it is also true that the five states of the Union with the highest proportion of families below the official poverty line are all in the South. So there is still a lot of ground to make up.

Second, the southern average is low because, in large part, the family income of the black population is low—as we saw in Chapter 2, about 60 percent of that of the whites. One of the South's greatest needs is to bring out black skills and to provide development capital for black businesses. The black population suffers from a chronic capital scarcity (more precisely, from a lack of black capital and a chronic reluctance of the white financial community to make such capital available), and unless black business can obtain the necessary investment funds, it has little chance of improving its contribution to the regional economy.

So the problems remain. Some writers have spoken of the basic southern problem as that of "nationalization," by which they mean the process of bringing the South to throw off the "burden of southern history" (the phrase is Vann Woodward's), and to join the rest of the nation on a footing of equality and mutual acceptance. It is over forty years now since John Hope Franklin, himself a distinguished black historian, expressed the hope that one day the South might become so integrated into the larger life of America that it would be nothing more than "a tattoo on the arm of the nation." There has been progress, but that time has not yet come. Perhaps we can at least say, however, that now the arm belongs to the nation and that it is playing its part in the activity of the whole body.

---

[13]The present writer's use of the word *partisan* to describe Metrolina's map is a reaction to the fact that, on the map, the urban–industrial growth of the southern Piedmont is shown as halting abruptly at the South Carolina–Georgia state line. It is as if Atlanta, with its 2.9 million metropolitan area population, did not exist—or Georgia's half-million industrial workers. Actually, everything that is true of the Carolina Piedmont is also true of the Georgia Piedmont.

16
Appalachia

The National Park system: Cade's Cove in the Great Smoky Mountains National Park, on the Tennessee–North Carolina border. This park encloses the largest mountain mass of the Appalachians, including the summit, Clingman's Dome (6643 ft, 2026 m). Its annual number of visitors is higher than for any other national park.

Negative: Appalachia, a Problem Region

The argument of the last chapter was that North and South, although they remain separate in history and outlook, are gradually growing together because they have common economic interests. As ties are created, a major role is played by programs of the federal government, which, because they apply equally to all parts of the nation, have a unifying effect.

We must now consider, however, another form of government action of a quite different character—a program that is specifically confined to one area, but that has the same significance for North and South as other programs: it binds them together because the area to which it applies links North and South in the same way that a rivet or dowel binds together two separate pieces of metal or wood. The program is one of regional development, and the region is called Appalachia.

Strictly speaking, Appalachia is not one of the southern regions of the United States, but then neither is it northern: it lies with one end in each. It takes its name from the Appalachian Mountains, which run through the North and the South, and the definition of its boundaries, as we shall see later, is itself a problem. But Appalachia does possess some southern characteristics; some of its problems—the distinctive problems the regional program was created to solve—are southern problems, shared by this region with the wider South. The center and north of Appalachia have their problems too, but on the whole they are merely special cases of the difficulties faced by the economic core, and reviewed in Chapters 11 and 14. The northern end of Appalachia actually lies within the core.

As its name suggests, Appalachia is first and foremost a physical region—that area of the United States within which the Appalachian system dictates surface, structure, and resources. As we saw in Chapter 1, the system extends in its essentials all the way into northeastern Canada, but it is to the part of the system that lies west and south of the Hudson and Mohawk corridors that the term Appalachian is generally applied. Here the breadth of the system from east to west is greatest and its westerly province, in particular—the Cumberland and Allegheny plateaus—broadens to occupy much of northwestern Pennsylvania, most of West Virginia and eastern Kentucky, and parts of southeastern Ohio and Tennessee. It is within these borders that the characteristic Appalachian topography is most clearly developed.

But these physical features alone do not account for the way in which the name *Appalachia* has become familiar to the average American in the past two decades. Appalachia is today not simply the name of a physical region: it is familiar because it carries cultural and political significance. Culturally, it stands for a world apart, whose inhabitants are generally considered to live lives unrelated to the standards and concepts of other North Americans: "Understanding the people of Appalachia must be based on the recognition that their value patterns are at variance with the value patterns found in the larger American society."[1] Politically, its existence and its boundaries have been given form by the allocation of several billion dollars of funds, under the Appalachian Regional Development Act of 1965, to the region as defined for a program of public works. In the American war on poverty, Appalachia has been a prime target.

The situation in Appalachia today is the product of both geographical and historical forces. From colonial times onward, the American's clearest con-

[1] F. A. Zeller and R. W. Miller, eds., *Manpower Development in Appalachia* (New York, 1968), p. 26.

sciousness of the Appalachian system has been as a barrier to east–west movement. It is true that, once across the Blue Ridge, the pioneers from the coastal settlements found in the Great Valley (occupied by the Shenandoah and Tennessee rivers and their tributaries) an easy route southwestward, but to break out of it to the west or northwest was far from simple, and led into a bewildering maze of valleys and forested hills that even today baffles the stranger. Daniel Boone's famous Wilderness Road of the 1770s left the valley at the Cumberland Gap (one of the few points between the Tennessee gorge at Chattanooga and the Mohawk Gap far to the north where a wagon trail could be carried up onto the plateau) and cut across a narrow section of the hills to reach the Bluegrass Basin. The deeply dissected plateau effectively defied penetration, except on foot.

Whatever hindrance to movement the Appalachian system as a whole offered, therefore, the plateau formed a kind of sanctuary and for that very reason attracted some of the pioneers. For most pioneers, however, it was an obstacle to avoid; it could neither be easily crossed nor easily cultivated. Even after a century or more of commercial timber cutting and coal mining, communication remains bad. The heart of Appalachia was not simply at the end of the road to nowhere; in many cases, there was no road.

Physically, then, Appalachia possesses the unity of a single geological system, but this unity is purely conceptual. Landscapes, communications, and access all vary greatly within the region, and with them, economic conditions. It is probably for this reason that the concept of Appalachia as a region has been largely missing in geography texts in the past; the eastern edge of the system is linked with and forms a backdrop for the tremendous development on the Atlantic Coast; the southern end of the Appalachians shares much of the character of the Old South, and the heavy industry region around Pittsburgh is normally treated as part of the Manufacturing Belt. In fact, Appalachia can be regarded as forming part of several other regions. This leaves the nonindustrial plateau as the epitome of Appalachia—the most remote, most poverty-stricken part of the region.[2]

How has this situation arisen? Most explanations have tended to focus on the cultural aspect of development—on the way in which the people of the area, living and intermarrying in a world apart from the mainstream of American life, have either missed or resisted the march of progress. But real though this cultural isolation may be, it is possible (to judge by some of the literature on the subject) to exalt the culture-differential argument to the level of a mystique and to suggest that, given the same range of choices as other Americans, the Appalachian settlers deliberately chose a life of feuding and moonshining, and idling in between.

This was not, of course, the real situation. The sequence of developments in Appalachia has followed quite logically from principles we have already reviewed in Chapters 7 and 11. In eastern North America, generally, the normal sequence of occupance has been (1) the coming of hunters and trappers; (2) the advance of the settled frontier, accompanied by subsistence agriculture; (3) a transition from subsistence farming to commercial agriculture and the rise of market-oriented production; (4) at some period after the beginning of stage (2), the growth of nonfarm employment and the coming of industry.

In Appalachia, however, with severe physical limitations of slope and soils on agriculture, even subsistence farming was often difficult: only in the "coves" and on the narrow valley floors could a few fields be cleared, and even these were liable to flood and soil erosion. When the time came for the critical transition from stage (2) to stage (3), even areas much closer to a market than Appalachia found themselves under fierce competitive pressure. The Appalachian farmers stood no chance at all.

Many of them took the obvious step of leaving their poor farms and looking for work elsewhere. In the cities, however, the Appalachians have often found it hard to obtain regular jobs. This is partly because they tend in many cases to be seasonal workers: after a few months in the city, they return to the hills and to their families, perhaps coming back a year or two later. On the other hand, their irregular appearance on the labor market makes them available for the kind of temporary job that every big city generates, at low rates of pay and without security. Since they fit all too well into this category of what might be called the labor cushion, their standard of living is generally low.

On economic grounds, therefore, the area's people, if they chose to remain where they were, had lit-

[2]The neglect of Appalachia in the past has been fully compensated for in the last two decades by the attention given to it by scholars. Among such studies may be mentioned K. Raitz and R. Ulack, *Appalachia: A Regional Geography* (Boulder, Colo., 1984); and R. D. Mitchell, ed., *Appalachian Frontiers: Settlement, Society and Development in the Preindustrial Era* (Lexington, Ky., 1991).

tle alternative but to continue in the primitive conditions afforded by stages (1) and (2) of our sequence until the fourth stage began, if it ever did—until the hills yielded something other than agricultural produce. Why they chose to remain may indeed require explanation in cultural terms, but if they chose to remain, it is difficult to see what else could have happened.

In due course, the resources of the Appalachians attracted outside attention, and this largely nonagricultural region came to life economically. The first of these resources was timber, for Appalachia possesses the finest stands of hardwoods in the United States, and very large stands of softwoods, in addition. From the time of the Civil War onward, agents of outside timber companies were at work, buying the mountaineer's trees and often employing him to cut them. The price was low and the method calamitous in its disregard for such other regional resources as soil and wildlife.

The cattle dragged or "snaked" the heavy mass of wood down the hillside to the creek, and along its rocky bank to a collecting point. There it was left, with hundreds of others like it, to await the log run.

To add to the stream's volume the mountaineers worked together to build "splash dams" at intervals along the creek. . . . When a heavy spring rain filled the rivers and sent torrents flowing over the crests of the dams, the mountaineers were ready to follow their logs to the great mills. . . .

A charge of explosives ripped out the dam nearest the head of the creek and the unleashed flood surged down in a bubbling wall on the thousands of "sticks" cluttering the channel. . . . Like rising thunder the water and its cargo rushed downstream, gathering momentum and freight with each succeeding mile.[3]

But this destructive exploitation of the forest resource could not, by its nature, provide a stable or continuing basis for the regional economy of the period. Nor, it now seems, could the second of these exploitive activities, although the impact it made on the region

was profounder by far and more widespread. If there is any one feature that unites the parts of this far from uniform region, it is an involvement with coal mining. Indeed, it could be argued with fair hopes of success that the region was united by poverty precisely because it was first united by a commitment to mining. Of the eleven states affected by the act of 1965, seven were coal producers; if we omit Tennessee with its small production, the output of the other six states has been as listed in Table 16-1.

It was in the last quarter of the nineteenth century that coal mining spread into the Appalachian valleys. In most parts of the region, development waited for railway construction, although the growing industries of Pittsburgh were fed with water-borne coal. The Appalachian coalfield had a number of peculiarities. One of them was that the coal underlying the plateau was, in a sense, too easy to mine. Much of it could be reached by opencast methods (and this was long before any question arose of restoring the mined-out surfaces, of which, at the last count, Appalachia had some 380,000 acres [153,850 ha]—80 percent of all such unreclaimed strip-mined land in the nation), and in the early period and in boom years, small mines were scattered everywhere throughout the field and there was little or no incentive to organize production rationally in big mine units. Another peculiarity was that Appalachia possessed few towns. Its rural population, though large, was scattered in nearly inaccessible cabins in the hills; consequently, it was necessary to assemble mine labor and to accommodate the miners in camps run by the coal companies. Therefore, although mining profoundly affected the physical qualities of the land, it had an equally disturbing effect on Appalachian social life.

But at least mining represented employment, so long as the market for coal remained buoyant. When

[3]H. M. Caudill, *Night Comes to the Cumberlands* (Boston, 1962), pp. 67–68. This remarkable book, although written with the frankest partisanship by a native of the Cumberlands (he lost two male relatives by accidents in a log-rush, such as that described above), should be read by anyone with an interest in Appalachia, if only because it set off such a tidal wave of "discovery" of the region and well-meaning but sometimes wrongheaded efforts to help its inhabitants. (Caudill tells the story of this aftermath of his first book in *The Watches of the Night* [Boston, 1976].) For Caudill's proposals for the region, see later in this chapter.

Table 16-1. Appalachia: Coal Output (in million tons)

State	1965	1975	1985	1989
West Virginia	149.2	109.2	127.3	151.2
Pennsylvania	95.1*	89.3*	66.1*	69.0*
Kentucky	85.7**	140.4**	161.8**	158.6**
Ohio	39.3	46.2	36.0	32.5
Virginia	34.0	32.7	44.3	51.2
Alabama	14.8	22.4	26.4	28.5

*Includes anthracite.

**Part of Kentucky's production is from the state's western coalfield, which lies outside Appalachia.

Appalachian landscapes: A remarkable mining landscape northeast of Harrisburg, Pennsylvania, where coal has been mined along the strike of the Appalachian folds. View looking southwest. *John S. Shelton*

markets began to contract, the extent of the region's dependence on this one resource became tragically apparent. First in the depression years of the 1930s and then in the course of the postwar reorganization of the industry, unemployment became endemic.

Not only in North America, but also in Western Europe, coal producers have confronted the problems of falling demand and rising costs. They met the situation by closing uneconomical mines and mechanizing their remaining operations. In this respect, Appalachia has not been exceptional; in fact, it has adapted to the change remarkably well, certainly more successfully than, say, the British coalfields, if we judge by sustained production and comparative cost. But what has produced the Appalachian distress has been that there was nothing out-of-work miners could do. Rationalization presupposes the transfer of labor to other tasks. In Appalachia, with

its primitive agriculture, its sparse distribution of industry, and its services, such as education and health, far below average, the unemployed miner generally had only two choices—to return to his cabin in the hills and live on welfare or to leave the region altogether. In the single decade 1950–60, over 2 million people chose the second alternative.

The problems of Appalachia during the decades of the 1950s and 1960s were epitomized by maps like the one in Figure 16-1. Although some parts of the Appalachian region—as defined by the act of 1965—increased their labor force by 10 or 20 percent during the 1950s, in four adjoining economic areas of eastern Kentucky, the reduction in employment was between 35 and 44 percent. In adjoining West Virginia, employment in mining dropped from 134,000 to 59,000 in the same period (by 1972 it was down to 48,000), and the population as a whole was reduced

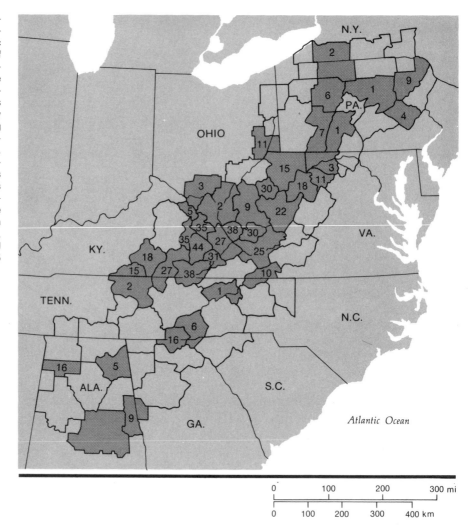

Fig. 16-1. Appalachia: Percentage decline in total employment, by state economic areas, 1950–60. This map is of considerable historical interest because it represents the situation that John F. Kennedy became aware of in his campaign for the presidency in 1960—one so distressing that he promised to take action to remedy it if elected. The Appalachian regional program described here is his memorial. In all the areas shaded, employment fell during the decade; percentage fall is indicated by the figure in each area. During this period, employment in the United States as a whole *rose* by 14.5 percent.

by 145,000. Throughout the 1960s, West Virginia lost over 25,000 people each year.

It is generally agreed that it was the plight of West Virginia that brought the problem of Appalachia to the attention of the nation. In 1960, when campaigning for the presidency, John F. Kennedy visited the state and pledged himself to deal with the situation if elected. His memorial is the 1965 Act. Meanwhile, Appalachia was growing on the consciousness of the nation in other ways. For one, an increasing number of Americans outside were finding in the region an outlet for their social awareness. In its provision of schools, health services, and dental care, the region lagged far behind the nation; infant mortality rates, doctors per thousand of the population, and number in high school all told the same story of submarginal conditions. Even before Kennedy discovered Appa-

lachia, medical personnel and other volunteers were quietly working weeks or months each year in the region.

The other way in which the region's problem became known was through the out-migration of its people. Most of them looked for work in the cities outside, but not too far from, the region's borders. Here they formed an element in the population almost as distinctive as the blacks, known as Appalachians.

Positive: Two Approaches to Regional Development

By the beginning of the twentieth century, the cultural and economic differential between the Atlantic Coast cities and the Appalachian plateau was probably greater than the one between the Atlantic Coast

Appalachia: Rural life. This photograph shows a cabin in Cade's Cove, in the Great Smokies, with its small clearing and traditional fencing. *National Park Service. Photo by Fred Bell*

and the mining or ranching West in the nineteenth century. It was upon this differential, and the attempt to get rid of it, that the Appalachian Regional Development Act of 1965 focused.

But those responsible for the program of the 1960s were not working in virgin territory. For inside the Appalachia defined by the Act, there was a region that had been the object of an earlier plan for regional development—a plan of the 1930s that was studied and emulated the world over. It was the Tennessee Valley. The Tennessee Valley Authority (TVA) had been set up in 1933 to do for the valley much the same as the regional commission for Appalachia was called upon to do for a wider area in the 1960s: relieve regional distress. In these two bodies we can, in fact, study the contrasting impact of two different approaches to the problems of a depressed area. And considering that the earlier TVA was faced by much the same types of terrain, people, and economy as the later Appalachian Commission, it is re-

markable how little similarity there has otherwise been between them. We shall consider each of them separately, and then attempt to assess the relative success achieved by each.

The TVA, formed in 1933, at a time when the fortunes of all rural North America were at a nadir, was created to work in one of the most depressed areas. It is easy to understand, therefore, how and why it has become a symbol of progress and an example to be copied throughout the world. At a time of deep depression, it showed that, with a little "pump priming" from outside, a poor and distressed region could achieve a new vitality and sense of purpose.

This being the case, it may seem strange that much of what has been said and written about the TVA in the past has been hostile to it. It is necessary, in fact, to explain this hostility before considering the example which the TVA offers, for otherwise the literature on the Authority is impossible to understand. The explanation is simply that, although control of the Tennessee River had been discussed by engineers for over one hundred years, the decision to create the TVA was a political one. It was made by Franklin Delano Roosevelt's Democratic administration in 1933, and for millions of Americans it was a decision prompted by party politics. It was therefore just as important for the Democrats to be able to show that the scheme was a success as it was for the Republicans to be able to demonstrate that it was expensive, unnecessary, or plainly unconstitutional.

The Tennessee Valley, early in 1933, epitomized rural America's most pressing problems. Low prices and uninstructed farming had undermined the valley's agriculture (cotton that year was selling at 5 cents a pound), and soil erosion affected runoff and drainage, so that the river was a real menace. On the one hand, its irregular flow and vast soil load made it almost useless for navigation. On the other, it presented an acute flood danger to the low-lying farms and cities not only along its own course, but also on the lower Mississippi, to whose flood crests it contributed.

This was the background to Roosevelt's decision to create the TVA. Clearly the menace of the river was only a symptom of the human problems of the valley. But the approach to these problems had to be indirect. Under the Constitution, as we have seen in Chapter 4, the powers of the federal government are restricted. The president could create an Authority to control the Tennessee, on the ground that it would be removing barriers to interstate commerce; all else

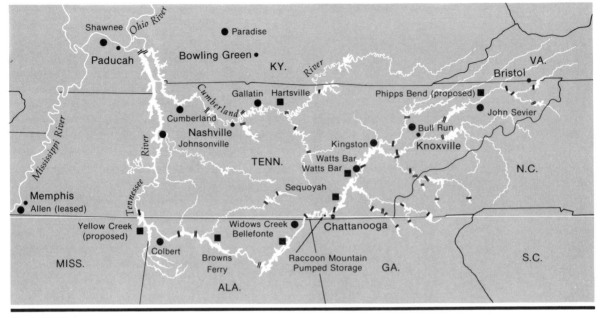

Steam Plants:
- ● Coal-fired
- ■ Nuclear

〓 Dams

Fig. 16-2. The Tennessee Valley and the TVA: Dams and power plants.

that he hoped for must grow out of that basic activity. Just how much might grow out of its one legitimate activity has been the great point of debate in the story of the TVA.

So it came about that the Authority was created with the dual mandate of flood control and navigation improvement. From the works constructed for these two purposes, it was to produce and sell electric power. It was in this somewhat backhanded way that the great Tennessee power development was initiated. The TVA was also given charge of a nitrate plant at Muscle Shoals, a relic of an earlier project, and so entered the fertilizer business and the sphere of agricultural improvement to which it made such an outstanding contribution in the succeeding years.

The area within which the TVA operates is some 40,000 sq mi (103,500 sq km) (Fig. 16-2). The Tennessee is formed by a number of rivers that rise in the Blue Ridge and Great Smoky mountains and flow in the corrugations of the Ridge and Valley country, where they merge to form the Tennessee proper. The main river follows the trend of the valley system to Chattanooga, where it turns west, cuts through the southern end of the Appalachian Plateau in a deep

gorge, and flows west and north to join the Ohio just before the latter joins the Mississippi. The area comprises parts of seven states.

The TVA has now been in existence for a full sixty years. In pursuit of its primary objectives, it has built some 20 dams (and coordinates the use of more than 20 others), and it has created for shipping a 9-ft (2.7-m) channel from Knoxville to the Ohio River, 625 mi (1000 km) away. It has about 25 million kw of generating capacity, with plans to double this, although the increase depends upon a decision to build a number of nuclear plants and that, as we saw in Chapter 6, is at present uncertain. So great has been the increase in electricity demand over five decades that nowadays four-fifths of the TVA's power is generated at thermal stations, and not at the dams. This increase in demand for power reflects the TVA's claims that, since it was set up, the number of manufacturing plants in the valley has increased fourfold, and that a "technology corridor" is developing between Knoxville and the atomic energy center of Oak Ridge that may soon seriously rival the "research triangle" of North Carolina.

Ironically, the TVA, which was set up to develop a

Appalachia: The Tennessee Valley Authority's Fontana Dam on the Little Tennessee River; at 480 ft (146 m) it is the highest dam built by the Authority. *TVA*

new source of power—hydroelectricity—has consequently become the largest single customer for an old one—Appalachian coal. It consumes 40 million tons a year. In fact, H. M. Caudill, ever watchful over the interests of his Kentucky miners, is severely critical of the TVA for having disregarded those interests and signed long-term contracts for the purchase of coal in which, using its immense bargaining power as a customer, it drove down prices and bought from the cheapest sources, regardless of either environmental damage or miners' welfare.[4]

D. E. Whisnant echoes this criticism of the poacher turned gamekeeper: "The Tennessee Valley Authority after an impressive twenty-year effort in comprehensive resource development succumbed to internal and external pressures, narrowed its functions,

[4]Caudill, *Watches of the Night*, pp. 60–61.

and committed itself to a power policy whose effect on coal-bearing central Appalachia was devastating."[5]

Flood damage along the river has been considerably reduced and, on a number of occasions, control of the Tennessee has lowered a flood crest further south by a few vital inches and so saved the levees. Yet it was obvious from the beginning that if the Tennessee was to be controlled, it would be necessary not only to build dams, but also to penetrate to the headwaters of the river and there to rectify the conditions responsible for the floods. So the sphere of the TVA's activities widened. As a complement to the program of dam construction and nitrate production, there was initiated, with the help of the Soil Conservation Service, an anti-erosion campaign in the valley farms. Gullying was checked and trees were planted on eroded hillsides, for otherwise the newly built dams would rapidly have silted up. This had led, in turn, to the TVA's participating in a detailed soil survey of its area. As a further by-product of the original construction, the Authority joined with the U.S. Geological Survey in the topographic mapping of the valley. Finally, where dams were built and reservoirs created, the shores were landscaped to create parks and to encourage a growing tourist trade to the "Great Lakes of the South." At the same time it was possible to campaign against the malaria that had for so long undermined the vitality of Southerners. Newly formed shorelines were engineered to avoid the creation of breeding places for mosquitoes, and swamps and standing water were sprayed.

The construction of the Tennessee navigation channel also enlarged the TVA's activities, for once it had achieved the 9-ft channel, the Authority set out to build up traffic on the route. Its economists attracted to the river a curious assortment of cargoes and in the process fought a number of battles over freight rates, which benefited the southeastern region as a whole.

It was in the sale of electric power that the TVA's interests found their widest extension, for it embarked upon a campaign to fulfill its creators' hopes by raising the whole standard of living in the valley. In this campaign it has been reasonably successful, for rural electrification and the use of machinery that

it makes possible have conferred many benefits. The scheme survived the years of political objection to its very existence, and has since had an enormous cumulative effect on land use, employment, recreation, transport, and industry.

It would be gratifying to be able to add that it had also been a triumphant achievement in regional planning on the grand scale, but that would be an overstatement. Its charter granted the TVA planning powers, but these were never seriously debated or developed—indeed in 1933, there could have been only a hazy idea of what planning implied in a regional context—and the Authority would have encountered stiff opposition, local as well as national, if it had tried to use them. So much activity on the Tennessee can only have been beneficial to the region as a whole, but the structure adopted for the TVA was not necessarily the best one. We must now consider the alternative of the 1960s.

The 1965 Appalachian Regional Development Act, as we have noted in Chapter 10, made the first application to a specific area of the principles of regional development that had been formulated in legislation during the period since 1961. In a sense, it was the first of the new regional plans and also the last of the old, ad hoc arrangements, which had brought into being federal schemes to aid the Tennessee and the Missouri valleys. Certainly, it could hardly be called a regional plan in the sense of being a comprehensive blueprint for the future of the region.

The act made available $1.1 billion of federal funds over a six-year period, to be spent on projects agreed upon between the federal government and the states concerned, all of whom would work together on a regional commission. Although $69 million of the allocation was earmarked by the commission for health centers, and smaller amounts for erosion control, vocational training, and other projects, by far the largest share ($840 million) was set aside for road building. Four-fifths of the total federal funds were for the construction of 2000 mi (3200 km) of "development highways."

There are two particular features of interest in this regional project. The first was the area to which it applied. The definition of Appalachia adopted by the act was a very broad one. Based on rather vague criteria, it comprised 373 counties in 11 states and was in fact the "biggest" Appalachia to emerge from any of the regional studies made in the past. Its bounda-

---

[5]D. E. Whisnant, *Modernizing the Mountaineer* (Boone, N.C., 1980 and 1986), p. 220.

ries did correspond in a rough way, however, with those of the physiographic Appalachian System. What happened was that because the regional commission was designed as a joint federal–state body, and because state governments were expected to make a contribution to projects undertaken, it was left largely to the states to decide what was and what was not Appalachia. The thirteen southern counties of New York, for example, were left out of the region, as originally delimited, but were included later. On the whole, the result has been to include within Appalachia any county that, from a physiographic point of view, could claim Appalachian affinities.

Within the area so defined, the regional commission found the following conditions: (1) A population of 17.2 million, at the 1960 census, with an increase during the 1950s of 2 percent, against a national increase of 19 percent; (2) an exodus of 2 million people between 1950 and 1960, and 1 million between 1960 and 1970; (3) only 31 percent high school graduates against a national average of 42; 5 percent college graduates (U.S. average, 8 percent); (4) 5.9 percent of the population on welfare, against 4 percent in the nation as a whole; 7.5 percent of housing dilapidated or dangerous, against 4.7 percent nationwide; (5) 33 percent of the population below current poverty level, against 20 percent in the nation as a whole.

But as might be expected in any region defined basically for administrative purposes, there were wide variations within its borders. It must, for example, have come as a shock to many people to discover that most of the area in which the TVA had been operating for the past thirty years was included in a poverty program. On the southeast, moreover, the region ended only just short of Fulton County, Georgia—a county that includes Atlanta and is one of the most remarkable growth points in the nation at the present time. In the north, virtually all indices of education, income, and welfare showed a sharp rise northward from the Pennsylvania–West Virginia state line—which is hardly surprising, since this Appalachia also includes the 2 million people who live in the Pittsburgh conurbation.

These same indices made it clear that the Appalachian problem had its focus in eastern Kentucky and West Virginia. The dilemma of the latter was particularly acute, because it was the leading coal producer and because the distressed area covered almost the whole state. There is no non-Appalachian section that, by its prosperity, could balance the poverty of

the problem area.[6] Eastern Kentucky, on the other hand, was statistically the most backward and the poorest section of all Appalachia. In 1966, when the average per capita income in the United States was $2963, the figure for Appalachian Kentucky was $1378.

Furthermore, the problem of Appalachia was essentially rural. Curiously enough, this did not mean that agriculture was the root of the trouble. Rather, it was the population classed as "rural nonfarm" that contained the hard core of the distressed. This element formed, in Appalachia, a proportion twice as high as in the nation as a whole. It was a nonagricultural rural population because agriculture, as we have seen, has never found much place in Appalachia, and commercial agriculture almost no place at all.

The unfavorable position of this rural population was made more acute because, up to this time, the relief programs the federal government had introduced had mainly been for the benefit either of the farmers or of the cities and their inhabitants. Since Appalachia's rural nonfarm population fell into neither of these categories, there was little assistance to which they could lay claim.

We now come to the second point of interest in the Appalachian program: the allocation of funds. Most of the original money was for highways. The hope of the planners was that the new roads would bore "development corridors" through the solid mass of this economically inert region and that, along the corridors at least, activity would be sparked off that would benefit the rest of the region. It was a variant of the "growth-pole" concept, a sort of "growth-axis."

The reasoning behind the commitment of so large an investment to the single purpose of road building was by no means clear, although it could be safely assumed that pressures on or within Congress played a part. Obviously, the coming of the new roads would yield some benefits: the question was whether

---

[6]It is worth recalling at this point that West Virginia owes its existence as a state precisely to the distinctiveness of its mountain population. Formerly a part of Virginia, it broke away during the Civil War, when Virginia joined the Confederacy, because it rejected the economic and political implications of Virginia's loyalty to the South. One feels, therefore, that there was a certain justice in taxing northern states in the 1960s to pay for a program of aid for West Virginia; had it not remained loyal to the Union in the 1860s, its problems today would be less acute than they are, for it would have lowland Virginia to support and subsidize it.

they would be worth $800 million when they appeared, and whether the benefits would accrue mainly to the inhabitants of Appalachia, or only to motorists in a hurry to get from, say, Washington to Chicago. There were some immediate doubters: "There are two reasons for taking the view that these development highways may not represent an effective route to economic growth. First, the highways do not attack the basic reasons behind Appalachia's lag. The second weakness . . . lies in the lack of care with which the system was planned."[7]

Nor were these doubts resolved at the midpoint of the original program, when the government's auditor remarked that "limited progress has been made toward the program objective of increasing accessibility to and through the Appalachian region."[8] All that the roads could do was to give Appalachia a better chance to compete on equal terms with other regions for what they were all trying to obtain—new factories, new employment, and more tourists. But obviously, the roads themselves could not guarantee any of these objectives. The main effect of the roads might, in fact, be not to bring industry into the area so much as to take people out.

Another thing about the Appalachian program was an obvious cause of concern, when contrasted with that of the TVA: the investment in the Tennessee Valley actually produced something—electric power—and power is revenue-producing. The TVA paid for itself. The roads of Appalachia would directly produce nothing; infrastructure creates only the potential for revenue, not the revenue itself.

It is now a full thirty years since the government's Appalachian program was launched. Like the last veteran from a long-forgotten war, the regional aid concept lives on here, when it has disappeared from other, once-familiar regions. Year by year through the 1980s, the Republican administration cut off funds for the regional projects, including that for Appalachia; in budget-trimming exercises, the regional programs were the softest of soft targets. Yet the Appalachian project has survived it all, through to the present day: the original $1.1 billion allocated by the federal government has been exceeded several times

over, and the individual states involved have contributed a billion dollars themselves. What has been the outcome?

On the whole, the doubters of the earlier years should be reassured. The commission can point to some early evidence of the success of the program: (1) In 1980, the population of Appalachia was 20.2 million, or 3 million more than in 1960. In every state in the region, with the single exception of Maryland, the Appalachian counties increased their population faster, between 1970 and 1980, than did the state as a whole. In Georgia, for example, the Appalachian section's increase was 35.5 percent, whereas that for the whole state was 19 percent. The state of New York lost population in the 1970s; its Appalachian counties gained 2.5 percent. (2) In contrast to the out-migration of the 1950s and 1960s, there was a net in-migration to the region in the 1970s of 1.1 million. (3) By September 1981, 1730 mi (2770 km) of highways had been completed out of a planned total of 3033 (4853 km), and funds had been commited for a further 667 mi (1067 km) (Fig. 16-3). (4) The number of families below the official poverty line had been halved since 1960.[9]

These are substantial achievements. Clearly, the original decision to concentrate on roads to open up the region was not capricious; the corridors have had their effect. But in practice, the balance of the program has changed considerably with the passage of time. We have already noted the original intention to spend four-fifths of the regional funding on highways. By 1981, out of a much larger sum, two-thirds had been used for roads, while the other one-third had been invested largely in the human resources of the region, which had received so little attention in the original allocation. The largest items on the "human" side have been a health program and vocational training, while water pollution control and the reclamation of mined-out land have been major environmental beneficiaries. Whatever the original

[7]J. M. Munro, "Planning the Appalachian Development Highway System: Some Critical Questions," *Land Economics,* vol. 45 (1969), 160–61.

[8]Report by the Comptroller General of the United States, *Highway Program shows limited progress toward increasing accessibility to and through Appalachia,* Washington, D.C., May 12, 1971, p. 9.

[9]It has also been claimed (*Appalachia: Journal of the Appalachian Regional Commission,* vol. 15 [1981–82], 8–17), although we are clearly dealing here with statistics of a different quality, that along the Appalachian highway corridors more than 400,000 new jobs have been created; that 60 percent of the 800 largest new plants are located within thirty minutes' driving time from one of the corridors, and that in 1980 nearly 800 million ton-miles of coal were moved over the new roads. There is no reason to suppose that the railroad companies are rejoicing over this last fact, but it is asserted that by using road transport the coal operators saved $7–8 billion.

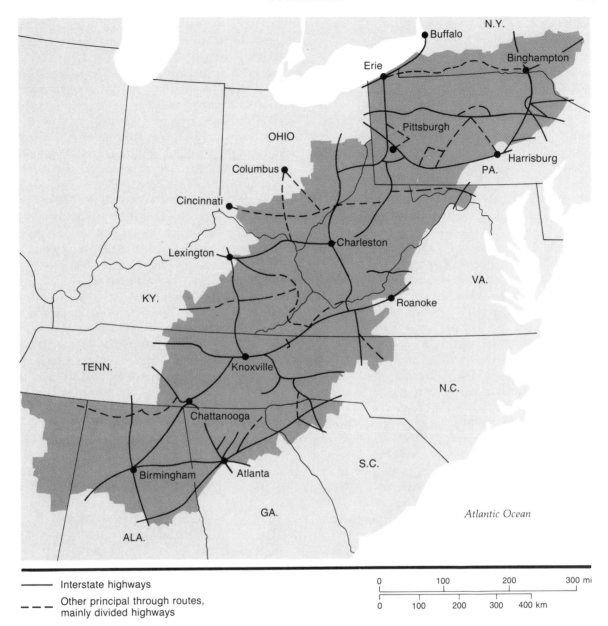

Interstate highways

Other principal through routes,
mainly divided highways

Fig. 16-3. Appalachia: The highway network, status in 1990. A large part of the regional aid provided for Appalachia was devoted to improving the highway network. Interstate highways and upgraded roads have certainly made a difference to regional accessibility, but some areas remain almost impenetrable. For a more detailed map of highways, see M. J. Bradshaw, *The Appalachian Regional Commission* (Lexington, Ky.), 1992, p. 125.

intention, the regional commission has given the program a human face.

One reason why this is so—a reason that is both a strength and a weakness of the original scheme—is to be found in the way in which funds have been allocated to the various projects. A commission consisting mainly of states' representatives can be ex- pected to favor state-sponsored or electorate-wooing projects, which will not necessarily add up to a uni- fied plan, and that is a weakness. The strength is that local needs and opinions can be considered. In this connection, there has emerged with the years a cen- tral role, which could hardly have been foreseen in 1965, for a planning unit known as the local development district (LDD). All federal grants-in-aid are routed through the LDD rather than being made by individual government departments to the state or the county or to individual recipients. The LDD has proved—unexpectedly perhaps—to be a useful level of organization, staffed by competent planners, but also near enough to the grass roots to respond to real needs.

We should not be surprised by the changes in direction that have occurred since 1965, for the region itself has changed. The most notable feature has been the resurrection of the coal industry. As it declined, in the 1950s and 1960s, Appalachia languished, but now that demand for coal has recovered and reached unprecedented heights, business in the coalfields is very different. The organization of the industry has changed: a few large mines, operated by major companies (which are often, in turn, parts of larger conglomerates) have replaced the many small operators of the past. The law has changed, particularly with the 1977 federal Strip Mining Control and Reclamation Act. Transport has changed, with the coal moving out over new roads and with plans for extra coal terminals on the coast at places as far afield as Charleston, South Carolina. Revenues from the industry have changed hands, with Kentucky, for example, charging a severance tax for the first time on each ton of coal mined, which augments the state budget, that is, it puts back into the coalfield areas, in the form of services, some of the wealth they have yielded to others over the past century.

As we have seen, then, the first fifteen years of the Appalachian program brought about some very marked changes in its economy and society. There remains, however, a question as to how permanent these changes may be. The doubts focus upon the area—the state—which has all along lain at the core of the Appalachian problem area: West Virginia. It is the only state to lie entirely within the region, and its welfare provides Appalachia's most direct criterion

of how successful the regional program has been. Between 1980 and 1990, West Virginia lost 8.0 percent of its population.

West Virginia is a coal state, and Table 16-2 summarizes its experience as such. Coal production and population have both fluctuated remarkably over the years. The Appalachian program achieved some real success in the 1970s against a background of sharply falling coal output, down to a trough in 1978. In spite of that fall, as we have already noted, the population rose after 1970; for the decade 1970–80 there was a net in-migration of 71,000 people. In the 1980s, by contrast, while coal production remained fairly steady there were worrying signs of out-migration once again—an estimated 53,000 between 1980 and 1985. In 1985, unemployment in the state was, by a wide margin, the highest in the nation: 13.0 percent overall against an average of 7.2; for male workers, 14.3 percent against 7 percent. West Virginia was forty-ninth out of the fifty states in income per capita, and this was not for the reason we have several times touched upon in earlier chapters—that racial minorities form low-income groups. West Virginia has only 56,000 blacks and a handful of American Indians: it comes close to being an all-white state.

So it would seem that the longer-term effects of the regional aid program have still to be assessed. Michael Bradshaw, in his official history of the Appalachian Regional Commission,[10] contents himself with asking and answering three questions:

1. Did the ARC fulfill its legislated charge?
2. Has it been able to change the human geography of Appalachia?
3. What lessons has it provided for regional development theory?

[10]M. J. Bradshaw, *The Appalachian Commission* (Lexington, Ky., 1992), pp. 122–43.

Table 16-2. West Virginia: Population, by Census, 1910–90

1910	1920	1930	1940	1950	1960	1970	1980	1990
1221	1464	1729	1902	2006	1853	1751	1950	1793

West Virginia: Coal Production, Selected Years, 1951–89 (in million tons)

1951	1954	1960	1965	1970	1975	1978	1980	1985	1989
163.3	116.0	119.0	149.2	144.1	109.3	85.3	121.6	127.3	151.2

On all three counts, Bradshaw gives the ARC cautious approval, but always with the provisos, first, that the ARC has not functioned alone—was not intended to function alone—but in concert with other bodies and programs; secondly, that while it can be credited with considerable achievements, it has succeeded least where the greatest need was, and the most program money was injected: in central Appalachia. The northern and southern sections had other programs, investments and enterprises, and the ARC has helped them to achieve development. Central Appalachia was and is the area that nobody wants, and there the ARC has fought virtually a lone fight against poverty and backwardness.

In the meantime, there are several other problems to be borne in mind, which have to do with the manner and method of handling regional aid, a policy of which the United States has, after all, limited experience.

One problem is the question: Should regional aid be spread uniformly or concentrated at particular points and, if concentrated, where? One sensible answer to this question is that it should be concentrated (this is the "growth-pole" approach) at centers where it will have the greatest multiplier effect. However, in the United States today, the fastest-growing centers are mainly metropolitan areas; indeed, B. J. L. Berry went so far as to say that "the basic regional distinction is that between self-generative metropolitan America, and the hand-me-down intermetropolitan periphery, condemned to progress characterized at best by lagged emulation and second-hand growth."[11] He found that growth rates generally rose for metropolitan areas up to about 1 million in population, and then leveled off. But metropolitan areas are precisely what Appalachia lacks. It has therefore been necessary to choose as growth-poles, or centers for regional projects, the far smaller communities with which Appalachia abounds. When this is done, the risk of failure is far higher; the chance of "self-generative" growth is small. Choosing growth-poles by sticking a pin in the map might give just as high a rate of return.

The second problem in Appalachia is how to put to use the one asset the region undeniably possessed—manpower. The Appalachian problem began with high unemployment; the trick is to try to change that weakness into a strength. To do so, it is

necessary that labor be retrained. The region has not offered the variety of labor skills other regions possess: an unemployed coal miner must be given new skills. At present, the number of workers being retrained is very small. To solve the problem, it has been estimated, between one-fifth and one-fourth of the present labor force should be enrolled in retraining programs each year. A large pool of labor, adaptable, healthy, and well-educated, would provide the region with a resource uniquely valuable in North America today: hence, the vocational training program that has come to play an important part in the commission's later planning.

The third problem is, paradoxically, a problem of plenty, not of deprivation. The revival of the coal industry may help to solve the unemployment problem, but it is liable to do so at the cost of further environmental damage. It seems as if Appalachia can either have unemployment or suffer damage, but it is bound to have one or the other, and for much of the present century it has had both.

The legislation of the 1970s, both by the federal government and by the individual states (Pennsylvania, for example, set very high standards in its Surface Mining and Reclamation Act of 1971) should help to control damage, but much will depend on the vigor of government enforcement and the cooperation of the coal companies. The last thing that Appalachia wants is another round of destructive exploitation of its resources.

TVA or Appalachian Commission: A Comparison

Both of the solutions for regional poverty we have been considering have yielded results in their time and context—the TVA in the depression years of the 1930s and the Appalachian regional plan in the 1960s, when demand for coal was low, and oil and gas were king and queen. If we try to establish which has been the more successful approach, then we can identify certain points of contrast, most of which might be felt to favor the TVA's style of solution: a direct versus an indirect approach; revenue-producing activity versus infrastructural improvement; different rates of return on capital invested. Given a very limited task, the TVA successfully expanded it to embrace many aspects of the regional economy. Given a rather broader brief, the Appalachian Commission has concentrated on a rather restricted range of projects.

[11]B.J.L. Berry, *Growth Centers in the American Urban System*, vol. 1 (Cambridge, Mass., 1973), p. 10.

On the balance of these factors, it is not surprising that, in 1961, before the 1965 Act brought the commission into being, Caudill was advocating the creation of a Southern Mountain Authority as a solution to Appalachia's most pressing problems.[12] The analogy with the TVA is clear; that was the power structure that, Caudill felt, was necessary to carry through the rehabilitation of the region. The TVA is a single, government-sponsored agency, with its own self-generated capital resources and its own income, whereas the Appalachian Commission is simply a coordinating committee that asks the appropriate government department in Washington to spend money held by it for individually chosen projects. It proceeds step by step and choice by choice, liable at any time to have its funds cut off by Congress. And by contrast with the TVA, it is specifically forbidden by its founding act to go into the electric power business. Even if it had wanted to emulate the TVA by producing power and, thus, revenue, it could not have done so; power development was not part of its remit.

These two regional organizations are one of a kind; neither is duplicated elsewhere. The TVA was created in the teeth of intense political opposition. It was challenged in the courts and fought by private interests. It produces power in competition with private power companies, and it has aroused sufficient hostility over the years to prevent the repetition of its particular formula anywhere else in the United States. There will never, it seems, be another TVA.

By contrast, the regional commission is a joint federal–state venture, and one that does not put government in competition with private business. It provides a formula that is politically acceptable—and so repeatable—in other regions. It offers a scrappy, piecemeal approach to regional development, but at least it is an approach unlikely to attract diehard opposition.

Both forms of organization can be criticized in detail. The TVA, criticized savagely in its early years, is now criticized for fundamentally different reasons. When it first came on the scene, generating hydroelectricity, it came as a new force, countering the regional power of the coal companies. Its interests and theirs were opposed to each other, and the Valley benefited from this fact. Now, however, that the TVA produces far more power in thermal generating stations than it does at its dams, this "balance of power" has changed; as the largest customer of the Appalachian coal companies, the TVA has, at least in some eyes, joined the enemy. It is this that leads Whisnant, a critic of practically everything that has happened in Appalachia since it was first settled, to comment:

The criticism TVA received during its first two decades, stemming by and large from political and ideological conservatives opposed to most forms of cooperative public enterprise, was qualitatively different from that which followed. Latter-day criticism, coming primarily from liberals and progressives, was in some sense the revolt of a child against its father, in which the child declared the father's values obsolete and his methods tyrannical.[13]

Whisnant also called the TVA "an idea whose time has gone" and he may be right: fifty years is a long time in the life of an agency created, as part of a blinding burst of government activity, by an innovating president in the depths of a depression. The regional commission approach has been rather cumbersome and largely experimental: in a sense, the program for Appalachia is only now finding an appropriate form as it winds down. The definition of the region was haphazard and the original program lacked the directness of the TVA's approach, for investment in either roads or people's welfare takes longer to mature than it does in electricity production or fertilizers.

It may, however, be that all this discussion of ways and means is secondary; that the most important problem has been, and remains, that of creating in a deprived region an atmosphere of activity and hope; a state of mind in which things get done, or what European planners call *animation régionale*. The TVA succeeded remarkably in generating this in its early years, whereas a regional commission is ill equipped to provide this kind of rousing leadership. Batteau[14] warns of the danger to the welfare of Appalachians of a "transformation of dependency" from the coal companies, which formerly dominated their lives, to the bureaucrats who may import their schemes and impose them on their unwilling beneficiaries.

The fact is that, for a century past, Appalachia has been to all intents and purposes a colony of the United States—not an overseas colony, but one in the nation's midst, with all the economic characteristics of other colonies. With the 1960s came a resolve

---

[12]Caudill, *Night Comes to the Cumberlands*, Chap. 22.

[13]Whisnant, *Modernizing the Mountaineer*, p. 49.

[14]Allen Batteau, ed., *Appalachia and America: Autonomy and Regional Independence* (Lexington, Ky., 1983).

that the colony should achieve its independence, like so many African and Asian colonies at the same period. And just as with those Third World countries, independence has brought a host of problems—the reshaping of economic relations with the former colonial power; the wish of the newly free to set their own goals; the danger of various forms of neocolonialism setting in. These problems come as no surprise: the surprise would be if there were none. In fact, the problems of formerly colonial Appalachia are greater than those of most of its African and Asian counterparts, which at least have a central government to speak for their peoples and set goals for them. Appalachia does not have one, and the Regional Commission never aspired to so lofty a role.

So the problem remains, and we leave to Whisnant its definition as:

[the] mistaken assumption that Appalachia had problems because it was not integrated into the larger economy, when in fact its problems derived primarily . . . from its integration into the national economy for a narrow set of purposes: the extraction of low-cost raw materials, power, and labor, and the provision of a profitable market for consumer goods and services.[15]

That is, as it happens, a very fair definition of what we mean by colonial status. It is where Appalachia is coming *from*. Where it is going *to* has yet to be seen.

[15]Whisnant, *Modernizing the Mountaineer,* p. 129.

17

The Southern Coasts and Texas

Texas: Dallas skyline. *Dallas Convention and Visitors Bureau*

In 1920, four centuries after the Spanish founded St. Augustine, less than 1 million people lived in Florida; it ranked thirty-second among the states in population. By 1990, the population had multiplied more than a dozen times, bringing Florida's rank to fourth. In 1920, Houston had a population of 138,000, about equal to that of Grand Rapids, Michigan. By 1990, its metropolitan area population had risen to more than 3 million, the tenth largest in the nation. Florida and Houston are representative of the Gulf Coast region. Why were population and economic growth so slow and why, once begun, have they been so spectacular? The answer to the first part of the question would seem to be that most of the area was poorly suited to the kind of plantation agriculture that developed in the inland Deep South. The answer to the second part is that the spectacular growth of the twentieth century occurred when the rest of North America discovered the unique resources of Florida and the Gulf Coast, especially oil and winter sunshine. The growth involved three major kinds of activities: (1) specialized production of subtropical crops for an increasingly large and affluent national market; (2) winter vacations and retirement living, especially for the people from the old Economic Core to the north, including peninsular Ontario; and (3) urbanization and industrialization, financed and fueled by development of the oil and natural gas resources of the Gulf.

The Physical Setting and Early Land Use

The southern coasts of the United States and the peninsula of Florida are subtropical in climate and vegetation. Brownsville, on the Texas–Mexico border, and New Orleans are at 26° and 30° N latitude, respectively, whereas the southern tip of Florida—Key West—is only one degree from the Tropic of Cancer. East of the Texas–Louisiana border, no part of these coastlands has less than 45 in (1125 mm) of rain per annum, a January mean temperature of less than 50°F (10°C), or a July mean of less than 80°F (27°C). The Florida Keys are frostless, the tip of the Mississippi delta almost so, and the frost-free season on most of the coast is more than 270 days (although the very infrequency of frost increases its economic impact when it does occur). West of the Texas border, rainfall diminishes rapidly to a coastal minimum of 24 in (600 mm) in the extreme southwest, but temperature conditions remain the same, and the Texas coastlands have a frost-free season of 300 days or more.

Under the climatic conditions of the humid subtropical coastlands east of the Texas–Louisiana border, the natural vegetation is luxuriant. To these climatic conditions, however, can be added another factor, which combines with the climate to give these coasts (and with them much of peninsular Florida and southeastern Georgia) their distinctive landscape. It is the low-lying sandy, swampy terrain on this gently sloping, lagoon-fringed coast. The combination of these circumstances creates a characteristic Gulf Coast landscape: flat, wooded plains (called appropriately, "flatwoods") interrupted by tree-filled swamps, with Spanish moss festooned on the branches of oak and cypress, and winding creeks that form a maze penetrated by no one but local fishermen and moss gatherers.

The flatwoods of the outer coastal plains east of the Mississippi delta held little attraction for nineteenth-century farmers. The generally sandy soils were infertile and subject to drought, despite high precipitation. Nor did areas of more fertile, silty soils have much appeal—they were wet and swampy. The outer coastal plain did not share in the cotton prosperity of the inland Deep South. Westward from the Mississippi delta, however, forests gave way to greenland. Here, in east Texas, Spanish settlers from

The Everglades of Florida: A scene on the Brighton Reservation of the Seminole Indians, near Lake Okeechobee. The swamp ecology of the subtropical Everglades is today threatened by extensive drainage and reclamation schemes associated with agricultural and residential development, but on the southwestern tip of the peninsula is the protected area of the Everglades National Park. *Florida News Bureau*

Mexico raised cattle in the eighteenth century. The cattle fared well and multiplied on the coastal prairies, even after being abandoned when the Spanish withdrew. Later, in fact, they became a major source of the stock that spread northward and westward over the Great Plains. But extensive grazing did not support many people, towns, or cities along the Texas coast in the nineteenth century.

Grazing of beef cattle is now important along the Texas–Louisiana coast and in Florida. In southern Texas, cattle are raised on the huge ranches that are typical of the dry West. The King Ranch at Kingsville, Texas, south of Corpus Christi, occupies about 825,000 acres (334,000 ha). It is famous not only for its size and long history, but also for its development of the Santa Gertrudis breed of beef cattle, based on the European Shorthorn and south Asian Brahman stock. These cattle, which are well suited to subtropical conditions, are important throughout the South, but especially in Florida, where they graze on improved wetland pastures interspersed with, or surrounding, the citrus groves that occupy the higher, better-drained lands. Wastes from both citrus and sugarcane processing are used for cattle feed.

Only in southern Louisiana did plantation agriculture gain a significant foothold. Rice was grown early along the lower Mississippi, but sugarcane soon replaced it as the major plantation crop. However, its area is restricted to the natural levees of the Mississippi and those of the bayous[1] west and south of Baton Rouge. The huge sugar plantations are mechanized now, but retain the use of black labor and much of the appearance of their pre-Civil War days.[2]

Specialized Production of Subtropical Crops

A second specialization, rice cultivation, was developed immediately to the west of the Louisiana sugar area, on geologically older coastlands with tight, water-holding soils. Rice had been grown earlier, using small-scale, labor-intensive methods, on the Mississippi delta. But in the late nineteenth century, Midwesterners who were experienced in large-scale, mechanized cultivation of grain brought and applied their technology to rice production. The rice is grown in huge, water-covered fields between low dikes over which combines can pass after the rice is mature and the fields have been drained. Airplanes are used to spread seed and to apply fertilizers or insecticides. Beef cattle are commonly grazed on the rice stubble. The outer coastal zone, just inland from the salt marshes, from central Louisiana to beyond Houston, is now a major rice- and cattle-producing area.

Sugar cultivation developed early, and rice cultivation late, in the nineteenth century; both have remained important. But the boom-like development of agriculture in parts of the Gulf Coast, especially Florida and the Lower Rio Grande Valley, did not occur until the 1950s, in response to the demand from northern American and Canadian cities for winter fruits and vegetables. Most spectacular has been the development of the Florida citrus industry. Orange trees were introduced by the Spanish settlers, but production was small and centered on the northern part of the peninsula until the twentieth century. Then, spurred by the demands of a rising national market, it was discovered not only that citrus could better survive occasional freezing temperatures further south, but that groves could be made to thrive on the dry sandy soils of the slightly elevated ridge stretching for 150 mi (240 km) along the central peninsula from north of Orlando to near Lake Okeechobee. This was made possible by heavy fertilization, supplementary irrigation, and sophisticated, intensive care. When the markets were in danger of being saturated as a result of extensive new plantings, new techniques were developed, first for juice canning and then for freezing of juice concentrate; pasteurized fresh orange juice is now marketed in waxed cardboard containers.

Florida normally produces three-fourths of the country's total crop of both oranges and grapefruit. Production is highly technical and generally large in scale of operation, with much vertical integration from grove to processing plant. Currently, the industry seems threatened not by overproduction and low prices, but rather by a lack of new, non-swampy land suited to citrus planting and by urban and recreational development, which is consuming thousands of acres of groves. It is well to recall that America's earlier premier orange-growing area, the Los Angeles Lowland, was virtually lost to such urbanization.

The same affluence and highly developed agribusiness that manages to put fresh grapefruit or orange juice on almost every American and Canadian breakfast table has stimulated the winter production of vegetables in the more subtropical areas. The most notable in the region we are discussing are southern Florida and the Lower Rio Grande Valley. In Florida, from Lake Okeechobee southward into the Everglades, the swampy muck, peat, and marl soils have been drained by a network of canals and ditches. Various vegetables, such as tomatoes, once considered suited only to garden-scale cultivation, are

---

[1]The French word *bayou* was applied to the sluggish streams that meander southward across the deltaic and coastal plains west of the Mississippi River. Several are former courses of the Mississippi or Red rivers—hence their natural levees are much broader than the streams now seem to warrant. This is the highly interesting Bayou or "Cajun" country populated by descendants of French-speaking Acadian people who were exiled from the Annapolis Valley, of present-day Nova Scotia, in 1755. They have retained their distinctive culture and French-based patois for more than two centuries, but the oil boom of recent decades, together with tourism, is making inroads.

See also R. C. West, "The Term 'Bayou' in the United States," *Annals,* Assoc. Amer. Geog., vol. 44 (1954), 64–74.

[2]On the geography of Louisiana, see F. B. Kniffen and S. B. Hilliard, *Louisiana: Its Land and People,* rev. ed., (Baton Rouge, 1988).

Florida: The citrus fruit industry. A view of orange groves in the Central Ridge District. *Florida News Bureau*

grown in huge fields, usually by very large-scale operators, who control production, processing, and marketing.

West of Lake Okeechobee, fields of vegetables give way to equally large fields of sugarcane. Whereas Louisiana's cane-producing area has been relatively constant, Florida's has expanded greatly since imports from Cuba were terminated. Florida's production is now twice that of Louisiana; on the Gulf Coast as a whole, production almost doubled between 1970 and 1985. The Gulf produces twice as much as Hawaii, but these two sources together barely exceed the production from sugar beet in the rest of the United States.

As might be expected, the rapid, large-scale reclamation of argicultural land by drainage south of Lake Okeechobee, together with the diking of the shallow lake in order to prevent overflow during storms, has brought new problems as well as solutions to some old ones such as flooding. At the southern tip of Florida, the vast Everglades grassland and Big Cypress Swamp, including Everglades National Park, are one of the continent's great natural preserves for subtropical plants, animals, and birds and, as such, one of its major tourist attractions. The Everglades flora and fauna depend upon a slow natural flow of water southward from Lake Okeechobee, but reclamation of farmlands has interrupted that flow and water that is in excess of farming needs is even pumped through canals into the Atlantic and Gulf of Mexico. The reduced water supply to the Everglades threatens plant and animal life, especially during periods of drought such as those in the late 1960s. The case is an example of an often-recurring dilemma: profit-oriented economic development versus wildlife-oriented preservation.

The humid Gulf Coast: Harvesting sugarcane by mechanical harvester. *U.S. Department of Agriculture*

The Lower Rio Grande Valley is the furthest east of the major irrigated districts of the West. Agricultural output is roughly balanced among three major specialties: winter vegetables, grapefruit and oranges, and field crops, expecially sorghum and cotton. Strangely, the dry, subtropical valley could be said to share one characteristic with Québec—most of the people are of Spanish-Mexican origin and speak Spanish, but much of the economy, including agriculture, is controlled by the English-speaking, Anglo minority.

One of the most unique, and possibly most futuristic, areas along the whole Gulf Coast is that included within a 50-mi (80-km) radius of Corpus Christi, Texas. In a continuous succession of huge fields of dark loam soils, each field barely discernible from the next, the flat landscape stretches to the horizon uninterrupted by tree or farmhouse. Sorghum and cotton are planted and harvested by workers and machines based not on farms, but rather in a few small, scattered hamlets. Managers of the huge operations may commute from Corpus Christi; owners of the land may live in Houston, Dallas, or elsewhere. Those who value family farming, or who appreciate a varied and interesting agricultural landscape, can only be alarmed at this possible model of the agriculture of the future.

Sunshine, Vacations, and Retirement Living

Basic to all these developments, however, is the one great natural advantage from which they all derive— the Gulf Coast's sunshine, which makes the region not only a leading producer of subtropical produce for a continent lying mainly in the temperate zone, but also a great resort area for the growing number of Americans who can afford to go south to avoid the winter.

The rise of the tourist industry has, indeed, been spectacular, particularly on the east coast of Florida, and especially around the metropolitan area of Miami, whose population of 42,000 in 1920 had risen by 1990 to 3.7 million within its CMSA. To turn these sandy beaches and coastal swamps into a string of thriving resort cities has involved a vast investment; vast, too, have been the rewards for those who participated in this astonishing boom, in which sandbars, suddenly appreciated as "palm-fringed," became valuable properties almost overnight. Further west, growth has centered on Tampa and St. Petersburg (population 2,068,000), but the whole Gulf Coast, through Pensacola and Biloxi to New Orleans, has gained from the vacation and retirement boom and so, further west, has the Lower Rio Grande Valley.

As has already been suggested, the boom in the southern resort industry must be seen as an expression of a rising standard of living, in that a growing number of Americans possess the economic freedom to move with the sun—farmers who fly their private planes south from the Wheat Belt; New York businessmen who conduct their business from Miami. The tourist industry caters for both summer and winter traffic: August and December are both peak months. In late summer, the resorts do, however, suffer from one undeniable drawback and that is the threat of hurricanes. Whereas one part of a hurricane's unpleasantness is that its habits are unpredictable, it can be said that these violent storms are most often generated east of the Antilles, move west, and strike the American coast several times a year, in late summer. In Florida south of Lake Okeechobee, there is a probabilty of severe hurricane damage one year out of every five. Southerners have learned to prepare for these emergencies, but these storms inevitably take a heavy toll of crops, orchards, and communication lines, while the high seas usually associated with their passage batter coastal settlements and endanger shipping.

Although the first and greatest attraction of Florida is its long coastline, not all its tourist attractions lie along the shore. Currently, Disney World and EPCOT near Orlando are the South's greatest draws, but in the world of nature, it is the Everglades that attract visitors—the wild maze of grass and swamp, water and wildlife in the southern end of the peninsula. But here, as we have already seen, is the paradox that bedevils all the touristic honeypots of the Western world: the more popular the natural attraction, the more difficult it becomes to maintain it. As larger and larger areas of Florida are taken over for housing and hotels or drained for agriculture, so the delicate balance of the Everglades ecosystem becomes more difficult to preserve. The water level falls; the edge of the swamp is filled in, and roads are cut through the wilderness. In the end, if the pressures of population growth and reclamation continue, the natural can only be maintained by artificial

Table 17-1 Florida: Population of Metropolitan Areas, 1970 and 1990 (in thousands)

	1970	1990
East Coast		
Daytona Beach MSA	169	371
Melbourne–Titusville–Palm Bay MSA	230	399
Fort Pierce MSA	79	251
Miami–Fort Lauderdale CMSA	1888	3193
Fort Lauderdale PMSA	620	1255
Miami PMSA	1268	1937
Palm Beach–Boca Raton–Delray Bay MSA	349	864
Jacksonville MSA	613	907
West Coast		
Fort Myers–Cape Coral MSA	105	335
Tampa–St. Petersburg-Clearwater MSA	1106	2068
Sarasota MSA	120	278
Pensacola MSA	243	344
Interior		
Orlando MSA	453	1073
Lakeland–Winter Haven MSA	229	405

means; one is reduced to pumping water into a swamp to keep it swampy.[3]

Along the coast, the urban sprawl continues (Table 17-1). From Coral Gables, at the southern end of Miami, northward through Fort Lauderdale, the built-up coastline extends for virtually 100 mi (160 km) to beyond Palm Beach, one urban area after another, with an aggregate population of more than 4 million. There is a new Megalopolis in the making.

Along the Gulf Coast, population has been increasing rapidly, particularly in Florida and Texas. In order to understand why this has been so, we must visualize two human tides breaking and meeting along this shore. One tide has been flowing from the north, and it represents the move to the Sun Belt which we referred to in Chapter 2. Americans have been retiring to Florida for decades; now they have been joined in their movement by younger people drawn south to the amenities of the Sunbelt, and the possibilities of employment that come with any in-

[3]The ecological issue was brought into prominence in the 1970s by such works as: R. F. Dasmann, *No Further Retreat: The Fight to Save Florida* (New York, 1971); W. R. McCluney, *The Environmental Destruction of Southern Florida* (Miami, 1971); and L. J. Carter, *The Florida Experience: Land and Water Policy in a Growth State* (Baltimore, 1974).

For a more recent report on the "plumbing system" of the Everglades, see M. L. Shelton, "Surface-Water Flow to Everglades National Park," *Geographical Review*, vol. 80 (1990), 355–69.

Boca Raton, Florida: The beach at Boca Raton, north of Miami, located approximately halfway along the 70–80 miles of continuous development which stretches along Florida's Atlantic coast between Miami and Palm Beach. *Florida Department of Tourism*

crease in population on this scale. In 1950, the population of the five states that share the Gulf Coast was 18.4 million. By 1990, it was nearly 41 million, which meant more than twice as many homes, cars, services, and opportunities for selling insurance (bearing in mind those hurricanes), and whole new leisure industries available to all, not just a privileged few.

The second tide has come from the south: it is nearly as numerous, although its exact dimensions are hard to discover. It consists of Spanish-speaking Mexicans, Central Americans, and Cuban refugees. And wherever two tides meet, as every mariner knows, there is likely to be rough water.

Table 17-2 offers the census statistics regarding this second tide, at least so far as the Southeast is concerned: Chapters 20 and 23 will carry forward the story. This table also reminds us that caught, in a sense, between the two tides is a third group, the blacks who have in many cases occupied the Gulf

Coast longer than the newcomers from either north or south. Their presence is a complicating factor in a city like Miami, where black and Hispanic are in direct competition for many blue-collar jobs. So far as the Hispanic newcomers are concerned, it is only right to point out that, although the present generation of immigrants may be "new," Spanish was the first European language of the Gulf Coast, Spanish was the first town (St. Augustine, Florida), and Spanish is the language of many place-names to this day. In a cultural sense Spanish-speaking peoples are recolonizing their former empire on the east and west Gulf.

Commerce and Industry

Quite apart from the growth of resort and retirement areas, there are other factors at work to bring urban and industrial development to these coastlands. One of them, as we saw in Chapter 8, is the attraction in

Table 17-2. Gulf States: Total Population, Black Population, and Population of Spanish Origin, Census of 1990 (in thousands)

	Total	Black Population	Population of Spanish Origin
Florida	12,938	1,760	1,574
Alabama	4,041	1,021	25
Mississippi	2,573	915	16
Louisiana	4,220	1,299	93
Texas	16,987	2,022	4,340

Metropolitan Areas of the Gulf States: Total Population, and Percentage Black and of Spanish Origin, Census of 1980

	Total (thousands)	Percentage Black	Percentage Spanish Origin
Houston CMSA	3101	18.2	14.5
Dallas–Fort Worth CMSA	2931	14.3	8.5
Miami CMSA	2644	14.9	23.5
Tampa–St. Petersburg MSA	1614	9.2	5.0
New Orleans MSA	1256	32.6	4.0
San Antonio MSA	1072	6.8	44.9
Mobile MSA	444	28.6	1.0
Corpus Christi MSA	326	4.0	48.5

Table 17-3. Gulf Coast Ports: Tonnage Handled, and International Tonnage, 1986

Port	Tonnage Handled (millions)	Percentage International Tonnage
Tampa	39.9	15.8
Mobile	37.6	18.1
New Orleans	149.1	55.3
Baton Rouge	77.2	24.2
Port Arthur	18.9	11.0
Beaumont	27.4	8.0
Galveston	8.0	6.3
Houston ship channel	101.6	49.8
Corpus Christi ship channel	50.1	27.4

any region or era of an expanding market, and the Gulf market is nothing if not expanding—there were over 10 million more people, in Florida and Texas alone, in 1990 than in 1970. New industry and services move into the region to cater for the needs of this growing population.

But there are more specific elements of growth, of which we can identify three:

1. *Port traffic and trade.* The long coastline of the southern United States possesses many harbors, although a number of them are artificial and are approached by way of rivers or ship canals. Through them the agricultural and forest products of the South and the southern Interior have always found their way to markets, for the most part, on the East Coast of North America or in Europe. But with the reorientation of U.S. trade, which we noted in Chapter 9, the importance of its trade links with the Caribbean and South American countries has increased, and so the role of its southern ports has become more prominent. The bulk of the foreign trade passes through New Orleans, Houston–Galveston, Corpus Christi, Tampa, and Mobile, but all the ports are involved in coastwise traffic (Table 17-3). Miami and,

to a lesser extent, Houston handle the Latin America air traffic from the U.S. South.

Dominating the central Gulf Coast, today as for the past century and a half, is the great port of *New Orleans*[4] (population 1,239,000). Thanks to its unrivaled position at the mouth of the Mississippi, New Orleans is ideally placed to participate in trade with the Southern Hemisphere, and construction of a ship canal has further increased its advantages. In addition, oceangoing vessels can penetrate up the Mississippi a further 100 mi (160 km) inland to Baton Rouge, whence the 9-ft (2.73-m) navigable channel of the Mississippi extends north to Minneapolis and (via the Illinois Waterway) to the Great Lakes, and also gives access to the Ohio channel and navigable Tennessee.

New Orleans' greatest days, admittedly, lie in the past. In the riverboat era between 1820 and the Civil War, before the westward-spreading railroads established a new, safe overland link between the Midwest and the Atlantic Coast, the Mississippi served as the great routeway for goods to and especially from the Interior. For all its length—down the Mississippi, through New Orleans, where transshipment took place, and around the Florida peninsula to the Atlantic ports—this route was, in an economic sense, a shortcut. Its importance grew as the Interior was opened up. "In the [1840s] the West had more marine tonnage than the entire Atlantic Seaboard, New Orleans alone in 1843 having twice that of New York, our greatest Atlantic port of the time." As early

[4]Peirce F. Lewis's study, "New Orleans—The Making of an Urban Landscape," in Adams, *Contemporary Metropolitan America*, vol. 2, 97–216, is so good that every reader should obtain a copy.

New Orleans in the riverboat days: An etching of the great southern port at the height of its commercial career in 1851. The modern city has expanded to Lake Pontchartrain (background) and has spread onto the southern bank of the river, but it has retained its geographic form and commercial preeminence. The Old French Quarter (right) retains many eighteenth- and nineteenth-century buildings and is the center of tourist interest. *New York Public Library*

as 1843 steamship tonnage on the Mississippi "was nearly half that of the whole British Empire, and it multiplied sixfold in sixteen years."[5]

The year 1859–60 was long remembered in New Orleans as "the best year on the river." But also in the 1850s, the railroads were already bringing the steamboat era to an end—even before the Civil War closed the Mississippi, and the Union blockade left the cotton bales lying and the grass growing on the quays of New Orleans. The revival of the port's commerce waited for the development of trade with Latin America and on the improvement of inland navigation. Today, its trade depends on a more harmonious balance between river and ocean traffic and between water, road, and rail transport than during the picturesque but hazardous Mark Twain phase of Mississippi navigation. Meanwhile, out of its colorful past—its background of French culture and its riverboat days—New Orleans has built up a carefully preserved reputation for Old Worldliness, which, in a continent where one city is much like another, is an asset worth millions of dollars annually in tourist traffic.

In the days of the riverboat, the produce of the western Cotton Belt was funneled through New Orleans to the outside world, and the capture and closure of the port was thus a primary objective of the Union forces. But since then, cotton production has spread west and has been supplemented, as we have seen, by other forms of farming made possible by irrigation. As a result of this westward spread of cultivation, and the growth of the oil industry, the western Gulf ports have been increasing their traffic, and it is in view of these developments that both New Orleans and the Texas ports have sponsored rival projects to improve water transport in the Southwest and

[5] J. T. Adams, *The Epic of America* (Boston, 1931), pp. 220–21. Mark Twain's comment on the subject was very much in character: "Mississippi steamboating was born about 1812; at the end of thirty years it had grown to mighty proportions; and in less than thirty more it was dead. A strangely short life for so majestic a creature."

The Intracoastal Waterway, west of Morgan City, Louisiana. *U.S. Corps of Army Engineers, New Orleans District*

thus draw off traffic, in the one case to the Mississippi, and in the other to the western Gulf Coast.

Although the ports of the Texas coast possess undoubted advantages of position, half a century ago they were hardly open to navigation. The alternation of sandbars and shallow lagoons along this coast has meant that almost all the ports west of New Orleans have been created only by costly dredging and cutting. Indeed, the two ports that handle the largest tonnage—Houston and Beaumont—both lie on the inland side of the lagoon fringe, 50 mi (80 km) from the sea, with which they are connected by deep-water canals. The ports, in turn, are interconnected by the shallower Intracoastal Waterway, which, throughout much of its length, makes use of the lagoons that are impassable to deep-water vessels. Traffic on the Waterway has built up rapidly: in most years, at least 100 million tons of freight are moved

along the section of the Waterway between Appalachee in Florida and the Mexican border (Fig. 17-1).

It seems probable that, in view of the growth of southwestern markets and the excellence of these ports' facilities, they will in time develop more extensive foreign connections. Around their docks and dredged channels are to be found the new industrial concentrations of the Southwest—petroleum refineries and chemical plants; smelters for imported ores, especially bauxite; and processing plants for deriving magnesium from sea water.

2. *Development and processing of resources.* From the beginning it was to be expected that the traffic through the southern ports would generate some processing industries. There are, first, those based on forest products: 15.3 million acres (6.19 million ha) of commercial forestland in Florida and 14.5 million (5.87 million ha) in Louisiana, an increasing propor-

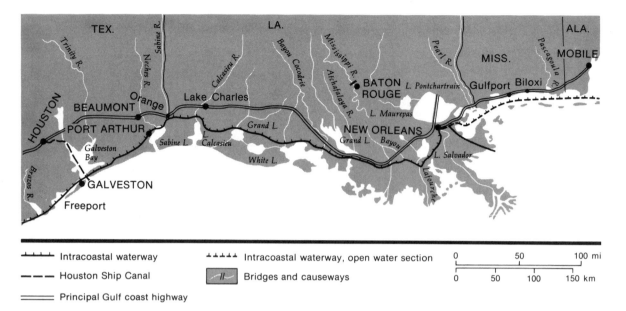

Fig. 17-1. The Gulf Coast: Major ports and the Intracoastal Waterway.

tion in plantations. But whereas the earliest forest products of the Gulf Coast were turpentine and ships' timbers, both production and consumption are today dominated by the big paper companies, whose interest is in pulpwood. Slash pine is most commonly planted, since it grows very rapidly to yield a quick crop of timber suitable for pulping.

The larger group of these resource-based industries, however, is associated with the Gulf's minerals. The discovery of oil at Spindletop, near Beaumont, Texas, in 1901 presaged the opening of North America's greatest oil and gas fields. Today, Texas and Louisiana rank with Alaska and California as the four major oil-producing states, while Texas and Louisiana account for 65 percent of the nation's natural gas output. The fields form an almost continuous coastal belt from the west bank of the Mississippi to beyond the Mexican border. Inland, they stretch in to Kansas and Oklahoma; seaward, they extend beneath the Gulf, and their working has led to not only technical problems on a formidable scale, but also a first-class political issue—the question of whether the state or the U.S. government should control tidelands oil.

As with oil, so with sulfur; the Gulf coastline has been merely an incidental barrier to mineral prospecting. The coastline sulfur deposits, already the largest known in the world, are being supplemented by further finds offshore from western Louisiana and eastern Texas, and this region supplies virtually the whole of North America's sulfur.

Associated with the same subterranean domes that contain the sulfur and petroleum deposits is a third mineral, rock salt, and this, too, is mined in quantity in Louisiana and Texas. The fourth major mineral resource of these southern coastlands is phosphate, a resource whose importance has been steadily increasing with the spread of the fertilizer habit among American farmers. Florida is the continent's leading producer.

Oil fields have seldom become the location of large industrial or urban centers, for their life span is too uncertain to attract the more costly forms of settlement. Where, however, some other factor creates a particularly favorable location within the vicinity of the fields, it is probable that the combination of circumstances will stimulate the growth of industries connected with the oil fields and create centers of importance. This is the background to the growth of the western Gulf Coast ports. They are the logical location for the refineries and chemical industries that accompany oil production. At the same time, they would never have achieved such importance as ports had it not been for their proximity to the oil fields. It is this combination of oil field and coastline (that is, of producing point and shipping point) that has brought into being the "Golden Crescent" (the Gulf Coast between Brownsville, Texas, and Pensacola, Florida), where 75 percent of the nation's petroleum-processing industry is located. Water transport extends the Crescent inland to Houston, while the lower Mississippi, between New Orleans and Baton Rouge, flows past a line of refineries and petrochemical plants from which imports and products can move north by river.

3. *Defense and space-related industries.* The southern coastlands of the United States have profited from another source of investment, and one independent of any local resource, such as forest or minerals—government spending on defense installations and missile and space research. With Cuba only 90 mi (144 km) from Florida, it is certainly to be expected that there should be a concentration of defense bases in the South; in any case, some bases, like those at San Antonio and Pensacola, have been there for a long time.

But it was, for the Gulf Coast states, a pure bonus that the federal government decided to establish NASA's headquarters in Houston and to use Cape Kennedy in Florida as the main launching site for rockets and spacecraft. Around these installations are the housing and the plants needed to make them operational. The government's space program thus brought an additional source of revenue to the states concerned and, in the case of Cape Kennedy, to an area that previously had almost no economic value at all.

Texas

The concluding section of this chapter must take account of one problem that confronts every writer who attempts to treat the geography of North America on a regional basis: what to do about Texas. If he or she adheres rigidly to regional subdivisions, then this vast state, which contains one-twelfth of the land area of the continental United States, must be divided into at least three major regions—the Great Plains, the Southeast, and the Gulf Coast. But while the writer's geographical conscience will have been satisfied in dissecting it, one of the most famous—and self-conscious—realities of American society, the Texas of the Texans, will have been destroyed. That its inhabitants are Texans first, and Americans second, has become a commonplace in the nation's reckoning and the nation's humor. That the state's individualism has been heightened by national rec-

ognition of it is equally commonplace. Expected to be separatists, Texans set out with gusto to live up to expectations.

Beneath this phenomenon, however, there lies a certain basis of historical fact and economic reality. Texas is the only part of the United States that was an independent republic before it joined the American Union. It was in 1835 that the 20,000 or so settlers who had entered Texas from the eastern United States revolted against the province's Mexican government; in the next year a republic was established and recognized. Only after this young republic had accepted annexation by a plebiscite in 1845 did Texas become part of the United States. The conditions of its entry into the Union were such that it retained somewhat more of its state sovereignty (for example, ownership of its public lands) than did those other western parts of the nation that began as sparsely populated territories.

The Spanish and Mexican origins of Texan land disposal also gave it a character of its own. Some of its Spanish land subdivisions resemble those of the French in Canada and Louisiana (see Fig. 5-2) rather than the American gridiron. In any case, the Mexican government was far more generous to (and realistic about) the earliest arrivals than was the U.S. government further north; the standard grant to a head of family was one league plus one *labor*—a total of 4605 acres (1864.4 ha)—while land remained so plentiful that right up to 1898 a Texan who had none could obtain 160 acres (64.8 ha) for the asking.[6]

Here, then, is the historical background to Texan self-consciousness. As a contributing factor, on the geographical side, we may note that the hugeness of the state (it is roughly 800 mi [1280 km] from north to south and from east to west, and there are 7–8 Texan acres [around 3 ha] for each inhabitant) give its people sense of both space and self-sufficiency. Perhaps most important of all, the great southwestern oil boom of the last eighty years created wealth and made available capital with an ease and speed relative to effort expended that can seldom have been equaled in history. In these circumstances, a belief that the dry soil of Texas has magical properties was perhaps understandable. Nor should the word *magical* lead us to overlook, as E. Cotton Mather has

---

[6]On the historical geography of Texas there is D. W. Meinig's fine *Imperial Texas* (Austin, 1969); T. L. Miller's *The Public Lands of Texas, 1519–1970* (Norman Okla., 1972); and the numerous papers in the geographical periodicals by T. G. Jordan.

Galveston, Texas: The legacy of the past. Before the advent of rail-roads and the oil era, Galveston was the chief port of Texas, a city of great wealth, beautiful homes, and many consular offices. Now, however, it has long been overshadowed by the growth of Houston, 40–50 miles inland. *Richard Reynolds / Texas Department of Commerce*

pointed out, the contribution of Texans to the solid business of settling the West:

It was Texans who initiated most of the great cattle drives a century ago to the northern railheads, it was Texans who pioneered the introduction of Brahma cattle in the United States, and it was Texans who were responsible for developing the first officially recognized American breed of beef cattle, the Santa Gertrudis. Texans opened most of the oil fields on the southern, central, and northern Great Plains. They initiated suitcase farming, they own and operate most of the migratory custom combines [which harvest Great Plains crops], they instituted the planting of winter wheat in the "spring wheat belt."[7]

Much of the wealth created by the oil fields has found expression, as we have already seen, in the growth of the metropolitan areas of the state. Of these, only San Antonio has a long history: it was the

[7]E. Cotton Mather, "The American Great Plains," *Annals,* Assoc. Amer. Geog., vol. 62 (1972), 237–57. Quotation from p. 257.

advance base for the Spanish thrust northeastward from Mexico. The "big three"—Houston, Dallas, and Fort Worth—all date from rather casual nine-teenth-century beginnings: Houston in 1836, Dallas in 1841, and Fort Worth, after a shaky earlier start, in the aftermath of the Civil War. The coming and crossing of cattle trails, roads, and railroads account for the sites of Dallas and Fort Worth. *Houston* was an anchorage on the Buffalo Bayou, well inland from the much earlier coastal settlement of Galveston. But on this hurricane-threatened coast, an inland loca-tion is a safety measure of a kind and, once the chan-nel across Galveston Bay and up the Bayou was dredged (in 1914), Houston became a well-placed shipping point for produce from the Texan interior. It was even, briefly, the state capital. But the city found its larger future in the oil industry, which for eastern Texas blew in at Spindletop, 80 mi (128 km) away in 1901.

Today, Houston's is the largest metropolitan area

Skylines of Texas: Pennzoil Place and The Republic Bank Center, Houston. *Copyright © Chas McGrath, courtesy John Burgee Architects with Philip Johnson*

in Texas. This is true not only in terms of its population, which rose from 806,000 in 1950 to 3.7 million within the CMSA (which includes Galveston and Brazoria), but also of its area. Houston, as we saw in Chapter 5, is the one major city in America that will have nothing to do with centralized zoning: its controls on land occupance are minimal, and its powers of incorporation of surrounding territory or jurisdictions are great. Consequently, it has spread hugely over the coastal plains. Houston claims that its policy offers better residential opportunities to low-income homeseekers than does that of other cities. If its population makeup is used as evidence (Table 17-2), there may well be some truth in this.

We have already noted Houston's leading position among the nation's ports; its industrial development

has been no less spectacular. Within some 50 mi (80 km) of the city is produced almost one-tenth of the nation's crude oil; it has more than a dozen refineries, as well as numerous plants producing chemicals and synthetic rubber; and it is the main steel-milling center of the Southwest. The availability of cheap oil and gas has attracted to the area numerous other industries, among which the manufacture of mining equipment is the largest employer. As port and as industrial center, there seems to be no immediate reason why the meteoric rise of Houston should not continue. And meteoric may well be the right adjective to use of a city that also possesses, halfway between Houston and Galveston, the Lyndon B. Johnson Space Center as the city's entryway to the universe.

The next three ranking metropolitan areas of Texas, Dallas–Fort Worth, San Antonio, and Austin, lie within the premier agricultural area of the Blackland Prairie and form a segment of the greater north–south line of hinge cities in the Great Plains, which connect the eastern and western parts of the continent.

Dallas and *Fort Worth* together form a CMSA of 3.88 million inhabitants. The disparity in size between them (Dallas is almost twice as large as Fort Worth) does nothing to mitigate the intensity of the famous rivalry between these two cities, whose centers are 33 mi (53 km) apart; proximity is a stimulant to the contest. From the geographical point of view, however, it is not their competitive similarities that are of real interest, but their differences. For these two cities, standing as they do close to the frontier between humid East and arid West (long-term records give Dallas a mean annual precipitation of 33.6 in [840 mm] and Fort Worth one of 31.6 in [790 mm]) divide between them those relationships with both regions that other hinge cities, Kansas City, for example, combine within a single metropolitan area.

Dallas belongs primarily to the agricultural humid East; it is a great cotton market, and looks especially northeast to the economic core. Fort Worth is a cattletown grown prosperous and industrialized; it looks westward to the ranges, and its stockyards are the largest anywhere south of Kansas City. But in reality, they form one urban concentration whose functions, like those of other Hinge cities in the western Interior, link it with both East and West; they share the quality of being Texan. For example, they share, in the rapidly growing suburban area between them, establishments that not only serve the whole region, but also represent the whole state, indeed, the whole Southwest. There is "the world's biggest airport," midway between the two cities, which has helped to make them the transport hub of the Southwest. And there are a number of federal agencies, like the Federal Reserve Bank, which have been established in Dallas as regional headquarters. This may well be because, to outside eyes at least, Dallas is cautious and conservative by contrast with the brasher boomtown of Houston. But those terms must all be interpreted by a strictly Texan scale of values.

Austin, further south, not only serves the flat Blackland Prairie agricultural areas to the east, and ranching and recreation in the hill country to the west, but is also growing rapidly as the seat of state government and the main center of the state university.

The climate of *San Antonio* is affected by its location between the humid area to the north and east and the arid lands to the south and west; geographically, it sits at the southern tip of the Blackland Prairie, the inner edge of the agricultural coastal plain, and the foot of the Balcones Escarpment, which ascends to the ranching country of the Edwards Plateau and Great Plains. Culturally, it links the Anglo East to the Hispanic Southwest.

San Antonio was the military base and capital of the Spanish and, later, the Mexican province of Texas, and all its early connections were with Mexico City. From San Antonio, there stretched the chain of mission stations that Spain thrust out northeastward to the Red River, to head off the French (see Chapter 20). It retains its military role to the present day; its air force and army bases employ thousands of military personnel and civilians. Throughout Texas, in fact, the armed forces represent one of the major employers, and a number of small towns owe their livelihood to huge bases like Fort Hood and Fort Bliss.

San Antonio's long history, varied cultural background, and proud possession of the Alamo make it an attractive convention and retirement center, a function encouraged by local enterprise; for example, in preserving Spanish missions or converting the course of the San Antonio River into the famous Paseo del Rio, or River Walk.

The story of Texas in the oil century has been, on the whole, one of dramatic success, a success reflected in its glittering cities. The exploitation of the state's oil and gas resources has made available a huge income for investment: it has made Texas one of the prime areas, not merely in the United States, but also in the world, for accumulating capital.

We must bear in mind, however, that there is another Texas—one of drought-ridden dirt farmers and jobless Hispanic Americans. We need to be reminded, too—as Texans themselves were in the 1980s—that oil sells on a world market, and in that market prices fluctuate. Oil and Texas have, on the whole, gone up, but sometimes oil goes down, too, and then Texas, like any other oil producer, is in trouble. We may smile indulgently at the way in which Texans think of themselves and their state as exceptional in every way, but with oil there are no exceptions to the law of supply and demand.

To return, in closing, to our theme of core and periphery: the whole of the South and Southwest has been, historically, a part of the periphery, dependent on the core, but it is evident that these categories of core and periphery must now be reconsidered. In two areas of North America, which geographically are entirely peripheral—the western Gulf Coast and California—core functions are being carried out and core characteristics are appearing. The center of gravity of the nation's population is moving south and west. Are the control functions of the economy following?

18
The Great Plains and Prairies

Agriculture on the Plains: The landscape of wheat farming, here seen at Wilcox, Saskatchewan. *National Film Board of Canada*

It is one of the splendid paradoxes of North America that the region of the continent possessing the least physical distinction has been the scene of some of the most dramatic episodes in the story of its human occupance. On the relatively flat surface of the Great Plains, which stretch from southern Texas north to the Arctic Ocean, have occurred the continent's sharpest clashes of group interest. And although these deceptively innocent-looking grasslands have attracted successive generations of settlers, nature has provided, both above and on their surface, hazards that time and again have forced the settlers to retreat before an environment they have failed to tame.[1]

For, in terms of human geography, the Great Plains have been a problem region—an area which forced upon settlers, even quite recent settlers, a respect for natural conditions, which even today cannot be treated carelessly. As the pioneers left the humid, fertile Midwest and moved westward across the central lowlands of North America, they were moving into an area where nature's control of human activities—by way of climate and soil—could not be mitigated. After a few decades of occupance, the region's growing problems were evident to the most casual visitor; they could be seen in duststorms, in eroded fields, and in abandoned farms. They served notice that, if the plains were to be settled, it would be necessary to accept the limitations of this fierce, but fascinating, environment and to develop a system of land use that was compatible with it.

This problem region does not exactly correspond in extent with the physical province of the Great Plains, as we defined it in Chapter 1, for the eastern edge of that province is an area of settled and generally prosperous farming that comprises the western parts of the Wheat, Corn, and Cotton belts. The regional problem is one of being at the mercy of the climate, for as we move west across the plains, the threat of a disastrous storm, slight in the Midwest, grows steadily greater and becomes serious after we pass the 98th or 100th meridian of longitude. Thus defined, the problem region comprises much of western Texas, the western parts of Oklahoma, Kansas, Nebraska, and the Dakotas, and the southwestern part of the Prairie Provinces. On the west, it terminates at the foothills of the Rockies, where the increasingly broken terrain removes the temptation to engage in unwise agricultural activity. On the east and north, the boundary between the region and the rest of the Great Plains fluctuates year by year; it is a boundary not of relief, but rather of risk; it divides relatively secure farming of the Agricultural Interior from the relatively hazardous farming of the dry plains.

Physiographically, then, the Great Plains comprise a landform region, extensive in area and relatively homogeneous in character. But by almost all other

---

[1] The Great Plains environment has attracted the attention of many writers. Although it is over half a century old, Walter Prescott Webb's *The Great Plains* (Boston, 1931) is unlikely ever to be displaced from its position at the head of the list, any more than those other fifty-year-olds, volumes 1 and 2 of the great *Canadian Frontiers of Settlement*, eds. W. A. Mackintosh and W. L. G. Joerg (Toronto, 1934 and 1938), will cease to be indispensable for the Canadian section of the region. For the latter, however, there are such recent studies as G. Friesen, *The Canadian Prairies: A History* (Toronto, 1984); and R. Rees, *New and Naked Land: Making the Prairies Home* (Saskatoon, 1988). For the more recent past, see B. W. Blouet and F. C. Luebke, eds., *The Great Plains: Environment and Culture* (Lincoln, Nebr., 1979); and M. P. Lawson and M. E. Baker, eds., *The Great Plains, Perspectives and Prospects* (Lincoln, Nebr., 1981) for the U.S. section; R. Allen, ed., *Man and Nature on the Prairies* (Regina, 1976); and J. R. Rogge, ed., *The Prairies and Plains: Prospects for the 80s* (Winnipeg, 1981) for the Canadian. There are good state and provincial atlases covering the region, and the presses of the region's universities deserve a special tribute for their enterprise in keeping up a flow of publications useful to the geographer.

The Great Plains environment: The margin of the Plains. The uplift of the Rocky Mountains contorted the otherwise largely horizontal strata underlying the Plains and created a series of west-facing scarps representing the tilted edges of these Plains strata. This pic-ture was taken southwest of Denver. Farther north, a similar valley between the scarps was dammed to provide storage for the Colorado River water brought by tunnel through the Front Range of the Rockies (see Fig. 18.5). *John S. Shelton*

criteria, it is not a region of homogeneity but, instead, a region of transition and differentiation: from a humid to a semiarid climate; from prosperous farming based on annual rainfed cropping to precarious agriculture based on irrigation, alternate-year cropping, or natural pasture grazing; from a pattern of relatively dense and continuous settlement to one that is sparse and spotty. Finally, it is an area in which, during the first century of white settlement, to gain a livelihood from the land was a difficult and precarious task.

The Physical Circumstances

The factors that govern the character of life on the Great Plains are primarily climatic. Since the plains lie in a continental interior, intermediate between the humid East and the arid West, their climatic regime reflects this transition.

PRECIPITATION

With the high wall of the Rockies blocking their western margin, the Great Plains depend for much of

Badlands of the Great Plains: An 1899 photograph by the U.S. Geological Survey of the Badlands at the head of Battle Draw, Washington County, South Dakota. *U.S. Geological Survey*

their precipitation on the northward intrusion of moist air from the Gulf of Mexico. There is thus a general decrease in the amount of precipitation from southeast to northwest across the plains; Abilene, Texas (32° N, 99° W), averages 25 in (625 mm) per annum and Oklahoma City (35° N, 97° W) 32 in (800 mm), whereas Miles City, Montana (47° N, 106° W), has 13 in (325 mm) and Medicine Hat, Alberta (50° N, 111° W), has slightly less than 13 in (Fig. 18-1). In the winter season, both rainfall and snowfall are light—a fact to which we must shortly return. Some 70 to 80 percent of the precipitation occurs between May and September, however, when the farmer and rancher need it most.

The rain-bearing Gulf air, upon whose intrusions the Great Plains depend, is not altogether reliable. The general trend of movement of this tropical maritime air is northeast rather than northwest, with the result that the amount of rain reaching the plains, and, in particular, the amount of the vital summer rains, varies greatly from year to year, the mean annual variation being as much as 25 percent of the an-

nual precipitation. This variation is greatest in the southern plains, where the rainfall is somewhat more plentiful, but where temperatures, and therefore evaporation, are highest. The dry northern plains have less rain, but their supply is slightly more reliable, for they not only lie closer to the source areas of the polar continental air, whose interaction with the humid Gulf air forces the latter to deposit its moisture, but also air from the Pacific passes over the region, and some precipitation does occur.

TEMPERATURE

The Great Plains experience the extremes of temperature characteristic of a continental interior. The northern plains on both sides of the international boundary have recorded the lowest winter temperatures in any populated part of the continent—between 50 and 60°F below zero (−45 to −51°C). On the other hand, summer temperatures in all parts of the Great Plains soar to maxima of over 100°F (38°C). In general terms, winter temperatures tend to vary with latitude. The southern edge of the plains

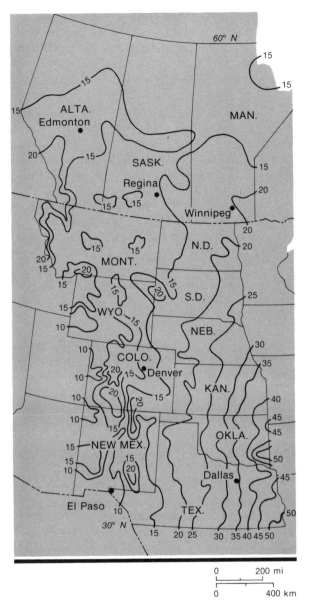

Fig. 18-1. The Great Plains and the Prairies: Annual precipitation (inches).

tain barrier and, being further warmed by compression as it descends the eastern slope of the mountains, brings a sudden spectacular increase of temperature to the foothills and western plains—a welcome if brief break in the intense winter cold of the area.

CLIMATIC HAZARDS

But the statistics we have so far considered tell only a part of the story of the farmer's struggle in the Great Plains. It is not the basic climatic conditions that are the menace of the region, but rather the climatic hazards that accompany them. These are of four kinds:

1. *Frost.* Since temperatures are affected by two or three different air masses, which influence different areas disproportionately, the length of the frost-free season varies greatly from year to year. On the average, its length is about 100 days in the southern Prairie Provinces and 240 days in central Texas, but over a forty-year period it has varied from 129 to 181 days at stations on the Nebraska–Colorado border, and from 89 to 172 days in western North Dakota, near the international boundary (Fig. 18-2).

2. *Hail.* The central and southern Great Plains are the areas of the continent most subject to hailstorms, and, infrequent though they may be, a single storm is sufficient to ruin crops over a wide area.

3. *Winds.* The winds sweep this flat, treeless area unchecked. There are several types of wind peculiar to the plains, and all of them are deadly in their effects. On the southern plains, the summer danger is provided by tornadoes, and by hot winds that blow from the interior and parch the crops, whereas in winter, this same area suffers from the visitations of the "Norther," a cold wind causing sudden drops in temperature and so representing a serious frost hazard.

But the "grizzly of the Plains," in W. P. Webb's phrase, is the blizzard. In winter the principal storm tracks cross the continent from west to east, approximately in the latitude of the international boundary. But occasionally, a surge of particularly cold air out of the north collides with warmer, moister air to the south—resulting in high winds and heavy drifting snow, followed by plummeting temperatures. The Great Plains, as we have already seen, generally have little snowfall, and ranchers usually leave their stock to winter outdoors. But the drifts prevent access to food supplies, so that the animals become victims of the blizzards. Serious economic loss often follows the passage of these violent winter storms.

has a January mean of 50°F (10°C), while at the Canadian boundary the figure is 0 to 5°F (−18 to −15°C). Summer temperatures, on the other hand, are governed partly by latitude and partly by altitude, the higher, western edge being, in general, cooler than the eastern edge, which is some 3000 ft (910 m) lower. However, this general pattern is disturbed by the effects of the Chinook. Dry air from the Pacific Coast, which is warmer than the prevailing winter air over the plains, crosses the Rocky Moun-

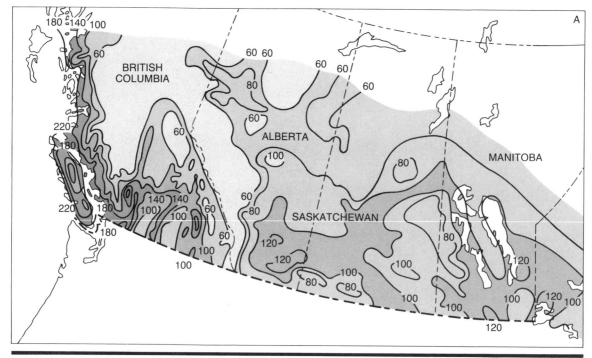

FROST FREE PERIOD
Average Length in Days

- 60
- 100
- 140
- 180

Fig. 18-2. Western Canada: (**A**) Frost-free season and (**B**) Growing season. The figures show the number of days in each season. For eastern Canada, see Figure 13.1. As a commentary on the figures shown on this map, most commercial crops grown in North America are in areas where the frost-free season is, as a minimum: for small grains, 90–100 days; for grain sorghum, 130–140 days; for corn cut ripe, 140+ days; for cotton, 180–200 days. Both maps are taken from the *National Atlas of Canada*, 4th ed. (Ottawa, 1974). © Her Majesty the Queen in Right of Canada with permission of Energy, Mines and Resources Canada.

Quite apart from the effects of these particular winds upon temperature and humidity, however, there is the ever-present wind erosion of the Great Plains. With little in the way of an obstacle above the surface, and with poorly consolidated parent material below it, the soil of the Great Plains falls an easy victim, once bared to the wind's action. In few other regions of the world does unwise cultivation receive such prompt and embarrassing publicity as when duststorms darken the skies above the Great Plains.

4. *Drought.* We have already seen that the principal source of precipitation for the plains is the unpredictable intrusion of moist air northwestward from the Gulf of Mexico and that this produces wide variations in rainfall from year to year. But what compli-

cates the problems of the settler in this area is that the years of subnormal rainfall tend to occur in groups, so that it is by no means safe to assume that a year of meager rainfall will be followed by one of higher than average rainfall. On the contrary, the records of the past half-century show that rainfall may be above or below average for a decade at a time. In Montana, for example, during the years 1906–16, the rainfall was above average every year (and was 125 percent or more of normal in six of them). Then between 1928 and 1937, it was below the fifty-year average for eight of the ten years, and for the three years 1934–36, it never exceeded 75 percent of normal.

Such a rainfall regime as this clearly increases the

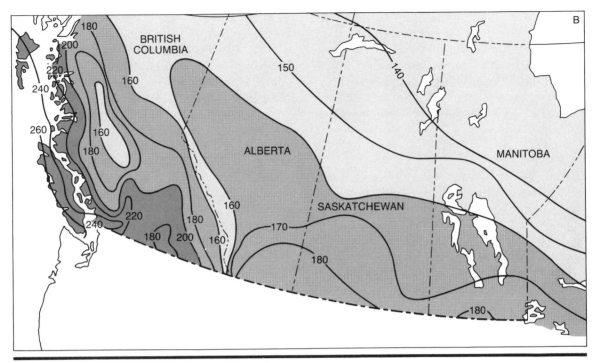

GROWING SEASON
Average Length in Days

160
200

Growing Season: Number of days with an
average temperature over 42°F (5.6°C).

difficulties of the farmer: the wetter-than-normal decade gives a false impression of humidity and fertility, and so lures settlers to the margins of possible farming. Then, the drier decade that follows robs them of any possibility of tiding over from one season to the next. In other words, on the Great Plains the average rainfall figure is doubly deceptive: it conceals year-to-year fluctuations, which resolute farmers can survive, but it also conceals decade-to-decade fluctuations, which they cannot.

In very general terms, the decade of the 1930s on the Great Plains was a bad one. Even without the general economic depression, which was causing acute distress throughout the nation, the drought years of 1934 and 1936 (when most of the plains had less than 75 percent of average rainfall) would have been serious; the coincidence of the events was catastrophic. The 1940s were years when rainfall was somewhat above normal, and farm yields on the Great Plains were high—happily for war-torn nations relying on North America for food. The 1950s, however, proved to be another decade of near-drought; on the southern plains, farmers and ranchers suffered a succession of drought years, and many gave up hope and abandoned their holdings.

But the 1960s again reversed the situation: they were a decade of generally average-to-better-than-average rainfall, and at the end of the decade, the plains wore an air of prosperity that they had all too seldom displayed in the century or so since the first settlers arrived. The sequence, however, continued: the 1970s saw one more swing of the pendulum. The mid-seventies saw the drought areas spreading eastward, and extending into the Corn Belt. The 1980s have brought both moisture and searing drought. But today's farmers, as we shall see, are much better prepared than their predecessors to survive the inevitable drought years.

SOIL AND VEGETATION

The general decline in the amount of precipitation from east to west across the Great Plains produces a fairly regular progression in the types of soil and vegetation found there. These soils fall within the general order of Mollisols and grade from Udolls in the eastern plains to Ustolls in the west and Borolls in the north: in the older terminology, from black earths in the more humid east, through dark brown chernozems to brown steppe soils on the dry western margins of the plains. As a result of glacial deposition over the northern part of the region, these soils are developed from a wide variety of materials, almost all recent, but most of them support a grassland vegetation whose character likewise varies with the amount of rainfall. The black soils support—or rather, supported, for most of them are now under cultivation—a tall-grass, sod-forming prairie vegetation, whereas in the semiarid west, the vegetation consists of shorter grasses, which form a mat cover in the moister parts and a cover of scattered bunch grasses in the driest ones.

The soil feature that reflects this progression is a layer of calcium carbonate and other salts accumulated in the subsoil, beneath which is a permanently dry zone. The depth of this subsurface layer varies with the amount of rainfall and with the length of the "wet" season. In the black earth belt, it lies between 30 and 40 in (75 and 100 cm) below the surface, and long-rooted grasses and cereals flourish. Westward, the salt layer rises nearer the soil surface as aridity increases, until at the western margin of the plains it is only 8–12 in (20–30 cm) down, and root development is only possible for short grasses. The point at which the depth of the layer reaches 25–30 in (60–75 cm) beneath the surface is an important dividing line: it is the line of change from black earth to brown, from tall, prairie grass to short, plains grass. In terms of human occupance, it is the line that divides the eastern Great Plains from the problem region to the west. Lying as it does between the 98th and 100th meridians of west longitude, it accounts

for the prominence of those meridians in every discussion of the settlement of the Great Plains.

But if this is the general pattern established by nature, it has been modified by humans. For reasons we will consider shortly, the "natural" vegetation of much of the Great Plains, especially in their southwestern section, is quite different today from that of a century ago. The character of the range vegetation has changed with use. In particular, changes have occurred in the balance between grasses and shrubs, and a tough, woody shrub vegetation of sagebrush or mesquite has spread over millions of acres of former grasslands, to change the appearance of the southwestern plains and to increase soil erosion.

Because the character of the plains vegetation is an index of the amount, reliability, and seasonal duration of rainfall, it serves also as a guide to the areas that may safely be cultivated and to those better left under the native grass cover. Most of the tall-grass prairie has been under cultivation for several decades without serious mishap. The area covered by shorter, grama–wheatgrass associations, on the other hand, represents an agricultural margin where cultivation is risky, whereas the plains areas covered by a grama–buffalo grass combination, or even more, by a mesquite–sage association, carry in their vegetation a warning cultivators ignore at their peril.

Land Use in the Great Plains

In the years between the end of the Revolutionary War (1783) and the 1850s, the frontier of settlement in the United States spread rapidly westward across the Mississippi Valley to the eastern edge of the Great Plains. From there it jumped, in the 1840s, across 1500 mi (2400 km) of intervening plain, mountain, and desert to the Pacific Coast, and not for almost half a century was progress made in filling the gap with permanent settlements. In Canada, the sequence of events was comparable. Beyond the wilderness barrier of the Canadian Shield small, farm-based colonies were established at the eastern edge of the Great Plains (in what is now the Winnipeg area) in the decade 1810–20. But for more than fifty years, these struggling colonies knew little expansion; the population of Manitoba at the census of 1871 was 25,000, and the West continued to be of interest as a source not of agricultural products, but rather of furs and, later, gold.

The reasons for this abrupt halt to the westward spread of settlement have become familiar to a later

The settlement of the Western Interior: *The Stone Boat* by Harvey Dunn. In a long series of paintings Dunn vividly recorded the lives and hardships of early settlers in the Dakotas. The "stone boat" was a sledge drawn by a team of oxen. Such stones were used for building on the treeless plains. *South Dakota State University*

generation as a classic example of the effect of environment upon human activities. For this was an area with an annual rainfall of 20–30 in (500–750 mm) and a flat surface, which, in theory, could have been cultivated, but which, in terms of the techniques available in the 1850s, might as well have been a desert. Indeed, so hopeless did the task of settling it seem to early travelers that the firm conviction took root that it *was* a desert, and it was treated as such, in the 1840s and 1850s, by American geography teachers and policymakers alike. As early as 1843, some scores of pioneer farmers had decided that prospects for agriculture in Oregon were better than those in the Great Plains and had made the hazardous transcontinental journey by the Oregon Trail; their reports, backed by the discovery of gold in California in 1848, encouraged others to follow. In the south, the Spanish-Mexican population had been equally unsuccessful in establishing permanent settlements on the plains. If the area possessed any virtue at all, in mid–nineteenth-century eyes, it was that at least it

was possible to cross it swiftly on the way to pleasanter places.

The environmental problems were threefold:

1. *The area was treeless.* It was not merely that to eastern minds land that would not grow trees was poor land, but also that in the 1840s there was in the West neither coal nor cheap iron and consequently no substitute for wood as a domestic necessity. There was no means of fencing land and no means of building homes; the earliest dwellings were sod houses, which were merely pits roofed with turf and a few precious timbers. The only substitute for wood as a fuel was buffalo dung.

2. *The sheer dryness of the plains found eastern farmers ill prepared,* their crops unsuited to the short wet season, and their farming techniques useful only in a more humid climate.

3. *The limited supply of available water raised its own problems;* rivers often flowed only seasonally, and waterholes were few and far between. Under the cheap land policies prevailing at the time, and even

more under the pre-emption policy, the earliest arrivals took up the lands adjacent to the water supply,
and so made valueless the waterless lands on the in-
terfluves and away from the waterholes. Only slowly
were legal measures adopted to control the use of
western water, and indeed, in spite of progress in
both legislation and provision of water, the problem
of waterless lands remains acute up to the present
day.

The farmer who had migrated from the humid
East was, therefore, simply not equipped to deal with
the plains environment; in this technical sense the
Great American Desert was a reality. But the pioneers in the United States (for here the story of settlement on the Canadian prairies tends to diverge
from that of the United States plains, and will be
treated separately in a later section) had also to face
another obstacle to westward progress—the Plains
Indians. We have seen in Chapter 2 how the horse
became available to the tribes of the Interior after the
arrival of the Spaniards in Mexico. Because this gave
the Indians a new mobility and, in particular, because the horse made it easier to hunt the buffalo that
roamed the grasslands, there took place a migration
of tribes from the surrounding forests to the plains,
to benefit from the new situation.

Thus, there occurred in the Great Plains an encounter between the two groups of newcomers, the
red and the white, which has become the most publicized culture clash in history. Indian resistance was
ferocious and, on the western trails, travel was safe
only in convoy, and sometimes not even then. Nowhere else, on a continent that had known a long and
tragic series of Indian wars, was the struggle so bitter
or the Indian strength so great as on the Great Plains.

If peace came at length, it was due partly to the development of superior fighting skills on the side of
the white people, with their famous Colt revolver,
and partly to the virtual extinction of the buffalo,
rather than to the triumph of sedentary agriculture.
The time for that had not yet come. Throughout the
history of white settlement of the plains, there have
been two principal claimants to the grasslands—the
rancher and the wheat farmer—and of the two the
rancher was there first. It took time for the frontier
farmers to develop the techniques necessary to bring
the plains under cultivation, and while they were
doing so the cattlemen had their heyday.

"The physical basis of the cattle kingdom was
grass," says W. P. Webb, "and it extended itself over
all the grassland not occupied by farms."[2] Livestock
grazing represented the most profitable means of occupying a grassland area, which, for the moment,
was useless to the agriculturalist. Its origin was in
Texas, where the early settlers had already developed a type of stock raising suited to the Great Plains
environment, for "in the final analysis, the cattle
kingdom arose at the place where men began to manage cattle on horseback. It was the use of the horse
that primarily distinguished ranching in the West
from stock farming in the East."[3]

Cattle-grazing in the Great Plains developed in
two stages. The first, shorter, but more romanticized,
was the open range stage in which the "cattle kingdom" was built up by two processes. First was the
process of delivering Texas cattle to eastern markets.
After some experimentation by Texas drovers for the
best routes, a pattern emerged. Cattle were driven
north in Texas, until the trail met the westward-
thrusting railroads that led to Kansas City, St. Louis,
or Chicago. There the drovers sold the cattle and the
buyers shipped them north and east by rail. As the
railroads continued west across the plains in the
1870s, so the shipping points moved west also.
Under these conditions, the grasslands simply served
as a great transit camp, providing forage along the
way for the northbound stock.

The second process was the spread of the open
range system from its original location in the southern plains to the whole of the unfarmed grassland
northward into Montana and Canada. Texas cattle
supplied not only meat for eastern markets but stock
for much of the Plains. By the mid-1880s, ranching
had taken hold in an area that, a little more than a
decade earlier, had been non-productive.

This early system was an unimproved and unim-
proving activity. Its basis was the natural range, open
and unfenced;[4] the stock was almost wild, and breeding to improve the strain was out of the question. It
is therefore not surprising that, once the plains began
to fill up, the cattlemen had to change their ways. The
need for change became acute when, in the unusually severe winters during 1886–88, thousands of

[2]Webb, *The Great Plains,* p. 207.

[3]Webb, p. 207.

[4]Since the grassland was still legally part of the public domain, it
was illegal to fence it; when some cattlemen tried to stake claims
by this means, troops were sent out to remove the fences.

cattle on the overstocked range were lost and many cattlemen went bankrupt. Fencing was made practical by the invention of barbed wire in the Corn Belt in the 1870s. Although at first opposed violently by open range cattlemen as infringing on their rights as well as injurious to their cattle, barbed wire was adopted by the newer, more progressive ranchers, since it made possible controlled breeding and improvement of stock as well as better use of pastures. Equally important to the new improved ranching was the development of the cheap, easily erected windmill. This made it possible, on the one hand, to raise water from greater depths and in more constant supply than by hand pump; and on the other, to provide a water supply at isolated and waterless places on the grasslands, so that cattle could be fenced into separate pastures instead of crowding around the few natural water sources and trampling the waterside areas.

The advance of the farm frontier across the eastern part of this debatable land occurred in the 1880s, whereas the more westerly farmlands were established in a series of advances and retreats lasting from the 1890s to the present day. The advance was made possible by a series of developments in technique and equipment, which opened the plains to the cultivation of grains. Most basic was the development, through a long period of research and trial and error, of dry-fallow farming (known in North America simply as dry farming). Essentially, dry farming consists of raising a crop of grain, usually wheat, every second year and, during the alternate years, cultivating the uncropped field to prevent weed growth and to conserve moisture. In early years, cultivation to produce a fine "dust mulch" on the surface was encouraged as the best method to conserve moisture. It soon became evident, however, that a fine mulch was of little use when it was blown into the next state by the strong winds sweeping across the Plains. Techniques and implements were gradually developed to cut weeds below the surface, leaving grain stubble only partially disturbed as erosion-resistant "stubble mulch." To protect further against wind erosion, wheat and fallow were arranged in alternating long, narrow strips perpendicular to the direction of prevailing winds. Drought- and disease-resistant or quick-maturing varieties of grain were imported or developed. The almost complete mechanization of farming increased the area one farmer could cultivate, and so made it possible for a family to live by farming extensive areas, even when yields were low.

Gradual development and improvement of the two land-use systems, sedentary ranching and dry farming, made possible the permanent occupation of the problem area. Ranchers and farmers came to divide the grasslands between them, the farmers taking the flatter and more humid east and the ranchers the drier, western areas and rough lands such as the Missouri Plateau, the Nebraska Sand Hills, and the Flint Hills of Kansas.

But "development" and "improvement" were only relative terms, in the conditions under which the Plains farmers were trying to establish themselves. These early farmers planted grain: between 1900 and 1910 the wheat acreage on the southern plains increased by 600 percent. They did this in face of the difficulty of securing from the all-powerful railroads satisfactory rates for transport to the flour mills, and under constant threat of catastrophe by drought, blight, and insect pests. In an area where yields of wheat might vary from 15 to 20 bushels per acre (37–49 per ha) in a good year to 5 bushels (12.3) or nothing in a dry one, they came to depend to a dangerous degree on this single, hazardous form of livelihood. Each series of wetter years tended to obliterate the memory of the preceding drought and to deceive a fresh group of newcomers.[5]

The effects of both ranching and wheat farming on the Great Plains soon became apparent; the ranching areas began to suffer from overgrazing and the wheat areas from unwise cultivation. Because the rainfall of the plains varied from year to year, the amount of range forage available also varied, and this, in turn, required that the numbers of stock grazed be adjusted to the condition of the range. But adjustment was seldom made in time; the temptation to graze the ranges to their peak capacity is always strong, and the grasslands deteriorated in consequence. The grasses grazed by the stock declined severely and were replaced by shrubs and plants that were either less sought after or not eaten by the stock. As the carrying capacity of the ranges in some areas dimin-

---

[5]"About eighty percent of the year-to-year variations in Prairie crop production can be accounted for by climatic factors, particularly soil moisture." W. J. Carlyle in J. R. Rogge, ed., *The Prairies and Plains: Prospects for the 80s* (Winnipeg, 1981), p. 23. In this connection, see also Cynthia Rosenzweig, "Crop Response to Climatic Change in the Southern Great Plains: A Simulation Study," *Professional Geographer*, vol. 42 (1990), 20–37.

ished, sheep and goats replaced cattle,[6] brush replaced grass, and the forces of erosion met less and less resistance. In 1936, when a famous government report, *The Future of the Great Plains*, was issued, it was estimated that the western ranges, as a whole, were 70 percent overstocked and that their carrying capacity had been reduced by more than one-half in eighty years.

For this deterioration, as Chapter 5 suggests, the individual rancher was not wholly to blame. If the ranges were overgrazed, it was partly because the government, failing to appreciate in time that the land policies of the humid East would not suit the dry West, prevented the rancher from securing, legally, a holding adequate to provide a living. Marion Clawson has written of an "institutional fault line" at 98° W—that is, a line to the west of which fresh policies and laws were needed, ones adapted to the thinning resources and the possibility of using land in large blocks or in common. Before such a policy had been formulated, much of the plains had been homesteaded in 160- or 320-acre (65- or 130-ha) blocks— as if they were tall-grass prairies in Iowa or Illinois.

The results of wheat cultivation were equally serious and far more spectacular. Removal of the natural grass cover on the dry plains, especially during the days of the "dust mulch" regime, exposed the soil to powerful erosive action. That the frontier of grain farming was too far to the west needed no demonstration in Dust Bowl years or when low prices made production, at 5 bushels to the acre (12.3 per ha), hopelessly uneconomical. The real problem has arisen in better years—especially in the humid 1940s, when the stimulants of wartime demand and support prices made grain farming of almost any standard profitable. It is in such years as these that the challenge of conservation farming has been hardest to face and the temptation to plant more wheat the strongest. To quote Marion Clawson again, "The real problem is not to put poor wheatland into grass—the real problem is to keep it there,

when unusually favorable weather and/or price years come again."[7]

Present Problems

TECHNICAL PROBLEMS

Today, most of the plains region wears a prosperous air, and the catastrophic events of the 1930s seem very remote. "The recovery of the Great Plains economy from 1938 to 1945 was as spectacular as had been its downfall," wrote Bailey; it was "almost miraculous."[8] This is not to say that hard times may not come again: climatically they are quite certain to do so, if the statistics do not lie. But the Great Plains farmers today are in a far better position to cope with them than their predecessors; they have been able to outgrow both the ignorance and the fears of earlier generations (Fig. 18-3).

In retrospect we can see that, given the variability of climatic conditions on the Great Plains, from year to year and from decade to decade, most of the problems that defeated so many earlier settlers were the result of three things:

1. *Lack of reserves.* When settlers homesteaded a 160-acre (64.8-ha) holding on the Great Plains, they were likely to start with minimal resources of capital and equipment and to stake everything on their success. If they were unfortunate enough to hit a dry spell in which to begin, they had nothing to fall back on (although some colonizing agencies, such as the railroads, did re-equip settlers on their own lands if they lost their first or second crop); they clung grimly to "their" land until they had nothing left and then suffered total defeat. It was this lack of reserves that caused so many personal tragedies among the first waves of settlers.

2. *Lack of technical alternatives.* The farming practiced by the first plains settlers was the only kind most of them knew—the mixed farming they had

[6]This practice was especially marked on the Edwards Plateau in southern Texas, which became the main sheep- and goat-raising area of the United States. Goat ranching was developed, especially after 1920, to provide mohair for automobile upholstery. Between 1920 and 1959, while the number of sheep and lambs in the United States fell overall from 40 to 34 million, the numbers in Texas rose from 2.5 million to 6 million. Since that time the totals have fallen again, but of the 11.06 million sheep and lambs in the United States in 1987, 2.06 million, or 18 percent, were in Texas.

[7]M. Clawson, "An Institutional Innovation to Facilitate Land Use Changes in the Great Plains," *Land Economics,* vol. 34 (1958), 75.

[8]W. R. Bailey, "The Great Plains in Retrospect," *Journal of Farm Economics,* vol. 45 (1963), 1092, 1095.

For some idea of what the Great Plains had to recover *from,* see P. Bonnifield, *The Dust Bowl* (Albuquerque, 1979), and J. N. Gregory, *American Exodus: The Dust Bowl Migration and Okie Culture in California* (New York, 1989). The core area of the Dust Bowl was Texas north of Amarillo; the Oklahoma Panhandle; eleven southwestern countries of Kansas; the southeasternmost two countries of Colorado; and the northeastern tip of New Mexico.

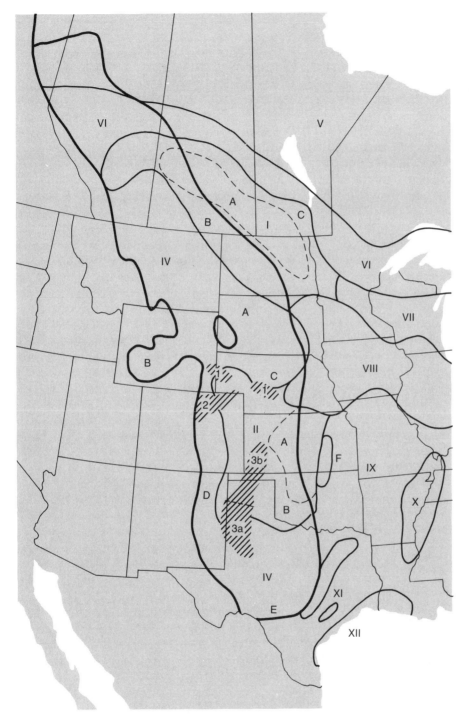

Fig. 18-3. The Great Plains: Regional patterns. Apart from the physical limits of the Great Plains, the map identifies the following land-use regions: I Spring Wheat Region; A Intensive, B Dry farmed, C Mixed farming. II Winter Wheat Region; A Intensive, B Dry farmed. III Irrigated Regions; 1. Platte River Valley, 2. Colorado Piedmont, 3. High Plains, (a) older intensive area, (b) area of expansion. IV Livestock Ranching Region; A Missouri River Plateau, B Wyoming Basin, C Nebraska Sand Hills, D High Plains, E Edwards Plateau, F Flint Hills. V Nonagricultural. VI Northern Fringe. VII Dairy Region. VIII Corn Belt. IX The South. X Mississippi Alluvial Valley. XI Texas Blackland Prairies. XII Texas–Louisiana Coastal Plain.

carried on back in the East or the Midwest.[9] But the drier conditions of the plains enforced changes of practice; yields were lower, conservation measures were necessary, and some of the midwestern crops could no longer be relied upon. To learn the new techniques and find alternative crops took time—too much time for many of the first settlers to survive until they did so. The outcome in terms of land use was likely to be either an attenuated mixed farming or an unwise concentration on wheat, neither of them immune to a cycle of dry years.

3. *Inflexible administrative and financial arrangements.* As we have already seen, the plains were settled within a framework of law and finance that had developed in the humid East. The most obvious expression of this eastern thinking was the retention of 160 acres (64.8 ha) as the basic size of government land grant. This figure of 160 acres first appeared in legislation governing disposal of land in what is now Indiana in 1804. It was a feature of other land acts up to and including the Homestead Act of 1862, and, despite the fact that settlement had by this time reached the borders of the dry plains it was left unchanged—even after Major J. W. Powell's report in the late 1870s had urged the necessity for a working unit in the dry West of at least 2560 acres (1036 ha). The concept of the homestead as a holding of what might be called "eastern-family size" went unmodified in legislation virtually until 1969; in some legislation, in fact, it survived into the 1980s.

The outcome was to tie the fortunes of thousands of settlers to one particular quarter section of land on which, in a region of uncertain climate, their whole future depended. It is little wonder that in the 1950s, Kraenzel wrote "It is not the fact of semiaridity that causes the difficulties in the Plains, but the fact of an unadapted culture that does so."[10] What was needed, then, was administrative flexibility: "Management principles for the Great Plains farmer must be such that he can shift quickly and roll with the punch," and he must have legal and financial freedom to do so.

If we consider each of these three problems, in turn, it becomes apparent that, in most respects, Great Plains farmers or ranchers of today are far better off than their predecessors. It is not so much a question of prosperity as of security and stability.

1. *Lack of reserves.* The provision of reserves may be considered at three overlapping, reinforcing levels: farm or ranch, community, and state and national. For better or worse, the result of decades of trial, error, and change is farms and ranches that are increasingly few in number, but larger in scale of operation. Also, farmers and ranchers have access to a knowledge of improved kinds and varieties of crops, breeds of livestock, and such improved techniques as dry farming and irrigation. The ability to provide reserves such as moisture retained in the soil, water stored in ponds, or feed reserves for winter or drought, is of utmost importance. The larger scale of operation and greater profits make possible the building of reserves of finance and credit and, so, the ability to survive years of drought or crop failure. Many farmers and ranchers have invested in nonfarm enterprises, as a financial cushion. Increase in scale may also facilitate diversification: a single ranch may, for example, raise beef cattle and grow wheat and sorghum or irrigated and dry-fallow crops—each on a scale large enough to be worthwhile. Such diversification is possible, however, only if the terrain, soil, and climate permit it. In many areas of the Great Plains, they do not.

Towns and cities in the Great Plains are small and dispersed. The smaller ones, as we shall see, have faced problems as the farm population has declined. Yet the small trade centers, the economic and cultural services they offer, and the network of roads and communications that connect them with the farms and ranches are much better developed today than during the first half-century of settlement.

The pervasive role of federal and state govern-

---

[9]The movement onto the Great Plains was generally not the sort of long-distance migration that resulted in Ukrainian peasants abruptly appearing in Saskatchewan or Scandinavians in the Dakotas (although it is true that some of the colonization societies, especially in Canada, produced this effect). Most of it represented short-distance moves by farmers whose previous experience was certainly American, often in borderline areas. Thus in Ole Rolvaag's classic, *Giants in the Earth,* his heroes moved only from southern Minnesota to eastern South Dakota: yet, even this short move presented difficulties of adjustment that taxed them to the limit. This book and its sequel, *Peder Victorious,* are well worth reading in this connection. By the same token, it has been calculated that in the main post–Civil War migration into Kansas (1870–90), of 526,000 new arrivals 75.3 percent were white Americans, 3.3 percent were black Americans, and only 21.4 percent were foreign-born whites: many of them, of course, would have had a previous home or homes in America.

For another Rolvaag-type study of Scandinavian settlement, see R. C. Ostergren, *A Community Transplanted . . . 1835–1915* (Madison, Wis., 1988).

[10]C. F. Kraenzel, *The Great Plains in Transition* (Norman Okla., 1955), p. 287.

ments in American society, in general, and in agriculture, in particular, has been discussed. Nowhere is such a role more important than in the Great Plains. Universities, experimental stations, and extension systems have been established. There is direct aid for irrigation, farm ponds, and soil conservation, as well as roads, county airfields, and crop insurance or disaster relief. Federal credit and marketing programs are fundamental. At the regional and national levels, also, are many professional farm and ranch organizations through which farmers and ranchers provide mutual support and also lobby for improved government support and services.

2. *Technical alternatives.* Many alternatives have been mentioned as critical in the development of improved ranching and dry farming. Other technique developments include irrigation, herbicides, and new or different crops such as grain sorghum or sunflowers. All help to provide greater stability and security. Some, such as irrigation, will be mentioned further in the discussion of areal patterns of land use.

One that has been mentioned, but not defined, is the water- and soil-conserving pond. With government encouragement and aid, thousands of such ponds have been built since the 1930s, not only in the Great Plains, but also throughout the humid East. In the East, the primary purpose is to reduce erosion and flooding, but the ponds may also provide water for livestock or for irrigation. In the Great Plains, water for cattle is most important; thus, natural runoff water in the ponds may be supplemented by water obtained from wells powered by windmills.

3. *Institutional arrangements.* As we have seen, the Great Plains farmers have gone a long way toward freeing themselves from the peculiar anxieties of their predecessors. In overcoming the third problem, inflexible institutional arrangements, they have been less successful. Most of the Great Plains area is in private hands, unlike the remoter West, where federal or local governments are often the majority landowners. Consequently, the status of private owners and the pressures on them are of vital concern for the welfare of the region. The best recipe for the region as a whole may well be inapplicable (or unacceptable) to the individual farmer.

What is to be done to ease these pressures? One thing is that government, banks, and creditors must recognize that, in a region in which the climate is unreliable, there will be runs of good and bad years. In bad years, it may be necessary to accept delay in repaying credits advanced in good ones, and tax assessments must take account of the *average* value of these lands. The other thing is to try to provide some kind of reserve of land within each district, so that in bad years ranchers and farmers need not feel trapped on a few burned-up acres. We have already seen that they have been successful in creating reserves of fodder and water; this would be a reserve of land.

On the Canadian Prairies, such a reserve already exists in the form of Community Pastures established under the Prairie Farms Rehabilitation Act. There are over 2 million acres (810,000 ha) of these on the Prairies, and they represent an unusual and apparently satisfactory balance of interest and control between the federal government, the provinces, and the local farmers.[11] They make it possible to take account of short-term, local changes in farm needs—and that, rather than an elaborate program of government intervention that most farmers would reject out of hand, is what the region mainly requires. There are plenty of other areas on the Great Plains to which the community pasture idea might usefully be extended, and the U.S. government has, in fact, been sponsoring a policy of identifying "set-aside" areas to form such reserves.

SOCIAL PROBLEMS

The technical advances put into effect on the plains in the past thirty years have affected the region's social geography. In very general terms, economic improvement (as reflected in larger farms and increasing mechanization) has been bought at the price of social disintegration. The new land use implies a greatly reduced labor force, and in a region so lacking in cities as the American plains (the Canadian situation will be considered in the next section), the effect of reducing farm labor requirements is to provoke out-migration. In the five Great Plains states between North Dakota and Oklahoma, the decade of the 1950s saw the farm population decline from 1.9 million to 1.3 million, and this figure of 600,000 loss from the farms corresponded almost exactly with the

[11]The provinces select and acquire the land (usually areas abandoned because of tax delinquency or severe erosion). They then lease it to the federal government, which puts the area in order for receiving stock and provides services on the pasture (including, for example, bulls-in-waiting). Allocation of the grazing privilege is then made by the elected committee of a Grazing Association formed locally, and grazing is charged at a low rate per head of stock and per day. For further details see *PFRA: The Story of Conservation on the Prairies,* Canadian Dept. of Agriculture Pub. No. 1138 (1961).

figure for out-migration from these states (595,000) during the intercensal period 1950–60. Between 1960 and 1970, a further reduction of the farm population occurred, to 966,000, and once again the decade was marked by a heavy out-migration from the five states, amounting to some 406,000. The 1970s saw the out-migration continue in four of the five states, but at a slower rate (Table 18-1) as the rural exodus worked itself out, and in Oklahoma there was a considerable in-migration, though not into agriculture—the number of farms there continued to fall, as it did in all the other states (Table 18-2).

Since 1980, out-migration has been light and, once again, Oklahoma defied the general trend.[12] In fact, since 1980, the rural population in four of the five states has risen, reversing a trend (Table 18-1b) that goes back several decades. Once again, however, this is not a rise in the number of farmers, but a part of the population movement (1) into rural areas where housing is cheap, and (2) into employment on the Great Plains—for example, the oil industry or coal mining—that does not require workers to live in town.

[12]Oklahoma is included in this discussion and in Tables 18-1 and 18-2 because it is undeniably a Great Plains state. But it differs from the other four states listed in a number of ways: (1) Its settlement history is unlike theirs. Before 1889, it was known as Indian Territory and was used, to put no finer point on it, as a dumping ground for Indian tribes displaced from other areas desired by whites. In 1889, white settlers were allowed to enter this territory, too, and it was admitted as a state in 1907. (2) It is an oil and gas state, and this has marked its economy and occupational structure in the same way that these minerals have set their mark on Texas. One consequence of this is that Oklahoma has what the other Great Plains states do not: two metropolitan areas, Oklahoma City (976,000) and Tulsa (733,000), which are, thanks to oil, miniatures of Dallas. (3) With its southerly location and its oil–gas interests, it forms part of the Gulf Coast hinterland of the Texan ports and cities, rather than looking toward the cities of the Midwest or the Mississippi Valley.

Table 18-1. Great Plains States: Population Migration Balance, 1960–85 (in thousands)

	1960–70	1970–80	1980–85 (est.)	1985 (est.)
North Dakota	−93	−31	−3	685
South Dakota	−92	−41	−15	708
Nebraska	−74	−47	−28	1606
Kansas	−140	−71	−11	2450
Oklahoma	−7	+230	+137	3301

Great Plains States: Rural Population, 1950–80 (in thousands)

	1950	1960	1970	1980
North Dakota	455	410	344	334
South Dakota	436	413	369	370
Nebraska	704	645	571	582
Kansas	912	850	762	788
Oklahoma	1094	863	819	990

In a farming area, the steady reduction in numbers of farm units shown in Table 18-2 means both empty farmhouses and reduced demand for service centers. On the plains, many communities are so small to begin with that they cease to exist altogether in any social sense, and the lower levels of local government, such as school districts, become extinct; there are neither children nor schools to administer.[13] It may well be argued that this is a way of eliminating waste in government, but the effect of the changes on

[13]The decline of community consciousness and communal activities on the Great Plains throws into sharper relief the communal existence of certain groups that, usually on religious grounds, have settled and farm together in the closest social units. An example of such groups is provided by the Hutterites, communities of whom farm both in the Dakotas and on the Prairies. It so happens, however, that these close-knit groups are now tending to break up, and individual holdings are becoming more common.

Table 18-2. Great Plains States: Number of Farms and Average Farm Size, Selected Dates, 1930–87

	1930		1950		1959		1969		1982		1987	
	No. of Farms*	Average Size*	No. of Farms	Average Size	No. of Farms	Average Size	No. of Farms	Average Size	No. of Farms	Average Size	No. of Farms	Average Size
North Dakota	78	513	65	676	55	755	46	930	36	1104	35	1143
South Dakota	83	545	66	719	55	805	46	997	37	1179	36	1214
Nebraska	129	391	107	471	90	528	72	634	60	746	60	749
Kansas	166	308	131	416	104	481	86	574	73	642	69	680
Oklahoma	204	194	142	300	95	378	83	434	73	446	70	449

*Farm numbers are in thousands; farm sizes, in acres. For hectares, divide by 2.47.

the generation caught in the transition is not something that can be lightly dismissed.

The fate of many small communities on the plains and prairies hangs in the balance. What must be particularly enervating is that their fate often depends entirely on outside factors. One common example of these factors nowadays is likely to be the decision of a railroad to close an unprofitable branch line. When that happens, the grain elevators beside the tracks are no longer served, and the community loses one of its basic central functions. Or the railroad may decide, for economy, to concentrate its shipping on the larger elevators only, which are better suited to the capacity of the "block" trains—the long, single-commodity loads—now commonly used. Another factor may well be the decision of the state or county government to change the siting of public service units, such as hospitals or high schools. As consolidation of these services is forced upon the government, much will depend for the community on the disposition of these central services. Some governments may try to keep alive the smaller towns by distributing services as widely as possible; much as the original state–provincial governments allocated their first three service functions to rival communities in the early years—the capital, the university, and the jail. A neighboring government may decide to focus all such services in a single center and let other communities die. Uncertainty is in the air these communities breathe today.

THE MISSOURI VALLEY

We have now considered the kind of problem confronting settlers on the plains and the suggested solutions. In the heart of the Great Plains, however, there lies an area that invites a far more comprehensive solution; a region where every known measure of conservation and agricultural adaptation could be applied as part of a unified regional plan—the Missouri River Basin. Any account of the future of the plains must necessarily include a consideration of developments in the Missouri Valley.

The basin of the Missouri embraces the whole of the northern Great Plains of the United States, south to and including northern Kansas and northeastern Colorado. Since the 1930s, it has been repeatedly urged that the Missouri Valley should be treated in the same way as the Tennessee; that an MVA should be established to parallel the TVA and that it should develop a plan for the whole basin (Fig. 18-4).

Such a plan actually exists (though an MVA does not)—the Pick–Sloan Plan of 1944. Among its features have been the construction of reservoirs and irrigation projects, flood control, improvement of navigation, and land management. Over the past forty years, this plan has been implemented, project by project, as Congress voted installments of the necessary funds.

In theory, such a plan should provide the best possible solution for the problems of the Great Plains. It should make available central irrigated areas, strategically located, as a firm base for farming and ranching; it should ensure the best disposal of the available water supply, and bring additional benefits, in the form of cheap electric power or agricultural education, to isolated plainsmen. Yet it is difficult to be enthusiastic about the Missouri Valley development to date. This is because, although the Tennessee Valley formed an intelligible and satisfactory unit for development, the Missouri Valley does not. In reality it falls into two parts, whose natural characteristics are different and whose interests, in consequence, conflict.

The southeastern tip of the Missouri Basin receives more than 40 in (1000 mm) of rain per annum. Through it flowed the silt-laden river, often in flood and always difficult to navigate. The eastern end of the basin was interested, therefore, in flood control, in navigation improvement, and in hydroelectricity. Much of the western part of the basin, however, receives less than 15 in (375 mm) of rain per annum, and the interests of its population are in water for irrigation and stock reservoirs. In short, the upstream section of the river has too little water and the downstream section, periodically at least, has too much.

As a result, two separate plans for the valley were drawn up, one—the Sloan Plan—by the Bureau of Reclamation, representing the interests of the dry western section, and the other—the Pick Plan—by the U.S. Corps of Engineers, which, because it is responsible for flood control and navigation, focused its attention on the eastern section. To resolve the dilemma of choosing between two plans, which were far from being compatible, the government adopted the peculiar expedient of combining the two plans, and in 1944 Congress approved a Pick–Sloan Plan (which was simply the Pick Plan united to the Sloan Plan), in what Rufus Terral has succinctly described as "a shameless, loveless, shotgun wedding."[14]

[14]On the Missouri Valley projects, lively opinions were expressed in their early stages by R. Terral, The Missouri Valley (New Haven, Conn., 1947); R. G. Baumhoff, The Dammed Missouri Valley (New York, 1951); and H. C. Hart, The Dark Missouri (Madison, Wis., 1957).

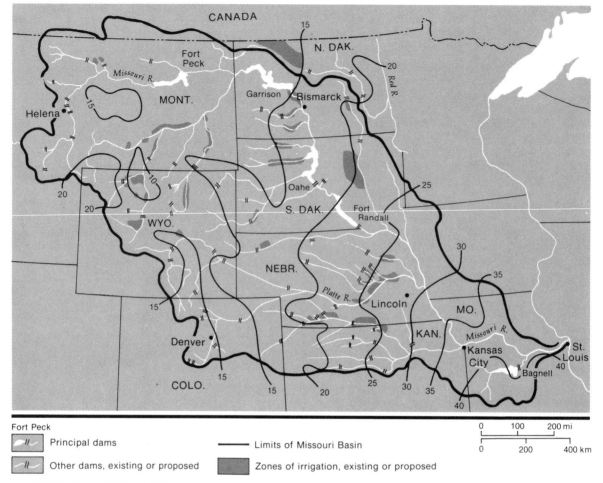

Fig. 18-4. The Missouri Valley. Solid lines represent the annual precipitation (inches). Notice how the Missouri Basin narrows to a "funnel" at its southeastern end, thus increasing the problem of flood control.

The wedding occurred in 1944, and it was not until five years later, when the Pick–Sloan Plan had already advanced through its early stages, that the Department of Agriculture produced its parallel program of agricultural improvements.

There is no doubt that the expenditure of several billion dollars on the Missouri Basin has brought many benefits. Individual projects have brought prosperity to one community or another. But this can hardly be called a truly regional solution. The construction of a few very large dams, for example, is probably not the most economical way of providing irrigation water for a dispersed farm population that has shown (as some critics predicted years ago) a marked disinterest in paying for water supplied by the government, preferring to sink wells on the farm. There are conferences of basin states and interagency committees, but no MVA—and then there are the ordinary inhabitants of the basin for whom, unless they live very close to the river indeed, the plan and all its works are largely irrelevant. Figure 18-5 reminds us that water is in any case limited.

Present Regional and Urban Patterns

1. *The wheat belts.* The two wheat belts were referred to in Chapter 14 as being parts of the great cropland triangle of North America and extensions of the relatively intensive, rain fed, annual-cropping kind of agriculture characteristic of the Corn Belt. But, like the Great Plains land-use region as a whole, the wheat belts also are areas of transition. The core of each belt, in which cropping is annual, and intensity and productivity are greatest, is toward the eastern,

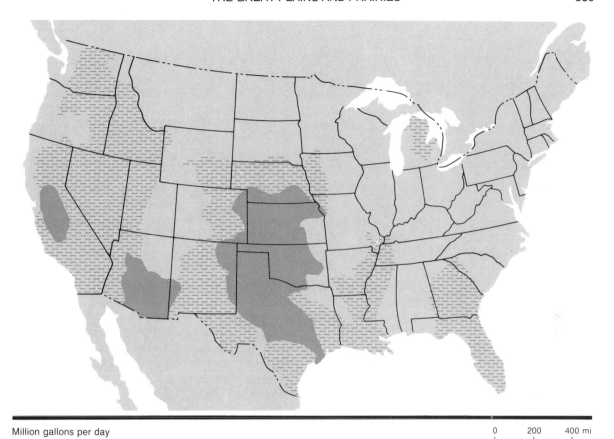

Million gallons per day

Moderate (21–500) Critical (500+)

0   200   400 mi

0   200   400   600 km

Fig. 18-5. United States: Groundwater overdraft; that is, the excess of withdrawals of water over recharges. This map is also relevant to discussion in Chapters 20 and 23 of this book. (Reproduced by permission from M. T. El-Ashry and Diana C. Gibbons, eds. *Water and Arid Lands of the Western United States* [Cambridge and New York, 1988], Fig. 9-1.)

humid margin (Fig. 18-3). Toward the western, more arid margins, annual cropping gives way to alternate years of crop and fallow; the risk is higher and the average yields are lower and more land is used for livestock grazing. McPherson and Greeley counties in Kansas are representative of the two levels of land use (Tables 18-3 and 18-4). Note the contrast in density of population, areal size of farms, and percentage of land that is in crops, in irrigation, and especially, in summer fallow.

In the winter wheat region, the crop is planted in September and October and reaches the height of lawn grass before winter, during which it lies dormant. It grows again during the maximum rainfall seasons of spring and early summer and is harvested in June, before the onslaught of the high temperatures and drying winds of mid- and late summer. Beef cattle, a basic part of the regional farming system, graze the young wheat in both autumn and

Table 18-3. McPherson County, Kansas: Agricultural Statistics, 1987

Population (1980)	27,000
Percentage change, 1970–80	+8.4
Area of county (ac/ha)	573,000/232,000
Percentage in farms	93.1
Percentage in cropland harvested	49.5
Percentage of this irrigated	8.5
Percentage in summer fallow	12.3
Value of agricultural products sold	$78 million
Percentage from livestock products	65.9
Number of farms	1357
Average size (acres/hectares)	393/159
Average value of land, buildings, & equipment, per farm	$268,000
Average value of agricultural products sold, per farm	$57,000
Percentage of operators whose principal occupation is farming	57.2

Table 18-4. Greeley County, Kansas: Agricultural Statistics, 1987

Population (1980)	2,000
Percentage change, 1970–80	+1.4
Area of county (ac/ha)	501,000/203,000
Percentage in farms	94.9
Percentage in cropland harvested	41.3
Percentage of this irrigated	12.9
Percentage in summer fallow	37.5
Value of agricultural products sold	$81 million
Percentage from livestock products	79.7
Number of farms	294
Average size (ac/ha)	1617/655
Average value of land, buildings, & equipment, per farm	$754,000
Average value of agricultural products sold, per farm	$277,000
Percentage of operators whose principal occupation is farming	73.5

spring: If the fields are very dry or very wet, they may be damaged by trampling; usually, however, controlled grazing for a few weeks aids wheat growth by causing the plants to "tiller," or to send up additional shoots. The post-harvest stubble provides still further pasturage. Wheat farmers who do not raise cattle often lease their fields to farmers who do, and to nearby ranchers.

The spring wheat region, extending as it does across the international boundary and along the transition zone between steppe and forest, exhibits a much greater diversity of farming than the winter wheat–sorghum–cattle region. There is the same general decrease in intensity and productivity, from a more humid core southwestward toward the more arid fringe. But, even within the core, there is more diversity of crops. They include, in addition to wheat, some corn, flaxseed, and sunflowers in the southeast and oats and barley throughout. Only in southern Saskatchewan does the cropping system approach monoculture of wheat, as it does in central Kansas. In the transitional zone along the humid margin, looping from the flat Red River Valley in the southeast to the foothills of the Rockies in the northwest, agriculture is still more diversified. Crops include all of those mentioned, plus potatoes and sugar beets in the Red River Valley and rapeseed in the northwest. Livestock, including hogs, poultry, and beef and dairy cattle, utilize hay and grain crops; their products supply the rapidly growing cities of the subregion, especially Winnipeg in the east and Calgary and Edmonton in the west.

2. *The irrigated areas.* Only two irrigated areas within the Great Plains are sufficiently extensive to be treated as regions. The Colorado Piedmont, as the name suggests, lies at the foot of the Front Range in Colorado. Old alluvial soils are supplied with water from the South Platte and its tributaries, including the St. Vrain, Big Thompson, and Cache le Poudre. Additional water from the upper Colorado River system is brought by tunnel under the Continental Divide (Fig. 18-6). The intensive farming produces such cash crops as wheat, sugar beets, and vegetables, and more important, alfalfa and feed crops to fatten cattle and lambs from ranches in the surrounding plains and mountains, as well as to feed dairy cattle. Increasingly, the meat and dairy products are used to supply Denver and the smaller cities of the area.

Unlike most major irrigated areas of the West that rely upon water taken directly from the streams or water pumped from alluvium in the stream valleys, the High Plains irrigated region pumps water from underground aquifers. The older, intensively cropped sector, centering on Lubbock, Texas, was developed after 1920 when improved equipment, especially the powerful turbine pump, became available,[15] along with electric power from oil and natural gas. In the early decades, the area concentrated upon cotton production, supplemented by wheat. Since 1950, grain sorghum has challenged cotton for supremacy. It is used to fatten beef cattle in huge feedlots, a method that has come to threaten the longtime supremacy of cattle feeding on family farms in the Corn Belt. However, intensive, irrigated agriculture in the region itself is now threatened by a rapidly falling water table and increased expense of pumping. The water is being removed from the aquifers much faster than it can accumulate—an example of what may be called "robber economy"; that is, agriculture that robs the land of the resources upon which the economy depends.

Throughout the 1970s, then, the irrigated acreage on the Great Plains rapidly increased. Then came the 1980s and the era of overpumping, and a reversal of the trend. In the United States as a whole, the decade saw a reduction in irrigated acreage of 2.6 million acres (1.05 million ha). Within the Plains region, the

---

[15]On another old-established irrigation area on the Plains, see J. E. Sherow, *Watering the Valley: Development Along the High Plains Arkansas River, 1870–1950* (Lawrence, Kans., 1990).

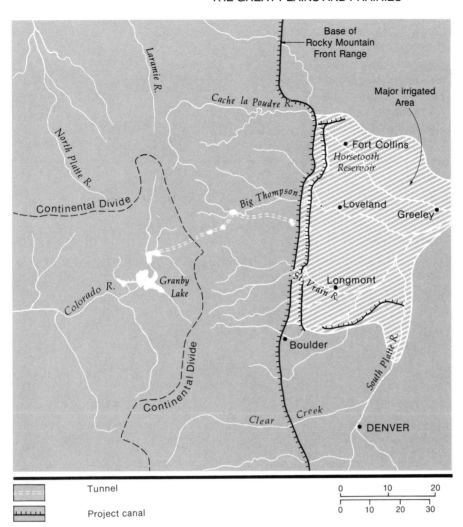

Fig. 18-6. The Colorado–Big Thompson Project: Water transfer across the Continental Divide from the Colorado River to the Piedmont, where population and water demand have increased very rapidly in recent years.

chief reductions occurred in the Texas Panhandle, Nebraska, and western Kansas.

Irrigated farming is important along many of the rivers that cross the Great Plains; these include the Arkansas, the Platte, and the Missouri and its tributaries. The largest and oldest of such irrigated areas are along the Platte and North Platte, especially near Kearney and Scotts Bluff in Nebraska. Irrigated farming helps to supply produce to these and other small cities along one of the oldest and still most important routeways across the plains. Other riverine-irrigated areas are often smaller and dispersed along a stream, like beads on a string. A typical, repetitive, north–south sequence of land use in the Great Plains is (1) patches of irrigated farming in the valley of an eastward-flowing stream; (2) beef cattle grazing on the rough lands of the dissected valley sides; and (3) dry farming of wheat on flatter lands of the interfluvial upland. Such a dispersal of small irrigable areas helps to make possible the use of the surrounding areas of rough or dry land by providing dependable feed crops or by supporting small trade centers. By contrast, a large irrigated area, such as that near Lubbock, Texas, constitutes an island of prosperity, but does not provide security for dry farmers or ranchers far from its borders.

3. *The livestock ranching areas.* Most areas of the Great Plains, outside the wheat belts and irrigated and dry-farming areas, are devoted to livestock ranching. Except on the Edwards Plateau, already mentioned (p. 382n6) as the leading sheep-ranching and only major goat-raising area of the continent, most ranches raise beef cattle. These cattle are then sold, at a year or eighteen months old, for fattening

Agriculture on the Plains: A beef cattle feedlot. The Rocky Mountain Piedmont and High Plains have become a principal location for this type of intensive stock raising: irrigated land supplies the necessary feedstuffs. *U.S. Department of Agriculture*

and finishing—usually to farmers in the irrigated areas or the Corn Belt. A family-operated ranch, with an economically feasible minimum of at least one hundred breeding cows, must necessarily be large in area. Fremont County, in the Wyoming Basin, is representative of livestock ranching areas (Table 18-5). However, the averaging of data from its large, dispersed ranches together with that from the small, irrigated farms along the Wind River tends to distort the data from both. Even so, the average holding is more than 2500 acres (1000 ha) and livestock (mostly cattle) accounts for four-fifths of the value of products sold. Although less than 2 percent of the total area is cropped, the irrigated area is much more than 100 percent of the cropland harvested. This is so here and in other western areas because irrigated pastures are not included in the category of cropland har-

vested. It should be remembered, however, that beef cattle or sheep are raised primarily on natural grassland; the small areas of irrigated crops or pasture on ranches are mainly supplementary.

The Nebraska Sand Hills and the Flint Hills in Kansas (outside the Great Plains) are exclusively beef-ranching areas. In both areas, plowing for dry farming has been avoided, in the one because of sandy soils and in the other because of thin, cherty ones. By contrast, the rough, dissected Missouri River Plateau does have scattered areas of irrigated farming and dry farming of wheat. The relatively flat Judith Basin in central Montana may be considered an outlier of the spring wheat region.

4. *Urban patterns.* As suggested in Chapters 14 and 17, the major metropolitan centers that serve the Great Plains are located, with the exception of Den-

Table 18-5. Fremont County, Wyoming: Agricultural Statistics, 1987

Population (1980)	39,000
Percentage change, 1970–80	+1.4
Area of county (ac/ha)	588,300/238,200
Percentage in farms	41.9
Percentage in cropland harvested	16.7
Percentage in irrigation	137*
Percentage in summer fallow	0.3
Value of agricultural products sold	$40 million
Percentage from livestock products	77.1
Number of farms	908
Average size (ac/ha)	2714/1099
Average value of land, buildings, & equipment, per farm	$451,000
Average value of agricultural products sold, per farm	$44,000
Percentage of operators whose principal occupation is farming	55.0

*For explanation of this figure, see accompanying text.

ver, Calgary, and Edmonton, outside the region. The Great Plains have, at least until recently, looked eastward, and the cities that serve as hinges linking the Great Plains to the economic core lie in a north–south line between the two regions. They include, from north to south, Winnipeg, Minneapolis–St. Paul, Omaha, Kansas City, Dallas–Fort Worth, and San Antonio. Even the secondary cities, state capitals, and state universities in the tier of plains states are, with few exceptions, located at or beyond the eastern margin of the Great Plains landform region. They include Grand Forks, Fargo, Sioux Falls, Lincoln, Topeka, Wichita, Tulsa, Oklahoma City, and Austin. It is these two groups of cities through which the agricultural and mineral products of the Great Plains (and much of the remainder of the West) pass, eastward to the Mississippi and beyond, or southward to the Gulf Coast, with a return flow of goods and services. The central and western parts of each of the states in the tier are an economic, political, and cultural outback.

The four states to the west are divided into eastern plains and western mountain regions, and the directional relationship of core and outback is reversed. In Montana and New Mexico, the core, with its larger cities, state capitals, and state universities, is in the mountain region. In Colorado and Wyoming, the core is at the margin between plains and mountains (although Wyoming lacks a clear distinction between plains and mountain section as well as a distinct

core). Again, there is a north–south line of cities at the eastern base of the mountains and western edge of the plains. Dominated by Denver, it includes Great Falls, Billings, Casper, Cheyenne, Colorado Springs, and Pueblo. Even the positions of Calgary and Edmonton in Canada are comparable, although Edmonton functions also as the gateway to the Canadian North.

Aside from the explosive growth of Denver (1990 CMSA population 1.8 million) as an industrial and financial center, the most rapid current urban growth within the American Great Plains is related to the opening of massive opencast mines to produce low-sulfur coal, most of which moves eastward in long trains. Most affected are the small cities near the Bighorn Mountains, including Billings, Montana, and Sheridan, Buffalo, and Gillette, Wyoming. The boom at Rock Springs, in the western Wyoming Basin, has been sparked by exploitation of coal, oil, and uranium.

The Prairies of Canada

The Canadian part of the Great Plains region, although not separated from the American part by a natural divide, must be treated individually for, in several important respects, it differs from the area south of the Canadian border. These differences are best grouped under four headings.

PHYSICAL DIFFERENCES

In many respects, the natural conditions of the Canadian Prairies are indistinguishable from those in the United States; there is the same generally flat relief; the same gradual rise toward the west, interrupted by a number of low scarps; the widespread cover of glacial deposits, and a climate that resembles that of the plains to the south in its continental characteristics and its summer rainfall. But the most clearcut feature of the American plains—the regular east-to-west progression from humid to arid conditions and from tall-grass to desert-shrub vegetation—is missing from the Canadian plains.

On the Canadian plains, the distinctive zones of climate, soils, and vegetation, which run from north to south in the southern plains, curve to run almost at right angles through the southern part of the Prairie Provinces. Thus, the better-watered plains, or subhumid prairies, form a loop round the northern end of the more arid zone (the area with 12 in [300 mm] or less of rain per annum), and the east-to-west

Cattle ranching in Alberta. On the whole, land use in this province is divided between crop raising on the smoother, moister lands and ranching on the drier and rougher areas, like the foothill region shown here. *Travel Alberta*

progression that is seen further south is replaced by a northeast-to-southwest progression of similar character. The dry heart of the prairies is thus an elliptical area some 300 mi (480 km) from east to west and 200 mi (320 km) from north to south that runs across the Saskatchewan–Alberta boundary somewhat north of the Canadian border. This area is better known as Palliser's Triangle—after the surveyor who, in 1857–60, presented a series of reports on the Prairies,[16] in which he described as unfit for agricultural settlement most of Saskatchewan and Alberta between the 49th and 51st parallels. Surrounding this dry area on the northeast and north are zones of somewhat higher rainfall, which correspond to the short-grass plains; between these plains and

[16]See I. M. Spry, *The Palliser Expedition* (Toronto, 1963).

the northern forest is the zone that corresponds to the tall-grass prairie—the Park Belt, a zone that receives up to 20 in (500 mm) of rain and that supports the densest rural population on the Prairies.

DIFFERENCES IN THE SEQUENCE OF SETTLEMENT

The settlement of the American Great Plains was a logical outcome of the occupation of the Mississippi Valley; the frontier of settlement moved west without a geographical break, with the cattlemen in front and the farmers following behind, as their means permitted. In Canada, there was no such regular sequence. The first settlers on the Prairies—the Selkirk colonists of 1812–13—arrived via the Hudson Bay, and many of the later arrivals had trekked north from the United States, following the Red River Valley. Between 1812 and 1870, there was virtually no westward advance; the whole of the empty West was the preserve of the Hudson's Bay Company, which discouraged agricultural colonization, and not until the Prairies were sold to the Canadian government after federation did a movement westward begin, a movement that can truly be said to have gained momentum only after the construction of the Canadian Pacific Railway in 1885.

When Prairie settlement did finally become a reality, its pattern was somewhat different from that across the border. For one thing, by the 1880s, some of the technical problems of plains agriculture had been overcome and their lessons learned. In this respect, Canada reaped the benefits of experience gained the hard way in the United States: "American experience contributed greatly to the opening of the Canadian 'dry belt,' since Yankee knowledge of dry farming antedated settlement of the Canadian West. . . . Experience on the American plains was highly valued by Canadian officials in their search for suitable settlers for the West."[17]

There was, in fact, a mass movement across the border from the United States to Canada. It began just before 1900, at a time when the prospects of obtaining land had plummeted in the midwestern states almost to zero, and was promoted by the Canadian government, which wanted the Prairies settled and which set up offices across the border, issuing pamphlets with titles like *The Last Best West.* Between 1896 and 1914, 590,000 Americans crossed into Canada. "In truth, the movement was one of the greatest land rushes in the North American experience."[18] As Sharp, the historian of this movement, remarks, "The mass migration into the Canadian West was the last advance in the long march that had begun on the Atlantic seaboard. . . . It was a movement brought about by a desire for cheap land, the same desire that had activated the earlier agrarian waves to the south."[19] The majority of the newcomers were from the northern Midwest and the Spring Wheat Belt. So alluring did Canadian land appear that the U.S. government modified its own land laws to try to offset the attraction of Canadian prospects.

But it is necessary to add that this was a transitory phase. In 1910 a drought year checked the movement, and World War I killed it. In any case, many who crossed the border in search of land recrossed it a few years later. "Nearly two-thirds of the American residents who migrated to the Prairies returned to the United States shortly thereafter."[20]

The first phase of settlement saw two divergent lines of advance. In this phase, which lasted from 1872, soon after confederation, to about 1900, the Canadian government granted grazing leases on big areas of the dry prairies. These leases, however, were subject to cancellation if the land was needed for agriculture; in other words, a cultivator would have official priority, and the ranchers obtained security of tenure only when and where areas were declared unsuited to agriculture.

Meanwhile, the cultivators were advancing the farm frontier, and the railroads were extending their lines northwestward along the Park Belt.[21] They were feeling their way along what was effectively a corridor, with its walls represented by aridity to the south and frost to the north. Before they could penetrate these, new varieties of wheat had to be developed (and a major breakthrough occurred when the Marquis variety became available in 1911). But even with these new varieties, the climatic hazards were such that the government encouraged Park Belt farmers to practice mixed farming and not to rely too heavily on wheat. Today's agriculture in the Park Belt is the lineal descendant of this early, government-sponsored mixed farming.

When, therefore, the great boom in prairie settlement took place in the first years of the twentieth century, it was a more restrained affair, in land-use terms, than that on the Great Plains. The distinction between ranchlands and wheatlands was clearer, and there were natural restraints on wheat growing. This did not, however, prevent some of the same mistakes being made on the Prairies. As the population of Saskatchewan and Alberta increased by 500 percent in the first decade of the century, settlers pushed out into the Triangle during a series of wetter-than-average years, displacing the ranchers. Then came the inevitable: dry years followed and the farmers retreated. Around Lethbridge, for example, the average wheat yields for the years 1911–21 were:

---

[17]P. F. Sharp, "The Northern Great Plains: A Study in Canadian–American Regionalism," *Missouri Valley Historical Review,* vol. 39 (1952), 72–73.

[18]K. D. Bicha, *The American Farmer and the Canadian West, 1896–1914* (Lawrence, Kans., 1968), p. 11.

[19]P. F. Sharp, "When Our West Moved North," *American Historical Review,* vol. 55 (1955), 287.

[20]Bicha, *The American Farmer,* pp. 140–41.

[21]See W. J. Carlyle, "Farm Population in the Canadian Parkland," *Geographical Review,* vol. 79 (1989), 13–35.

20, 16, 18, 6, 43, 34, 7, 5, 13, and 9 bushels per acre. It was the familiar story of advance and withdrawal: of all the lands "entered for" in this region under the Homestead Act, only about one-half ever became the settlers' property. The rest were abandoned before the four-year occupance period was up.

There was one other factor in the early days that made life on the Prairies somewhat more secure than on the U.S. plains to the south. The Canadian government took a more realistic view than the U.S. government of the size of farm needed in the dry West, and although it only offered the same 160-acre (64.8-ha) homestead, it encouraged settlers to acquire extra land. This could usually be obtained from the railroads, whose land grants gave them alternate sections over most of the best prairie land. The companies were generally glad to sell the tracts of land to homesteaders on adjoining sections, usually for a modest price such as one dollar an acre. By the end of the prairie boom, 70 percent of the farms were over 200 acres (81 ha) in size, and some settlers who would otherwise have been defeated by drought or frost were able to hold on because of their extra acreage.

For all these reasons, the pattern of land use that developed in the Prairie Provinces was slightly more in harmony with the weather than that of the plains further south. In the years since then, the difficulty of

marketing the wheat crop (and the need to have, in consequence, other sources of farm income) has had the same effect of discouraging overdependence on wheat (see Fig. 18-7). Thus, although there are areas that, owing to their roughness or their aridity, are primarily ranching areas, the wheat areas are by no means without livestock. Oats and barley are grown for stock feed; the Prairie Provinces produce most of the Canadian output of these two crops, production of which has greatly diminished in the eastern provinces in recent years. As settlement has consolidated and, in particular, as the cities of the Prairies have increased in size, local markets for agricultural produce have grown relative to the previously dominant export market. Dairy products and meat now figure prominently in Prairie output, and certain other specializations have grown up to offset dependence on grain, such as the cultivation of rapeseed.

In spite of the somewhat greater degree of harmony that has characterized farming in the Canadian section of the Great Plains, however, the area suffered much the same fate as did the American section in the hard years between 1930 and 1935. The conjunction of low, depression-hit prices and drought during these years brought the Prairie farmers to despair and bankruptcy and obliged the federal government to intervene on their behalf. The Prairie Farms Rehabilitation Act of 1935 made available

Fig. 18-7. Canada: The Prairie Provinces. The diagram shows (a) the production of wheat in the three Prairie Provinces, 1926–87, and (b) the farm price of wheat over the same period (in constant-value, 1981 dollars per bushel). The diagram raises the question as to how any region, or its individual farmers, could hope to operate successfully in so violently fluctuating a market. (Source: M. Fulton, K. Rosaasen, and A. Schmitz, *Canadian Agricultural Policy in Prairie Agriculture*, [Ottawa, 1989].)

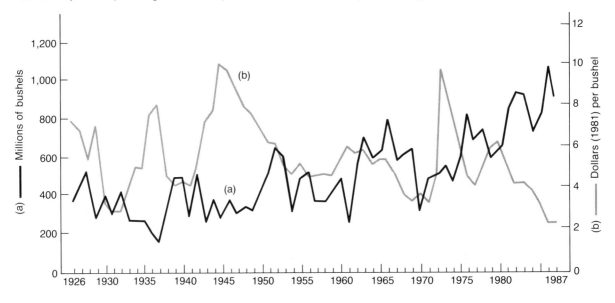

Table 18-6. Prairie Provinces: Agricultural Statistics, Census of 1986

	Manitoba	Saskatchewan	Alberta
Provincial population (thousands)	1071	1010	2375
Number of farms	27,336	63,431	57,777
Area in farms (million acres)	19.0	65.7	50.9
(million ha)	7.7	26.6	20.6
Percentage of provincial area in farms	11.8	40.7	31.1
Area under crops (million acres)	11.1	32.8	22.7
(million ha)	4.5	13.3	9.2
Percentage of farmland under crops	58.4	50.0	44.7
Area of unimproved farmland (million acres)	5.7	16.0	19.0
(million ha)	2.3	6.5	7.7

funds for repairing the ravages of drought and erosion. With the passage of time, the scope of PFRA work has been extended, in parallel with the work of government agencies in the United States, to cover a program of instruction in conservation, a program of irrigation works, and, as we have already seen, the administration of a system of community pastures. Small irrigation projects abound on the Prairies; for the most part, the larger schemes are to be found along the South Saskatchewan River in southern Alberta and Saskatchewan. Between 1970 and 1980, the irrigated area in these two provinces' schemes increased from 620,000 acres (250,000 ha) to 1,450,000 acres (587,000 ha). There is no lack of irrigable lands in this area, but the estimated cost of developing them is very high.

As Table 18-6 shows, the three Prairie Provinces between them had over 66 million acres (27.0 million ha) of cropland at the 1986 census, out of the Canadian total of 76.1 million (31 million ha). Manitoba, the first-settled of the three, had fallen well behind Saskatchewan and Alberta in area of farmland and volume of output: on the other hand, a larger proportion of its farmland is under crops than in the other two, and less falls into the depressing category of "unimproved" which, in 1981, covered 30 percent of all Canadian farmland.

DIFFERENCES IN SPACE RELATIONSHIPS

If the sequence of land settlement has been different on the Prairies of Canada from that on the plains of the United States, the simplest explanation of that fact lies in the difference in location of the two areas in relation to the more settled, eastern parts of the two countries. In the United States, as we have seen, settlement spread westward without a break from the Mississippi to the plains because there was no geographical obstacle to such a logical extension. In Canada, on the other hand, there existed a most formidable obstacle—the southward extension of the Canadian Shield, which interposed a barrier in the form of 1300 mi (2080 km) of forested wilderness between the settlements of Ontario and the site of the future city of Winnipeg.

When the westward spread of settlement was resumed on the Prairies in the years after 1870, it was, from the Canadian viewpoint, settlement in a world apart. A new natural environment was encountered not, as in the United States, after a period of adaptation in the subhumid eastern plains, but abruptly, and the new settlements were only tenuously connected with Canada proper. Many of the early settlers, in fact, were not from Canada at all, but rather arrived either direct from Europe, by way of Hudson Bay, or from the United States (where movement northward down the fertile Red River Valley offered a less hazardous prospect than movement westward into the dry plains). Only in 1855, with the opening of the Soo Canal, was a practicable route established to link the Prairie settlements to the East. The earliest railroad line into Winnipeg reached the city in 1878 from the south, and only in 1885 was the transcontinental rail link with Ontario completed. Not until the Trans-Canada Highway was completed did there exist a modern road link; before that time it was easier to "cross" this part of Canada by making a southward detour into the United States. Connection across the Shield was virtually connection through a foreign country.

The break in the sequence of settlement has also had an effect upon the present distribution of population and the urban pattern. We have already seen in Chapter 14 that the Great Plains of the United States lie within the hinterland of a line of cities located to the east of the region itself—Minneapolis, Kansas City, and Dallas–Ft. Worth, in particular. On the Plains themselves, there are no large cities between the 98th meridian and the Rockies except Denver, which owes more to the fertile Piedmont zone than it does to the plains; and its economic domain stretches west rather than east. Just as the surface of the Great Plains slopes east to the Missouri and the Mississippi, so the economic "gradient" has run the same way; the density of settlement increased from west to east; the products of the plains flowed eastward; cattle from the ranges have been fattened and finished in the Midwest; and, in turn, manufactured goods needed on the Plains have been distributed from midwestern factories and warehouses.

The commercial situation of the Canadian Prairies is different. Although there is the same contrast in density of settlement and communications network between the Dry Belt and the Park Belt as there is between the western and eastern plains in the United States, the economic "slope" ends abruptly at the edge of the Shield. The Canadian section of the Midwest, with its markets and its industries, is separated from the eastern edge of the plains by a vast area that is virtually uninhabited, and so neither provides markets nor generates traffic for the railroads that cross it. Across this empty area all goods leaving or entering the Prairies must be ferried, either literally, by lake steamer or, economically, in the sense that they are carried by railroad lines whose maintenance must be borne by consumers at either end.[22]

The obstacle presented by this "ferry service" was particularly serious because the Prairies depended for their livelihood on exporting bulky agricultural produce, and nearly all the exports (like those of the Great Plains) moved east. To reach their markets, they must first reach Montréal. Two things about this situation are worthy of comment. One is that the Prairies developed their own processing industries, whereas the plains in the United States remained largely nonindustrial.[23] These processing industries

were established further west in Canada than in the United States to reduce the bulk of the exports making the long journey east: in this sense, the movement of industry (especially meat-packing) onto the Great Plains, which we noted earlier in this chapter, represents the Plains belatedly catching up with the Prairies—belatedly, because their problem was never so acute.

This industrial development has served another purpose also—the supply role in relation to the northern frontier of Canadian settlement. The nature of this frontier will be examined in Chapter 22; for the moment, we need only note that the Prairies' industries must supply both the West and the North and that the Northlands frontier is continually expanding, and must be equipped and maintained from the nearest available supply bases—the cities of the Prairies.

Industry first came to the Prairies at Winnipeg. It is at the foot of the Prairie slope and at the western end of the route-bridge over the Shield: it is the funnel through which all east-bound produce must pass to leave the Prairies. Winnipeg quickly became the focus of the railroad network, and the lake city of Thunder Bay serves as its outport. It became the distributing point for goods received from the east and a great manufacturing center, in its own right. In 1991, its population was 652,000.[24]

In due course, however, manufacturing spread to other cities, too, as the agricultural processing industries reached out toward their raw materials. They have now been followed by a number of other national industries, and smaller centers have increased their manufacturing. Regina (192,000) makes steel in electric furnaces, and Saskatoon (210,000) is developing a high-tech specialization. Only in Saskatchewan is agriculture a more important employer than industry. In Alberta, in any case, there have been great changes that have stimulated industrialization, as we shall see in the next section (Table 18-7).

The other factor of interest about the space relations of the Prairies is that, although they were opened up and settled as an eastward-looking region, they could develop, and have developed, a second face—toward the west. Once the railroads penetrated the Rockies, everything west of Winnipeg was closer to the Pacific Ocean than the Atlantic, and cargoes began to move west rather than east, even

[22]As we have already seen in Chapter 6, the Canadian government recognized this economic fact and, changing only the simile used by the present writer, offered a "bridge" subsidy to the railroads operating across the gap.

[23]Except, of course, in their most southerly parts, where the presence of the oil fields (as in the Prairies) and proximity to the Gulf Coast combine to give the Texas plains some industry. It should, in fact, be re-emphasized that although what has been written about the "economic gradient" from west to east across the Great Plains holds true for most of the region, proximity of the Gulf Coast to the southern end of the plains gives to that part of the region a similar north-to-south "gradient" toward the coastal cities and factories.

[24]See J. Silver and J. Hull, eds., *The Political Economy of Manitoba* (Regina, 1990).

Table 18-7. Prairie Provinces: Leading Industrial Groups, by Employment, 1987

Alberta		Saskatchewan		Manitoba	
1. Food & kindred products	12,086	1. Food & kindred products	3,858	1. Food & kindred products	8,631
2. Fabricated metal products	8,581	2. Printing & publishing	2,817	2. Clothing industries	7,312
3. Printing & publishing	8,387	3. Machinery industries	1,813	3. Transportation equipment	6,300
4. Wood industries	6,082	4. Fabricated metal products	1,510	4. Printing & publishing	5,320
5. Machinery industries	5,842	5. Electrical and electronic products	1,475	5. Fabricated metal products	3,927
6. Chemicals & chemical products	5,403			6. Machinery industries	3,424
7. Nonmetallic mineral products	4,311			7. Electrical & electronic products	2,740
8. Refined petroleum & coal products	3,548			8. Primary metals	2,270
9. Electrical & electronic products	3,363			9. Wood industries	2,211
10. Primary metals	3,190			10. Furniture & fixtures	1,989
Total provincial industrial workers	78,220	Total, provincial industrial workers	19,772	Total, provincial industrial workers	54,031

when they were bound for Europe. But now a further change has taken place. Not only is it the overland distance that influences shippers to use the Pacific ports, but also some of Canada's most important bulk-cargo markets are in Asia. Grain from the Prairies, potash from Saskatchewan, and coal from British Columbia are all bound for the Far East. In gravity model terms, the pull of the West has greatly increased, partly due to comparative costs of transport, and partly to location of markets. The westward route to Vancouver becomes the natural one, and the role of Winnipeg as the original "funnel" is reduced relative to that of Edmonton and Calgary, the gateway cities on the trans-Rocky routes.

THE EXPLOITATION OF PRAIRIE MINERALS

The fourth factor that gives to the Canadian section of this great region a distinctive character is the presence of important mineral resources, especially oil. The existence of large deposits of coal, both bituminous and subbituminous, has been known for many years, but local demand was limited and external markets were out of reach. When such markets were eventually found, in the 1970s, they were in Asia and so coal joined the other west-moving commodities the Prairies now export via the Pacific Coast. The reserves of bituminous coal are huge, and present policy is to export this and use the bulkier subbituminous coal for domestic needs.

The second Prairie mineral to join the train for the West is potash, of which Saskatchewan has important deposits. Production is increasing rapidly, as it is mined to supply an apparently insatiable demand in Japan, China, and Southeast Asia.

But the main changes have come to the Prairies through the exploitation of oil and natural gas. It was in 1947 that the first major field came in, south of Edmonton, and although Saskatchewan, Manitoba, and British Columbia all came to have a share in the development, Alberta has in recent years been accounting for 85 percent of Canada's oil and gas production.

Alberta's economy forty years ago was based on agriculture, ranching, and forestry. The impact of the discovery of oil, consequently, has been considerable. Supply industries rapidly appeared and population increased. The most dramatic expression of this change in the economy is to be seen in the skylines of Edmonton and Calgary. These two cities have shared the development: Edmonton, the provincial capital, developing oil refineries and petrochemical industries; Calgary, the former cattletown, becoming the financial and commercial center. Their growth has been spectacular: Edmonton, with a 1986 population of 785,000, has outstripped Winnipeg; Calgary is smaller (671,000), but growing faster. The province has grown wealthy on severance taxes imposed on its mineral shipments.

We considered in Chapter 6 some of the economic and political implications of western Canada's rise as an oil and natural gas producer. In the years between 1947 and the opening of the Alaska North Shore fields twenty-five years later, Alberta's oil held an obvious strategic importance, because it was the only major oil source in the northwest of a continent where most of the production was in the south or southwest. During World War II, the northwestern part of the continent was oil-deficient; so much so that the wartime governments went to the length of developing a small known oil field at Norman Wells, at 65° N on the Mackenzie River, and of piping the oil from there 400 mi (640 km) over the wild Mac-

Commerce of the Prairies: The Canadian Pacific Railroad's classification yards at Winnipeg, into which funnel the grain and other goods gathered from the Prairies farther west. *CP Rail System*

kenzie Mountains to a refinery at Whitehorse. Nothing that has happened in the years since then has lessened the importance to Canada and the United States of their Northwest: quite the reverse.

It also gave Canada its first opportunity to reduce its dependence on foreign petroleum: indeed, to enter the market as a producer. This strengthened the hands of the Canadian government; it also strengthened the hands of the province in its dealings with Ottawa. It brought power, financial and political, to a West that formerly had no chance of counterbalancing the overwhelming power of Ontario and

Québec.[25] Here on the oil fields we meet again a phenomenon we last encountered in Texas—the impact of oil on capital formation and investment resources; the ability of an oil-rich area to generate finance and industry, and to take over a measure of the control of its own economy. If the federation weakened, the

[25]"The prairies are growing in size and wealth but the major economic decisions are still being made at corporate head offices in Toronto and Montréal, while the major political decisions come from Parliament Hill in Ottawa." R. K. Semple, "A Geographical Perspective on the Prairies: The 1980s," in Rogge, *The Prairies and Plains*, p. 121.

province could go its own way. If capital is to be raised for developing the great, empty Northlands, this might well be where that capital could be raised. For the first time, the Prairie Provinces could begin to decide their own destinies within the confederation.

But we must conclude on a rather less clarion note. Alberta's known and conventional oil reserves are running low, and will not last many more years. If energy production is to continue into the twenty-first century, "known" reserves must be (1) extended by continuing prospecting, and (2) replaced, as an energy source, by natural gas, which is available in quantity and not yet at peak production. The second adjective, "conventional," reminds us that, around Fort McMurray in northern Alberta, there are the huge reserves represented by the oil-bearing Athabasca tar sands. The technology of oil extraction from them is now well understood, and the plant is constructed: it is simply a question of prices—the world price of oil, and Canada's internal price. In the high-priced 1970s, it paid to exploit the tar sands. At the moment, it does not. And whatever powers the western provinces of Canada have acquired since they entered the petroleum business, the power to set world oil prices is not one of them.

19
Mountain and Desert

The Desert: Death Valley, looking south from Furnace Creek. Note the large alluvial fan spreading out from its apex at the left center of the picture and the conduit carrying water to the irrigated area on the right. *Spence Air Photos*

The Region and Its Character

Where the foothills of the Rockies rise to break the monotony of a thousand miles of plains, the westbound traveler enters a region of remarkable natural splendor. From the Front Range of the Rockies to the crest of the mountains of the Pacific coast is some 900 mi (1440 km) along a line from Denver to San Francisco; some 400 mi (640 km) on the line from Calgary to Vancouver. Spread across this great area is a magnificent variety of scenery, a wide range of environmental conditions, and a thin scattering of natural resources available to a population whose average density is less than 5 per sq mi (2 per sq km).

Both the variety and the splendor of the region are products of the relief. It is a region where the mountain ranges occupy only a part of the area; it is a region of basins and plateaus also; some of the latter lie at altitudes—7000 ft (2130 m) or more—above the tops of the highest mountains east of the Mississippi and high enough to have been glaciated. Some of its most interesting sections are the downfaulted valleys in the southwestern margins of the region—Death Valley and Salton Sea—where the earth's surface sinks to more than 200 ft (60 m) below sea level. In the southeastern section, the Colorado Plateaus often rise steplike above one another, to give the observer the impression of being surrounded by mountain ranges.

This physical variety, in turn, affects the climate. Although certain generalizations are possible (the frost-free season, for example, is seldom more than 120 days and is often less than 100), relief exerts the main control on temperature and precipitation, and overall description becomes impossible. The northwestern parts of the region are under the influence of Pacific air and have a winter–spring rainfall maximum, whereas the southeastern part has a summer maximum and receives much of its rain from violent thunder showers; but everywhere there are local conditions of rain and rain shadow produced by relief. The main rain-bearing winds west of the Continental Divide blow from the Pacific, bringing 80–100 in (2000–2500 mm) of precipitation, including very heavy snowfall, to the mountains along the coast. Immediately east of these mountains, there occurs a change of dramatic suddenness: the rain shadow falls across the adjoining plateaus and basins, and the dense forests of the mountain slopes are separated by only a narrow transition belt from the desert scrub of areas where annual precipitation averages 8–10 in (200–250 mm). Eastward the surface rises again, and the rainfall increases, especially in the Rockies; on their higher western slopes, a precipitation of 30–40 in (750–1000 mm) is again to be found. In a manner that is even more marked, because the west-to-east distance is shorter, the rainfall of southern British Columbia varies from 150 in (3750 mm) in the coastal mountains, down to less than 10 in (250 mm) in the Okanagan Valley, and up again to 30 in (750 mm) or more in the eastern Cordilleras. In both Canada and the United States, the Rockies cast their own rain shadow over the plains that lie to the east of them, beyond their forested slopes.

Climate, in turn, affects vegetation, which varies with both the amount of rainfall and the time and duration of the season. A regular sequence of vegetation zones can be distinguished, both horizontally and vertically; that is, a sequence of zonal changes generally holds true between low elevations and high, and between the heart of the desert and the better-watered lands surrounding it.

The zonal changes are generally from scrub through grass and transitional woodland to forest. At one end of the scale, in the lowest and driest areas, is found desert shrub vegetation, which covers the floor of much of the Great Basin. In the lowest areas, including the beds of former lakes, vegetation is sparse and consists of alkali-tolerant shrubs, such as shadscale and creosote bush. From this vegetational nadir, an increase in rainfall produces a sagebrush–grass combination. In the southern section, there is sufficient late summer rain to produce semidesert grasslands, which provide valuable, year-long grazing. Elsewhere in the intermontane region, however, the lack of summer rain restricts the growth of grasses, and the value of the vegetation for grazing purposes varies inversely with the amount of sagebrush.

Beyond the sagebrush–grass zone there is generally a belt of transitional woodland—an area where increasing rainfall encourages the growth of small, scattered trees. In the southern Rockies and the high plateaus, this zone is represented by a pinyon–juniper–grassland association and is encountered generally between 4000 and 6000 ft (1215 and 1825 m); the trees themselves have little value except for pinyon nuts, but the grasses that accompany them provide forage for stock. In California, a belt of transitional woodland occupies the lower mountain slopes. It is the tangle of manzanita and other shrubs known as chaparral. Above the transitional woodlands are the forests—open at their lower limits and so used for grazing, but becoming denser at higher elevations until they thin out into the alpine meadows above the treeline.

These introductory statements apply to the whole West between the Rockies and the high mountains of the Pacific. In so large an area, however, there are inevitably subregional contrasts based on cultural or economic differences within it; therefore it seems best to isolate two of the "corners" of the region and to deal with them separately. These are the southwestern corner, with its highly distinctive cultural associations, covered in Chapter 20, and the Northwest, lying in relative isolation behind its barrier ranges, which is covered in Chapter 25. Both these subdivisions of the West are well recognized by geographers, but neither of them is divided from the remainder of the region by any physical change in landscape other than those gradual changes brought about by increasing latitude. All share the quality of being "western," and the great variety of scenery and climate that forms the basis of work and recreation within the Mountain West.

Land and Livelihood

In such an area as this, where much of the land is extremely dry or rugged, the ways in which a livelihood can be made are strictly limited. Consequently, population is sparse and scattered. But scattering is not random—rather, the map of population can be explained by the locations of the various opportunities for employment as they have been discovered and developed. The major occupations have been the fur trade, mining, agriculture (irrigated farming, dry farming, and ranching), forestry, inter-regional transportation, and tourism–recreation.[1] The fur trade although fascinating in its history and geography and important for the knowledge of the region it generated, left no permanent settlements.

MINING

Many of the explorers and earliest settlers of the Mountain and Desert region were miners. Apart from the communities on the Pacific coast and the Mormons at Salt Lake, miners made up the bulk of the population west of the 98th meridian in 1855. Before permanent agriculture spread further, many a mining town had reached its gaudy heyday and was already on the decline—such mining camps as Virginia City on the famous Comstock Lode in Nevada, which was discovered in 1859. At various times a town of up to 20,000 inhabitants before its peak in the 1880s, Virginia City has since declined to become almost a ghost town, with a population of 600; it remains on the maps by grace of the tourist trade alone. Few of these early mining camps have had a continuous career from their establishment until the present day; even fewer, such as Butte, Montana, have remained to become towns in their own right. The history of the western mineral industry has been one of precipitate change.

Apart from the obviously temporary character of an extractive activity, these changes have been due to several factors. One of these has been the progressive discovery of new mineral ores. Almost without exception, the earliest miners were seeking gold, and so flocked to California after 1848 or to British Columbia and Colorado after 1858. After the first rush, the miners came to realize that, even if a fortune was not to be made in gold, there was an expanding market for silver and lead, copper and zinc; and this led to new beginnings in many mineralized areas the goldseekers had too hastily abandoned.

---

[1]Among the best, though not necessarily the newest, accounts of the advance into the West of the various frontier groups are R. A. Billington, *Westward Expansion* (New York, 1949 and revisions); and J. A. Hawgood, *The American West* (London, 1967). For Canada, consult the *Canadian Frontiers of Settlement* series, W. A. Mackintosh and W. L. G. Joerg, eds.

The western mining industry: (1) Early development depended on transport like this twenty-mule team on its way to a minehead in 1892. *U.S. Borax*

Another and continuing factor of change has been the discovery of new uses for the rarer minerals. The exploitation of tungsten and molybdenum, for example, waited on the demand for electrical goods and tougher steels; and the use of molybdenum brought into prominence a valley high in the Colorado Rockies, where, at the town of Climax, 11,300 ft (3436 m) above sea level on the Continental Divide, one-half of the world's supply of the mineral used to be mined. Even more recently, the continent-wide search for uranium has set off a fresh burst of prospecting and has brought to light, besides supplies of the mineral itself (notably in the region we are considering), many deposits of other minerals.

A third factor of change is the impact of new mining techniques, and especially the effect of enlarging the scale of production. The copper mines of Upper Michigan were put out of business by the opening of mines in Montana and Utah in the 1880s. There, the opportunities for open-pit mining and a larger scale of working created a decisive advantage, despite the fact that the percentage of copper in the ore was but

a fraction of that in Michigan. Much more recently, the petroleum crisis of the 1970s led to a burst of activity in mining the oil-bearing gilsonites and shales of northwestern Colorado and southeastern Utah. Their existence had been known for a long time; the only problem was that of extracting the oil economically. When the price of oil rose in the mid-seventies, extraction became worthwhile. Unfortunately for the companies involved, the price of oil then began to fall, and they were no sooner in production than they ceased to be competitive, and operations had to be suspended. But change and suspension are characteristic of the western mining and oil industry. Depending on price and demand, mines and wells may open and be shut down. The symbol of the oil worker is the mobile home.

The most important features of the pattern in the 1990s can be briefly summarized.

Copper has for a long time been the region's most valuable single metal. Arizona has been the leading producer, followed by New Mexico, Utah, Montana, and Nevada. In British Columbia, copper is mined in the south, especially around Kamloops and, further north, in a number of locations on the plateau; it accounts for more than half the value of metals produced in the province. But the recent history of copper illustrates how mining fortunes can fluctuate in the West. In 1980, the mines at Butte in Montana closed after a full century of production: the smelter at nearby Anaconda had already been closed down. In 1982, there were closures—blamed on foreign competition—in southeastern Arizona. The ores are still there, to be worked again as and when market conditions change, but American copper production in the 1980s varied widely from year to year.

Silver was the source of many a fortune in the West, especially on the Comstock Lode in Nevada, around Virginia City: the lode was at its most productive in the 1860s and 1870s, and is estimated to have yielded minerals worth $500 million. Today, as then, the Mountain states of the United States dominate production: the leading producer is now Idaho.

In British Columbia, the principal silver mines are around Kimberley.

Gold was what every miner dreamed of finding. There were the periodic incentives of strikes in California, Colorado, British Columbia, and, ultimately, Klondike in the Canadian Yukon.[2] Then the gold fever died down, and production remained small until, with the price of gold rising, it once more be-

[2]"Thus was completed a northward osmosis that had been going on since the rush to California, a kind of capillary action that saw restless men with pans and picks slowly inching their way along the mountain backbone of North America from the Sierras to the Stikines, up through Arizona, Colorado, Nevada, and Idaho, leaving behind names like Leadville, Deadwood, Pikes Peak, Virginia City, Cripple Creek, Creede, and Tombstone; up through the wrinkled hide of British Columbia, through the sombre canyons of the Fraser and the rolling grasslands of the Cariboo to the snowfields of the Cassiars, at the threshold of the sub-Arctic." Pierre Berton, *The Klondike Fever,* quoted by F. Walker, *Jack London and the Klondike* (London, 1966), p. 44.

The western mining industry: (2) The Kennecott copper mine at Bingham, west of Salt Lake City, Utah. *Salt Lake Area Chamber of Commerce*

came worthwhile to prospect and produce. Nevada, Utah, and Montana in the Mountain West, together with the Black Hills of South Dakota, are today's producing areas in the United States. In Canada, British Columbia is a gold producer, as it was in 1858, but sources in Ontario and Québec Province are today more important.

We have already referred to *molybdenum* and the early concentration of production of this useful metal in the mountains of Colorado. Its value, however, stimulated search and production elsewhere, in Arizona, Idaho, and New Mexico, and particularly in British Columbia, where mines were opened in remote corners of the province, and on Vancouver Island. In the early 1980s, the price began to fall, and a number of the more recent ventures have been shut down.

Zinc and *lead* are other metals mined in the mountains. Zinc comes from Colorado and Idaho, and from southern British Columbia and Vancouver Island, although larger Canadian supplies are to be found in the Northlands, and in Ontario and New Brunswick. Lead is often found in association with zinc and silver, and Idaho and Colorado are, as with zinc, major U.S. producers. The southern British Columbia zinc and lead ores are refined in one of the world's largest smelter complexes, at Trail on the Columbia River, where chemical by-products add to the value of the refined metals.

Among the region's nonmetallic minerals, the most important is *coal*. In earlier years, output from the western fields—in Wyoming, for example—was limited by the fact that the only local market for the coal was provided by the steam locomotive: in fact, the railroads were responsible for the original development. But, as we saw in Chapter 8, great changes have taken place in the demand for coal and in the methods of working it. The western coal industry, using strip or open-cast methods, has enormously increased its production. In 1972, Wyoming's output was under 11 million tons. In 1989, it was 171 million (Fig. 19-1). Much of this, admittedly, comes from the Great Plains section of Wyoming and the other producer-states, but in Canada the rapidly increasing output is from mines situated along the spine of the mountains, on the Alberta–British Columbia border. On the Alberta side, subbituminous coals and lignites underlie the prairies, with the harder bituminous coals and coking coals beneath the foothills of the Rockies. On the British Columbia side, mines line the southern end of the interprovincial border, with what is claimed to be the world's largest open-pit

coal mine at Sparwood, high on the western flank of the Rockies, below the Crow's Nest Pass. Other B.C. mines lie east of the Continental Divide, south of Dawson Creek. Provincial production for 1989 was 25 million tons.

Petroleum and *natural gas* are found in almost all the states of the region. Wyoming is the largest producer of petroleum, and New Mexico of natural gas. In British Columbia, the northeastern corner of the province, east of the Continental Divide, forms part of the oil-producing Prairie fields. This is the Peace River country, and it is linked by pipelines to the coast at Prince Rupert and Vancouver, as well as to the settlements in the southeast of the province.

Phosphate rock is worked extensively in southeastern Idaho, to supply an ever-increasing demand for fertilizers. This same demand has encouraged the production of *potash*, especially in New Mexico and Utah. In addition, the floor of the Great Basin in Utah and Nevada, with its many former lake beds, is rich in all kinds of salts.

Mining played a key role in the establishment not only of ephemeral towns, such as Virginia City, Nevada, but also of permanent ones. It has been crucial to the economies of such small cities as Kimberley and Trail, British Columbia (lead); Coeur d'Alene in the Idaho Panhandle (silver, lead, and zinc); Butte and Anaconda, Montana (copper); Ely, Nevada (copper); Helper, Utah (coal); Leadville and Grand Junction, Colorado (molybdenum, uranium, oil shale, lead, and zinc); and Bisbee and Douglas, Arizona (copper). The role of mining in larger metropolitan areas, such as Denver, Salt Lake City, Phoenix, El Paso, Albuquerque, or Spokane, is indirect, but significant; these cities perform processing, transporting, and supply and marketing services. They also manufacture mining equipment. Provo, south of Salt Lake City, services the large Geneva Steel works nearby. The plant was constructed under government direction during World War II to supply steel for the shipbuilding and aircraft industries on the Pacific Coast. It draws coal from the Castle Valley, across the Wasatch, and ore from southwestern Utah.

World War II disclosed to Americans both the strategic importance of many minerals whose names were scarcely known to them and also the possibility of obtaining almost all these minerals, at a price, in the Mountain West. In consequence of wartime need, many low-grade deposits were worked in the West, in operations that, after the wartime emergency, were too expensive to continue. We must

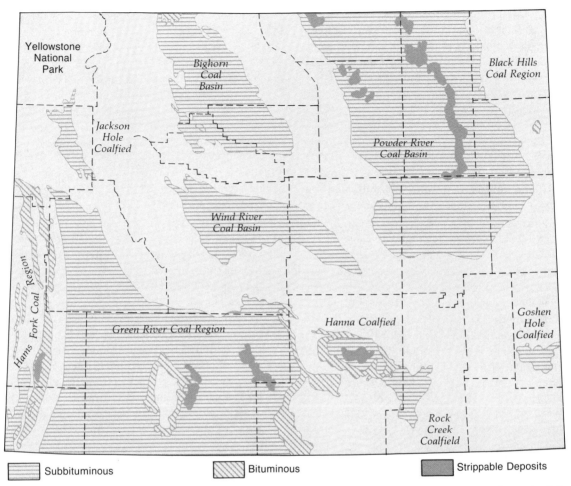

Subbituminous Bituminous Strippable Deposits

Fig. 19-1. Wyoming: Coal-bearing regions. These regions make up some 40 percent of the state's area and, it is claimed, contain nearly 24 percent of those coal resources of the United States that lie less than 6000 ft (1825 m) beneath the surface.

therefore think of part of the western mining industry as being in suspension; the deposits (and, in many cases, the plant) are available in case of strategic need or a rise in price, but for the present, it is simpler and cheaper to import from higher-grade sources abroad.

Not only the distribution, but also the character of mining operations in the West has changed. Large corporations have replaced the highly individualistic operator[3] of earlier years, and this has had the effect of increasing the stability, both geographically and

economically, of the modern operations. Whereas the early miner worked and perhaps looked for only one mineral, scientific mining has made possible the working of mixed ores found in association; thus, with a wider range of resources to draw upon, the life of the mining community is prolonged and made more secure. The coming of the larger mining unit has also made possible new and more costly prospecting methods. It was only with the availability of the Geiger counter and the search for radioactive minerals that the western mining industry reverted to the free-for-all of its earlier years; that once again, a century after its opening, the Mountain and Desert region became an El Dorado for the individual treasure seeker.

IRRIGATED FARMING

Farming in the intermontane areas is mainly oasis farming, and the oases are, for the most part, made

[3]This phrase was penned several years before the author discovered how gruesomely appropriate it is. A prospecting trip by Colorado miners in 1873 produced the only successful prosecution for cannibalism ever brought in a U.S. court. The five-man expedition became trapped by winter snows in the Rockies and when spring came only one man, a certain Alfred E. Packer, emerged. Prosecution followed. Obviously, to call an industry in which an operator eats his partners "individualistic" is to err on the side of understatement.

by man. Some dry farming is carried on; most important is the Columbia Basin wheat area, discussed in Chapter 25. The only other significant dry farming is some on the mountain bench lands, on the margins of the Utah and upper Snake River irrigated areas.

Irrigation was carried on in the south of the region by the Spanish settlers, and by the Indians before them. The first major agricultural development, however, was that of the Mormon community that founded Salt Lake City in 1847 and turned the arid alluvial lands at the western base of the Wasatch Mountains and high plateaus into green fields.[4]

Irrigation farming in the Mountain West may be considered at three levels of scale and organization: (1) the large, relatively self-contained irrigated farming regions, such as the Columbia River Basin, the Snake River Plain, the Wasatch Piedmont, the Salt–Gila valleys, and the Imperial Valley; (2) scattered small irrigated patches that support small trade centers and provide a base for the use of surrounding dry areas: examples are found along the Humboldt River in Nevada or the upper Colorado River and its tributaries; and (3) irrigated fields and pastures on individual ranches that augment natural grazing lands of low productivity.

Along the U-shaped course of the Snake River across southern Idaho are three large irrigated areas. The eastern section stretches upstream from American Falls and Pocatello past Idaho Falls; the middle section is centered on Twin Falls; the western section lies in the valleys of the Boise and Payette rivers, well above the deeply incised Snake River and near the cities of Boise and Payette. As in almost all irrigated areas in the West, alfalfa and other crops, such as barley, are grown to feed sheep and beef cattle from surrounding ranches and, near cities, some dairy cattle. Sugar beets are also important, both as a cash crop and a feed crop (using the plant tops and the pulp, obtained after processing). The cash crop that distinguishes the Snake River oases, however, is the famous Idaho baking potato. Additional cash crops in the westernmost area are alfalfa seed, vegetables, and fruit, especially prunes and sweet cherries. The western base of the Wasatch and high plateaus in Utah boast a diversified agriculture. Sugar beets, fruits, and vegetables are grown to supply the urban markets, and alfalfa to feed livestock, the most important product. Dairy products from this area go to Salt Lake City, Ogden, Provo, and smaller cities.

Other irrigated areas are small and scattered, although highly important for local trade centers, which they help support; examples are Grand Junction on the upper Colorado, Reno and Carson City at the eastern base of the Sierra Nevada, and Winnemucca and Elko along the Humboldt River routeway across the Great Basin. The Grand Junction area is an important producer of peaches.

LIVESTOCK RANCHING

Ranching in the Mountain West differs from that of the Great Plains in two major respects: (1) much of the land used is owned by the federal government rather than by ranchers, and (2) much of the ranching is carried out on plateaus, or in dry basins surrounded by rolling foothills and mountains.

As we have seen, federal ownership is a result not of a particular sociopolitical philosophy, but rather of the course of political and economic history. Such federally owned lands are often the dry, rugged, and remote lands of low productivity that were not homesteaded or purchased because they could not support a livelihood. Most of the mountain areas, and the highest parts of the Colorado Plateau, being better-watered and forested, were placed in national forests and are administered by the U.S. Forest Service, which was established early in this century. The more extensive, lower and drier areas, including much of the huge Great Basin, the lower parts of the Colorado Plateau, and almost all of the desert to the south, were left in federal ownership but largely unsupervised until 1934.

As ranching spread into the area, the ranchers chose sites—called base properties—that could provide water, not only for household and livestock use, but that could assure pasturage and some forage crop acreage, often irrigated, for supplementary feeding during the critical dry or cold seasons. Invariably, such sites were at the base of a mountain range near small streams that flow into the lowland basins. Situated as they were, these base properties allowed ranchers to utilize the surrounding lands of lower productivity, that is, to drive cattle or sheep into the mountains to graze upon woodland or alpine pastures in midsummer, and into the dry basins in winter, when rains revive the pasture and snow may provide water for livestock. Many ranchers have been able to extend their base properties to include suffi-

[4]D. Worster, *Rivers of Empire: Water, Aridity and Growth in the American West* (New York, 1985).

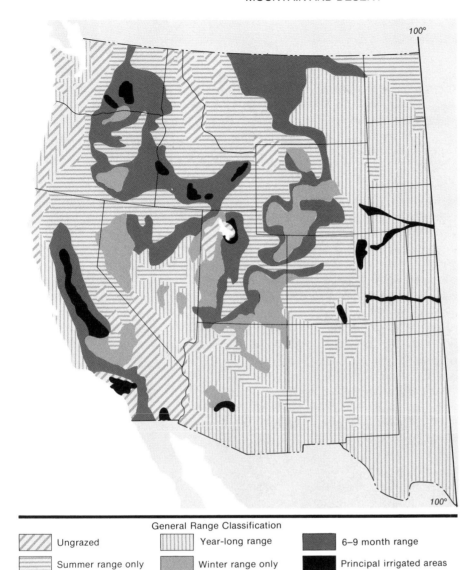

Fig. 19-2. The Western United States: Seasonal land use.

General Range Classification

Ungrazed	Year-long range	6–9 month range
Summer range only	Winter range only	Principal irrigated areas

cient foothill lands to provide pasturage in spring and fall (Fig. 19-2).

Summer grazing rights in the mountains are leased from the U.S. Forest Service, and the right for winter grazing in the basins from the Bureau of Land Management (BLM). Until the BLM was established by the Taylor Grazing Act (1934), grazing in the vast basin areas had been carried out by an open-range system, like that of the Great Plains half a century earlier, that is, a given range was used by whoever could control it. Such competitive use was bound to lead to serious overgrazing and deterioration of the range. The BLM, like the Forest Service, has at-tempted to map and scientifically appraise the range-lands. Today, in order to preserve and maintain their quality, its Grazing Service, like the Forest Service, assigns ranchers grazing permits and levies fees, cal-culated in animal unit months—the estimated amount of pasture that will support one cow, or its equivalent (such as five sheep), for one month. The Bureau assigns permits through local committees of ranchers on the basis of two major criteria: (1) own-ership of a base property that allows efficient use of nearby federal lands and (2) prior history of use of the range. Such a system provides for stability and continuity.

The system is not without problems, as W. Calef found in his study of grazing permits in operation,[5] since the Bureau has never had sufficient staff to map and appraise the rangelands adequately. It has always been difficult to make the necessary range forecasts and very difficult to get the local ranchers' committees to agree to reductions in numbers of stock to be grazed. The 1963 congressional hearing on the grazing law was told quite frankly by witnesses that ranchers would not be able to cover their costs if proper stocking rates were insisted on. Thus, public lands are overgrazed and eroded, as are private lands, although in some areas the ranchers do not use their range, but rather hold onto their right to do so, as a means of keeping their prior use status or as a reserve, "just in case."

The use of different grazing lands is a kind of transhumance: In spring, a flock of sheep may graze, and the lambs may be born, on the rolling foothills, preferably a part of the base property, so that the lambing may be more easily supervised. In June, the flock and the new lambs are herded up the slopes, following the retreat of the snow and the new growth of pasture, reaching the high alpine meadows in July. There, several herders, with a chuck-wagon "home" and each with one or two horses and dogs, move the flock of perhaps 2000 sheep and lambs slowly, but systematically, over the meadows, protecting them from such predators as coyotes and bears. Once or twice a week, a pickup truck from the ranch brings food and supplies and perhaps takes sick or injured lambs back to the ranch. In late August or early September, the herders begin to guide their grazing flock back down the slopes.

The increasing cost of satisfactory labor, the low returns from much of the rangeland, and the competing uses for such land, for example, for recreation, have tended to bring a decline in the old style of ranching and, especially, in sheep raising. The vast rangelands have contributed only a small fraction of the total feed requirements of the country's cattle and

sheep. As more livestock are raised on irrigated farms, that fraction is becoming smaller still.

FORESTRY

Lumbering and the forest-products industry are carried on in many parts of the mountains. In general, the forests become denser and commercially more valuable toward the colder North so that lumbering increases in relative importance through western Montana and northern Idaho until, in British Columbia, it overrides all other occupations. The forests of the Rockies are, for the most part, pine, spruce, fir, and larch, and in the United States less than one-third of the forest area represents virgin growth. South of the border, the lumber industry is most fully developed around such centers as Coeur d'Alene and Lewiston in Idaho and Missoula in Montana (where the adjacent mining areas have in the past provided an immediate market for timber), but there are important outliers of the industry in the upper Colorado Basin and in the highlands of central Arizona (Table 19-1).

Much of the forest in the United States is federally owned and lies within national forests; most of that in British Columbia (94 percent) is owned by the province. Rights to harvest timber are sold to private companies, usually under long-term lease in Canada and by cutting contracts in the United States. The use of the forestlands, especially in recent years, has

Table 19-1. Western United States: Areas Lying Within National Forests, and Areas of Commercial Timberland, by States, 1985

	Area Lying Within National Forests	Area of Commercial Timberland	Timberland as a Percentage of State's Area
Arizona	11.3	3.9	5.4
California	20.5	16.3	16.3
Colorado	14.4	11.3	17.0
Idaho	20.4	13.5	25.6
Montana	16.8	14.4	15.5
Nevada	5.2	0.1	—
New Mexico	9.3	5.5	7.1
Oregon	15.6	24.2	39.3
Utah	8.0	3.4	6.5
Washington	9.1	17.9	42.0
Wyoming	9.2	4.3	6.9

Note: The area within national forests in these states is in most cases larger than the area of commercial timberland because the National Forest is merely an area *containing* forest: it may also contain—as in Nevada—large areas of non-forested land it administers.

Numbers in millions of acres; for hectares, divide by 2.47.

[5]W. Calef, *Private Grazing and Public Lands* (Chicago, 1960). The Forest Service charges more for grazing on its section of the public domain, and generally succeeds in keeping a closer control over users than does the BLM. This is not because of any carelessness on the part of BLM officials, but rather because the Bureau has operated for years on a shoestring, a fact that, in turn, can be traced to political struggles over its existence and functions.

On this topic, see also W. Voight, Jr., *Public Grazing Lands: Use and Misuse by Industry and Government* (New Brunswick, N.J., 1976).

Tourism in the Mountain and Desert region: (1) Lake Louise, Alberta.
National Film Board of Canada

caused considerable conflict in the United States. Because the forest and mountain lands are largely coincident, they are valued not only for timber, but also for their scenery, as a wildlife refuge, and for recreation; the forest itself is important for water catchment and conservation. There is general agreement that, over much of the area, timber should be harvested on a sustained yield basis. The principal issue is *how* the timber should be harvested. Timber companies claim that it is most efficient to cut all the timber within a given area at once (clear-cut), and then to reseed. Conservationists, appalled by the longlasting scars such a practice leaves on the landscape, and especially the loss of soil and water and the disturbance in the balance of nature, favor selective cutting.

In British Columbia, the forest-products industries grew up along the coast, where the forests were accessible and the timber could be moved by water along the coastal inlets and channels. With increasing accessibility to the interior, however, and the spread of roads and power lines, and with a steady demand for lumber from the treeless Prairie settlements to the east, the industries spread inland. It is estimated that 46 percent of the province's area is covered by productive forest.[6] We shall consider British Columbia's forestry further in Chapter 25.

[6]The critical factor in rate of exploitation along the accessible edges of Canada's forests is that 80 percent of the productive forest is on provincial crown lands, where the cut can be monitored and controlled.

413

Tourism in the Mountain and Desert region: (2) Lake Mead. *Las Vegas News Bureau*

TOURISM

Whereas commercial tourism in the West may be said to have begun in 1872 with the opening of Yellowstone National Park, its modern development was based on the automobile and the extension of the region's road network. Banff and Jasper, Yellowstone and the Grand Canyon have become places familiar to millions who visit each year these products of glacial erosion, volcanic activity, or desert weathering. The distinctive cultures of the Spanish Southwest and the Indian reservations attract other thousands. Tourism has been responsible for the opening of large sections of the region, which previously were both economically valueless and inaccessible. Lacking the profitable resource base of the lowlands, the Mountain states have made capital out of their scenery. Ironically, the state whose economy is most dependent upon tourists and recreation, and the two

largest resort cities of the region, Las Vegas and Reno, depend not only upon outdoor scenery, but also upon gambling and glamorous indoor entertainment.

Las Vegas and Reno are year-round resorts, but over much of the Mountain West there are, in effect, two separate tourist industries—the summer and the winter trades. In summer, with the passes over the highest mountains open, tourist traffic flows into the remotest parts of the region, by car, on horseback, and on foot; only the hottest desert areas in the Southwest are closed to it. In September, however, the roads over the mountain passes begin to close, and with this closure—only the major routes are kept open—the service population withdraws, leaving whole sections of the mountains empty of inhabitants until the following spring. The winter tourist traffic then begins, making either for such centers as

Banff, Sun Valley, or Aspen, which have boomed with a remarkable increase of interest in skiing, or for Arizona, which, like Florida and southern California, attracts visitors by offering them a January mean temperature of 45 to 50°F (7 to 10°C) and clear, dry weather. Then, too, is the season to explore Death Valley and the deserts of the Southwest.

Tourist facilities are constantly being improved; new settlements spring up, and roads are cut further into the wilderness. Such roads as the Banff–Jasper Highway, the "Going to the Sun" mountain road in Glacier National Park, and the roads above the 10,000-ft (3040-m) contour in the Colorado Rockies are, besides being a testimony to engineering skill, a token of the force of modern tourism in opening the remoter West.

But it is axiomatic that in a region that relies as heavily as this one does on attractive scenery and the

Yellowstone National Park: Old Faithful Geyser. The picture highlights the general conflict between preserving nature and encouraging tourism; each year, 2 million visitors come to witness this geyser's eruptions. *Wyoming Travel Commission*

appeal of the wild, two things will happen. One is that the honeypots—the exceptional locations like Yellowstone or the Grand Canyon—will become overused; that is, their popularity will destroy the very qualities that made them famous. Unless the National Park Service is prepared to operate some kind of rationing or quota system, it is difficult to see how the visitors can be kept away: success threatens to overwhelm nature. The other thing is that a certain number of those who first visit these areas as tourists decide to return as settlers. The Mountain states have been experiencing a rapid rise of population; Colorado, the best example, increased its population by 38.7 percent between 1970 and 1980, and by 14 percent between 1980 and 1990. Some of this in-movement has been generated by environmental attraction, and some by mineral developments. Denver has become an important business center, sharing some of its new features with such cities as Houston and Calgary.

Transportation and Cities

In few regions of the world is there so large a proportion of the population involved in the transport business as in the Mountain and Desert region. If to the running of regular communications we add the service of the millions of tourists who annually invade the region, then we account for the livelihood of almost the entire population of some areas.

There are two main reasons for the importance of the transport industry:

1. It was transport, in the form of the western railroads, that dominated the settlement period in the region, as we saw in Chapter 9. It was the railroads that led the way for all but the hardiest pioneers into the West; that had land to sell and the means of reaching it; that located the towns; that were, in short, the agents of civilization in the West. Their primacy was incontestable until the 1930s brought a comprehensive road network over the region.

2. As a region of sparse population and scattered resources, the Mountain West has always been a geographic barrier between areas of denser population and more intensive activity; a barrier zone in which few Americans had business to transact and across which they were eager to travel as quickly as possible. But to maintain travel and communications across an empty region requires almost as large a staff and administration as to maintain it across populous areas, so that, although the regional population in the early years was very small, a large part of it consisted of the railroadmen who manned the division points and the lonely section posts along the tracks. The towns were located primarily for the convenience of the railroads and only gradually did they develop functions that linked them with their surroundings. They were simply the piers of the transport bridges that spanned the empty West, strung out in east–west lines along the routes of the Canadian Pacific, the Union Pacific, or the Santa Fe.

When road travel began, the same pattern was repeated. The roads, built to cross the area rather than to serve local settlements—which were, in any case, few—ran for scores of miles through uninhabited areas. Along them, therefore, there sprang up service points for motor traffic that duplicated those of the railroads. What the towns were to the early railroads, the "rest-stop" service clusters have become to the roads, with this distinction: with the increase in size of the locomotive and especially with the coming of diesel haulage, there is less and less for the railroad towns to do. Meanwhile, the business of the road service-points and the number of such points is still increasing (Fig. 19-3).

Most of the routeway cities that have grown in size owe a part of their growth to the kinds of land use and livelihood we have discussed, especially to irrigated farming that supports relatively dense populations. A few cities, such as Butte, Anaconda, and Trail, have had their economic base in mining and in mineral processing. Albuquerque in New Mexico (481,000), on the Santa Fe Railroad and the main highway routes that skirt the southern end of the main ranges of the Rockies, and El Paso, Texas (597,000), on the Southern Pacific route at the Rio Grande crossing, are routeway cities that have become regional centers serving military bases, small irrigated farming, or widespread mining areas. Even Las Vegas, Nevada (741,000), the gambling resort, had its beginnings as a routeway city (Salt Lake City to Los Angeles) and mining center. Most of the larger cities benefit from several economic bases, including transportation. Two of the rail lines that cross the northern part of the region within the United States converge at Missoula, Montana, before crossing the Rockies, to emerge from the mountains at Spokane, Washington (356,000). Both cities have become regional centers for forestry, mining, ranching, and irrigated farming. Pocatello, Idaho, on the Oregon Short Line route from Salt Lake City to Portland, serves the upper Snake River irrigated district and

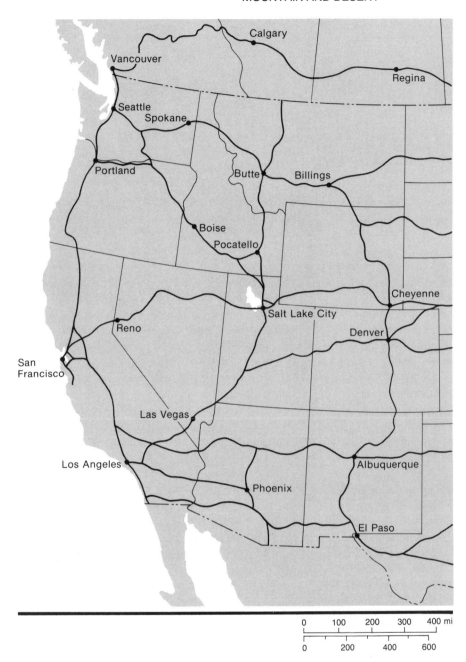

Fig. 19-3. Highways in western North America: The Interstate System and the Trans-Canada Highway. Construction of the system began in the 1950s, and all but a few short sections have been completed.

phosphate mining and ranching nearby. Reno, Nevada, serves mining and irrigated farming districts at the base of the Sierra Nevada, but depends on its casinos and resorts; it is on the oldest transcontinental rail and road route across the West.

Two cities dominate the whole of the Mountain and Desert region, from the Canadian border to northern New Mexico. One of these cities is Denver (1,800,000), the other Salt Lake City (1,072,000).

One is in the oasis belt of the Rocky Mountain Piedmont, the other in the center of the irrigated lands at the foot of the Wasatch Mountains; the hinterland of one lies mainly to the east of the Continental Divide, the other to the west. Although Denver actually lies outside the region, in all that concerns the economic activity of the Mountain West, the two cities are rivals—as railroad and road centers, miners' and stockmen's markets, manufacturing cities, and distribution points. Denver is larger, but Salt Lake City, because of its religious significance for the Mormons, who have settled far and wide west of the Rockies, exerts a powerful influence over its hinterland. In Canada, Edmonton and Calgary share the functions and something of the situation of Denver in servicing the mountain region.

Development or Preservation?

The interior West was for a long time literally passed over, a land rejected or unclaimed in favor of more economically attractive areas. As such, much of it was left in public ownership. In the twentieth century, its resources—minerals, water, forests, irrigable lands and lands suitable for grazing, and, especially, its scenery—have become more valuable and more coveted. The challenges to the federal government to release the land and its resources for development by private enterpreneurs become ever more vigorous.

There has developed, consequently, a double controversy about the West; more particularly, about the public domain. The first point of argument is should there be federal or private development? The second is should there be development at all, or preservation?

Since the U.S. federal government owns 727 million acres (294 million ha) of land, or 32 percent of the nation's land area, and most of that is in the West, there is an obvious case for arguing that it should release land for private development. That argument, however, implies that, under federal ownership, the land is *not* being developed, whereas we have already seen that the government actively exploits—or permits the exploitation of—most of the useful resources of the domain by timber-cutting and grazing permits. This first argument simply boils down to the question of a decision about the kind and rate of development: a development for individual interests, or a development overseen by a kind of referee—the federal agencies—which will try to balance all the interests in an area for the benefit of all.

But the second argument challenges the idea that development should occur at all. In modern North America, with a population predominantly urban in character, there is an obvious need for the physical relief offered by recreational space and outdoor life. This felt need is given an emotional edge by the knowledge that much natural wealth and beauty has already been destroyed, and there are very strong pressure groups which urge that the preservation of

parts of the West as nature reserves should be undertaken as a matter of urgent policy.

Certainly the evidence in favor of action is depressingly real. It might be imagined that to suburbanize a desert was an impossible task; yet the Americans have succeeded. Some of the most attractive, and some of the most arid, parts of the West have been sold off in small lots and, with their beauty parlors and their real estate agents, these provide plenty of ammunition for the preservationists.

What has happened, of course, is that the highways that have opened up the region's splendid scenery to outside view have also had the effect of exposing that scenery to the full force of tourist traffic from a largely urban-based population[7] for whom the weekend habit and the long vacation have become second nature. Furthermore, that population enjoys a total mobility up to, say, 500 mi (800 km) at any given weekend. It has become fashionable to use this private mobility to live in the desert and work in the city or, at the least, to own a second home out in the wilds. The rising price of desert and mountain building lots over the past decade reflects this developing life-style.

The concept of "wilderness" has in these circumstances taken on for some Americans the qualities of a religion.[8] It is a religion which already has its law—the Wilderness Act of 1964, which provides for the setting up of reserves. One of its prophets, quoted with approval by a former secretary of agriculture in the United States, explains the concept as follows: "Wilderness is an anchor to windward. Knowing it is there, we can also know that we are still a rich Nation, tending to our resources as we should—not a people in despair searching every last nook and cranny of our land for a board of lumber, a barrel of oil, a blade of grass, or a tank of water."[9]

The idea of protecting and preserving nature is not new: the United States already has over 75 million acres (34.4 million ha) of national and state parks and Canada has no less than 108 million acres (43.9 million ha). But to press for extension of these areas is

---

[7]Bureau of Reclamation figures tell us that, in 1989, such places as Lake Mead and Glen Canyon (the latter of which was, not many years ago, virtually inaccessible), in the heart of the desert and plateau country, attracted twelve-hour day visits by 7.0 and 6.4 million tourists, respectively.

[8]On the idea of wilderness, see R. Nash, *Wilderness and the American Mind* (New Haven, Conn., 1967 and revisions).

[9]Stewart Udall, *The Quiet Crisis* (New York, 1963), p. 181.

aggravating to westerners who are denied other use of the lands insulated by reservation. On these nature reserves wildlife increases, and the surrounding ranch and farmlands may suffer its depredations. What is more, such nonuse, or "single use" as it is perhaps fairer to call it, contravenes the basic principle with which we began: that in the West the important economic point is that only by multiple use, the combination of all possible forms of resource value, can these lands pay for themselves. The old "single-use" problem of the West arose from such activities as wheat monoculture and overgrazing. Today's problem is different in detail, and moreover it is willed on the region from outside (for the pressure comes from city-based groups). But it is the same problem in essence: how to make the Mountain and Desert region a valued and valuable part of the nation.

20
The Spanish and Indian Southwest

Pueblo country: Acoma Pueblo and Katzimo, the "Enchanted Mesa"
west of Albuquerque, New Mexico. The view is toward the northeast.
Photo by Dick Kent

Regional Identity

Most of the statements made in the last chapter about the Mountain and Desert region apply to the whole of the great American West—statements about its economy, its emptiness, or its problems over land and water use. But it is justifiable and, in fact, necessary to look beyond the generalizations, to examine more closely the character of individual parts of the West, and here and in Chapters 23 and 25, we are concerned with doing precisely that. Between them they cover the peripheries of the West: southern, coastal, and northern. In each of these three areas, individual factors combine to produce a distinctive regional character, superimposed upon the basic quality that comes from being western.

The first of these peripheral areas is the Southwest. Its distinctiveness is accepted by regional specialists, yet hardly at all in the general consciousness of the nation.[1] Perhaps this is because of its shape and position: it is a narrow periphery of the nation that stretches all the way from southern California to eastern Texas, with an extension into Florida. It has never *acted* as a region or section and, to most Americans further north, it is an alien margin to their mainly English-speaking subcontinent.

On what grounds, then, can its regional identity be justified? The argument is primarily a cultural one. Within the United States this region possesses a marked concentration of two earlier cultures and their artifacts, cultures that, in contrast with the blended "American" culture that predominates over most of the remainder of the national territory, are relatively unmixed and resistant to change. These are the American Indian and the Spanish.

CULTURAL IDENTITY OF THE INDIAN LANDS

Although Indian peoples have at some time occupied the whole of the North American continent, there are valid reasons for regarding their occupance of what is now the southwestern part of the United States as in some ways the most remarkable of their achievements. Certainly that occupance is of great antiquity, when contrasted to the time-span during which other Indian peoples further north had occupied the lands where settlers encountered them. At least since the time of Christ, there have been settled Indian tribes in the Southwest; so that at the period of the first contacts with Europeans, this region of the continent was supporting the densest Indian population anywhere north of Mexico—in an area, we should note, of meager rainfall and limited resources of game. At least since the third century A.D., these Indians have been building permanent structures, reports of which lured the Spaniards northward in search of "cities," to which they gave the Spanish name that has been attached to them ever since— pueblos. Among their other characteristics, these same Indians practiced irrigation agriculture; in a historical sense they were the forerunners of the Spaniards, the Mormons, and the Bureau of Reclamation. Finally, the southwestern region today contains by far the largest area of Indian reservations in America north of the Rio Grande: of some 50 million acres (20.2 million ha) of tribal and Indian trust lands in the United States, 20 million (8.1 million ha) are in Arizona and a further 7 million (2.8 million ha) in New Mexico (see Fig. 20-1). These two states have

---

[1]In recent years, however, it has been well studied, as a spate of books testifies. Among these may be noted J. W. House, *Frontier on the Rio Grande: A Political Geography* (Oxford, 1982); O. J. Martinez, *Troublesome Border* (Tucson, Ariz., 1988); L. C. Metz, *Border: The US–Mexico Line* (El Paso, 1989). These are general studies: other titles will be found in the sections on Spanish settlement and on water problems.

Fig. 20-1. The Southwest: Indian reservations. Notice the curious overlap of the Navajo and Hopi reservations, which has resulted in considerable intertribal friction.

everyday life of at least 10 million Americans of Spanish-Mexican origin.[2]

The Southwest is the only large area of the United States (as opposed to inner-city districts of limited extent) where people buy groceries or offer services in a language other than English. By this criterion, Los Angeles is the largest city of Spanish culture north of Mexico City; at the 1980 census, there were over 2 million persons of Spanish origin in its metropolitan area, with many more in adjacent cities, in San Diego, and in the counties east and south of San Francisco Bay. In a cultural sense, the southwestern states of the United States form an area of overlap between the two macro-regions of Spanish and Anglo-Saxon culture—between the Hispanic and the Anglo, as they have long been identified.

Within this Hispanic region, whose extent we shall consider in the next section, society and landscape have been stamped with many features whose origins lie in the Spanish Empire to the south—or for that matter back in Spain. Among the elements of contrast between the colonizing work of Spain and that of its northern rivals were the Spanish attitude to the Indians, the Spanish land policy, the conscious, centralized, planned imperialism of Spain (however haphazard its application at the fringes of the empire), and the relative neglect of cultivation by the European colonists, who tended to concentrate on ranching and to depend on the native population for food crops. Even though these attitudes and habits were modified in time, their application for two and a half centuries following the first settlement (in 1598) of what is now the Santa Fe area has left a distinctive mark on these southwestern lands.

It is possible to argue that either of these cultures by itself would give the Southwest a marked regional personality. In practice, the two reinforce each other in giving the area its distinctiveness, for each had taken over much from the other before any third party—in the form of the Anglo culture—appeared on the scene. Even after 1850, however, when the whole of the Southwest had become American territory, there was a real sense in which it remained a

an Indian population of about 270,000 and form the core of a distinctively Indian Southwest.

CULTURAL IDENTITY OF THE SPANISH LANDS

In any account of the settlement and use of the West as we know it today, however, pride of place must be given to the Spanish pioneers. It is easy for Anglo-Americans in other regions to forget that Spanish settlement of the continent predates the Pilgrim Fathers by a full fifty years. The Spanish approach was from the south, initially to the Gulf Coast and Florida (where the town of St. Augustine was founded in 1565) and then overland from northern Mexico to the Rio Grande Valley. Although it is now well over a century since political control of the Southwest by a Spanish-culture government in Mexico City was terminated by the treaty of Guadalupe Hidalgo (1848) and the Gadsden Treaty of 1853, the cultural influence of Spain remains. It is found in place-names and architecture; it is recalled by the importance to the region of mining and ranching—both introduced by the Spaniards—and it is a part of the

[2]The U.S. Bureau of the Census identifies "Persons of Hispanic Origin." For the United States as a whole, the 1990 census reported a resident population under this head of 22,354,000, a remarkable 53 percent increase over the figure for 1980. As we have already seen in Chapter 2, however, and shall see again in this chapter, it would be naive to take this number seriously: all the census takers have recorded is the number of Hispanics willing to stand up and be counted.

The Southwest and the
Spanish influence: Church
of San Xavier Del Bac, at
Tucson, Arizona. *Arizona
Department of Tourism*

region apart. In his perceptive analysis of the West, D. W. Meinig has pointed out the essential character of western settlement as being based on the growth of a series of nuclear areas:

Although folk colonization is always selective and uneven in area, in the East the general tide of settlement was relatively comprehensive and local nuclei and salients in the vanguard were soon engulfed and integrated into a generally contiguous pattern . . . holding to the same scale, the pattern in the West is a marked contrast; several distinct major nuclei so widely separated from one another and so far removed from the advancing front of the East that each expands as a kind of discrete unit for several decades, only gradually becoming linked together and more closely integrated into the main functional systems of the nation.[3]

Meinig recognizes six such major nuclei and a number of minor ones. If we accept this analysis, then the Southwest is constituted by the development of two of the six nuclei—those of what he calls "Hispano New Mexico" and southern California, with three minor nuclei as links or outliners: the El Paso area, the Phoenix area, and (although outside Meinig's immediate consideration) the San Antonio nucleus in Texas. Between these nuclei and those to

the north—Denver, Salt Lake City, and San Francisco—there is a cultural borderland to be crossed.

Regional Boundaries

If we accept the fact that the Southwest is a region distinguished by its cultures and isolated to a degree from its neighbors, then there is no reason to expect that it will have clearly defined boundaries. Only if the cultures themselves are related to, or depend on, environmental factors shall we find a correlation with natural conditions. Perhaps to the northwest, but there alone, can we speak of a natural boundary in the form of the Grand Canyon, a gash so wide and deep as to have formed a barrier to north–south movement, which is even now bridged in only two or three places along a 350-mi (560-km) stretch in Arizona and Utah.

To what extent do environmental factors create a unity in the Southwest? O. B. Faulk argues for the unifying effect of low rainfall: "The overriding geographic feature of the Southwest is aridity. The economy of the region, the outlook of individuals, and the philosophy of the state governments are based on the limited availability of water and the fight to secure more of it."[4]

[3]D. W. Meinig, "American Wests: Preface to Geographical Interpretation," *Annals*, Assoc. Amer. Geog., vol. 62 (1972), 159–84; quotation from p. 160. See also Meinig's other fine works on this region—*Southwest: Three Peoples in Geographical Change, 1600–1970* (New York, 1971); and *Imperial Texas* (Austin, 1969).

[4]O. B. Faulk, *Land of Many Frontiers: A History of the American Southwest* (New York, 1968), p. 3.

It is true that virtually the whole of the region as Faulk delimits it receives less than 16 in (400 mm) of rainfall a year. And it is possible to argue that the Spanish Empire in Central America was essentially an empire of semiaridity (the parallel with the Spanish Meseta is evident) and that it terminated where this ranch country gave way to more humid lands that would one day be cultivated. In Texas and Oklahoma, this was approximately true. But it would be difficult to argue from this a conscious decision or causal association.

Topographically, the Southwest runs across a number of the physical regions of the continent; these were identified in Chapter 1. It penetrates into the southern fringes of the Rockies and, east and west of the Rio Grande, it crosses the belts of plateau and lowland desert that extend south into Mexico. It spreads up over the Mogollon Rim where the latter marks the southern edge of the high Colorado Plateau, and reaches within a few miles of the Gulf of California and across the troughs of the Pacific coast to the sea.

A clearer idea of the extent of the region can be gained, however, from an understanding of the diffusion patterns of the two major cultures, the Indian and the Spanish.

THE INDIAN CULTURES AND THEIR EXTENSION

As we noted in Chapter 2, the Indian cultures of North America are extremely varied; they have also no more been static in their distribution than those of the European in-comers. In the Southwest, however, we have to visualize a number of groups occupying the same lands over a period of hundreds of years, and there developing the remarkable agricultural civilization of the pueblos. They were evidently dislodged periodically by drought or invasion, but they did not move far; they were sedentary peoples cultivating corn, beans, and squash and practicing irrigation and such crafts as basket making and pottery. They formed two relatively stable blocks, one on the southern part of the Colorado Plateau and around the headwaters of the Rio Grande, of which the Hopi and Zuni Indians are the survivors, and the other along the border between Mexico and the United States, where such tribes as the Pimas and Papagos once enjoyed a high level of civilization, probably transmitted to them from the great culture hearths of Central America.

To the east and the northwest of these areas of rel-

ative cultural stability, conditions were much more fluid. East of the pueblos were the tribes of the Great Plains, for whom life was transformed in the seventeenth century by the introduction of the horse. These tribes, among whom the Comanche were to play the largest subsequent role, had always been nomads, using dogs for haulage; now, although still nomadic, they were transformed from pedestrians into equestrians. They could hunt buffalo and consequently enrich their culture with a host of new artifacts; they could also, when the time came, form a highly mobile striking force against the Plains traders and settlers.

Northwest of the pueblos, there began to be felt, in the thirteenth century, the pressure of a wave of newcomers from the far north. These migrant Indian peoples, who were different in most ways from the sedentary Pueblo Indians, were the forerunners of the Apaches and Navajos: they came, we are told, "almost empty-handed," but they proved to be "cultural vacuum cleaners",[5] they adopted crafts and customs from the people they overran. After they had acquired some knowledge of farming and metalworking from the Pueblo Indians, they went on to learn stock raising from the Spaniards, and, when the time came to settle the Navajo on their present reservation in 1866, it was as pastoralists with stock provided by the government that they were established.

Whether we go back a thousand years or consider the situation of Indians today, we are justified in regarding this southwestern region as critical to their development and their future. Unlike the 170,000 Indians who today live in Oklahoma,[6] the southwestern peoples have a long history of occupance of the region in which they live and, in some cases, of the very lands they occupy today.

---

[5] A. Marriott and C. K. Rachlin, *American Epic: The Story of the American Indian* (New York, 1970), p. 60. See also J. M. Goodman, *The Navajo Atlas* (Norman, Okla., 1982).

[6] The point of contrast between Oklahoma and the Southwest is, as already noted in Chapter 18, that the former was established by the U.S. government as "Indian Territory," in which the remnants of a large number of tribes (eventually almost seventy) could be collected, in order to clear them off lands desired for white settlement and exploitation. The two main groups so collected were (1) from the southeastern United States, a group that included the so-called Five Civilized Tribes (Cherokee, Chickasaw, Choctaw, Creek, and Seminole), who were moved to Indian Territory between 1829 and 1842, and (2) from the Great Plains after the Civil War. Plains Indians from Wyoming in the north, to Texas in the south, were brought together in what is now western Oklahoma.

THE SPANISH CULTURE AND ITS EXTENSION

For the Spanish government, represented by a Viceroy in Mexico City, the region we are considering formed a part, but only a part, of the northern frontier of the empire. That frontier was of enormous length; as Caughey put it, "In 1789, in fact, when Washington was inaugurated President, the United States was confronted by the reality of a Spanish Southwest that began at the Georgia–Florida frontier . . . and wound up north of Nootka on the Pacific." Our viewpoint for the moment is from the opposite side of this Spanish–American frontier, and under these circumstances, it need not surprise us to find that the government in Mexico was preoccupied more with holding a vast military frontier than with encouraging the colonization of new lands, which could only have the effect of straining the Spanish defenses still further. As Spain advanced north, there was a period when, because of the configuration of the North American continent, it was almost literally true that every mile of advance northward meant a doubling of the length of the frontier to be protected. To quote Caughey again, "When Spanish occupation did occur . . . in every instance it was more for the sake of erecting defenses for Mexico and the Caribbean than because of the intrinsic attraction of the new lands. . . . Imperial policy did not call for building up much more than a token occupation of these northern borderlands."[7]

Initially, "defenses" meant local protection against Indian hostility, such as the uprising that drove back the Spaniards from the Rio Grande Valley colonies in 1680. But on the larger scale, a more serious threat was posed by the French thrusting west from the Mississippi Valley and the Russians pressing south along the Pacific coast. If, as was to be the case, the Spanish advance northward was three-pronged (for their advance into Arizona was halted only a short distance north of the present international boundary), then the eastern prong was designed to block French expansion into Texas, and the western prong was an effort to secure California ahead of the Russians. (Meanwhile, far away to the east, in Florida, yet another chain of missions was being pushed northward to block the English, whose colonies—the Carolinas in the 1660s and Georgia in 1732—were gradually encroaching on the eastern end of the Spanish sphere of influence and threatening the oldest of all their North American settlements, St. Augustine.)[8]

The initial expansion of the northern frontier of the empire was logical enough: it was due north up the Rio Grande into the heartland of the Indian Pueblo civilization, where Santa Fe was founded in 1610. In the late seventeenth and the eighteenth century, it was the turn of the Texas frontier, whereas the main expansion west into California may be dated from the founding of San Diego in 1768. The general pattern of expansion was the same in each case: it was spearheaded by priests who established missions, around which they gathered Indians, and where they encouraged the cultivation of gardens and fields, introducing European fruits and plants in the process. The priests were followed by military detachments, so that the frontier became a chain of mission stations and military posts (presidios). As colonization took place, civil settlements (pueblos) were founded. Settlement policy favored big land grants to ranchers, who introduced horses, cattle, and sheep from Spain and herded them over the semiarid grasslands between the islands of cultivation represented by the missions. New Mexico became famous for its sheep, and the whole culture of the cowboy as we have come to know it can be traced to Spanish origins.

Eventually, the northern limits of the Spanish sphere of influence were set not so much by the advance of French or Russian colonization—the threat from both these sources proved ephemeral—as by the expansion of the young republic of the United States. Mexico gained its independence from Spain in 1821 and spent the first quarter-century of its life as a nation wrestling with the problem of its relations with its expansionist neighbor, independent since 1776. Already, under Spain, the problem had arisen of Anglo penetration and settlement of Texas. After 1821, traders from the Midwest began at once to appear at Santa Fe (Spain had been careful to keep them out). In California, the effect of the Gold Rush of 1848 was cataclysmic for the thin veneer of His-

---

[7]J. W. Caughey in M. Jensen, ed., *Regionalism in America* (Madison, Wis., 1963), pp. 174, 176.

[8]A selection of references on Spanish settlement in the Southwest is: D. W. Meinig, *Imperial Texas* (Austin, 1969); J. F. Bannon, *The Spanish Borderlands Frontier, 1513–1821* (Albuquerque, 1974); J. E. Officer, *Hispanic Arizona, 1536–1856* (Tucson, 1987); G. R. Cruz, *Let There Be Towns: Spanish Municipal Origins in the American Southwest, 1610–1810* (College Station, Tex., 1988); and A. W. Carlson, *The Spanish-American Homeland: Four Centuries in New Mexico's Rio Arriba* (Baltimore, 1990).

panic culture in the central and northern parts of the state. War and treaty came and went, and at the end Mexico had lost the whole of the Southwest. Culturally, the Hispanic world had lost eastern Texas and central and northern California, for in those areas the tide of Anglo settlement had been overwhelming: California was culturally and commercially divided in two. But over the remainder of the Southwest, although the political sovereignty had changed, the cultural legacy of Spain remained to challenge or to enrich the new political masters.

The eventual international boundary between Mexico and the U.S. Southwest was established only in 1853. In the east, it followed the Rio Grande, but in the west, it marched across largely unoccupied and unsurveyed land, and it owes its present course to Mexico's insistence on a land bridge between her mainland territories and Baja California and a hazy belief by some U.S. senators that the United States needed, for a transcontinental route, the strip of land it bought through the Gadsden Treaty of 1853.

These, then, are the limits of the Southwest. If we now pause to inquire whether the area we have identified forms in any sense a coherent region in the life of the United States today, we have to reply in the negative: distinctive it may be; coherent it is not.

The northern border of the Spanish Empire ran in a rough arc of a circle whose center was Mexico City; the arc extended from Texas to California. The major lines of movement radiated out from the center, but along the circumference itself, movement was difficult, slow, and infrequent. Thus, the early settlements on the upper Rio Grande found their natural lines of commerce running down the river to the south; in this sense, the opening of the Santa Fe–Missouri trade route after 1821 represented an about-face for New Mexico's commerce.[9] Administratively, the only links between Texas at one end of the frontier and California at the other were via Mexico; the project for the short-lived Atlantic and Pacific Railroad (later built as the Southern Pacific) was still

some decades in the future. And after the thrust of the missionaries into Arizona, in the last years of the eighteenth century (their main mission of San Xavier del Bac was founded in 1700), Arizona remained a sparsely settled area of the borderlands, a gap in the frontier, and the forms of Hispanic society, as they developed there, were quite different in New Mexico, Arizona, and California: "The date of immigration and settlement, the attendant cultural concomitants, geographic isolation, natural resources, the number and kind of Indians among whom they settled, and many other factors resulted in not one Spanish-speaking people but several."[10]

Fragmented though this Hispanic population has been, and small in its beginnings—in 1848, there were between 75,000 and 80,000 Spanish-speaking inhabitants of what is now the U.S. Southwest—it survived and grew and made its mark on several hundred thousand square miles of U.S. territory. The parallel with French Canada, with its 65,000 French speakers in 1763, is a suggestive one. The Hispanics, without the cultural protection the French Canadians enjoy in the educational and legal systems of Québec, have held their ground at least as well and, after 130–40 years of government and education under an alien system, today show a higher level of group consciousness than probably at any time in the past.

Land and Livelihood

The Spaniards entering the Southwest brought with them their livestock and various European grain and fruit crops to add to the range of native plants cultivated by the Indians. Irrigation was known to these Indians, and to Spain since Roman and Moorish times, so that ranching and irrigation agriculture, two of the activities we have already seen to characterize the West as a whole, grew out of a merging of the Spanish and Indian economies. The third economic base was mining: "Every Spaniard who came north . . . no matter whether he was a soldier, a missionary, or a civilian, had some hopes of discovering mineral deposits. The dream of quick wealth . . . permeated every facet of Spanish endeavor in the provinces after they were permanently settled."[11]

In terms of minerals, the Southwest did not prove

---

[9]It is interesting to notice (as D. W. Meinig does in *Southwest*, pp. 38–40), that the idea that New Mexico's principal link would remain with Old Mexico rather than with the American Midwest—which lay at the other end of a journey across the width of the Great Plains—underlay the first project of the railroad era: the concept, initiated by Anglo interests, of a north–south line to be called the Denver and Rio Grande.

The railroad was eventually built, but in the process it lost its planned direction and grew as an east–west stem. The first railroad into northern New Mexico was the Santa Fe line, which followed the trail from Kansas City.

[10]R. W. Paul, "The Spanish-Americans in the Southwest, 1848–1900," in J. G. Clark, ed., *The Frontier Challenge* (Lawrence, Kans., 1971), pp. 31–56. Quotation from pp. 33–34.

[11]Faulk, *Land of Many Frontiers*, p. 79.

a rewarding area for Spain, as did some areas further south: the major mineral discoveries, particularly of gold in California, came later. But it was the Spaniards who led the way in prospecting and in techniques.

These, then, have been the traditional means of livelihood in the Southwest. We must now trace their development in more recent times.

MINING IN THE SOUTHWEST

Among the fifty states, New Mexico ranks seventh in the value of its mineral production. In the days when copper was king, Arizona also ranked high on the list of mineral producers but, as we have already seen, its copper production has declined in recent years. In New Mexico, by contrast, a wide variety of minerals are mined, and the oil fields of the Permian Basin of Texas extend across the state line, so that natural gas and petroleum are the leading minerals by value. In a similar way, the Colorado coalfield extends across New Mexico's northern boundary.

The mining industry has felt the effects of strikes on the Indian reservations, strikes that have encouraged a greater awareness of the needs and potential of the Indians. Whereas they have oil, gas, coal, and ores, they obviously have not had the technical means to exploit these minerals. It is to be hoped that, in the future, the Indians will themselves benefit to a greater degree from their resources.

RANCHING IN THE SOUTHWEST

The Spaniards introduced their livestock to the region, but it was with an influx of Anglos after 1870 that the ranching industry developed in its modern form: New Mexico had only 57,000 head of cattle in 1870, but nearly 350,000 in 1880. The dates are significant: these were the years of the "cattle kingdom" (Chapter 18), and the natural rangelands of the Southwest attracted the cattlemen. It was during this same period, in the years after the Civil War, that the Indian tribes were "pacified" and settled on reservations. With cattle challenging sheep on the open ranges, the latter became the particular staple of the reservations, where, as the largest tribe, the Navajo built up on their arid lands after 1866 an economy based on sheep raising and weaving. The Indians thus became the heirs of a tradition of livestock raising in the New World that was, as Paul has noted, "medieval if not biblical in aspiration,"[12] whereas the

[12]Paul, "The Spanish-Americans," in Clark, Frontier Challenge, p. 36.

Anglos developed the range cattle and feedlot industry of the modern Southwest. Overstocking and erosion have been serious problems on the reservations, just as they have been, less justifiably, on the richer grasslands elsewhere.

IRRIGATION AGRICULTURE IN THE SOUTHWEST

Throughout the Southwest today, most crops are grown under irrigation. The Hopi Indians have developed to a fine art the techniques of dry farming, but most of the commercial agriculture of the region is concentrated in the oases. These are to be found in the valleys of the principal rivers and their tributaries—Rio Grande, Pecos, Gila, Salt. But the supply of water in these rivers is inadequate for the irrigated acreage that has been developed.[13] The time came when the Rio Grande was reaching the border of Mexico at El Paso virtually dry, and the U.S. government was obliged to construct a federal storage project on the New Mexico section of the river to resolve an international argument. On the Arizona side of the Continental Divide, the situation today is much more acute. Large-scale irrigation in central Arizona was made possible by the construction of the Roosevelt and Coolidge dams (built in 1911 and 1928, respectively), so that by the early 1930s the flow of the Gila and Salt was fully utilized for irrigation purposes. But the demand for water continued to rise, both because of the rapid increase in Arizona's population and the remarkable postwar expansion of irrigated cotton, which in 1952 reached a peak of over 650,000 acres (263,000 ha). Effectively, Arizona was converted from a mining and ranching to a mining and cotton state.

With this conversion the increased demand for water had to be met from the only alternative source available besides the rivers—groundwater reservoirs. It is estimated that the annual water "capital" accruing to Arizona is about 3 million acre-ft (3.67

[13]On southwestern water shortage and some of the international complications of this situation a selection of titles is: M. L. Comeaux, *Arizona: A Geography* (Boulder, Colo., 1981), Chap. 6; J. L. Westcott, "Impacts of Federal Salinity Control on Water Rights Allocation Patterns in the Colorado River Basin," *Annals*, Assoc. Amer. Geog., vol. 76 (1986), 157–74; I. G. Clark, *Water in New Mexico: A History of Its Use and Management* (Albuquerque, N.M., 1987); D. J. Eaton and J. M. Anderson, *The State of the Rio Grande/Rio Bravo: A Study of Water Resource Issues Along The Tex/Mex Border* (Tucson, 1987); F. A. Schoolmaster, "Water Marketing and Water Rights Transfers in the Lower Rio Grande Valley, Texas," *Professional Geographer*, vol. 43 (1991), 292–304; and the annual reports of the Bureau of Reclamation, Washington, D.C.

billion m³), of which two-thirds come from stream and river diversions and the remaining one-third from groundwater recharge. But by the mid-1960s, the estimated consumption in the state was 6.5 million acre-ft (7.96 billion m³), which meant that the groundwater supplies were being drawn down at the rate of some 3.5 million acre-ft (4.28 billion m³) each year. This also means that the water table falls and that recovery of groundwater from wells becomes more expensive. The true danger here is depletion of the groundwater reservoirs.

After the cotton boom of the 1950s, the acreage under the crop fell off for a time, only to rise once again to 631,000 acres (255,000 ha) in 1980 and fall back to 415,000 (168,000 ha) in 1985. In the mean-

time other crops, such as winter and spring vegetables, had been introduced, so that total demand increased. This agricultural demand, moreover, is focused on the Phoenix oasis, where in 1990, 2.1 million of Arizona's population of 3.7 million were concentrated.

We shall refer again in the next section to the growth of Phoenix. For the moment, our concern is with the shortage of water in central Arizona, an area that lies within the basin of the Colorado and is subject, therefore, to the terms of the Colorado River Compact (see Chapters 5 and 23). With the entire flow of the Gila and Salt already committed (see Fig. 20-2), the only way to obtain more water for Arizona is from the Colorado, and the state has been to the

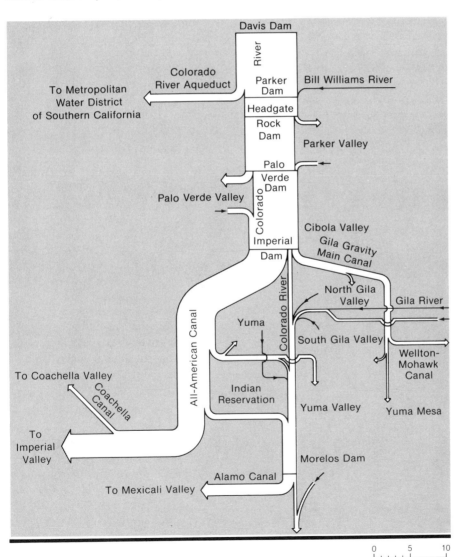

Fig. 20-2. The Colorado River: Distribution of water below Davis Dam, 1961–63. Note the very small residual flow reaching the Gulf of Mexico (bottom). The Central Arizona Project (Fig. 20-3) now under construction, will divert water from behind Parker Dam. The reader may wonder where the extra water is to be found. It is hoped to increase the dependable flow of the Colorado by regularizing works higher upstream. Upper-basin states, however, also have hopes and plans to use more of the Colorado water. In any case, the desirability of huge dams and reclamation projects is being questioned, especially by those concerned with environmental quality.

courts several times to try to secure more favorable treatment under the compact. Its best hopes for the future are now pinned on its sponsorship of a scheme known as the Central Arizona Project. This has been designed to tap the Colorado at Parker Dam (see Fig. 20-3) and to carry water—that is, if the water is present in the river and unclaimed by any of a number of potential litigants—to the Phoenix area, where it will reinforce the flow of the Gila and Salt rivers, and then on to Tucson. The chances of such water being regularly available will be improved if some additional storage can be provided higher up the Colorado, but this involves the construction of dams, which is the object of determined opposition by conservationists and environmentalists.

The Bureau of Reclamation, however, has meanwhile been pushing forward the Central Arizona Project step by step. In 1989, two new areas became irrigable, although take-up by farmers of the irrigation water was slow. In 1991, half a million acre-feet

Fig. 20-3. The Central Arizona Project: The map shows the project as it will be when completed, but progess has been slow and funds have been voted by Congress only for parts of the project.

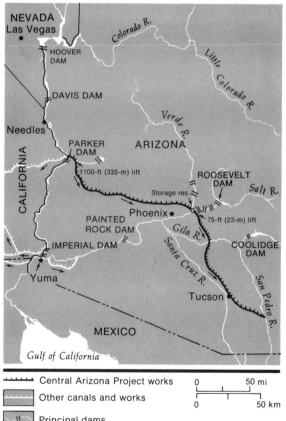

Central Arizona Project works

Other canals and works

Principal dams

0   50 mi

0   50 km

of water were drawn from the Colorado, with an intended 800,000 acre-feet for 1992, to allow filling of newly built reservoirs. Already, in its 1989 report, the Bureau had been able to claim a crop value of $1,714 per irrigated acre ($4,233 per ha) in Arizona.

But the question that always has to be asked about western irrigation must be: Is this the best use for this water? Should it not be reallocated to other uses (see also Chapter 23)? Of the Tucson area, for example, A. W. Wilson wrote:

> Domestic, industrial, and mining uses consume 50,000 acre-feet annually, which is a substantial portion, and perhaps all, of the annual recharge. A small area of irrigated agricultural land, about 14,000 acres, uses on the average an additional 42,600 acre-feet of water annually. Agriculture in the Tucson area employs less than 1500 out of a total labor force of some 70,000. In other words, about two percent of the workers are supported by the use of almost half the water consumed.[14]

The question, therefore, becomes one not simply of a physical shortage of water supplies, but also of resource allocation of a scarce commodity among a number of possible uses and users.

AMENITY AND TOURISM

Shortage of water is only one aspect of the impact on the Southwest of its attractiveness to tourists and settlers in modern America. It is now nearly forty years since E. L. Ullman drew attention to the importance of amenity as a factor in regional growth.[15] The Southwest has been the region of the United States that has most vividly illustrated his case during the succeeding three decades. The tourist traffic began soon after the Santa Fe Railroad (which linked southern California with the Midwest in 1885) completed a branch line to the very rim of the Grand Canyon. From Albuquerque westward to Los Angeles, it gave access to one scenic marvel after another, while further to the south the Southern Pacific provided the link to the old Spanish settlements west of El Paso. This tourist traffic has steadily increased, but it has also understandably resulted in an influx of settlers, people who have chosen to live year-round in an environment they previously enjoyed on vacation.

The principal amenity factors of the Southwest are its climate and its relief: the relief ensures variety of

[14]A. W. Wilson, ''Urbanization of the Arid Lands,'' *Professional Geographer*, vol. 12, no. 6 (1960), 7.

[15]See Ullman's article under this title, *Geographical Review*, vol. 44 (1954), 119–32.

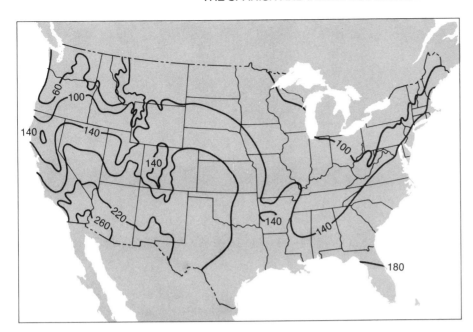

Fig. 20-4. The United States: Hours of sunshine in December. The map underlines the climatic attraction of the Southwest, particularly as compared with its most obvious rival for the winter tourist trade, Florida.

climate. In the winter, it far outdistances its obvious rival—Florida—in total hours of sunshine (see Fig. 20-4). Yet not only does Tucson have a January mean of almost 50°F (9.8°C) and a minimum of 77 percent of possible sunshine (65 percent in Miami), but only a few miles to the northeast it has the Mt. Lemmon winter sports area. On the other hand, the plateau country offers relief from the heat of summer, with July means generally between 68 and 70°F (20 and 21°C), and cool nights. Relative humidities are low and the Southwest's first and continuing attraction as a home area is for those who have rheumatic and bronchial complaints.

Only the spread of air-conditioning, however, has made possible the tremendous population growth within the Phoenix metropolitan area. At Phoenix, the July mean temperature is 90°F (32.2°C), and maxima of over 100°F (37.8°C) regularly occur in the summer months. In 1940, it was a small city of 65,000 inhabitants. By 1990, the city had passed 850,000, and the metropolitan area 2.1 million. Tucson had 36,000 inhabitants in 1940; by 1990, its metropolitan area counted 667,000. The single decade of the 1960s brought nearly 200,000 newcomers to Phoenix and 50,000 to Tucson. That same decade saw the emergence of Arizona as an industrial state, with the value of manufacturing overtaking that of agriculture by 1970. The California aerospace industry has spilled over into Arizona, and Phoenix has, in addition, a hot rolling mill for steel and a large food-products industry. New Mexico, for its part, remains one of the least industrialized states of the United States, but its now-historic connection with the early work on nuclear fission led to large government investments during and after World War II, and a slow but steady growth in manufacturing is occurring in Albuquerque and downriver, at El Paso, Texas.

Regional Mixture

The population of today's Southwest is made up, as we have seen, of three main components; in their order of appearance, Indian, Hispanic, and Anglo. In addition, there are black minorities in most of the cities. The next section of this chapter examines briefly the characteristics of each group and the extent to which these diverse elements have become a homogeneous regional population.

In the early stages of European occupance, a mixture of Spanish and Indian blood became general. The Hispanics of the Southwest today are largely the product of this mixture; their language is Spanish, but their pride is much more in their Mexican and Indian background than in their remote connection with Spain. It might therefore be supposed that this group would form a valuable social binding force, reaching out to the Indians on one side, to the Caucasians on the other. But this is far from being the case. The policy of gathering Indians on reservations set up a barrier of official segregation on one side of

Phoenix, Arizona. In the census of 1940, the city of Phoenix had a population of 65,000 and its metropolitan area, 186,000. By 1984 the corresponding figures were 853,000 and 1.8 million. The city of Phoenix has managed through annexation to spread itself over 386 sq mi (1000 sq km) of the former Salt River oasis and its irrigated farmlands. This makes it comfortably larger, territorially, than either New York City or Chicago. *City of Phoenix, Public Information Office. Photo by Bob Rink*

the "center," whereas further political developments erected obstacles just as formidable on the other. Those Hispanics who found themselves citizens of the United States after the 1848 treaty suffered the immediate disadvantage of a change of legal system (affecting particularly their title to land), of government in a foreign language, and of invasion by alertly commercial Anglos, who quickly came to dominate all economic outlets. With no comparable commercial experience, and penalized by a system foreign to them, the Hispanics were either submerged or impoverished, so that of the old Hispanic heartland R. W. Paul could later comment, "By the opening of the twentieth century the high, dry lands of rural New Mexico began to stand forth as a cultural island of poverty, illiteracy, and premodern customs."[16]

With the competitive advantage clearly in the

[16]Paul, in Clark, *The Frontier Challenge*, p. 37.

hands of the Anglos, the scene was set for a deep social cleavage to develop between Anglo and Hispanic. In other words, where Hispanic and Indian had formed a relatively homogeneous society (although one that had certainly had its social gradations), the coming of the Anglo proved highly disruptive. For our present purposes, the expressions of this disruption—or segregation, as it came to be—which chiefly concern us are those we noted when considering the relationship of the black to the white American—residential location and job opportunities. Most southwestern cities have their Mexican-American districts, which compare with the black ghetto: cities with both Hispanic and black populations have two ghettos. A city like El Paso exhibits clearly the effective segregation in housing and employment (and, consequently, income), within the city (Table 20-1). In one respect, in fact, the blacks are better off than the Hispanic. They do not have to

Table 20-1. Cities of the Southwest: Percentage of Blacks and Spanish-Origin Population in Total Population, Census of 1990

Area	Percentage Black	Percentage Spanish Origin
El Paso MSA	3.7	69.6
Phoenix MSA	3.5	16.3
Tucson MSA	3.1	24.5
San Diego MSA	6.4	20.4
Los Angeles CMSA	8.5	32.9
City of El Paso	3.4	69.0
City of Albuquerque	3.0	34.5
City of Phoenix	5.2	20.0
City of Tucson	4.3	29.3
City of San Diego	9.4	20.7
City of Los Angeles	14.0	39.9

choose between speaking Spanish or speaking English. The difficulties begin at school. The five southwestern states have between them probably 2 million Spanish-surname pupils. But not more than 10 percent of their schools have bilingual programs; consequently, most pupils of Spanish origin are being taught throughout their schooldays in a foreign language, which simply ensures that they will emerge as backward by contrast with their English-language peers. The rest of the story is contained in Table 20-2, which effectively shows for each group listed the point at which education ended. Of Spanish-origin persons over 25 in 1985, for example, 13.5 percent never got beyond the fourth grade. Only 7.3 percent of Spanish-origin female students completed four years of college, compared with 16 percent of white females, or 11 percent of blacks.

Economically, this is a very serious handicap indeed. Some occupations in the Southwest have

become traditional to the Mexican-American population, such as railroad track-laying and irrigation ditch maintenance, but the obstacles to rising beyond them are formidable, and a steady immigration from Mexico ensures that there is always competition for them. To improve this depressed status in the community and broaden vocational opportunities through educational programs is the object of Chicano[17] pressure on the state and federal governments.

This is, however, very far from being the whole story. When we are considering the status and welfare of the Spanish-speaking population of the Southwest, we must distinguish between its two parts: the "legals" and the "illegals," for the impact of the second upon the first is very marked. As House reported in *Frontier on the Rio Grande*,[18] there was not a large volume of cross-border migration before World War II and, indeed, in some years the outflow from the United States to Mexico exceeded the inflow. But wartime labor needs called for the import of a large number of Mexican workers, many of whom were then compulsorily repatriated in the

[17]The term *Chicano* for Mexican-American is derived from "Mexicano" and is accepted for common use by Mexican-Americans generally. Their growing group awareness is further indicated by their use, in referring to themselves, of the term *La Raza* (the race) and by the identification of their variant of the Spanish language as *Pocho*. J. W. House (see n. 19) further explains that *Chicano* tends to be identified in the minds of middle-class Mexican-Americans as associated with the radical culture coming out of the Border *barrios* (p. 103).

[18]J. W. House, *Frontier on the Rio Grande: A Political Geography of Development and Social Deprivation* (Oxford, 1982). This book deals equally clearly with the problems of river flow and water division on the Rio Grande between the United States and Mexico, and those of the population flows across the border.

Table 20-2. United States: Termination of Education, Percentage of All Persons 25 Years Old and Over, 1985

	Elementary School			High School		College		Median School Years Completed
	0–4	5–7	8	1–3	4	1–3	4+	
White	2.2	4.3	6.5	11.5	39.0	16.5	20.0	12.7
Male	2.3	4.6	6.3	10.8	35.3	16.7	24.0	12.7
Female	2.1	4.1	6.6	12.1	42.4	16.4	16.3	12.6
Black	6.2	8.6	6.2	19.2	33.9	14.8	11.1	12.3
Male	7.8	9.0	6.0	18.7	31.9	15.3	11.2	12.3
Female	4.9	8.3	6.3	19.6	35.5	14.3	11.0	12.3
Spanish Origin	13.5	15.5	8.7	14.3	28.4	11.0	8.5	11.5
Male	13.6	15.8	8.2	13.9	27.2	11.5	9.7	11.6
Female	13.4	15.3	9.3	14.7	29.5	10.6	7.3	11.3

The Mexico–U.S. border: The official crossing point at El Paso, Texas. Compare this with the next illustration. *U.S. Customs Service. Photo by Bill Mason*

1950s. A legal limit to immigration was established in the 1960s, but, at 20–40,000 a year, it accounted for only a fraction of the number of Mexicans who wished to settle in the United States. Out of this quota system grew the great illegal flood of the years since then. The final section of this chapter will deal with this subject.

The same characteristics of separate, low status and growing group awareness mark the Indians on the reservations, where 400,000 remain up to the present time, although nowadays there is no *legal* obstacle to their leaving. In the Southwest, the pursuit by the reservation Indians of either their traditional agriculture or their acquired pastoralism for long went on in an economic vacuum: with arid lands, few roads, and almost no elements of a modern infrastructure, they subsisted apart from the mainstream of American commercial life. Well-meaning attempts to provide employment in the kind of occupations common to the rest of the country met with little success.

Three things have acted to change this situation:

the growth of tourism, the construction of roads, and the discovery of minerals on Indian lands. The result of these three factors has been a much greater awareness both of the handicaps and of the advantages of the Indian in contemporary America. The handicaps are formidable—a backlog of educational and occupational disability going back to the multiple deprivations of the Indian in the nineteenth century, as well as ill health and a high mortality rate. The advantages are legal rights to the new-found minerals, a tender spot in the nation's conscience, and an exotic culture to draw visitors; younger Indians, however, do not appreciate the side-show aspects of the last.

There is growing self-awareness among younger Indians and it focuses on the question of treaty rights. In a series of court actions they have called for compensation or renegotiation arising out of many of the treaties made between the tribes and the U.S. government, for rights to land or minerals take on a new importance when it becomes a question no longer of simply herding sheep on an arid plateau, but of a fortune in minerals or in real estate to be gained from these same dry lands, once so useless and now so coveted.

One particular question of treaty rights has not so much brought the Indians of the Southwest into conflict with the federal government as it has set them against each other. As Figure 20-1 shows, the original treaties provided for a curious arrangement by which the Hopi Indian reservation was within the larger Navajo reservation, and the two tribes were to have "common use" of a large area. With the increase in numbers, both of people and livestock, the two tribes have clashed over the occupance of this area. With claim and counterclaim, it seems clear that only a program of relocation, paid for by the federal government, is likely to ease the pressure. Even then, the question will remain Relocation to where?

A Most Remarkable Border

If there is one subject more than another in North American geography which, in recent years, has provoked a flood of publications, it is the United States–

Mexico border. There are probably two reasons why this is so. One is that the volume of traffic across the border has enormously increased, and the other is that, unlike the majority of the world's political boundaries, opinions are clearly divided about the role which this particular border should play, and about the futures of the two areas, the American and the Mexican, which lie on either side of it.

As we have already noted in Chapter 2, the United States–Mexico border has become a frontier where the First and Third Worlds meet across a line. "Along its entire length," comments House," the U.S.–Mexican boundary is one of the most remarkable and abrupt culture contact-faces in the world. . . . Nowhere else in the world are there such steep economic and social gradients across an international boundary." And since the contrasts are actually visible across this line, it can be added that nowhere else in the world, probably, is there such an incentive to cross the line, if only temporarily, to a land of plenty.

This border is in all respects remarkable. One has only to stand on Scenic Drive above El Paso and look

The U.S.–Mexican border on the Rio Grande at El Paso, Texas. The photographer is on the U.S. side, looking across into Mexico. The month is November. Between the two countries flows the "great river," here fordable without the necessity of getting one's feet wet. At this point there are no obstacles to movement in either direction. Compare the previous illustration. *J. H. Paterson*

across to Ciudad Juarez in Mexico to be aware of the two worlds. On the El Paso side there are half a million people enjoying the clear Texas air, reasonably free from pollution and with a good level of car maintenance. On the Ciudad Juarez side, there is the perpetual smog that comes from domestic fires, poorly maintained transport, and a much larger population ponded back, so to speak, behind the frontier; people waiting for the opportunity to cross over, or living where they can do so, in crowded conditions with much poverty.

That people on the Mexican side should want to look for work, and perhaps settle, on the American side of the boundary is perfectly understandable. Why, then, should there be disagreement about this? It is because the U.S. Immigration Service, on one hand, is charged with keeping the borders of the United States inviolate; with knowing who crosses them, and with deciding whether he or she may remain in the country. For the Immigration Service, in other words, the border is a matter of legality and control. On the other hand, a large part of the population of the U.S. borderland is only too glad to keep the border porous because they want to employ, for low wages, the garden boys, maids, and laborers who can cross the border from Mexico and who, provided they stay out of trouble, are generally left in peace to earn a living from these jobs. For these people, both American and Mexican, the border is nothing but a nuisance.

The two photographs reproduced here of the border at El Paso–Cuidad Juarez epitomize, in a way, this confusion of ideas. One is of the official crossing point over the Rio Grande, with all the panoply of flags and officials. The other, taken only a mile or so away, is of this same Rio Grande in November—dry, unfenced, and unguarded. It is agreed by all that there are far too few border guards to seal the whole boundary line. At the same time, it is also agreed by all that there must be *some* way of preventing the whole population of northern Mexico from crossing at will into the United States. How to achieve a balance between these two ideas is something that would involve roundtable agreement between federal and state/provincial governments, municipalities, and many local interests—a most improbable

scenario. For the present, it is easier to let things stay as they are—to turn back some border crossers and admit others, and let the sleeping dogs of the law lie.

There is, however, an interesting development that meets the problem halfway, by providing Mexicans with American employment, but without leaving Mexico. It is possible for U.S. firms to set up assembly plants in Mexico's border areas, where duty-free zones have been established for goods assembled within them. These plants are known as *maquiladoras.*[19] At a recent count, they employed over 300,000 workers. They are, for the most part, outliers of U.S. industry no different from overseas plants in Asia or Africa, which perform a similar function, but here there is no sea to cross—only a few hundred yards of Mexican territory.

The advantage to the U.S. firms involved is that of cheap and plentiful labor, although it must at once be added that minimum wages are laid down by the Mexican government and that these are not, on the whole, sweatshops. The advantage to the Mexicans is that the border area has seen a dramatic increase in population in recent decades, as her people are drawn north by the magnet of American prosperity, and the *maquiladoras* have offered employment, ever since they began to multiply around 1965, to several hundred thousand of this thronging population.

The Mexicans, however, see this as a *transitional* stage in their industrial development: it fills gaps not only in employment but also in industrial earnings, to go some small way toward balancing the huge outflow of funds that, legally or illegally, have passed from Mexican capitalists into U.S. banks and trusts over recent years. One day, the Mexicans hope, their industry will be their own, and dependence on their larger neighbor for this large amount of employment will cease. What effect the newly announced Canadian–U.S.–Mexican free-trade agreement (see Chapter 8) will have on the *maquiladora* phenomenon remains to be seen.

[19]On questions of legal and illegal entry and on the rise of the *maquiladora*, a virtual one-man library has been created by Ellwyn R. Stoddard at El Paso, with a large number of papers scattered among several periodicals. As a starting point, the reader might look at a more substantial example of his work: E. R. Stoddard, *Maquila: Assembly Plants in Northern Mexico* (El Paso, Tex., 1987).

21
The Northern
Atlantic Coastlands

The Nova Scotia coastline: A view of the Cape Breton Highlands National Park. *Nova Scotia Department of Tourism*

The Region

We are concerned, in this chapter, with Newfoundland, the three Maritime Provinces of Canada, and the adjacent, long-settled, but thinly populated portions of northern Maine, the Gaspé Peninsula, and the shores of the Lower St. Lawrence. Nova Scotia, New Brunswick, and Prince Edward Island have been long and widely known as Canada's Maritime Provinces. Since Newfoundland joined Canada in 1949, the term Atlantic Provinces is generally applied to the four units. Labrador is a part of Newfoundland province, but, in its physical character and economy, it is more a part of Canada's North, and is treated as such in this book as well as in the thinking and terminology of most Canadians. The region is centered on the Gulf of St. Lawrence and lies beyond the northeastern extremities of the economic core region of the two countries. It has a long history of settlement, but its economic development has been slowed by handicaps of position and physical geography.[1]

Although the history of European settlement in Atlantic Canada is as long and as distinguished as that in Massachusetts or Virginia, in population growth and in economic development this region has lagged far behind the lands to the west and south. At the time of Canada's confederation in 1867, more than one-fifth of all Canadians lived in the three Maritime Provinces. Since then, the population of

Canada has multiplied by a factor of 5.6 as compared to less than 1.2 in the Maritimes, whose population now comprises 9 percent of that of Canada. During the last three decades, the long, slow growth in Atlantic Canada has quickened somewhat, yet population in the four Atlantic Provinces increased during 1951–81 by 38.1 percent, only about one-half the 73.8 percent increase for Canada as a whole. This has resulted from a difference not in birthrates, but in migration. Most immigrants to Canada after 1867 bypassed the Atlantic region en route to the interior, and were even joined by people who had earlier settled in the Maritimes.

THE PERIPHERAL–FORELAND POSITION

The physical handicaps of the area—which in any case are shared by much of New England—cannot by themselves explain the slow growth of Atlantic Canada. The explanation must also take into account, first, the region's position in relation both to the rest of Canada and to the continent as a whole and, second, its political and cultural fragmentation.

To the south, the Virginians and New Englanders of the colonial period found their expansion westward from the Atlantic coast obstructed by the Appalachians and the New England ranges. As we have seen, these dammed the flow of migrants long enough for a relatively dense population, with an early growth of cities and industries, to be produced. Even when settlement moved beyond the Appalachians, the Atlantic Seaboard cities became the nodes of the developing economic core and the gateways to Europe. Further north, however, the St. Lawrence gave easy access to the continental interior. The Maritime Provinces lay not astride the main line of westward movement, but rather to one side of it; and their forested and often infertile lands offered little incentive to costly clearance and settlement, when contrasted with the vastly superior, easily accessible

---

[1]For a rather fuller account of the region than that given here, it is hard to improve on the two chapters devoted to it in L. D. McCann, ed., *Heartland and Hinterland* (Scarborough, Ont., 1982). Chap. 6, "The Maritimes," is by Graeme Wynn, and Chap. 7, "Newfoundland: Economy and Society at the Margin," by Michael Staveley. As explained in an earlier footnote in this book (see Chap. 13, n. 4), there is a later (1987) edition of McCann's book: use it for preference. See also E.E.D. Day, ed., *Geographical Perspectives on the Maritime Provinces* (Halifax, 1988).

areas of Lower Canada to the west. There was, in other words, not the same geographical compulsion about the occupance of the Maritimes—and still less of interior Newfoundland—as there was about that of the coastlands further south. Many of the early settlements looked seaward to the fishing grounds rather than landward, and, significantly enough, the earliest agricultural areas were not forest clearings, but rather coastal marshes reclaimed by diking.[2]

The position of Atlantic Canada became that of a foreland reaching toward Europe, but peripheral to the developing economic core of Canada and the continent as a whole. As a foreland, the region enjoyed certain advantages, most of which have disappeared with time and changing technology. In the days of sailing ships, a landfall only 1800 miles (2880 km) from Ireland, more than 800 miles (1280 km) closer than either Montréal or Boston, was encouraging to exploration and to potential settlement. St. John's (Newfoundland), Halifax (Nova Scotia), Saint John (New Brunswick), and smaller maritime ports carried on a flourishing trade with Great Britain, the Mediterranean lands, colonial America, and the islands of the Caribbean, and built many of the ships that carried the cargo. The foreland position was important in defense—for example, Fort Louisbourg, built by the French near the easternmost point of Nova Scotia, changed hands repeatedly during the conflict for control of the region. During both World Wars, ship convoys were assembled at Maritime Atlantic ports, especially Halifax, and, in World War II, Gander in Newfoundland was a takeoff and refueling point for military aircraft bound for Europe. Gander continues to serve as an international airport, though with a diminishing amount of business.

In the twentieth century, the U.S. economic core region expanded westward from the Boston-to-Baltimore seaboard to the lower Great Lakes, where it merged with Canada's developing economic core in the Ontario Peninsula and St. Lawrence Lowland. Atlantic Canada's position has become even more peripheral. Oceangoing ships bypass the Maritime ports en route to Montréal and, since 1959, penetrate as far as Thunder Bay, to be as close as possible to the destination or source of the goods hauled. Jets now pass overhead on their nonstop flights to the great urban centers.

PHYSICAL CHARACTER AND FRAGMENTATION

The handicap of remoteness is intensified by the physical character and fragmentation of the Northern Atlantic Coastlands.[3] A glance at the map makes the statement obvious. Newfoundland, Anticosti, Prince Edward Island, and even Cape Breton, the northeastern part of Nova Scotia, are all islands. Cape Breton was joined to the mainland by a causeway, with a road and a railroad, only in 1955. Both the region's mainland and the islands are cut into by arms of the sea. Even mainland Nova Scotia is almost an island—Annapolis Royal, the site of the first settlement in 1604, is about 50 miles (80 km) across the Bay of Fundy from Saint John, but they are almost 600 miles (960 km) apart by road.

Geological structure and landforms add to the fragmentation. The Maritime Provinces continue the New England mountain system, which is related in turn to the older, eastern half of the Appalachians. The Appalachians extend into northern New England and adjacent Québec in southwest–northeast trending groups of ranges: (1) the Green Mountains of Vermont and the linear Notre Dame Mountains of Québec and (2) the White Mountains of New Hampshire and the highland continuation northeast to Mt. Katahdin in central Maine. Beyond the Maine boundary, alternate uplands and lowlands and arms of the sea tend to fragment the region. The main relief features are determined by a number of granite batholiths and other igneous intrusions. The highest of these are the Shickshock Mountains, which form the backbone of the Gaspé Peninsula, whose highest elevation reaches 4160 ft (1248 m); the Cape Breton Highlands in Nova Scotia; and the Long Range Mountains in western Newfoundland. All three of these highlands rise abruptly from their seashores; their steep cliffs and headlands form some of the most spectacular scenery in eastern North America. A second group of broader uplands, lower in elevation, occupy north-central New Brunswick, most of

---

[2]"The dyked marshes of the Bay of Fundy gave somewhat the same grudging tolerance to primitive animal husbandry as the St. Lawrence region did to cereal husbandry." V. C. Fowke, *Canadian Agricultural Policy* (Toronto, 1946), p. 3.

[3]"Fragmentation of the environmental resources has been a critical factor in the use of natural resources in the Atlantic Provinces from the time of the first settlement to the present. In farming, mining, forestry, and in the fisheries, small operations developed that were highly individualistic. . . . This legacy from the past has hindered the formation of larger economic units compatible with the modern competitive demand for primary resources." W. A. Black and J. W. Maxwell, "Resource Utilization: Change and Adaptation," in A. G. Macpherson, ed., *The Atlantic Provinces* (Toronto, 1972), pp. 77–84.

Nova Scotia, and much of Newfoundland. Finally, still smaller but conspicuous uplands outline parts of the shores of the Bay of Fundy: the Caledonian Hills east of Saint John, the Cobequid Hills along the north shore of Minas Basin, and North Mountain, which separates the Annapolis Valley from the south shore of the Bay of Fundy. North Mountain is a narrow, 120-mi- (192-km)-long ridge of Triassic rock geologically related to the ridges in the Connecticut Valley and the Palisades of the New York City area.

Lowland areas not covered by arms of the sea consist of relatively small plains between the uplands. The largest is in eastern New Brunswick, with a continuation along Northumberland Strait into Nova Scotia. Unfortunately, much of this plain is poorly drained or covered with poor soils. The largest plain suitable for agriculture is Prince Edward Island; its moderately fertile red soils are derived from sandstone and shales. Other, smaller lowland agricultural areas are the middle valley of the St. John River and its Aroostook River tributary along the Maine–New Brunswick border, the Vale of Sussex (Kennebecasis River) northeast of Saint John, the long, narrow Annapolis Valley between North Mountain and the Nova Scotia upland, and the low tidal flats around Minas Basin and Cobequid Bay, the northeastern arms of the Bay of Fundy.

The climate of the region is the result of two very different influences, the first continental, the second maritime. A mid-latitude position on the eastern side of the continent leads to massive invasions of polar continental air in winter. The region also lies in the path of cyclonic storms that follow the Ohio Valley and Great Lakes storm tracks. On the other hand, the influence of the sea in coastal areas is marked. Winters are cold, although moderated near the sea; summers are cool throughout. Precipitation is ample and well distributed throughout the year. The chief climatic handicaps are the short growing season (especially in inland and upland areas) and infrequent sunshine in the summer.[4] The southern coasts are often fog bound (for some seventy days in the year), and consequently, the moderating influences of the seas on this long, indented coastline are offset by the fog.

Settlement and Economic Development

The population of the Atlantic region is highly diversified in origin. During the sixteenth century, fishing fleets from western European countries operated from ships and summer bases along Newfoundland coasts until fishermen from the British Isles gradually began to establish permanent settlements. During the following century, beginning in 1604, the French settled at Port Royal (now Annapolis Royal) and around the Minas Basin arm of the Bay of Fundy and then spread throughout Acadia, as the region was then known.[5] A struggle for control of the region followed the founding of Halifax by the British in 1749, with the British finally prevailing in 1763.[6] English, Scottish, and Irish settlers helped to turn the tide. There followed influxes of Germans, mainly to Lunenburg south of Halifax, of British Loyalists who chose to move north rather than remain in the new American republic after 1783, and of the Scots and the Irish, who, entering in their thousands in the nineteenth century, accepted the hardships of the pioneer life in preference to the famine and eviction of their homelands. Some black refugees from the United States settled in Halifax during the War of 1812. Today, about one-third of the population of the three Maritime Provinces is of English origin; the French account for about one-fifth, being a majority in northern and eastern New Brunswick,[7] and the

---

[4]The difference between inland and maritime locations is illustrated by data for Edmundston, in the upper St. John Valley, and Charlottetown, Prince Edward Island.

		Edmundston		Charlottetown	
		Jan.	July	Jan.	July
Average temperature:	°F	9.5	65.1	20.5	66.8
	°C	−14.5	18.4	−6.4	19.3
Average precipitation:	in	3.0	4.0	4.1	3.1
	cm	7.6	10.2	10.4	7.9

[5]The geography of settlement in the Maritime Provinces was a subject which the late A. H. Clark made peculiarly his own. See his *Three Centuries and the Island* [Prince Edward Island] (Toronto, 1959), and *Acadia: The Geography of Early Nova Scotia to 1760* (Madison, Wis., 1968). He seems to have a successor in G. Wynn: see, for example, the latter's "A Region of Scattered Settlements and Bounded Possibilities: Northeastern America, 1775–1800," *The Canadian Geographer*, vol. 31 (1987), 319–38; and his contribution to McCann, *Heartland and Hinterland*.

[6]In 1755, the British authorities at Halifax, fearing disloyalty, rounded up thousands of French settlers at Grand Pre, put them aboard small ships, and exiled them abroad. Many found their way to the bayous west of New Orleans and created the cultural island known as Cajun (for Acadia) Country.

The French retained a foothold near, if not on, the continent—the islands of Miquelon and St. Pierre, just off the coast of Newfoundland, where they continue until today.

[7]In New Brunswick, 33.5 percent of the population claims French as its mother tongue. Apart from Québec (82.8 percent), no other province has a figure above 5 percent.

Scots and the Irish for perhaps one-eighth each. By comparison with the Canadian average, the population of the Atlantic region has a much larger rural component than is seen in other provinces, although the farm element of this rural population is actually smaller than for the nation as a whole; many of those classified as farmers are fishermen or woodsmen as well.

We have already noted that the early advantages of the foreland position of the region were soon outweighed by the region's remoteness from Canada's economic core. In a similar way, the diversified resources of the region—of sea, forest, mine, and soil, and materials for manufacture—were an inducement for early settlement but, ultimately, proved too limited in quantity or quality to sustain population growth and economic development. Let us look at each of the major resources and their related occupations in turn.

FISHING

Cod, haddock, halibut, flounder, sole, and redfish are caught offshore the Maritime Provinces; cod and lobster, inshore, especially along the coasts of Prince Edward Island and southern Nova Scotia. Fishing is the major industry in Newfoundland, and it ranks high in the other Maritime Provinces, with some 50,000 fishermen involved; 17,000 workers are employed in processing plants. In inshore fishing, which is the more important, one or two men put out in a boat and return to port daily. Offshore fishing involves larger vessels and longer stays at sea, in some cases making use of a depot ship, where the catch can be stored or processed.

Since the 1950s, there have been marked changes in the fishing industry. Governmental loan programs at both the federal and local level, as well as technological advances, have made possible a general overhaul of the industry—improving port facilities and automating vessels and streamlining the processing and marketing of the catch. The bulk of the catch is processed into fresh or frozen fillets or fish sticks, and the waste into meal for animal and pet food and fertilizer. This is in marked contrast to early centuries. Then, the catch was dominated by cod, which was cleaned, salted, and dried on wooden platforms, called "flakes," along the shore. A food staple for poor people and slaves, the product was marketed mainly in the Caribbean and Mediterranean.

One of the industry's major problems concerns competitive fishing on the offshore banks, especially by large foreign vessels whose huge trawl nets literally scrape the sea bottom. Through international Law of the Sea conferences in the 1970s, the Northwest Atlantic Fisheries Organization was established in 1979; and through other negotiations, territorial jurisdictions were established, proportional shares of the catch agreed upon, and conservation measures taken—for example, the Grand Banks was closed to cod fishing during 1980 in an attempt to replenish the resource. Canada and the United States extended their offshore fishing limits from 3 mi (4.8 km) to 12 mi (19.2 km) in 1970 and to 200 mi (320 km) in 1974. Measured as they are from irregular coastlines, the limits were bound to overlap, and the two nations have had to negotiate about their conflicting claims. The remaining, and larger, problem is that of fleets from much further afield, whose size and ruthless thoroughness threaten the livelihoods of the Canadian fishermen.

The Atlantic fisheries account for more than four-fifths by weight of the total Canadian catch, although in value terms the proportion is only two-thirds. Cod accounts for over 40 percent of the catch on this coast. In value terms, lobsters, scallops, and other mollusks and crustaceans account for a similar proportion. The lobster fishery is most important around the shores of southwestern Nova Scotia, Prince Edward Island, and Northumberland Strait. Cod are fished everywhere along the coast, as well as out on the Grand Banks. In weight of catch, Nova Scotia and Newfoundland bring in about 40 percent each but, in value, the Nova Scotia catch, particularly its lobsters, amounts to almost one-half of the total for the region.

AGRICULTURE

Except in Newfoundland, most of the early settlements in the region were based on agriculture. Yet only a few areas were able to maintain themselves as important agricultural districts, especially after the opening of better-endowed areas in the Ontario Peninsula in the 1820s and the western Prairies after 1890 (Fig. 21-1).

The best farming area in the Northern Atlantic Coastlands is Prince Edward Island, as the data in Table 7-4 make clear. One-half of its area was in farms and more than one-quarter in crops in 1986. The entire island enjoys gentle relief with red podzolic soils (Spodosols) of moderate fertility developed from sandstone and shales of the Permian age. But the province does not escape the handicaps of both its peripheral position and its fragmentation.

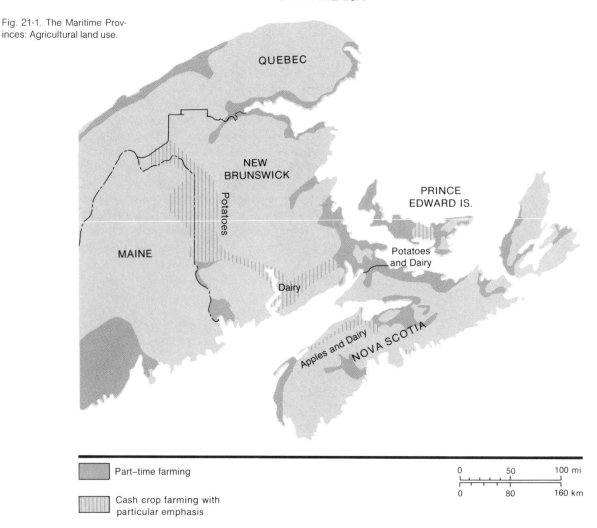

Fig. 21-1. The Maritime Provinces: Agricultural land use.

Part–time farming

Cash crop farming with particular emphasis

The local market is limited to the approximately 127,000 people of the island, a number that has increased only recently after a century of little growth. All surplus production, and most farm supplies, must cross Northumberland Strait by ferry. In keeping with the modest quality of soils and growing season, Island farmers specialize in producing high-quality potatoes especially for use as seed. Otherwise farming is diversified and oriented toward local markets, with emphasis on livestock—dairy and beef cattle, hogs, sheep, and poultry—and forage and grain crops, such as hay and oats, to support them. Some wheat and vegetables are also grown.

Farming is moderately developed in three valleys of the region, the Annapolis, the St. John, and the Vale of Sussex. The middle St. John River Valley and its extension across the border includes the well-known potato-growing area of Aroostook County,

Maine. As in Prince Edward Island, farmers on both sides of the border produce high-quality potatoes, including seed varieties. They have been affected by two trends. The first was the gradual decline in potato consumption during the century after 1850 as people replaced some of the starches and grains in their diets with dairy products, meat, vegetables, and fruits. The second, after 1950, was a revival in potato consumption, but as baking potatoes from Idaho and, especially, as processed and packaged French fries and chips. Farmers in the St. John–Aroostook area have found it difficult to compete for the new markets against large-scale producers and processors (often vertically integrated) in irrigated areas such as the Snake River Plain, Red River Valley, or Central Wisconsin Sand Plain. Yet they have shown that they can do so, in spite of distance to market, if only the U.S. government will leave them free to enter the

U.S. on reasonable terms. One producer of frozen French fries in New Brunswick is among the largest in the world—and its principal customer is in the United States.

The long, narrow Annapolis Valley is separated from the Bay of Fundy by the ridge of North Mountain and bounded on the opposite side by the South Mountain edge of the Nova Scotia upland. Its production of apples, especially for export to England, reached 8 to 9 million bushels (about 185,000 metric tons) per year early in this century. Time and competition have brought change, however—production is now about one-third of its former amount. Part of the crop is still exported to Britain, Europe, and Iceland, but the major market is now in Halifax and other regional cities. Old orchards and varieties have declined; new ones and new methods are replacing them. The modern, cooperatively owned cold storage and processing plant at Kentville stores, packs, processes, and markets much of the crop. Valley farms also produce dairy and poultry products for

Halifax and the regional market. The agricultural area continues from the northeastern end of the Annapolis Valley along tidal flats around Minas Basin and Cobequid Bay. Such lands as these were diked and farmed by the French settlers early in the seventeenth century. Dairying is most important, with some production of small fruits and vegetables.

The Vale of Sussex is a long valley northeast of Saint John, lying between the Caledonian Highland to the southeast and lower ridges to the northwest and drained by the Kennebacasis and Petitcodiac rivers. Dairy farming, based on hay, oats, and pasture, supplies the markets at Saint John and Moncton.

The areas so far described possess an agriculture with recognizable commercial outlets and local specializations. Together, however, they cover only a small part of the region: there is then the rest of the "farmland." About this remainder, two sets of statistics are interesting. One is given in Table 7-4: 69 percent of Newfoundland's farm area, and 62 percent of Nova Scotia's, is unimproved. In practice, it is prob-

The Annapolis Valley, Nova Scotia: A view across the fruit orchards.
Nova Scotia Department of Tourism

ably indistinguishable from the bordering forest or the scrub of abandoned fields. Since in relatively well-farmed New Brunswick (59 percent of farmland unimproved) the number of farm units had fallen in 1981 to only 15 percent of the number in 1951 (Table 21-1), there are abandoned fields in plenty.

The other interesting statistic is the high proportion of farmland in hay and fodder crops: 70 percent in Newfoundland, 60–65 percent in Nova Scotia, and 55 percent in New Brunswick. Taken together, these figures suggest that, outside the areas of specialized farming, agriculture is neither intensive nor profitable: little of the land is cultivated, and the part that is grows grass. In fact, most of the small farmers double as fishermen or lumberjacks.

FORESTRY

The Northern Atlantic Region was mostly forested at the time of settlement, and most of the area is still considered forestland. The volume and rate of growth of timber generally are highest in the interior southern areas with better soils; they are slower with increasing latitude, on higher, rocky land, or near the seacoast; that is, where trees must endure a short growing season and high wind. About 90 percent of the area of New Brunswick and northern Maine is productive forestland; the proportion falls to about three-quarters in Nova Scotia, one-third in Newfoundland, and one-fifth in Labrador. In the northern areas, black spruce and balsam fir trees are used mainly as pulpwood. Coniferous softwoods are also the dominant species in the southern interior, but there they include white pine and hemlock and are mixed with such deciduous hardwoods as sugar maple and yellow birch. Consequently, Maine, New Brunswick, and Québec support the lumber as well as the pulpwood industry. In fact, Maine was the first

area of the continent to enjoy (or suffer) a lumber boom—its great stands of white pine, after having supplied masts for British and American sailing ships for two centuries, were decimated in the mid-nineteenth century, after which the lumber boom shifted successively to the Great Lakes, the South, and finally the Pacific Northwest. Sawmills are scattered along the south shore of the lower St. Lawrence, along rivers such as the Penobscot, St. John, and Miramichi, and through much of Nova Scotia. Most of the mills are small and supply local and regional needs.

In Canada, as a whole, about 90 percent of forestland is owned by provincial governments and leased to private companies for cutting on a sustained-yield basis. This is the practice in Newfoundland. But in New Brunswick, more than one-half of the forestland is privately owned, in Nova Scotia more than three-quarters, and in Prince Edward Island almost all. The major forest product in the region is now paper, or pulp for papermaking, and comes from a relatively small number of very large plants, rather than from many scattered small ones, as in the case of lumber. Among the largest are the newsprint mills at Corner Brook and Grand Falls and a linerboard mill at Stephenville, all in Newfoundland. They account for the only sizable urban settlements in that province outside St. John's. The largest concentration of pulp and paper mills is in northeastern New Brunswick, with plants at Chandler, New Richmond, Dalhousie, Bathurst, and Newcastle. Other large mills are at Baie-Comeau on the north shore of the St. Lawrence, Edmundston on the upper St. John River, Saint John, and the northern and eastern coasts of Nova Scotia. Some mills manufacture only chemical or mechanical pulp, some paperboard, several of the largest newsprint. Much of the newsprint is exported to the United States (Table 21-2).

Table 21-1. New Brunswick: Agricultural Statistics, 1951–86

	1951	1961	1971	1976	1981	1986
Number of farms	26,431	11,786	5485	4551	4063	3554
Area in farms (million acres)	3.5	2.2	1.3	1.15	1.08	1.01
(million ha)	1.4	0.9	0.5	0.46	0.43	0.41
Number of farmers reporting off-farm work	—	5825	2328	1829	1801	—
Percentage of farmers reporting farm sales of less than C$5000	—	—	66.5	58.0	46.7	35.7
Area under crops (thousand acres)	712	483	322	339	323	320
(thousand ha)	288	195	130	137	131	129

Corner Brook, Newfoundland: View of the paper mills. The impor- inhabitants, Corner Brook is Newfoundland's second largest settle-
tance of this industry can be judged by the fact that, with 23,000 ment. *Corner Brook Pulp and Paper, Ltd.*

Table 21-2. New Brunswick: Categories of Employment, by Percentages, for 1975 and 1986, with Estimates for the Year 2000

	1975	1986	Estimate 2000
Agriculture	2.68	2.25	1.9
Other primary activities	4.91	4.12	3.4
Manufacturing	15.63	13.48	11.5
Construction	8.93	5.99	3.4
Transport, communications, & other public services	10.71	8.61	6.8
Commerce	19.64	19.10	18.6
Finance, insurance, & real estate	3.57	4.12	4.7
Services	25.89	33.33	39.8
Public administration	7.59	8.61	9.6
	100.00	100.00	100.00

Source: M. Beaudin and D. J. Savoie (eds.), *New Brunswick in the Year 2000,* Canadian Institute for Research on Regional Development, Moncton, 1989, pp. 50, 56.

MINING

In 1989, the four Atlantic Provinces produced minerals with a value of almost C$2 billion. While this figure may appear relatively insignificant alongside that for oil- and coal-rich Alberta, it makes a very substantial contribution to the economy of the region we are considering, and that in spite of a long history of mine closures.

Yet mining in the northern Atlantic Coastlands is highly localized. Coal is mined in northeastern Nova Scotia along the coast near Sydney.[8] There, the underground galleries of long-exploited mines follow the coal seams that dip beneath the sea, making production difficult and costly. A major market has been the local steel mill, which, like the coal mines, is struggling to survive in competition with larger,

[8]See H. Millward, "A Model of Coalfield Development: Six Stages Exemplified by the Sydney Field," *The Canadian Geographer,* vol. 29 (1985), 234–48.

newer plants in Canada's economic core. Both the coal mines and the steel mill have been taken over by provincial Crown corporations in order to modernize them and keep them in operation. Some coal is also mined at Pictou and Springhill, south of Northumberland Strait, and at Minto, in southern New Brunswick (Fig. 21-2).

Iron ore for the Sydney steel mill was mined on Bell Island in Conception Bay near St. John's, Newfoundland, with galleries also beneath the sea, until ore brought more economically from the Québec–Labrador border area forced the Bell Island mines to close. The Québec–Labrador producing area is shared by Newfoundland province, but, like the great hydroelectric power development at Churchill Falls, it lies outside the Northern Atlantic Coastlands region.

There are currently three large, nonferrous metal mining centers within the region. Ores mined at Bathurst in northern New Brunswick yield zinc, copper, lead, and such precious metals as silver and gold;

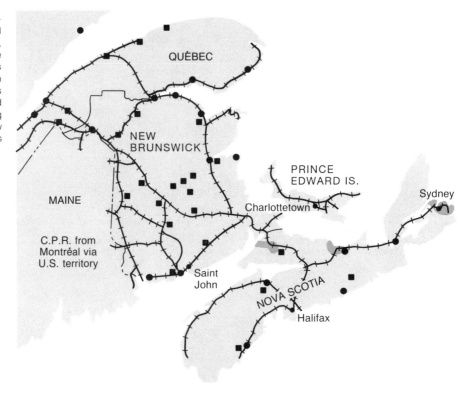

Fig. 21-2. The Maritime Provinces: Coalfields, pulp and paper mills, and rail network, status in 1987. As elsewhere in North America, the rail net is being steadily reduced in mileage, with the branch lines closing first. In Newfoundland (not shown here), the long cross-island line has now gone: there are no railroads left there.

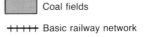
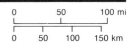

● Pulp or pulp and paper mills

■ Saw mills, capacity in excess of 6 million board ft.

Coal fields

+++++ Basic railway network

they are processed at a local smelter and refinery. A mine and smelter at Murdochville in the Gaspé Peninsula has been producing copper, with some silver, gold, and bismuth among other metals. The third center is in central Newfoundland at Great Gull Lake; its products are copper, lead, and zinc. Together the three centers account for about 7 percent of the value of mineral production in Canada. Each development may loom large in the economy of its sparsely populated area and be an important earner of income for the region. Other mining enterprises include production of potash in New Brunswick and gypsum in southwestern Newfoundland and Nova Scotia.

One mineral development that has, for some years, promised great things for the Atlantic Provinces is drilling for petroleum on the Grand Banks, off the Newfoundland and Nova Scotia coasts. Exploration has been going on since 1966 and two fields, Hibernia and Ventura, have been located. In 1985, the federal government in Ottawa and the Newfoundland provincial government signed an Atlantic Accord regulating the joint exploitation of offshore energy and revenue sharing. The same terms were available to Nova Scotia.

All that is needed now is a sufficient rise in oil prices to make the whole venture pay. Exactly as with the tar sands in western Canada, the resource is available; the question is simply whether the costs of exploitation will be covered by the world price of oil. For the costs of this particular venture are enormous. It is difficult enough, as the oil companies know, to produce oil from Europe's stormy but relatively shallow North Sea. It is going to be very much more hazardous and costly to produce it here, while avoiding storms, icebergs, and the wreck of the *Titanic*. Already, in 1982, one large oil rig had sunk, with the loss of eighty-four crewmen. One fears that, as has so often been the case in the past, this may be yet another occasion when the region grasps at a bright hope, while it foreknows disappointment.

TOURISM

For the tourist industry, the peripheral location of the Northern Atlantic region is both handicap and blessing. One of the handicaps is distance—from Toronto by road through Canada to Gaspé is over 1500 mi (2400 km), to Halifax over 1900 mi (3040 km); from Boston to Cape Breton Island, over 1000 mi (1600 km). The journey to Newfoundland is even longer and more costly. It is, however, only a little more

than 200 mi (320 km) from Portland Maine, to Yarmouth, Nova Scotia, by ship.

The remoteness from centers of population may also be viewed as a blessing. Although the coastland scenery, especially in the Gaspé and Cape Breton, is superlative, it is the historic and unspoiled character of the more remote coastal villages that is most attractive to many visitors. Even more than in such honeypots as Yellowstone or Yosemite, such attractiveness is destroyed if too many tourists come to see it. But there is more to the dilemma. The villagers have no desire to remain poor and quaint in order to please tourists—they would rather abandon their rowboats and cod-drying flakes for labor-saving equipment or a job in town and abandon outdoor baking ovens (as in the Gaspé) for an electric stove or a microwave. Although the quaintness is rapidly disappearing, the small size and beautiful settings of the villages endure.

In this region, as in other economically underdeveloped areas, such as Appalachia or the Indian Southwest, change seems to proceed through the following sequence: (1) the area is bypassed by progress and retains its old ways; (2) at first a few, and then many tourists discover the area and find its people, folkways, crafts, and landscapes appealing; (3) the people of the area discover the tourists, abandon their old ways and crafts, and devise imitation folkways or import machine-made souvenirs to entertain or sell to the visitors; (4) the more discerning among both the natives and the tourists get to know each other better, and the conservation or even revival of some of the genuine old ways and crafts occurs—hence the basis for a more stable and mutually beneficial exchange is established. All four stages of the sequence can be observed currently in this region—one hopes that stage three will be limited in both area and duration.

The Character and Role of the Atlantic Provinces Region

The character of this region depends on three factors.

1. *Geographical fragmentation.* To describe the Atlantic Provinces as "a region" is already to take liberties with their real geography. In 1990, their aggregate population was 2.3 million, or under 10 percent of the Canadian total, and this population was spread over islands and peninsulas covering 16° of longitude and 8° of latitude. In such an area, no real regional concentration is possible; there is no central

Halifax, Nova Scotia, in the mid-eighteenth century. *R. Short, 1759, The Town and Harbour of Halifax in Nova Scotia as They Appear* *from the Opposite Shore Called Dartmouth. Courtesy, Public Archives of Nova Scotia*

location to offer the opportunity of metropolitan growth. Rather, there are lines of penetration and development—axes of movement, and local concentrations of farming, industry, or services.

One can attempt to apply here the core–periphery model; to argue that the core of Newfoundland is St. John's and the Avalon Peninsula, and that the core on the mainland stretches from Halifax around the head of the Bay of Fundy to Saint John. But to argue in this way it is first necessary to find and define a core. In the Atlantic Provinces it is much more a choice between areas remote and more remote.

Halifax (320,000) is the largest city, with the widest range of functions in the region. It became the terminus of the Intercolonial Railway (now Canadian National) from Montréal via the Matapedia Valley and eastern New Brunswick in 1876. It also is a major port, naval base, and airline node. Its industry includes shipbuilding, oil refining, and auto assembly, and the manufacture of machinery and electronic equipment, and food and other consumer goods. But its rapid growth, 25 percent in the 1970s, and 8 per-

cent in 1986–91, was most closely related to its development as the provincial and regional center for government, finance, and other administrative and commercial services.

Saint John, the major node near the western end of the corridor, grew rapidly after the Canadian Pacific Railway across Maine was completed in 1890. Like Halifax, it is a major port and regional commercial center. Manufactures include food, paper, fabricated metals, and transportation equipment. Fredericton, at the western end of the corridor, serves as provincial capital.

Truro, at the head of the Chignecto Bay, and *Moncton* in southeastern New Brunswick, are trade centers along the corridor. Moncton, a major railway node, competes with Saint John as a center for eastern New Brunswick and, to a degree, includes Prince Edward Island within its hinterland.[9] Two branches from the central corridor may be recognized. One reaches

[9]Railway cars and motor vehicles are ferried from Cape Tormentine to Borden on Prince Edward Island.

southwestward into the Annapolis Valley, with Kentville its major trade center. The other reaches northeastward to Trenton on Northumberland Strait and to Port Hawkesbury on the Strait of Canso. A causeway-bearing road and rail line were built across the latter channel in 1955. Although a lock permits ship passage, the closure created a deep-water harbor on the Atlantic side, where a power plant, oil refinery, pulp mill, and atomic heavy water plant formed a new industrial node.

The major port, governmental, and regional center for Newfoundland is *St. John's* (172,000) on the Avalon Peninsula at the far eastern tip of the island. Since the beginning of the sixteenth century it has been the major base for fishing the Grand Banks. It was, until the line's closure, the terminus of the 700-mi (1120-km)-long, narrow-gauge railroad across the island. Manufacturing is much less important than administration, trade, and services; in addition to fish processing, some consumer goods for the provincial market are produced. For the island as a whole, fishing provides the largest amount of employment, but in value terms it ranks behind mining and construction.

2. *Economic fragmentation.* If the region is fragmented by sea, mountains, and barren places, fragmentation also marks its economy. This is true in two senses. One is that, in an area where there is no focal point and where, in any case, contact with neighboring communities often involves a sea trip, the region's small population clusters have built up a high degree of self-sufficiency, relying on land, forest, and ocean for most of their needs.

This worked well enough until confederation in 1867, and the coming of the railways, after which the Atlantic Provinces found themselves part of the Canadian nation and the national economy. It was then that peripheral position and fragmentation became handicaps: three sparsely populated provinces and Newfoundland (which had a form of independence under Britain until 1949) formed nothing of an economic force; their small agricultural producers and manufacturers were in no position to compete with large-scale operators in the Canadian heartland. But the heartland was where the principal markets were to be found.

Taken as a whole, enterprises in the region were small-scale and labor intensive. By comparison with those in other parts of Canada and the United States, they were therefore high cost. In the isolation of the coastlands, this had been an acceptable price to pay

for independence and local supply. But now that goods from the heartland could enter the region cheaply, local industries could no longer compete. They declined, or were taken over by national concerns.

This leads us to the second sense in which one can speak of economic fragmentation in the Atlantic Provinces: they have what we can only call a dual economy. Wynn expresses this clearly:

> there is a great variation in the scale of operations; small family farms, private woodlots, and individual fishing ventures exist alongside large mechanized agricultural operations, corporate forest monopolies, and expensive freezer trawlers. . . . Economic dualism is a marked feature of the region, most clearly reflected in the juxtaposition of capital-intensive, technologically-complex industries tied directly to national or international markets with small-scale, family enterprises.[10]

The old and the new, the national and the local, exist side by side with the national, in the nature of things, gradually gaining on the local. What the interests outside the region want are its resources. What they do *not* want is the local method or scale of exploiting them. It is a classic colonial situation.

Out of these circumstances arise the facts (1) that income per capita in the Atlantic Provinces was, in 1950, only about 63 percent of the national average; since then it has crept up to some 80 percent, and (2) that during that same period the region had easily the highest unemployment rate in Canada and, seasonally, a rate that was double the national average. Even allowing for lower costs of living in these provinces, the record shows their inhabitants to have taken on, unwillingly, the role of poor cousins to the rest of the nation.

The government in Ottawa has certainly not neglected these poor cousins. The prime target of each of the national schemes of regional subsidy that we reviewed in Chapter 10, from the equalization payments of 1957 to the Department of Regional Economic Expansion (DREE) of 1969 and its successors, was these same provinces. National pensions and unemployment benefits help to spread the nation's wealth, and are acknowledged to have swayed many voters in Newfoundland to vote for incorporation with Canada in the referendum held in 1949. Without these federal transfer payments incomes in the region would, of course, be lower still. In New Bruns-

---

[10]G. Wynn in McCann, *Heartland and Hinterland,* pp. 158–61.

Sydney, Nova Scotia: The Sydney Steel Corporation's plant on Cape
Breton Island, North America's most isolated surviving steelworks.
Sydney Steel Corporation

wick, which may be taken as typical of the four provinces, more than 40 percent of provincial revenues
derive from transfer payments.

3. *External influences.* No region today exists in a
vacuum of self-sufficiency, and the prosperity and
future of the Atlantic Provinces depend on factors
beyond their immediate control. The most important
of these are: (1) the attitude of the Ottawa government, (2) the policy of the U.S. government, and (3)
the movement of world commodity prices.

Canada's federal government, as we have already
seen, has made transfer payments to the provinces
from the federal revenues under many different
guises. But these payments represent other people's
taxes, and there may well come a point where the political opposition to these subsidies obliges a govern-

ment to reduce or halt them. That, after all, is what
has happened to most of the regional subsidy
schemes in the United States. In any case, some of
these grants were earmarked for particular developments, which in turn means that decision making is
taken over by the granting agencies. By and large, as
Beaudin and Savoie put it: "There is now a realization in New Brunswick, as in the other provinces,
that self-sustaining economic development cannot
be imported. New Brunswickers themselves will
have to provide the knowledge, the skills and the
energy to create and manage new economic
activities."[11]

[11]M. Beaudin and D. J. Savoie, ed., *New Brunswick in the Year 2000*
(Moncton, New Brunswick, 1989).

Then again, some of the Atlantic Provinces' past difficulties have been caused by federal tariff policies. Every nation reserves the right to implement these, but their impact is likely to differ from region to region, and the provinces have no single loud political voice to speak for them in the formulation of tariff policy, compared with the Prairie farmers, the oil interests, or the manufacturers of Ontario.

But because of its location, the Atlantic region has to take into account not only Canadian tariff policy, but also American. As we have already suggested, northern New England and New Brunswick are very similar in character; potatoes are grown on both sides of the international boundary in the Aroostook–St. John Valley country, and for some of their products the provinces' natural market is in the United States. Yet access to this great market has been, to say the least of it, uncertain. Several times in the past a good start has been made on entering a particular U.S. market, only for a change of policy to close it once again.[12]

The third external factor giving character to this region's economy is the play of world prices. We have already noted that offshore oil fields represent a rich resource that can be exploited when the price of oil is right. Much the same external influence governs the output of other coastland minerals. And then there is the steel mill at Sydney on Cape Breton Island. We have already briefly reviewed the continuing saga of how this mill—situated, with a kind of geographical perversity, on the outermost tip of the mainland—has been kept in production under different managements. But with the mighty U.S. steel industry in grave difficulties (Chapter 8), and with plants closing in favored locations in Pennsylvania and Ohio, what serious prospect can there be for profitability in the far Northeast?

Choices and Chances

Given this situation, geographical and economic, what choices are open to the Atlantic Provinces?

1. *To accept full integration into the wider national economy,* with what that implies: the takeover of local industries by outside corporations, the closure of uneconomical units (uneconomical when judged by national or international criteria), and the loss of jobs,

which can only then be replaced by developing service functions such as tourism. This process would seem certain to lead to some concentration of activity in fewer and larger centers; looked at from the other end of the telescope, the disappearance of many small settlements, like the 300 communities closed down since World War II under Newfoundland's "centralization" program.

2. *To resist such changes, and try to preserve the region's traditional activities and culture.* This is, after all, what the tourists come to see. Perhaps tourism can be transformed into a year-long activity by creating in these provinces the kind of "Ecotopia" that already exists on the Pacific Coast, in Oregon and, for that matter, in northern New England. In a continent with fast, excellent communications, it may be that there are sufficient people to sustain a region on this basis: certainly the population of the three northern states of New England has shown steady increase and an in-migration in the past 15–20 years of people looking for an alternative environment.

For the Atlantic Provinces, however, this is hardly a realistic expectation. Nobody can commute to work from Newfoundland, and Canada, with its huge empty areas and sparse overall population, has plenty of potential Ecotopias nearer to Toronto or Montréal than this. Attractive as the provinces may be to summer visitors, their winter climate is harsh, and 2.3 million people cannot live by summer tourism alone.

3. *To develop regional capital and regional leadership,* so as to ensure economic expansion without the loss of all political and financial control. These communities, many of them among the oldest in Canada, are only too anxious to succeed in the world of the late twentieth century, but not on somebody else's terms; not the federal government's, and certainly not those of some multinational corporation.

The problem is, of course, to find some means of generating capital locally. New England did it with trade and textiles; Texas and Alberta have done it with oil; California with gold, oil, and a huge in-migration. Oil might one day provide the necessary lever here, too, if the offshore fields are brought into full production. Forestry and fishing are not sufficiently propulsive in character, and the entrepôt trade from the Atlantic to the heartland has had little long-term effect. Besides tourism, what other stimulus can there be?

Meanwhile, a steady out-migration continues from all four provinces.

---

[12]For details, see Wynn in *Heartland and Hinterland,* pp. 168–80. Then see his "A Province Too Much Dependent on New England," *Canadian Geographer,* vol. 31 (1987), 98–113.

In writing about this region, one is constantly struck by the parallels it presents with other areas. Our earlier mention of "colonial" status recalls our study of Appalachia in Chapter 16. Here are two regions with considerable natural resources but an isolated population, to some extent detached from mainstream North American culture. In each, there are outsiders only too ready to exploit the regional resources, but without necessarily consulting the wishes of the inhabitants, or having regard to their welfare or employment. We saw how a program of regional aid and development had effected changes in Appalachia, but also noted that the manner of bringing that aid had aroused much criticism.

Compared with the Atlantic Provinces, Appalachia is favored in one respect: it lies in the heart of the nation, surrounded by areas more prosperous and more fully developed. It is, in fact, a periphery surrounded by a core. Out of the way it may have been, in the days of the pioneers, but no longer—the development highways have seen to that. Today, Appalachia is a sensible central location for a manufacturer or wholesaler to settle.

The Atlantic Provinces do not possess this advantage: their location is peripheral, not central. Consequently, while an Appalachian-style program might open up new forest areas or mines, this would make little difference to the region's position within the *national* economy. As with Appalachia in the early days of regional aid, the roads might simply make it easier to leave for elsewhere.

But there is another parallel. If there are people in this region with long memories and Scottish ancestors (and there are plenty with both), they can hardly fail to be struck by the way in which history is threatening to repeat itself. For many of the Scots who came to Nova Scotia came because they, too, formed part of a society that was largely self-sufficient, albeit at a low level of living, and that was then invaded and dislodged by an outside economy of larger scale and greater capital resources. The coming of sheep to the Scottish Highlands late in the eighteenth century produced the Clearances, and the Clearances produced the emigration.

It is not a story that anyone would wish to have repeated in the New World.

22
The Northlands

The Canadian North: The midnight sun. Like northern Norway, this is a land of the midnight sun, so that although the summer and growing season both look short on the calendar, the hours of sunshine during the long days permit crops to be grown. The picture was taken at Great Bear Lake, about 66° N Lat. *National Film Board of Canada*

The Northlands of the continent, considered politically, are made up of Alaska, Yukon, and the Northwest Territories, with a total area of 2.1 million square mi (5.44 million sq km). If, however, we use the term to include all those areas that lie north of the limits of continuous settlement, then the area covered by this great region of North America is much larger, for it includes Labrador, much of Québec, and the northern parts of Ontario, the Prairie Provinces, and British Columbia.

A number of attempts have been made by geographers to achieve a clear definition of where and what is the Canadian North. L. E. Hamelin has been the most persistent, and he has proposed that, on the basis of ten variables, it is possible to rate any location in terms of its "polaricity" or, in the case of Canada, its "nordicité."[1] This polar index may then be used to fix an arbitrary but quantitative boundary around the Northlands. In practice, it might also be used for such purposes as fixing rates of pay or bonuses for workers undergoing the hardships and privations of northern employment, by relating wage rates to the scale of nordicité. On the basis of his index, Hamelin has seen the North as divided into four zones—the Pre-North, the Middle North, the Far North, and the Extreme North (see Fig. 22-1).

A similar exercise by Gajda[2] produced a three-fold division of the North which has found some general acceptance; it distinguishes between (1) the Near North, a zone in which a good deal of settlement, including agriculture, has already taken place—the "older pioneer zone"; (2) the Mid North, with a scatter of settlements, mostly mining communities, in a "new pioneer zone," and (3) the Far North, the "zone of strategic occupation," where economic activity is absent and the only settlement is military or political in character.

Whichever of these schemes of subdivision we may adopt, our interest is in recognizing that within an area so huge as the Northlands, there are degrees of activity and isolation; that the onset of northern conditions is neither even nor consistent all along the southern edge of the region; and that the North has always been a region delimited by individual perception rather than by specific latitudes or temperatures. To a farmer successfully cultivating the Peace River prairies, "the North" may well lie beyond the provincial boundary along the 60th parallel, whereas to an inhabitant of the southern United States, it would appear equally self-evident that even the southern boundary of Canada along the 49th parallel is within the margins of the frigid Northlands. J. W. Watson goes so far as to say, "To speak about *the* geography of the North is nonsense; there have been as many geographies as there have been illusions."[3]

Although the North that geographers have tried to define may lack unity in its physical or cultural character, it does possess a unity of role or function in terms of our original identification of Core and Periphery. There can be no question but that the North forms part of the Periphery. Its development is dictated by, and financed from, the economic core. It depends upon the Heartland not only for capital and expertise, but even for the supply of many of its everyday needs, not to speak of labor, and of markets

[1]L. E. Hamelin, "Un Indice Circumpolaire," *Ann. Géographie*, no. 422 (1968), 414–30. The ten variables suggested, each of which can be scored on agreed scales to give a composite total index, are latitude, summer heat, aggregate annual cold, types of freezing and icing, total precipitation, vegetation cover, accessibility other than by air, air service, population density and degree of concentration, and measure of economic activity. A map drawn on this basis appears on pp. 424–25 of Hamelin's paper: see also Fig. 22-1.

[2]R. J. Gajda, "The Canadian Ecumene: Inhabited and Uninhabited Areas," *Geographic Bulletin*, no. 15 (1960), 5–18.

[3]J. W. Watson, "The Role of Illusion in North American Geography," *Canadian Geographer*, vol. 13 (1969), 22.

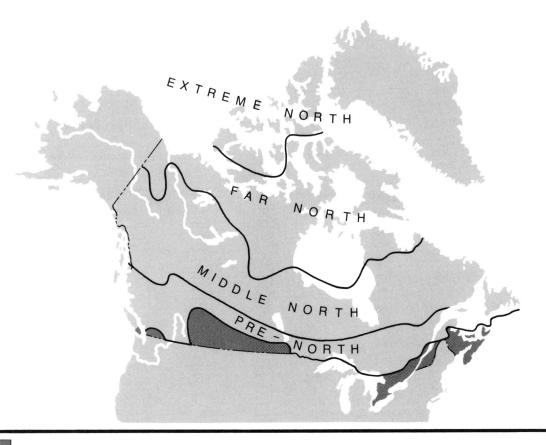

Principal Canadian Ecumene

Fig. 22-1. The Canadian North and its Zones.

for its products. It is in virtually every sense a dependent region. It is true that its native inhabitants, Indian or Inuit (Eskimo), have in the past maintained, and may in places still maintain, a life-style independent of the rest of Canada, but even they have on the whole accepted—or succumbed to—a creeping modernization that has left them, also, dependent on outside resources and outside technology.

Physical Conditions

Over so large an area, physical conditions vary widely. All types of relief are to be found, although even within this area, which most people think of as simply cold and barren, significant differences in climate and vegetation occur. We can list here only the major physical factors that govern the use of the Northlands.

RELIEF, CLIMATE, AND SOILS

If one is thinking in terms of the physical environment, it is probably simplest to regard the Northlands as comprising two separate regions: the Northwest and the Arctic. The differences between the two are in relief, climate, and vegetation. Most of the Northwest is made up of the northward extension of the Cordillera and the Pacific Coastlands and so is mountainous, whereas the mainland Arctic is underlain by the Laurentian Shield and is an area of gentle relief.

As we saw in Chapter 1, the climatic heart of the American Arctic is offset somewhat to the eastern side of the Northlands. In summer, the isotherms run from northwest to southeast, so that although parts of the Northwest have a short but agriculturally useful summer season, the Arctic receives much less heat in summer.

The effect of this climatic difference is seen in the vegetation of the two areas. The Northwest is a mountainous area in which (as in southern British Columbia and in Washington) lower slopes are forested and upper slopes are barren, but the Arctic is treeless tundra. The northern limit of trees, which runs diagonally across Canada from the mouth of the Mackenzie to northeastern Manitoba, forms a natural boundary between the two areas.

The soils of the Northlands are also varied. Besides the Spodosols (northern podsols) one might expect in these latitudes, there are the lacustrine clays and sands of former glacial lakes and areas where glaciation has swept away the soil. Where drainage has been interrupted and swamps, or muskegs, have been formed, there is often a peat cover several feet thick. This must usually be stripped off, or at least reduced in thickness and drainage channels cut, before the land can be brought under cultivation.

But all soils in these high latitudes have characteristics that make their use complex and difficult. Rates of decomposition and humus formation, for example, are very slow, so that in the surface layers organic matter is held "in cold storage." Freeze and thaw are responsible for constant rearrangement of soil-making materials and, indeed, for a continuous formation process of minor surface landforms and irregularities. Solifluction is widespread on bare surfaces of quite gentle slope, and gullying is extensive when, for example, the surface is disturbed by construction.

The major soil problems in the Northlands are, however, created by permafrost. Below the surface, the ground is perennially frozen to a depth that may well exceed 1500 ft (460 m) in the extreme north. At the surface, a narrow layer will thaw in summer: this is known as the active zone; in the Far North it has a depth of only a foot or so, but it deepens southward. The permafrost area, which is continuous in the north, becomes discontinuous in the south, consisting at its southern edge of a series of permafrost "islands." The presence or absence of permafrost depends on a number of factors—of vegetation cover, aspect, or relief—some of them highly localized in their nature, as shown in Figure 22-2.[4] Thus the presence or absence of permafrost also enters into the calculations of farmers, miners, and construction engineers as a major variable in their planning.

Although a permanently frozen subsurface has to be reckoned with, most of the problems actually arise because the active zone is not permanently frozen; it thaws in the summer season. One then has to deal with a surface layer that is iron hard in winter, but generally soft and waterlogged in summer, for the permanently frozen subsoil impedes drainage. This seasonal contrast is of great importance to an engineer planning to construct an oil pipeline from the Alaska oil fields southward or, for that matter, to build a sewage system for a northern settlement. A cultivator has equally to reckon with the interruption of both soil drainage and leaching processes (a hardpan tends to form at the lower limit of thaw) even when a soil can be found that will thaw to a sufficient depth to make crop rooting feasible.

What is the impact of physical conditions such as these on agriculture? Apart from the limitations imposed by relief and soils, it is widely assumed that the main determinant is the cold of the Northlands. But experience has shown that this is not wholly true. Although the frost-free season is very short and highly unreliable in its occurrence, some crops can nevertheless be raised in the Far North, where summer days are long, even if the summer season is short.

The main climatic limitations are lack of summer heat and, rather surprisingly, drought. The first of these, as we have seen, marks an important difference between the Northwest and the Arctic. Certain crops can be grown on the Lower Mackenzie, beyond the Arctic Circle, that would not mature in central Québec, 15° further south. At Dawson, in the Yukon (64° N), the mean temperature is above 50°F (10°C) for almost three months in summer, which is nearly a month longer than at Fort Mackenzie, in Québec (57° N), and only a little shorter than at Gaspé (49° N).

The second limiting factor is drought. Over the interior of the Northlands, precipitation is generally between 10 and 20 in (250 and 500 mm) per annum; only on the Alaska–British Columbia coast does it rise to 40, or even 80 in (1000–2000 mm). This means that the Canadian Arctic, with its cool summers, is subhumid in climate, whereas the Northwest is definitely semiarid, and irrigation has as much relevance for its few cultivators as concern over frost danger.

Finally, not merely the amount, but also the seasonal distribution of precipitation, must be borne in mind. Not only does most of the Northeast have a short, cool summer, but its 10–20 in (250–500 mm) of precipitation arrive with a late-summer maximum,

[4]See R.J.E. Brown, *Permafrost in Canada* (Toronto, 1970), for further details.

A

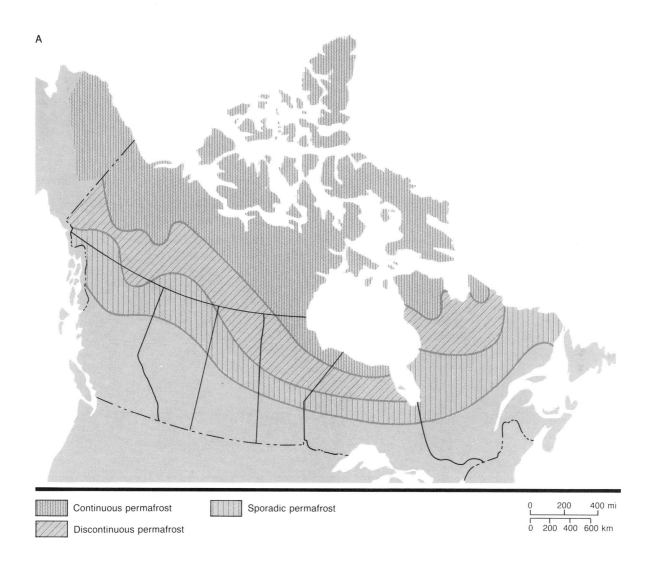

Continuous permafrost		Sporadic permafrost
Discontinuous permafrost		

0   200   400 mi

0   200 400 600 km

B

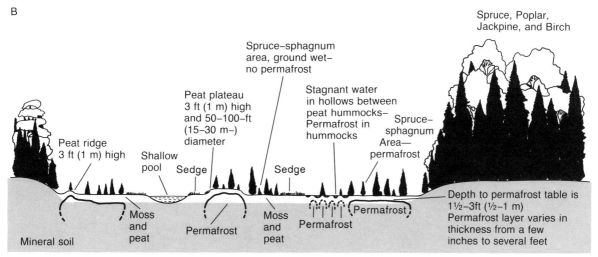

Spruce, Poplar,
Jackpine, and Birch

Spruce–sphagnum
area, ground wet–
no permafrost

Peat plateau
3 ft (1 m) high
and 50–100–ft
(15–30 m–)
diameter

Stagnant water
in hollows between
peat hummocks–
Permafrost in
hummocks

Spruce–
sphagnum
Area—
permafrost

Peat ridge
3 ft (1 m) high

Shallow
pool

Sedge

Sedge

Depth to permafrost table is
1½–3ft (½–1 m)
Permafrost layer varies in
thickness from a few
inches to several feet

Mineral soil

Moss
and
peat

Permafrost

Moss
and
peat

Permafrost

Permafrost

Fig. 22-2. Canada: Permafrost. Map **A** shows the areas affected by permafrost in the Canadian North, and is taken, by permission, from R. M. Bone, *The Geography of the Canadian North* (Toronto, 1992),

Fig. 2-7. Diagram **B** illustrates permafrost localities in the zone of discontinuous permafrost.

457

too late to be of much help to the would-be cultivator. Thus the Arctic, and in particular northern Ontario and northern Québec, suffers a triple drawback to land use: lack of soil, lack of sunshine, and lack of growing-season rainfall. Further west, in the Peace River country and the Mackenzie Valley, the amount of precipitation is no greater, and evaporation rates are higher, but the rainfall regime is of a Great Plains type and shows a growing-season maximum.

Future Prospects

In 1990, the population of the Canadian northern territories was 69,000, and that of Alaska 550,000. The population of all those census divisions of Canada that lie to the north of Hamelin's boundary line between his Pre-North and Middle North was approximately 1.25 million. Although it is difficult to compare Canada and the Russian lands directly, it must be recognized that the population of the Russian Northlands is many times larger than that of its North American counterpart. Therefore, questions arise: Why have so few settlers entered this great region? Under what circumstances might its population increase? Is increase in its settled population essential to its future development? The next task in this book is to attempt a brief answer to these questions.

THE PAST—WHY SO LITTLE?

There is no real difficulty in answering the first of our questions if we bear in mind the conditions under which Canada has been peopled—why a population largely composed of free and individual immigrants should have chosen on the whole not to strike out for an area so physically harsh and lonely as the Northlands; an area that supports a native population of Indians and Inuits amounting to a mere 100,000 or so. With few constraints to restrict their choice of settlement area, and plenty of space available, it would be surprising if many of the newcomers had chosen the North. The great in-migration of the years before World War I was basically farm-orientated; so, too, was the internal movement of French Canadians away from the crowded farmlands of Québec, but the Northlands repelled the farmer. On the other hand, the second great immigrant wave, in the period since 1945, flowed into a changed Canada—a country in which industry had developed, cities were booming, and the whole fabric of employment was being altered by the growth of service occupations.

Less than ever before was there any incentive for the average person to settle on the fringes: the cities absorbed the immigrants, and for the few interested in farming, there were empty farms and agricultural frontiers to occupy even in the southernmost reaches of the country. In the absence of any constraint imposed from outside, or of an unprecedented increase in the nation's population, Canadians would be likely to prefer comfort and community in the south to the hard and lonely life of the Northlands.

THE FUTURE—HOW MUCH?

In temperate and long-settled lands, it is normal to find that agriculture forms the basis of continuous settlement, and that other occupations grow like a pyramid from that base. But we have already noted that physical conditions limit the growth of agriculture, and, as we shall see in the next section, the overall prospects for agricultural expansion are not good. It therefore follows that the basis for continuous settlement is unlikely to exist: if it takes place at all, settlement will almost certainly not be continuous, but rather concentrated—at strategic locations, along routes, or around mineral deposits. These concentrations are likely to be more and more widely spaced as they spread northward into higher latitudes.

We must now consider the probability that development will occur, and see what form it is likely to take.

Agriculture in the North

Despite the dwindling significance of agriculture within the Canadian economy as a whole, it will be well to start with its prospects in the North. On the basis of the physical limitations we have already considered, we can recognize that some regions of the Northlands offer positive prospects and that others, in the foreseeable future, offer none at all. In northern Québec and Ontario, for example, despite the growth of mining settlements, potentialities appear to be very slight. Although the eastern Shield has adequate rainfall, it has neither adequate soil nor summer heat. Even the Clay Belts, to which the farm frontier advanced several decades ago, are not much more favored; they are somewhat less sterile than the ice-swept surface around them, but their cultivation is not easy. Before they can be farmed, they must usually be cleared of either forest or peat, the danger of frost is always present, and few crops are grown besides hay and a little grain. Further west and north, the situation is different only in detail; in northern

The Peace River region: Farmland in the Peace Valley in British Columbia. *Government of British Columbia*

Manitoba and Saskatchewan, a combined land use of forestry and tourism offers far better prospects than agriculture, and although estimates of cultivable soils run as high as 5 million acres (2 million ha)—along the Liard and the Slave and in the northern valleys of British Columbia—reconnaissance work has been sketchy and the climatic data on which the estimates are based are limited and suspect.[5]

This being the case, generally, along the northern frontier of agriculture, the contrast presented by the Peace River region of Alberta and British Columbia is all the sharper. Not only does this region possess millions of acres of potentially cultivable land to add to the 5 million mentioned above; it actually has 4.5 million acres (1.8 million ha) under cultivation and is now a farming frontier. Part of the reason for the lack

[5]To show how wildly tentative such estimates can be, it is worth quoting those recorded by H. A. Johnson and H. T. Jorgenson, *The Land Resources of Alaska* (Seattle, 1963). Estimates of potential agricultural land in Alaska have been as follows: in 1940, 42.6 million acres (16.8 million ha) were thought to be cultivable; in 1946, this figure was revised downward to 2.9 million acres (1.2 million ha),

and in 1954 to 0.8 million acres (324,000 ha). The actual cultivated area is minute, and Alaskan agriculture, despite growth in local population and demand, consists of only about 300 farms, half of which are too small to be commercial. See further, J. R. Shortridge, "The Collapse of Frontier Farming in Alaska," *Annals,* Assoc. of Amer. Geog., vol. 66 (1976), 583–604.

of agricultural progress northward elsewhere in Canada is probably that any of the few Canadians who are interested these days in pioneer farming set out automatically for the Peace. Between 1971 and 1986, the area in farms there increased by 18 percent, and the average farm size rose from 667 acres (270 ha) to 820 acres (332 ha). The population increased to keep pace with the farm area. The Alberta government was anxious to encourage settlement, and made land available: take-up has varied from year to year, but it is felt that, given good harvests and the substantial government loans which are available, the province may hope to dispose of as much as 150,000 acres (60,000 ha) each year.

These changes reflect a combination of both expansion into new lands and retreat from poor ones, as well as enlargement and improvement of existing holdings. Improved land increased significantly during both census periods. Under the structure of provincial law, it is usually possible to obtain extensions to existing farms for little more than the cost of clearing them, and that, to a farmer in, say, Illinois or Iowa must sound like paradise indeed. Part of the increase in farm area on the Peace, however, represents genuine pioneering: the provinces issue several hundred homestead titles each year, and all through the region one sees the work of cutting, burning, and clearing forest going on. The southern edge of the region is fully settled: its leading crops are barley (which

now occupies double the area of wheat) and rapeseed, which proved a runaway success, but which then fell out of favor for a time, before increasing its acreage once again in the early 1980s. Beyond this firm base of prosperous farms, there is a thinning out of agricultural settlement and finally the northern forests. The Peace River country is the gateway to the North.

There are, of course, special conditions here that help the farmer. In terms of soils and topography, the region is an extension of the Prairies rather than a representative section of the Northlands. Relief is gentle and access relatively easy by road and rail. The big Peace River farms operate no differently from those of the main prairie lands further south. They can thus bypass one of the main problems of northern farmers everywhere else—how to market their produce.

Yet even along the Peace, some farmers fail. They are mainly the homesteaders and they fail because they cannot survive the long initial period before they have a productive holding large enough to yield a livelihood. This has been the case with many—perhaps the majority—of the French-Canadian settlers who came in from Québec in groups, settling 40–50 acres (16–20 ha) apiece.[6] A few produce honey or

[6]R. Breton and P. Savard, eds., *The Quebec and Acadian Diaspora in North America* (Toronto, 1982).

Peace River Country: Grain elevators at railside, Fairview, Alberta. Here, close to the northern edge of commercial grain farming, the size and number of the elevators testify to output, quality, and demand for the product. *J. H. Paterson*

work farms that others have abandoned, but the rest have left the land to seek work in the city or on oil rigs.

What are the lessons to be learned on the Peace River?[7] First and most important, northern farming is susceptible to peculiar pressures. In the North, lands suitable for agriculture are widely scattered, mostly on old lake beds and river terraces. To the isolation of the region there is added, therefore, the isolation of the homestead. This isolation and its counterpart, cost of transport and service, are the great economic handicaps in every northern project. They have been aggravated by the method of settlement. Where settlement has been on a homesteading basis, each newcomer has claimed a promising quarter section without reference to the location of other settlers; thus, the cost of services is increased by the unnecessarily wide scatter of individual homesteads.

The second lesson to remember is one of cost—the cost of clearing the ground (for much of the potentially cultivable land is under forest). This means that either the early years for the farmer will be a long, drawn-out struggle, clearing an acre or two at a time, with only modest chances of success, or there must be capital to invest before a start can be made—which is the same sort of problem that an irrigation farmer in the dry West would face. The North suffers, however, an additional handicap: whereas, on irrigated lands, higher yields will help to pay for the initial outlay, yields on northern lands will not be appreciably higher—indeed, they may be lower—than elsewhere.

If agricultural settlement is to occur, it must clearly be governed by certain principles. (1) Settlement should be made not at the whim of the individual cultivator, but rather in blocks, as the need for it arises from increasing population pressure or enlarging demand for food. Each of these blocks must be large enough to support the cost of services. (2) Just as the settlement block must be large enough to provide essential services, so the individual unit must be large enough to give the settler a reasonable chance of success. We have seen how the restriction to one particular quarter section was the ruin of many an earlier homesteader in the United States. Today in Alberta, it is possible to obtain a 1000-acre (405-ha) homestead, and to lease in addition up to 5000 acres (2025 ha) of rough grazing. What is more,

the residence requirements have also been made more realistic: here in the North it is accepted that many homesteaders will occupy their holdings only seasonally and that they will require the freedom to take other jobs in winter. All this the provincial government accepts as part of the necessary framework, within which settlement should take place. (3) Blocks of new lands should be settled only in strict relationship to transport and to market opportunities—perhaps the clearest lesson to be learned from northern settlement thus far. In the Peace River country, that "laboratory" for northern experiments, when the first good lands were opened up early in the present century, the mistake was made of trying to grow wheat commercially at distances of 50–60 mi (80–100 km) from a railroad. This proved impossible; only when the railroad was carried north could this wheat compete with crops grown 10–20 mi (16–30 km) from a railroad in the central Prairies.

On the other hand, the few places where agriculture has proved successful in the North have been places where a market, and in particular, a local market, has been available. The Matanuska Valley in Alaska was linked by road and rail with Anchorage, which it supplied; and after initial misjudgments were corrected, it settled down to a sensible routine of local supply of fluid milk and potatoes. The Ontario Clay Belts supply various mining and forestry communities with food. In short, the only justification for expanding northern agriculture in the future, away from the frontier of continuous settlement at least, will be to supply a specific market within or close by the locality where cultivation is possible.

Whether the provincial and federal governments ought to encourage farmers to attempt settlement even on this limited basis is very much open to question. It is even doubtful whether the provinces ought to continue to sponsor new homesteading (as opposed to enlargement of existing units) in the Peace River country, where the chances of succeeding are so much greater than elsewhere on the margins. One thing that seems clear is that by itself agriculture cannot support the range of services to which any sector of the Canadian population in the 1990s feels itself entitled. These services can attain an acceptable level only if other activities, such as mining or forestry, are present to help to foot the bill. On this ground, if no other, therefore, the idea of agricultural development, independent of the growth of other economic activities, should be firmly suppressed by the governments responsible for these Northlands.

---

[7]These are dealt with by W. Hamley, "The Farming Frontier in Northern Alberta," *Geographical Journal*, vol. 158 (1992), 286–94.

Mining in the North

To date, the largest part of all northern development has been the product of mining activity. The Precambrian rocks of the Shield contain a great variety of metals; the younger formations that underlie the northern end of the Great Plains Province yield coal and petroleum; and in the western mountains lies the gold that blazed the name of Klondike round the world in the 1890s.

Some of these mineral deposits have, over the years, been important on a world scale—for example, the nickel-copper ores of Sudbury, Ontario, once the source of some 90 percent of the nickel supply of the world outside Russia. Other deposits are only now revealing their riches; other areas showing that where one mineral has been worked, there are others to follow. The opening up of the Québec–Labrador iron ores, for example, proves to have been only the first step in exploiting the riches of the Labrador Geosyncline, which is now found to contain not only iron but copper, lead, manganese, nickel, and zinc, in a mineralized belt extending from eastern Québec, all the way north to the west shore of Ungava Bay.

At the same time, new sources replace old ones of the same mineral. In nickel production, Thompson in Manitoba has become second only to Sudbury as a source of the metal, and the radium-uranium beds of Great Bear Lake have been worked out and eclipsed by those of northern Saskatchewan and of the Elliot Lake area, on the north shore of Lake Huron, midway between Sudbury and Sault Ste. Marie.

Each recent decade has brought its developments. The 1950s saw the opening of the Labrador iron mines, the founding of ore towns—of which Schefferville was the first—and the railway links with Sept Iles and Port Cartier on the St. Lawrence shore. The 1960s saw the opening of both operations and townsite at Pine Point, on the south shore of Great Slave Lake, where the newly constructed railway line and Hay River road (since extended to the Mackenzie Valley) were available to haul out valuable lead-zinc ores, whose existence had been known for years, but whose remoteness had prevented their useful exploitation.

Also belonging to the late 1960s were the oil strikes along the North Slope of Alaska. Between the Alberta strikes, which began in 1947, and those in Alaska, there was a great leap forward, both geographically and technically. It was in March 1968, after several years of fruitless drilling, that oil was

struck in the Prudhoe Bay area and Alaska's future, as *Time* magazine commented, "lit up like a pinball machine." The technical problems of merely supporting the drilling crews were gigantic; those of moving the oil once it was flowing were almost insuperable. Yet in a period of increasingly tight domestic supplies, there could be no doubt that the United States needed the Alaskan oil—that the problems had to be surmounted. The difficulty was to find a feasible way of moving the oil to market under Arctic conditions, and Alaskan Arctic at that, which meant not only taking into account low temperatures, permafrost, and the fragility of the tundra ecosystem when disturbed, but also earthquake risk. It took the oil crisis of late 1973 to overcome environmentalist opposition to the construction of a specially designed pipeline, a pipeline that, four years later, began carrying oil to the warm-water port of Valdez on the southern coast. By 1980, Alaska had become second only to Texas among the oil-producing states of the United States. By 1985, its production was 75 percent of that of Texas, and 20 percent of U.S. production. Its role as a gas producer, however, was more modest: it accounted for only 2 percent of national output.

Meanwhile, the rise in oil prices in the mid-1970s drew fresh attention to the Athabasca tar sands. The existence of these oily sands had actually been known since the late eighteenth century, and small-scale experiments had been made to extract the oil from an area where reserves were estimated to run to an astronomical 1 trillion barrels. In the seventies, the prospect for "synfuels" appeared to be good. The sands lie around Fort McMurray in the northern Peace River country—an area accessible over existing routes. In places, the sands could be reached by removing an overburden of only 50 to 100 ft (15 to 30 m).

By 1978, two major operations had been established; in one, the Canadian government was a main participant. All seemed fairly set for a new northern development that would usher in the next round of expansion on the Peace River frontier. But then the price of oil began to fall; the shadow of the "energy crisis" of the seventies faded; production was cut back, and the whole project was reduced to little more than a standby operation. It remains to be seen when and whether it will be revived.

In an area so huge and so little known as the Northlands, mineral discoveries are likely to continue in the future as they have in the past. It is al-

Oil in the Arctic: Drilling from an artificial island (see below) constructed in the Beaufort Sea off the coast of northeastern Alaska.
Standard Alaska Production Company

ready clear where the major exploratory effort of the 1970s and 1980s has been made: along the Arctic coasts and under the waters of the Beaufort Sea. The Alaska oil strikes of the late 1960s resulted from the work of the oil companies in the Far North, at the same time that it became clear that reserves in the Alberta fields were running low. Even if, by the standards of the 1970s, oil prices remained low, the Canadian companies needed at least to know where future oil supplies could be found, against the time when the price rose again sufficiently to make drilling worthwhile.

It is now clear that there is plenty of oil in the Arctic, beyond the edge of the Canadian Shield (see Fig. 1-4). The technical problems of extracting it and transporting it are, however, horrendous. To drill for oil beneath an ocean that is frozen for much of the year is difficult enough. But to do so beneath one that thaws briefly is even worse: the circumpolar ice—the polar pack—moves slowly clockwise, and the land ice remains *in situ*, but thaws at the edge, which means that one shears off from the other, and ordinary drilling rigs are crushed like matchsticks between the two.

To overcome this problem, the oil companies have created artificial islands, either by dredging them up or by sinking a circle of caissons to surround the rig. With these, and a sentry posted to drive off polar bears, it is possible to drill. Meanwhile, exploration has continued among Canada's Arctic islands beyond the 75th parallel. How oil or gas would be transported south to market from these High Arctic fields—by pipeline or by icebreaker—remains to be seen.

But while exploration continues and mines open, other mines close—in a recent ten-year period, eighty-two of them. Perhaps the most notable recent casualties have been a number of the Québec–

Labrador iron ore operations, including the town of Schefferville, which has been abandoned—Schefferville, perhaps the most complete, and completely planned, of all the northern mining communities of the past thirty to forty years. Either a reduction in demand, or a fall in price, can do this to a mine and its supporting community. So the question is bound to arise as to whether the size and duration of the mining operation will justify building a permanent settlement.

North America is littered with mining ghost towns, the remains of temporary communities. But in the Mountain West, where most of them are situated, these towns were built of local timber, cheaply and quickly and, when the mines were exhausted, it was no great loss if the settlement was abandoned to its ghosts. In some cases, the whole town was dismantled, and moved with no difficulty to a new mine head. But this is hardly a treatment to be recommended for the Scheffervilles of the Northlands. To create a town at all in the northern wilderness calls for a huge investment in construction, insulation, and services. It is not something to be lightly undertaken, or lightly abandoned. Construction will probably involve bringing every plank and bag of cement from a thousand miles away: it will certainly involve problems of insulation, sewage, and servicing peculiar to the northern climate. Only when the life expectancy of the mine is long or when, better still, the mineralized area contains a variety of different deposits, can the construction of a permanent settlement be justified. In other words, mining alone is likely to provide the North in the future with little more than it does at present—a series of scattered communities whose existence begins and ends with the mine.

Hydroelectricity in the North

A further possibility of development in the Northlands arises out of their wealth of waterpower potential. The impact of waterpower on a region's economy is already vividly illustrated by developments in areas that adjoin the Northlands on either side of the continent—the Columbia–Snake basin and the St. Lawrence Valley. In detail, the impact of such development within the North itself can also be observed in the Kitimat scheme on the coast of British Columbia. Here the flow of one of the Fraser's tributaries has been reversed by damming, the water is diverted to an underground power station at Kemano, and the power generated there is transmitted to Kitimat, on one of the coastal fiords, where an aluminum smelter (see p. 516) has been brought into operation.

The next generation of projects is already in being. In the Northwest, there is the Bennett Dam on the Peace River, just to the west of the Alaska Highway. In the Northeast, the harnessing of the power of the Churchill Falls in southern Labrador is now complete, and it contributes most of its 5.2 million kw capacity to the Québec grid. On the east shore of Hudson Bay, the La Grande complex (see Chapters 8 and 13) is coming into being: one of its stations alone, La Grande 2, exceeds the capacity of Churchill Falls. Beyond these schemes lie other possibilities on still more remote rivers—the Yukon and the Mackenzie. If such schemes are ever to be realized, the problems of financing and marketing of power will have to be overcome, and neither is likely to be easy. But it would seem certain that in a continent facing both an energy shortage and pollution problems, the use of this "clean" source of power will be extended in the future.

Apart from the brief period of construction, such schemes will not, of course, do much to increase the population of the Northlands. With improvements in transmission techniques, it is likely that nothing but power generation will take place in the North: utilization will occur in the already settled south.

Forestry in the North

One obvious user of electric power within the region is the forest-products industry. Throughout the forested zone of the North there are large numbers of saw mills, and a discontinuous forestry frontier is moving gradually northward as the development of access roads and water transport brings fresh areas of timber within economic range. Experience elsewhere has shown that the forest-products industry is self-sufficient when it can support pulp and paper mills, for these represent its more stable, long-term element. This takeoff point was reached in Alaska in the late 1950s, and the number of mills in the state has increased steadily since. The industry is a very large consumer of electric power, and offers clear growth potential in the Northlands.

Although there is no doubt that the forest-products industries will expand in the future or that they will make a valuable contribution to the region in terms of employment and road building, they will certainly not expand free from constraints. These constraints are of two kinds. One is the vigilance of conservation interests, who will be concerned to make sure (1) that great areas of forest resources are not given over to private hands without strict provisos about cutting and replanting, and (2) that the pulp industry is not allowed to pollute the northern environment.

The other constraint applies not only to forestry, but to oil and mineral development. For all these northern products, the main markets, the main financial interests involved, and the main management are not Canadian but American. There is very little doubt that the quickest way to open up the Northlands would be to throw the border wide open for the entry of companies and investors from the United States. The chronic lack of capital a nation of 25 million people is bound to experience when confronted with development potential on the northern scale can most simply be met by drawing on American resources.

But if this is the quickest method, it is not necessarily the one Canadians will wish to adopt. We have already seen the extent of Canada's concern over the large part played in its industry and economic affairs by U.S. interests. American investment means that the profits from that investment are all too likely to end up south of the border. Against the advantages of exploitation in the immediate future, Canadians must therefore balance two other considerations. (1) It is one thing to see a pulp mill or oil refinery constructed and giving employment locally, but another to know that local employment is all that the plant is providing; that all the value added by manufacture and the profits from the operation are being drained off outside the region and, indeed, outside the nation. (2) The resources being exploited today will, in a resource-hungry world, be worth more tomorrow and even more ten years hence. In other words, if Canada accepts for the present a slower rate of northern development, one in line with its own capabilities, the value of its products when they are actually marketed will be that much greater. What is more they will be Canada's own, to dispose of at will. These are decisions that call for the wisdom of Solomon.

Strategic Considerations in the North

The Northlands have come into their own in the air age and with the rise of the former USSR as a military power on the other side of the polar wastes. The knowledge that Alaska was Russian until Seward bought it—in the face of fierce criticism—for $7.2 million in 1867 is a lingering nightmare in the Amer-

ican mind. Members of the armed forces make up a considerable part of the population of Alaska, and the presence of numerous civilians is also attributable to strategic needs. Of these needs, the wartime Alaska Highway, running northwest from Dawson Creek, British Columbia, was the most obvious geographical expression. Since World War II, the United States and Canada have collaborated in the construction of the continent's radar defenses against transpolar attack.

Such measures tend to open up the North, but involve little permanent settlement outside a few major bases. If, however, we extend the term "strategic" a little to cover a wider field, then the general development of northern communications, which will certainly serve a strategic purpose, is likely to have the most direct effect on increasing the region's population in the future. At least the converse is true: there can be no hope of successful expansion without extension of communications, and this extension should preferably precede any attempt at further settlement.

Transport and Tourism in the North

In the early days of northern penetration, movement was relatively simple only in summer, when the rivers and lakes were open for water transport. But developments in transport are slowly reducing the inaccessibility of the Northlands and that, almost by definition, means increasing the flow of tourists. There are new roads into the North; an extended network of air services; even a few new railways, at a time when the rail net over the rest of the continent is contracting. Railways were built to bring iron ore from the Québec–Labrador mines to Sept Iles and Port Cartier on the St. Lawrence. Great Slave Lake has been linked with the settled lands to the south. Altogether, over five thousand kilometers of railway carrying significant mineral traffic were built in Canada between 1950 and 1975, nearly half of them serving mines in the Shield. What these railways would carry when the mines close is not clear.

Road corridors across the Northlands are gradually being opened up. It is possible to travel by road to Inuvik, at the mouth of the Mackenzie, via Whitehorse and Dawson City and, one day soon, a road will follow the Mackenzie all the way upriver from Inuvik to Fort Simpson. Mining developments have ensured road links to Lynn Lake and Thompson in northern Manitoba, although Churchill is still best reached by rail.

It has already been suggested that new roads and routeways mean more tourists. They probably do. But we must be careful not to exaggerate this expectation. It is significant that, in the most recent geography of the Northlands at the time of going to press, Robert Bone does not mention growth of tourism among his prospects, and barely lists "tourism" in the index of his well-informed study of the area.[8] Tourism there is likely to be, but not *mass* tourism; certainly not enough, in the foreseeable future, to form the basis for regional centers in the Far North, or for a regional economy. Distances are great, and scenery is empty and awesome, rather than dramatic.

Characteristic of the region has been the use of air transport to carry mining equipment to new projects or to assist in the export of precious metals, furs, and fish. Yet most of this northern transport network has to be laid out in acceptance of the fact that, since it is economically "strategic," its justification is not to be sought in terms of immediate financial returns, but rather in the long-term development the coming of transport makes possible within the nation.

This being the case, we should adopt a realistic view of what the transport network of the future will be like. Clearly, there will never be a dense network of roads over the whole region; moreover, the tourist demand of the future will encourage the preservation of stretches of untouched wilderness as a major attraction. What we may expect to see is the development of particular areas, served by appropriate transport routes. Yet these areas will be separated from each other by empty stretches, much as the Prairies and Ontario are at present separated by the desolate north shore of Lake Superior with its single road and two railroad lines.

It is important that development should be concentrated in the Northlands in order to reduce the mileage of costly transport routes that would have to be maintained. Indeed, this principle of concentration holds good in all respects in northern development: with a limited amount of investment capital to spread over 2.5 million sq mi (6.47 million sq km), to distribute it too widely would automatically be to do too little everywhere.

THE FUTURE—HOW PERMANENT?

The last several sections of this chapter have been taken up with an assessment of the likely answers to two of the three questions with which we began: Why has there been so little development in the North up to now? Under what circumstances may we expect development in the future? We now come to the third question: Does development necessarily imply permanent settlement in the Northlands?

If it does, then we must note that future settlement will probably have to be very different in character from that of the past, because the settlers themselves will be different. The lonely, rough-living prospector will be followed by the technician and scientist, and for these northerners, settlements will have to be planned in advance rather than allowed to grow up in the old, haphazard fashion of, say, a Dawson City in the gold rush days. Facilities will have to be provided and families catered to. It may well be that failure either to realize this or to provide the new type of settlement has contributed to the slowness and hesitancy of the Canadian occupance of the North. As Trevor Lloyd remarked:

The type of person who will be needed in the new north will be expensive, because he will be a technical specialist in a world where mechanization will be very advanced. He and his family will need and demand conditions as good as or better than those found in southern suburbs. . . . Such communities will be expensive, but their citizens will be well paid and able to afford them. There will be no place for the shoveller of snow or the guardian of a trapline.[9]

As Lloyd stresses, all this will be expensive. It is therefore worth pausing to consider whether it is necessary for "technical specialists" actually to "live" in the North. Given the long-standing attachment of the Northlands to air travel, and the speed of such travel today, it would seem reasonable to suggest that much of the truly northern development we have been considering might be carried out without construction of an indefinite number of miniature Torontos for the workers and their families beyond the 60th parallel. The workers would "live" in Montréal or Winnipeg and spend three or four working days a week in the North: they would be commuters on a grand scale (and some already are). The cost of commuting 500–1000 mi (800–1600 km) would be heavy, but it would not be borne by the individual, and it would be weighed against the cost of the alternative—a northern town complete with the amenities suggested by Lloyd and having a life expectancy of perhaps twenty-five to fifty years. Schef-

[8] R. M. Bone, *The Geography of the Canadian North* (Toronto, 1992). However, help is at hand: see W. Hamley, "Tourism in the Northwest Territories," *Geographical Review*, vol. 81 (1991), 389–99.

[9] In J. Warkentin, ed., *Canada. A Geographical Interpretation* (Toronto, 1967), p. 590.

Yellowknife, Northwest Territories. Gold was first discovered here in 1896, but it was not until 1938 that the first large-scale production from modern mines led to the growth of the town. By 1945 Yellowknife's population numbered 3,000. Today it is over 12,000, for the mining town is also the capital of the Northwest Territories, with government offices, an arts and cultural center, and ample sports facilities. *Government of the Northwest Territories, Department of Culture and Communications. Photo by Tess Macintosh*

ferville, the iron ore town, lasted a little over twenty-five years. What is to be done with it now?

However, there is one other factor. Whether the technical specialist commutes in this fashion or becomes a permanent settler in the North, there is one section of the northern population that has no choice, and that is the indigenous element, which is "permanently" part of the Northlands. It is noteworthy that none of the future possibilities we have been considering offer any obvious role to the Indians or Inuits. On the contrary, the forecast by Lloyd which we have just quoted will, if valid, have the effect of excluding them from the future development of the Northlands. We cannot leave the subject of northern prospects without pausing to ask what has created this anomaly.

In Canada as a whole, the Indian and the Inuit populations represent only 1 percent of the total. But this proportion increases sharply as one moves north. In northern Saskatchewan, it is almost 50 percent; in northern Québec it is more than 50 percent, and in the Northwest Territories it is 65 percent. Moreover, the fringes of this indigenous population are blurred by the existence of the métis, or half-

Indians, as a virtually separate group; their status is distinctive, but their economic condition usually approximates that of the Indian rather than the white Canadian. Both Indians and métis in the North occupy treaty reservations, in most cases selected for their potentialities as hunting grounds. A few groups are successful cultivators; some others are marginal farmers, but the majority, insofar as they have a regular basis of livelihood, are hunters, trappers, and fishermen.

There is no particular reason why a thin scatter of such groups should not survive in their traditional life patterns in this vast wilderness, if they choose to do so. But there are two complicating factors. One is that not all these groups are scattered; there are actually some considerable concentrations of Indians in the North, one thousand, or two thousand, of them living in and around a small settlement that acts as a miniature service center at the end of a dirt road. Not all of them can hunt and fish under these circumstances; economically, concentrations like these assume other employment—in industry or service occupations. But in practice, there are none. The "town" Indian lives on welfare.

It is a cold statistic that the average income of the Indian and the Eskimo is one-sixth . . . the national average, that the infant mortality rate is more than four times the national average, that the life expectancy of those who survive infancy is decades shorter than that of the white man, that the percent of the uneducated element in the labour force, mainly Indians and Eskimos, is twenty-four times higher than the national average.[10]

That was written nearly thirty years ago. We can bring the figures up to date to some extent. In 1986, the Inuits had an infant mortality rate three times the national average. Their illiteracy rate was well over twice that for Canada as a whole. Their suicide rate was over three times the nation's. The native peoples of the North as a whole had a birthrate more than three times greater than the national average (2.8 against 0.8).

To put no finer point on it, much remains to be done, to bring the Native Americans into the circle, economic or social, of Canada as a whole. But what should be the aims and the methods? To create employment for such groups, a number of agencies and

programs have been launched. But here we come to the second complicating factor. Although it may be possible—given funds and patience—to generate secondary and tertiary employment for the Indian population in such remote places as Wabasca, or Fort Chipewyan, or Slave Lake, the quicker method of dealing with this problem is to train the indigenous population to do the sort of work available in existing centers. Yet the first effect of training them is likely to be that the more enterprising members of the community will take the earliest opportunity to leave for the south. It is unrealistic to invest in manufacturing plants in the places just named if the markets for the manufactured goods are back along the same third-class road by which the raw materials are brought in. The question to be decided is one that has confronted developers in many other parts of the world (Appalachia, for example)—whether to take the work to the population or to bring the population to the work. In the Northlands, the case for maintaining by artificial stimulus the native population in its present harsh environment of poverty is a weak one.

There is, however, another dimension to this problem. The effective exclusion of the native Canadian peoples from modern economic life was neither inevitable, nor was it the result of some natural lack of skills or enterprise. It was brought about by the overwhelming numbers and technical resources of the incoming white population. Although the present circumstances of the native peoples are the product of numerous treaties between themselves and the newcomers, and although the new (1983) Canadian constitution fully recognizes the Indian-Inuit rights, it has to be said that those treaties were not made between equal contracting parties. The imbalance between the two sides created long-standing grievances among the tribes about the disposal of wealth that white Canadians extracted from lands regarded by native peoples as their own, while they themselves stood and watched.

Generations of Indians and Inuit have nursed these grievances. What may be said to have brought them out into general consciousness was (1) the setting up in the 1940s by the U.S. government of an Indian Claims Commission to consider specific grievances, as we saw in Chapter 2, and (2) the sudden starburst of wealth that fell upon the tribes in Alaska, following the oil strikes of the late 1960s. With this to highlight native claims, the U.S. government negotiated a Native Claims Settlement in 1971, as a result of which those natives have received hundreds

[10]*Report of the Advisory Commission on the Development of Government in the Northwest Territories,* vol. 1 (Ottawa, 1966), p. 172, quoted by M. C. Storrie and C. I. Jackson, "Canadian Environments," *Geographical Review,* vol. 62 (1972), 309–32.

of millions of dollars in cash and oil royalties, and clear title to millions of acres of land.[11]

This naturally served as an incentive to the Canadian tribes to press their own claims, and it led to the establishment of a native claims procedure in 1974. Since then, three major claims—or, rather, claims covering three major areas—have been settled: the first in northern Québec–James Bay in 1975, the second in northeastern Québec in 1978, and a third covering the western Arctic (1984). But, unfortunately, other claims, small or large, were held up by the prolonged negotiations on these major cases, and the temperature of the discussions has risen. For, as Usher, Tough, and Galois have written: "land struggles have been revived with determination, and even militancy, by aboriginal peoples throughout Canada over the last 20 years."[12] Whatever the outcome, however, there is general agreement that, this time around, the Native Americans must be offered a fairer deal.

To begrudge them their new prosperity would be both vindictive and absurd: they have so much catching up to do. But it does raise the intriguing question as to what these northern peoples are to do with all that wealth—at least if they remain in their northern environment. That, of course, is an important "if." They will presumably be able, once they have their settlement, to live where they like, and to afford good health care and education for their children. What they will find difficult, in the foreseeable future, is to invest their huge new revenues *in the North*. Tourist facilities might usefully absorb a small part of the total sum. As for mining, they will be well advised to stay out of it, let others take the risks, and simply collect the royalties. But other possibilities for investment, or even conspicuous consumption, on the lower Mackenzie are hard to visualize.

Many Indians have left the North, driven out by a harsh environment, poverty, and the false hopes of the towns further south. It would be a sad outcome if the righting of ancient wrongs, and the joys of wealth at last, were to have no more than the same effect as poverty has had: to detach the tribes from their roots and their ancestral home, to become mere wanderers among the other 99.9 percent of the inhabitants of Anglo-America.

Alaska

In dealing with the Northlands and their future, we must treat Alaska separately, because it is a part of the region where the future has, so to speak, already arrived. Over the past thirty years, it has been possible to watch the impact on this section of the North of statehood and oil strikes, as well as the negative impact of the 1964 earthquake. During the decade 1980–90, Alaska's population increased by 36.9 percent (second only to Nevada's), and reached 550,000. This population had the highest per capita income of any state, and the lowest proportion of its general revenue provided by the federal government. By contrast, Canada's Yukon and Northwest Territories are sparsely inhabited, run by the Ottawa government as federal territories and, up to the present, are largely without oil.

All this has become a reality in spite of the fact that, in Alaska, the physical obstacles are immense. Apart from the general problems of settling and exploiting the cold North, Alaskans have to contend with mountains, ice fields, and long, deep fiords, with the chief obstacles concentrated in the southeastern corner of the state; that is, precisely in the part nearest to the rest of the United States.[13]

Quite apart, however, from the political independence conferred on Alaska by its grant of statehood in 1959, and the financial freedom secured for it by its oil revenues, it was always reasonable to expect that, as the "spare room" of a nation of more than 200 million people, it would attract more development than the huge and empty Canadian North, owned by a nation only one-tenth that size. The resources that could be brought to bear on Alaskan development projects were vastly greater, taxpayer for taxpayer or mile for mile, than those available to the Canadians on their side of the international boundary.

A further advantage for Alaska has been the possession of an ice-free coastline and the wealth it represents. This wealth takes two main forms: fisheries and coastal forests. The fisheries, the largest part of

[11]See J. B. Haynes, "The Alaska Native Claims Settlement Act and Changing Patterns of Land Ownership in Alaska," *Professional Geographer*, vol. 28 (1976), 66–71.

[12]P. J. Usher, F. J. Tough, and R. M. Galois, "Reclaiming the Land: Aboriginal Title, Treaty Rights and Land Claims in Canada," *Applied Geography*, vol. 12 (1992), 109–32: see also, in the same issue, pp. 133–45, Evelyn J. Peters, "Protecting the Land Under Modern Land Claims Agreements."

[13]Among works on Alaska may be mentioned Melody Webb's *The Last Frontier* (Albuquerque, N.M., 1985), which has a good bibliography. The periodical *Alaska Geographic* contains a good deal of valuable material.

Fig. 22-3. Alaska and the
Lower 48: A comparison of the
states' sizes, based on an
equal-area projection.

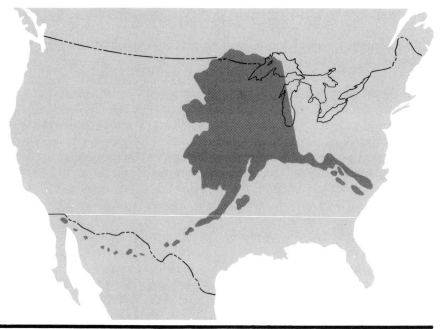

Equal Area Projection

which are represented by the salmon catch, brought in $590 million in 1985. The coastline also gives access to more than 5.5 million acres (2.2 million ha) of commercial forestland,[14] an area immediately available for development. These sources of income existed before oil started to flow from the North Slope fields and, properly managed, may well last much longer. Meanwhile, Alaska's severance taxes for 1985, mainly on oil, amounted to $1.4 billion, second only to Texas, with $2.1 billion, but thirty-one times as many inhabitants.

So we may readily agree that Alaskan development has been, and is likely to be, much more rapid than that of the Canadian Northlands. But we are still far from certain what the nature of that development will be. More than that, there is already a certain geographical impertinence implicit in treating Alaska as a single unit, as if its fate and future will inevitably be uniform. Point Barrow lies beyond latitude 71° N, whereas the southern tip of the state,

just north of Canada's port of Prince Rupert, is 55° N. From the state capital Juneau to the westernmost of the Aleutian Islands is a span of more than 50° of longitude—the same span as that from Portland, Maine, to San Francisco (Fig. 22-3). Even with a single state government responsible for this huge area, no one can predict, still less require, that it be developed as a unit.

Thus, the future of Alaska will be decided by the interplay of five separate forces: (1) the federal government; (2) the state government; (3) private development interests; (4) preservationist interests; and (5) the original population of Indians, Aleuts, and Inuits (Fig. 22-4).

The federal government is involved because it still (in some sense yet to be defined) "owns" 86 percent of Alaska. Its decisions are clearly critical to the area's future, in many ways much more critical than those of the state government. The pressures upon the federal departments concerned come from various lobbies in Washington, but the national interest, as we saw in Chapter 6, demands, as a bare minimum, an energy policy and a resource policy, and it is up to the federal government to formulate these. It was Congress that ultimately authorized the Alaska pipeline, and it is in Washington that the decision must be taken as to what proportion of the nation's resources

---

[14]Land that is (1) producing or physically capable of producing usable crops of wood, (2) economically available now or prospectively, and (3) not withdrawn from timber utilization. In Alaska this definition is important: the total forest area in the state is no less than 120 million acres (48.6 million ha). Much of this is too remote to meet condition (2) above, whereas other areas are excluded, under condition (3), by various federal measures.

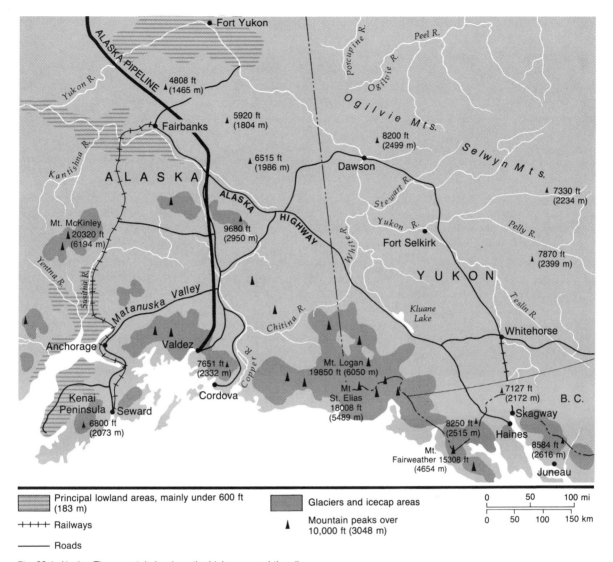

Principal lowland areas, mainly under 600 ft (183 m)

Glaciers and icecap areas

++++ Railways

▲ Mountain peaks over 10,000 ft (3048 m)

—— Roads

0   50   100 mi
0   50   100   150 km

Fig. 22-4. Alaska: The mountain barriers, the highways, and the oil pipeline.

should be used or, alternatively, set aside by the formation of parks and preserves; that is, what proportion of its resources can a wealthy society afford to tie up in the interests of present amenity and future reserves?

The state government is involved because when Alaska attained statehood in 1959 it received powers of taxation and resource control, and because the federal government, as a kind of christening gift, presented it with 103 million acres (41.7 million ha) of public land, to be selected by the state itself. The selection is very far from complete; yet the state has benefited already by such windfalls as the $900 mil-

lion it received from the auction of oil leases around Prudhoe Bay, and that before a single barrel of oil had actually been produced. No less than any of the other forty-nine states, Alaska is empowered to make laws controlling its own resources, and to press the federal government to reduce its role within the state.

Private interests are involved because journalists have long since run out of clichés about treasure chests full of resources within Alaska and, just when these clichés were beginning to sound a little too familiar and complaints were being heard that Alaska was greatly overrated (the 1964 earthquake did not exactly help), the North Slope oil fields were brought

The Alaska pipeline, which carries oil from the North Shore fields to the ice-free port of Valdez. The pipeline is insulated and positioned above ground (1) to avoid disturbing the permafrost surface below it and (2) to permit free passage for Alaska's wildlife, like these bears. *Chevron Corporation*

in to make good the earlier claims. Alaska is full of forest and mineral resources and, if they are to be exploited, there is no question of keeping commercial interests out; the only question is the extent to which either the federal or the state government may wish to control them.

Preservationist interests are involved because they see Alaska (and we are driven back again to clichés) as the Last Frontier, or the Last Chance. The word *last* implies that if the wrong policies are adopted, there will never be an opportunity to remedy them: the United States will simply have run out of areas to experiment upon. But just because the U.S. government still controls 86 percent of the land, policy enforcement should be practicable enough, if only the initial policy decision is right.

It is the preservationists' special concern that the natural equilibria in Alaska are very easily disturbed; that the tundra, in particular, has a natural replacement rate so slow that once the ecosystem is upset, vegetation and surface may well be destroyed. An area in which a single bulldozer or drill can cause irreparable damage obviously requires special treatment. Of equal concern to the preservationists is the wildlife of the area, and much of the argument about

the oil pipeline, both before and after its construction, has focused on 600,000 caribou, which annually migrate to their feeding grounds across the line.

Last, the native Alaskan population is involved because, although white Alaskans have tended to come and go (there was, for example, quite an exodus after the 1964 earthquake), the 60,000 natives remain: this is their true home. A rather academic series of arguments about "ownership" of Alaska had dragged on for years—adding one more example to a worldwide list of cases where concepts of ownership are differently understood by different cultures—when the 1968 oil strikes suddenly enhanced the value of this empty land by billions of dollars.

Coming quite soon after the 1959 grant of statehood, the oil strikes involved the federal government, the state, and the Native Americans in a threeway discussion about who should own what. As we have already seen, the federal government endowed the state with millions of acres of land at statehood, and by a 1971 agreement it made over further millions to the tribes. The status and wealth of these native Alaskans have thus been transformed in a very short time: let us hope that it has not been too short, and that they are able and willing to make the necessary adjustments.

What have military posts, oil, and earthquake produced in Alaska today? The population has more than doubled since 1960, but the turnover of population during that time has been great. Many have come, and a much smaller number have stayed. Bearing in mind experience on earlier frontiers, we are bound to notice how much of Alaska's employment has been temporary—how much is still strictly seasonal—and how, throughout the early 1980s, it consistently showed one of the highest unemployment rates of any of the fifty states. But then again, paradoxically, for those who did have employment, all was well: average personal income per capita is the highest for any state.

It is this curious mixture of poverty and wealth, opportunity and failure, incentive and disincentive, that makes the future of Alaska so hard to predict.

Ample notice has been served on all concerned that the environment is endangered, whatever use is made of it. The main area of dispute is between those who see Alaska primarily in terms of resource base and those who see it as a refuge. We need not be surprised if there is a divergence of viewpoint between these two groups. Among the latter group, however, there is still a second question to answer: A refuge for whom or what? For the 60,000 descendants of the original inhabitants, to use as they see fit? For world-weary Americans from other states, who may wish to keep in perpetuity one area, at least, where they can get back to the wild? Or for 600,000 caribou and sundry moose and reindeer, who should be left undisturbed and who, although they may not be able to speak for themselves, have plenty of enthusiasts to speak for them? It is not a question the author of this book will attempt to answer.

23
California

The Yosemite Valley on the western flank of the Sierra Nevada in central California. So spectacular is this valley, with its sheer rock faces and waterfalls, that as long ago as the 1850s it became one of the earliest preservation projects of ecologically minded Americans; in spite of this, the press of visitors to the valley has robbed it of some of its calm grandeur. *Ansel Adams*

The regions of North America we have considered in Chapters 15 through 22 are all, in some sense, tributary to the economic core; those of the South by virtue of the long-continuing legacy of the Civil War, and most of the remainder because of remoteness, sparsity of population, or late development. But when we come to the regions of the Pacific Coast, we cannot adopt the same simple division between core and periphery. Certainly, the original settlement here was from core areas in the east or the Mexican south. Certainly, too, there was a period of acute dependency on the East, when virtually every article needed on the West Coast had to be carried across the continent, or transported by sea and mule train via the Isthmus of Panama or around Cape Horn. We have only to read, for example, the account of the California coast in R. H. Dana's *Two Years Before the Mast* (1840) to realize how empty, barren, and unproductive the Pacific Coast appeared to early arrivals.

Yet the fact is that from these unpromising beginnings has developed a regional economy to challenge the strength of the East; more, itself not only to become a focus of those functions that characterize the core, but also to reverse the historic flow across America from east to west, and to transmit eastward all manner of cultural, religious, and political influences that have shaped the life of the nation. In the 1990s, California is not Periphery, but is itself Core. In this chapter, we must try to trace this transformation.[1]

California has been a name to conjure with for more than a century. Since the gold discoveries of 1848 brought in the Forty-niners, the state has seldom been out of the news for long. Yet it is probable that nothing would surprise the old-timers more, with their memories of dry grasslands and scanty food supplies, than the information that the California they knew had become the home of nearly 30 million Americans, the site of great cities, and one of the world's richest agricultural areas, and that the value of the gold they so laboriously won had long since been eclipsed by that of the crops and the manufactures produced within the state.

Their surprise would probably increase on learning that the intensive development of today has spread even to those areas which to them were most forbidding. The deserts of the southern Central Valley have become an area of cotton and vegetable farms, and the dry scrubland south of the Tehachapi Mountains is now the site of a conurbation whose population is more than 12 million. Even in the Mohave Desert there are now cities, holiday homes, and research centers. Only Death Valley remains largely untamed.

For modern California is, to a considerable extent, the creation of the Americans who have chosen to make their homes there. In particular, it is the water engineer who has made the present measure of settlement possible, by transferring water from rugged mountains to plains and from rivers to fields.

The physical obstacles that confronted the pioneers were certainly formidable. To reach their new

[1]There has been a steady flow of texts on the geography of California since World War II. Among the more recent may be mentioned L. Jelinek, *Harvest Empire* and *California Land* (Sacramento, 1982 and 1984, respectively); C. S. Miller and R. S. Hyslop, *California: The Geography of Diversity* (Palo Alto, 1983); and C. Jurmain and J. Rawls, eds., *California: A Place, A People, A Dream* (Bedford, N.Y., 1986). There are also two atlases: M. W. Donley et al., *Atlas of California* (Culver City, 1979); and D. Hornbeck, *California Patterns: A Geographical and Historical Atlas* (Palo Alto, 1983). See also D. W. Lantis, R. Steiner and A. E. Karinen, *California: The Pacific Connection* (Chico, Calif., 1989).

homes by an overland route, future Californians had first to cross the 7000-ft (2130-m) passes of the Sierra Nevada and find their way down through snow-drifts, forests, and canyons to the Central Valley. Once they were west of the Sierras, they entered a region where the characteristic landscape consisted of grass-covered hills, lying dry and brown under the summer sun like old, wrinkled canvas, with shrub-sized trees on their higher slopes and desert scrub on the valley floors.[2] The Spanish, who were the first settlers of European background in California, were ranchers, and when the territory was ceded to the United States in 1848 the only agriculture was in the immediate neighborhood of the Spanish (later Mexican) religious missions, which had been founded in the eighteenth century.

The climate of California, as we have seen in Chapter 1, is generally classified as "Mediterranean," but this is a serious oversimplification that conceals important differences between, for example, the northern and southern ends of the Central Valley or between the coastal hills and similar elevations at the foot of the Sierras. Along the coast, mean temperatures increase and rainfall decreases with fair regularity from north to south. However, over the coastal waters of the Pacific, which are cooled by the south-flowing California Current, fog is frequent in the immediate vicinity of the coast, especially in summer. Thus, San Franciscans spend many sunless summer hours, and the July mean temperature is held down to 59°F (15°C), giving the city an annual temperature range of less than 10°F (5°C).

[2]M. G. Barbour and J. Major, eds., *Terrestrial Vegetation of California* (New York, 1977).

In the Central Valley, conditions of both temperature and rainfall show much wider extremes; indeed, the climate is more accurately classified as continental. We may illustrate this by comparing the climatic data for three Valley stations with those for stations in the same latitude on the coast. In each of the three pairs listed in Table 23-1, the Valley station is given first, with the coastal station corresponding to it following.

Rainfall diminishes from north to south through the Central Valley and, at the southern end, true desert is encountered. At the same time, the eastern side, nearer the Sierras, is generally wetter than the western. Both on the coast and in the Valley the summer months are almost rainless: Bakersfield receives only 0.6 in (15 mm) between May and September. Summer temperatures are high, and winter frosts limit the growing season in the Valley. In all this, the effect of the coastal hills in shutting off marine influences is marked, for the figures in Table 23-1 reveal that at Stockton, inland from the only break in the hills (at San Francisco Bay), the July temperature is moderated and the growing season is lengthened by the entry of maritime air.

The coastal hills and the Sierra Nevada turn inward at their southern end to enclose the Central Valley, and south of their junction lie the plains on which Los Angeles has grown up. Such a southerly position within the state means higher temperatures and lower rainfall than on the coast further north, but the low elevation of the plains permits air from the Pacific to flow inland without hindrance. This carries the moderating influence of the ocean into southern California, but it also brings Los Angeles its now-famous smog. Maritime air spreads inland until

Table 23-1. Climatic Data for Selected Central Valley Stations (a) and Coastal Stations (b) in California

Station	Growing Season (days)	Av. Annual Precipitation		July Mean Temperature	
		In	mm	°F	°C
Northern End					
Redding (a)	278	37	925	82	27.8
Eureka (b)	328	37	925	56	13.3
Center					
Stockton (a)	287	14	350	74	23.3
San Francisco (b)	356	20	500	59	15.0
Southern End					
Bakersfield (a)	277	6	150	83	28.3
San Luis Obispo (b)	320	22	550	64	17.8

The California missions: Santa Barbara, one of the twenty-one missions in the chain built by Spanish missionaries in the late eighteenth century. This mission church, which was rebuilt in 1820 after earthquake damage, lies within the present city of Santa Barbara. *Santa Barbara Conference and Visitors Bureau*

it is blocked by the mountains further back, and being very stable, it lies over the metropolitan area, where the exhaust fumes of millions of vehicles create below it a dense and irritating fog with a dangerously high ozone content. It is a singular climatic irony that denies to the inhabitants of the two great California cities a sight of the sun that drenches the rest of their state.

The great barrier of the Sierra Nevada lies across the path of moisture-bearing winds off the Pacific, its crestline rising from 4000–5000 ft (1230–1520 m) above sea level opposite the northern end of the Central Valley to 9000 ft (2736 m) in the south, with individual peaks thrusting above this level to 13,000–14,000 ft (3950–4260 m) and Mt. Whitney, highest of all North America's mountains south of Alaska. Consequently, precipitation is very heavy, and since it falls mainly in winter, produces an annual snowfall of up to 400 in (1000 cm). This snow cover forms an invaluable reservoir for central and southern California during the rainless lowland summer. East of the mountains, there is an abrupt decline in precipitation, reflected in a swift transition from alpine pasture through forest to scrub and desert. From the mountains only a few ribbons of vegetation run out into the basins of eastern California and Nevada, where east-flowing streams penetrate only a short distance before losing themselves in desert sinks.

The Settlement of California

Such was the region penetrated by the Spaniards, the first white men to reach it. They thrust northward from Mexico, first with isolated expeditions and then, in the eighteenth century, with the same kind of planned advance that they had used in Texas and, far to the east, in Florida and Georgia. They established a chain of missions, from San Diego in the south to San Francisco in the north, following the line of the coastal hills. These were backed by civil settlements and by military posts, for the Russians were beginning to move southward from Alaska, and it was important for Spain to forestall them. Around the missions, the priests then strove to settle the Californian Indian population, to preach the Christian gospel and to demonstrate the cultivation of crops.

It is estimated that, at the time of the missionaries' arrival, there were about 250,000 Indians in California, most of them in broad, well-watered valleys like those of the Sacramento and Klamath. By the time the area passed into U.S. hands, in 1846, the estimate is that only 100,000 remained, but even that is far in excess of the 1900 count, which was just 15,500.[3] Despite the best efforts of the missionaries, the Indian population had died off or moved away and, in any case, those who survived had proved reluctant to become sedentary converts.

Whatever the missionaries may have failed to transmit in terms of religion, however, they certainly made up for in terms of agriculture. In the gardens and fields of the missions they introduced the crops—largely Mediterranean in origin—which were one day going to make California farmers rich. While secular Spanish and Mexican settlers used the grassy hills for ranching, and sold hides and tallow to ships calling at the coast, like the one on which Dana served, the missions laid the foundations of California agriculture.

Into this Indian–Mexican region there erupted, in 1846, first the forces of the U.S. government to claim California for itself and then, within months, a wave of miners looking for gold. They poured in, either overland or by sea to San Francisco, which rapidly became the commercial focus of the region: the old administrative and supply connections from San Diego south to Mexico City had lost their importance with the political takeover. The goldfields were in the foothills of the Sierra Nevada, east of what is now Sacramento, and it was for central California that the first transcontinental railroad aimed, fighting its way through the passes to link the Pacific Coast with the East in 1869.

More railroads followed, and out of the competition between them was born the southern California boom. Heads of families were encouraged to prospect for a California home by the offer of a return ticket from midwestern cities to Los Angeles and

[3]By the 1990 census, the state's Indian population had risen once more to 242,000.

back for prices that, at the height of the rate war of March 1887 between the Santa Fe and the Southern Pacific railroads, amounted to a free ride. The railroads brought a host of settlers to the dry southern plains, where commercial orange growing was established before the end of the century. By World War I, therefore, there were two population nuclei in California, with the newer southern one rapidly overtaking the older-established region around San Francisco Bay (Table 23-2).

Since World War I, California has experienced a continuously high rate of immigration from states further east. Over the years, however, the character of the immigrants has varied. In the 1920s, when roughly 1.25 million people moved into the state, the majority were Midwesterners of adequate means looking for new opportunities in pleasant surroundings. These newcomers tended to settle in the cities. Then came the depression of 1929–33. The 1930s again brought over a million immigrants to California, but they were of very different character from those in the previous decade. Many of them were destitute farmers from the depression-ridden, drought-hit Agricultural Interior—the "Okies" (from Oklahoma, one of the worst Dust Bowl areas) immortalized by John Steinbeck in *The Grapes of Wrath*. Unlike their predecessors, they sought work on the land. And fortunately for them—at least in the short term—California was looking for farm labor when the tide began to flow most strongly, in 1935. California agriculture had been dependent from the first on foreign labor—Chinese, Japanese, Filipino, or Mexican. But during the early 1930s, the state had been encouraging the repatriation of Mexican workers. For anyone willing to accept wages at the lowest level—and the Dust Bowl migrants certainly fell within that category—there was field work to be had. Furthermore, these Southwesterners knew about cotton, a crop California's farmers were planting in ever-larger quantities in the late thirties. So the

Table 23-2. California: Population of the State and of Principal Cities, Census Years, 1860–1990, (in thousands)

	1860	1870	1880	1890	1900	1910	1920	1930	1940	1950	1960	1970	1980	1990
State	380	560	865	1213	1485	2378	3427	5677	6907	10,586	15,717	19,953	23,668	29,760
City of San Francisco	—	—	—	299	343	417	507	634	635	775	740	716	679	724
City of Los Angeles	—	—	—	50	102	319	577	1238	1504	1970	2479	2816	2967	3485
City of San Diego	—	—	—	16	18	40	74	148	203	334	573	697	876	1111

Note: For the populations of the metropolitan areas surrounding these and other California cities, see Tables 23-6 and 23-7.

state absorbed this migration, too. There were probably 300,000 of them from the Southwest as a whole; 100,000 from the state of Oklahoma.[4]

The 1940s brought new circumstances—war and the growth of industry. To man the war industries, some 1.5 million people entered the state; they in turn created a market that attracted other industries. Between 1940 and 1950, the population of California increased from 6.9 to 10.6 million.

The 1950s were like the 1940s, but on a larger scale. Three million people arrived in California from other states. Most of these postwar immigrants fell into two categories. They were either elderly people who were retiring to a warm climate or they were young, active people, impressed by the possibilities and the wealth of California. Almost all of them made for the cities, for by this time the state that had attracted first miners and then farmers was well launched on its industrial and commercial career.

The 1960s saw a continuation of the process of population increase. In the last years of the decade, California overtook New York as the most populous state in the Union, an occasion for rejoicing missed by nobody in the West. By this time the census recognized fourteen metropolitan areas in the state, two of which, in southern California, grew more rapidly over the period of 1960–70 than any other U.S. city except Las Vegas.

By the 1970s, with well over 20 million people in the state, the percentage rates of population increase began to fall off, and to appear less remarkable than those of Florida. But the absolute increase in numbers continued—for the decade of the seventies it was 3.7 million. New metropolitan areas were recognized (there are now nineteen in all), and the proportion of the inhabitants classified as urban, already highest in the nation, rose still further.

In the 1980s, California's population continued to increase. In recent years, however, two new sources of growth have been added to the influx. One is a considerable legal immigration from Mexico and Central America, and an even larger illegal immigra-

tion, as we saw in Chapter 20. Since California was Hispanic, culturally and politically, before 1848, there is a sense in which the Hispanic peoples are simply recolonizing their old homeland. One-third of the population of the Los Angeles CMSA is of Spanish origin (Table 20-2), and the proportion of Hispanic school children is much higher: as far north in the state as Salinas, in fact, that proportion is well over 50 percent.[5]

The other new source of growth, and one to which we have already referred in Chapters 2 and 20, is the Asian population, which includes thousands of refugees from the politically troubled lands of Southeast Asia. In large numbers, they have chosen southern California as their new home. Among the groups whose numbers have increased most rapidly in the past fifteen years are Koreans, Filipinos, Vietnamese, and Samoans.[6] Between them, they have changed the face of Los Angeles, "the new Ellis Island."

California's Agriculture: Organization

As the population of California has increased, so its agriculture has developed and changed. In 1848, California was cattle country; ranches occupied most of the lowlands. The gold rush created a local demand for food supplies, and, although cattle were raised in increasing numbers, there was a growing diversification to supply this market. Following the completion of the transcontinental railroad in 1869, California, like eastern states before it, enjoyed a wheat boom—by 1890 it had achieved the position of second wheat state in the Union.

It was in the last two decades of the nineteenth century that there began to develop the fruit growing that brought California more permanent wealth and fame. The construction of railroads in southern California and the development of refrigerated railcars made possible the shipment of fresh fruit to eastern markets, and the new enterprise, sedulously publicized by the railroads, provoked a tremendous land boom, as we have seen, in the arid surroundings of

[4]W. J. Stein, *California and the Dustbowl Migration* (Westport, Conn., 1973). It is worth adding that there was a similar, though numerically smaller, migration from the northern Great Plains into the Pacific Northwest; smaller because the effects of drought were less severe further north. But a whole generation of farmers in Washington and Oregon started life in the Dakotas.

For a more up-to-date reference than Stein, see J. N. Gregory, *American Exodus: The Dust Bowl Migration and Okie Culture in California* (New York, 1989).

[5]In the state as a whole, 26 percent of the 1990 population was of Hispanic origin: this was an increase since 1980 of 69 percent.

For some impression of Hispanic life in southern California, the reader might refer to Ricardo Romo, *East Los Angeles: History of a Barrio* (Austin, 1983).

[6]Samoa is a group of islands in the southern Pacific. The two largest, western islands of the group form the independent monarchy of Western Samoa. The easterly islands of the group form American Samoa, a territory of the United States since 1900.

Los Angeles, while other areas of the state developed their own fruit and vegetable specialties. California today sells a greater value of farm products than any other state (Iowa and Texas are its nearest rivals). It has some 33 million acres (13.4 million ha) of farmland and ranchland, the largest irrigated area of any state (8.6 million acres [3.5 million ha]), and is the major U.S. producer of about thirty crops and the only state producing several of these.

It is as a producer of fruit that California is most renowned, but its reputation as a fruit grower may make it necessary to counter at once an impression that fruit trees grow everywhere in the state, or that there is some simple reason why they do so. In reality, about one-quarter of the farm income is derived from fruits and nuts, and about 12–15 percent each from cattle, vegetables, and dairy products. Grapes and cotton are the leading single crops in terms of income. In area, orchards and vineyards occupy nearly 2 million acres (0.8 million ha), cotton and hay each about 1.5 million acres (0.6 ha), vegetables about 900,000 acres (364,000 ha), followed in order by barley, wheat, and rice.

But it is not merely the volume of its output that makes California's agriculture outstanding, nor the degree of its dependence on irrigation water, which is high—as it must be in any area with a winter rainfall maximum and an almost rainless growing season. It is also the organization of the farm business within the state. For in this respect California offers what is, perhaps, a model of the agriculture of the future.

It is a model that has developed out of "traditional" western farming over the past century and a half. Livestock ranching was followed by family farming, just as it was elsewhere in the West. Even today, the average size of farm unit in California is below 400 acres (162 ha), (see, for example, Table 23-3) and it has, moreover, fallen considerably since the start of the 1970s. The average area in crops on the state's 80,000-plus farms is only some 60–65 acres (25–27 ha). But the heavy capital investment represented by irrigation, the need to sell many of the state's specialty crops in nationwide markets (and the need to sell them fresh), together with the distances separating California from those markets all called for a special organizational effort: in particular, for combination among growers and shippers. Marketing agencies have been set up to provide quality controls, selling arrangements, and provide bargaining strength against the carriers of farm produce, from

Table 23-3. Stanislaus County, Central California: Agricultural Statistics, 1987

Population (1980)	266,000
Percentage change, 1970–80	+36.7
Area of county (ac/ha)	966,000/391,000
Percentage in farms	74.5
Percentage in cropland harvested	29.9
Percentage of this irrigated	108.0
Value of agricultural produce sold	$786 million
Percentage from livestock products	61.5
Number of farms	4630
Average size (ac/ha)	155/63
Average value of land, buildings, & equipment, per farm	$495,000
Average value of agricultural products sold, per farm	$170,000
Percentage of operators whose principal occupation is farming	50.9

whom they have obtained some favorable freight rates on commodities moving east (see Chapter 6). But inevitably this kind of production is dominated by a few large firms, usually canners or processors, which means that the part of Californian agriculture that deals with specialty crops is in many ways organized more like industry than agriculture. Where the small grower survives, he does so, if not under contract to a canning firm that tells him what to do, at least within a market overshadowed by the big producers.

There is another factor. Growing special fruit and vegetable crops has traditionally been a highly labor-intensive operation. For some crops, it still is, and we shall return to that point a little later. But since World War II, in one crop after another, planting and harvesting have been mechanized. It happened, as we have earlier seen, with cotton, a California field crop. It has since happened with such crops as grapes and tomatoes, although in most cases it has been necessary to modify the plant genetically to make it fit the machine. But such machines are highly specialized in their operation, and it would be out of the question for a small farmer to own a whole row of them. He is likely, therefore, to specialize in a single crop and leave the large machine inventories to farmers such as those in the Imperial Valley (Table 23-4), with their million-dollar investments.

Among the latter, there are still some family farms—farms that their operators have successfully capitalized and enlarged to the biggest size category. Many others in this largest group—12 percent in Im-

Table 23-4. Imperial County, Southern California: Agricultural Statistics, 1987

Population (1980)	92,000
Percentage change, 1970–80	+23.7
Area of county (ac/ha)	2,712,000/1,098,000
Percentage in farms	20.0
Percentage in cropland harvested	15.1
Percentage of this irrigated	101.0
Value of agricultural produce sold	$717 million
Percentage from livestock products	47.9
Number of farms	804
Average size (ac/ha)	675/273
Average value of land, buildings, & equipment, per farm	$1,464,000
Average value of agricultural products sold, per farm	$891,000
Percentage of operators whose principal occupation is farming	69.6

perial County—are, however, owned by corporations, either family or commercial. Some are huge, vertically integrated businesses owned by canning companies, or by multinational corporations with a side interest in wines or fruit. These farm units are supervised by managers, who may operate from airplanes, and communicate by shortwave radio with their foremen on the ground. They keep in touch with market brokers, equipment manufacturers, and agricultural experiment stations, much as any successful businessman would.

Whether or not one finds this vision of the future attractive tends to depend on the part of the country where one is. There are states in the Interior—North Dakota and Kansas—whose laws actually prohibit corporations from entering farming. The mystique of the family farm still exerts its power in an era when family farms are going out of business every day. California is, however, one of the few, the very few, states in the Union where a politician can speak against the family farm, and hope to survive.

We must now return to a topic that was mentioned earlier: the supply of labor for this Californian type of agriculture. With its great north–south extent, the state can produce a wide range of crops: temperate-zone crops in the north, provided that the water supply is adequate; tropical-zone in the south, provided that frost danger can be avoided. Most of the state's specialty crops are fruits and vegetables, and they were originally cultivated and harvested by hand. To

do this work, a large migrant labor force was therefore needed, season by season. The harvesters would start in the south and move north for the later ripening of the temperate crops, then return to the far south for winter-harvesting work.

In recent decades, this labor force has contained many different elements. In the 1930s, as we have already seen, the Okies and other migrants from the drought-hit Great Plains were only too ready to perform this thankless task. Filipinos and other Asian workers have, all along, provided some of the labor needed. In the 1950s, the U.S. government operated a policy of employing Mexican labor, the *braceros*, temporary workers who were housed in labor camps up and down the state, but in 1962 the program was suspended and the workers were repatriated.

The labor situation in the 1980s showed some marked changes:

1. *Many of the hand operations on fruit and vegetable crops have been successfully mechanized*, reducing labor requirements. This development was predictable, but it was also encouraged by organization of the workers, discussed next.

2. *The labor force became unionized*. With their low pay and temporary employment, the itinerant workers had minimal bargaining power. But with two decades of militancy behind them; with union negotiators and campaigns to stop the public buying particular Californian products, the workers and their unions have forced up wages and improved conditions.

3. *Much of the off-field work that used to be carried out in packing centers throughout California is now done in the fields,* eliminating the need for the packing sheds and their workers. In the lettuce fields, for example, the field crew cut the lettuce, put it in boxes sealed on the spot, and place the boxes aboard trucks which take them straight to cooling sheds. From there, the lettuce is shipped directly to market. This means, of course, that the field hand must be able to decide instantly whether the product is up to standard; it also requires a high degree of standardization of product across the field or farm.

4. *With all these changes, today's worker bears little resemblance to his or her predecessor.* The single migrant worker, living in labor camps, a few weeks here and a few there, and working for a pittance, has given way to a more settled, and certainly wealthier, work force. Most of the work camps have been closed. To take again the example of the lettuce workers in the "salad bowl" around Salinas, they

California: The great Central Valley at its southern end. The flat, irrigated floor of the valley contrasts sharply with the bare foothills of the Coastal Ranges, seen in the foreground. At the upper left-hand edge of the photograph lies the town of Bakersfield. *John S. Shelton*

have health insurance, family homes, in some cases pension plans, and a stable work pattern of eight months a year on the lettuce followed, if they wish, by four months winter work in the Imperial Valley. For this they may hope to make $25,000 a year. Most of the workers are Hispanics. The Okies would not have believed it.

California's Agriculture: Crops and Regions

Most of California's crops—including hay and pasture—are irrigated. Since there are three main irrigated areas in the state, it will be simplest to deal with the regional details for each of these three areas in turn. They are the Central Valley, the Imperial Valley, and the Los Angeles and Ventura Lowlands.

The largest irrigated area in California (or on the continent, for that matter) is the Central Valley. It may be divided into four subregions or, roughly, quarters, from north to south: (1) the Sacramento River Valley; (2) the combined Delta of the Sacra-

mento and San Joaquin rivers, a low, extremely flat area consisting of islands of silty soil protected from a maze of stream channels by man-made levees; (3) the San Joaquin Valley; and (4) a basin of interior drainage lying south of Fresno and the San Joaquin River and centered on Tulare Basin. Natural drainage northward is blocked by the great alluvial fan of the Kings River.

Agriculture in the Central Valley began with cattle ranching on Mexican land grants, with cattle driven up from Mexico and, following completion of the first transcontinental railroad in 1869, was dominated for two decades by dry farming of wheat. In the coastlands, the Spanish missions introduced irrigated agriculture, but major development did not come until after 1890. As in other parts of the West, early irrigation projects were small in scale; often they were begun by individual farmers tapping small streams and then enlarged, or amalgamated, as communities or private water companies developed larger and larger projects.

In the 1930s, the state turned to the federal Bureau of Reclamation to develop the Central Valley Project. It is bold and simple in basic design, but complex in detail. Two-thirds of the irrigable land in the 400-mi (640-km) Central Valley is south of the Delta. But two-thirds of the water for irrigation comes into the Sacramento Valley, brought mainly from Sierra Nevada snowfields by Sacramento tributaries, especially the Feather and the American; often much water is lost when the river floods in spring, running off to the Pacific through the Delta and San Pablo Bay.

The project design is based on two great water-conserving dams, two major canals, and a giant pumping plant. At the northern tip of the Central Valley, Shasta Dam and Shasta Lake store Sacramento River water, some of which is carried from the lower river across the Delta by the Cross Delta Channel (Fig. 23-1). Near Tracy, the water is lifted 196 ft (59 m) to flow southward for 120 mi (192 km) along the western side of the Valley in the Delta–Mendota Canal to Mendota, where it enters the San Joaquin River. The water of the latter is stored behind Friant Dam, to be carried southward in the Friant–Kern Canal along the eastern foothills where it is used in the dry southern basin of interior drainage. Water is stored behind smaller dams on the many streams that flow from the Sierras. Their alluvial fans coalesce to form a piedmont alluvial plain that slopes almost imperceptibly toward the flat floor of the Central Valley. Much of the irrigation water is pumped from the alluvium, which acts as a giant reservoir.

The Central Valley Project irrigates some 5 million acres (2.02 million ha). Yet by the 1950s, there were strong demands for more water, by both farmlands and cities. This time, with its greatly increased population and wealth, the state undertook in 1960 its own project, the California State Water Plan, to coordinate with, but also to augment, the Central Valley Project and other earlier developments. It included new storage facilities, most notable of which is the Oroville Dam on the Feather River, and the California Aqueduct, above and parallel to the Delta–Mendota Canal, which carries additional water southward along the west side of the Central Valley from Tracy.[7]

The Aqueduct supplies water to parts of the Central Valley that previously were without irrigation.[8] But this newer canal continues on to the Tehachapi Mountains, which enclose the southern end of the valley. There, pumps lift water nearly 2000 ft (660 m) across the mountain barrier to Lake Perris near Riverside. Subsidiary aqueducts carry water to farmlands and cities south of San Francisco Bay, to coastal areas near San Luis Obispo and Ventura, and to southern California.

The main features of the State Water Plan have been completed, but construction and expansion continue. One of the most contested features, still unbuilt, is the Peripheral Canal. It is designed to carry Sacramento Valley water around the eastern and southern margins of the Delta to the California Aqueduct near Tracy. This new canal is supported vigorously by southern California, which wants more water for its cities and farmlands, the former because they will lose to Arizona part of the water taken from the Colorado River when the Central Arizona Project is completed. Northern California, on the other hand, opposes the canal; among the more vocal opponents are large-scale farmers in the Delta and environmentalists who fear more interference with the ecologically fragile lands of the Delta—even though the canal is designed to protect the area from current overpumping for irrigation and consequent underground intrusion of seawater. Other opponents include residents and businesses in north-central California, whose water supply it is.

In the Central Valley, crop location depends, in part, on latitude. More important, however, is the

---

[7] The west side of the San Joaquin Valley is a loosely defined area with large corporate farms, mechanized cotton production, and oil fields that extend some 175 mi south from San Luis Dam near Los Banos to the Tehachapi Mountains south of Bakersfield. It has an average width of 25 mi along the western side of the valley floor.

Sparsely populated and arid, the region is mainly irrigated by deep wells, which are steadily depleting the groundwater resources. See G. W. Dean and G. A. King, *Projection of California Agriculture to 1980 and 2000* (Davis, Calif., 1970). See, in addition, J. J. Parsons, "A geographer looks at the San Joaquin Valley," *Geographical Review,* vol. 76 (1986), 371–89.

[8] In areas such as the western side of the Central Valley, part of the irrigation water is from the federal Central Valley Project and part is from the California State Water Plan. Even if the two sources are not distinguishable in the canals, they certainly are so in law, because federal irrigation water is liable to the acreage limitation clauses of the 1902 reclamation act as amended, but state water is not. The legal problem of what happens when federal and state water mix is very much like that of Shylock attempting to take his pound of flesh without drawing any blood. The problem here is particularly acute, since most of the irrigated land (and the farm produce from it) is held in very large units, and if the acreage limitation rule was ever applied to it even in its recently amended and more generous form, nothing short of an agricultural revolution would be necessary.

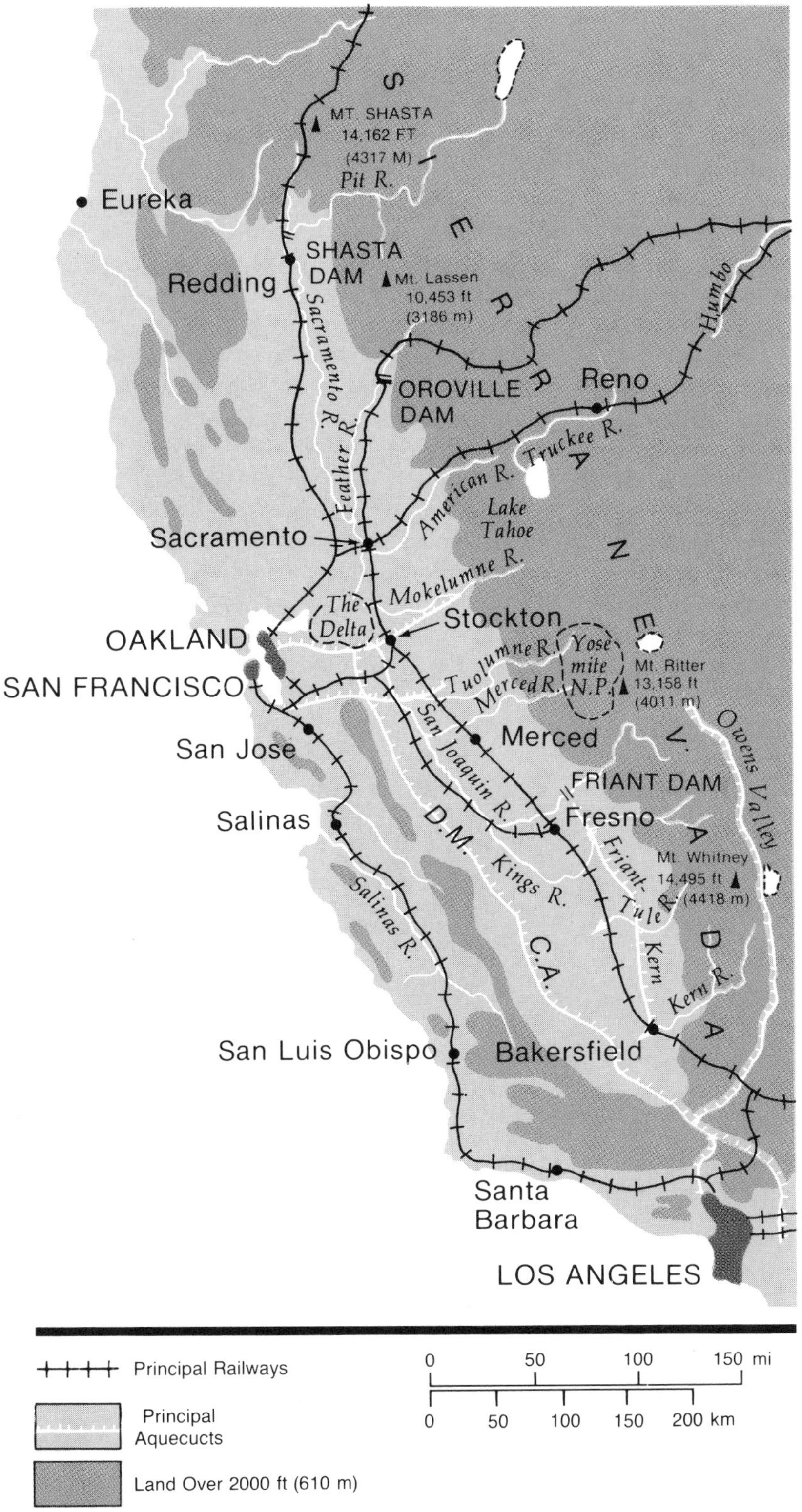

Fig. 23-1. Central California: Location map. C.A. = California Aqueduct, D.M. = Delta–Mendota Canal

Map labels:

S
MT. SHASTA
14,162 FT
(4317 M)
Pit R.

Eureka

Redding

SHASTA DAM

▲Mt. Lassen
10,453 ft
(3186 m)

Sacramento R.

Feather R.

OROVILLE DAM

Reno

R

Humbo

American R.

Truckee R.

Lake Tahoe

Sacramento

N

Mokelumne R.

The Delta

Stockton

Tuolumne R.

Yosemite N.P.

▲Mt. Ritter
13,158 ft
(4011 m)

OAKLAND

SAN FRANCISCO

Merced R.

E

San Jose

San Joaquin R.

Merced

V.

Owens Valley

Salinas

D.M.

FRIANT DAM

Fresno

Friant-

Mt. Whitney
14,495 ft ▲
(4418 m)

Salinas R.

Kings R.

C.A.

Tule R.

Kern

A

D

Kern R.

San Luis Obispo

Bakersfield

A

Santa Barbara

LOS ANGELES

Legend:

╫╫╫ Principal Railways

▦ Principal Aquecucts

▮ Land Over 2000 ft (610 m)

0   50   100   150 mi

0   50   100   150   200 km

485

vertical arrangement relative to terrain, soils, water availability, soil drainage, and growing season. In the San Joaquin Valley and southward, where the broader piedmont alluvial plain is on the eastern side of the valley (most of the water and alluvium have come from the Sierras), there is a progression from east to west. The rugged foothills of the Sierras are used mainly for livestock grazing. Smoother bench-lands, too high to be irrigated, support dry farming of wheat and barley. The better-drained, coarser soils of the upper and middle parts of the alluvial plain are irrigated and support orchards and vineyards. Deep-rooted trees and vines require good drainage; they usually escape frost damage because during cold snaps, the colder, heavier air flows downslope to the floor of the valley. Hardier field crops, such as cotton, rice, alfalfa, sugar beets, and vegetables, occupy much of the lower piedmont and the flat valley floor.

Prunes and pears are most important in the northern part of the piedmont alluvial plain, oranges near the southeastern margin. On the middle and upper levels of the San Joaquin plain are a continuous succession of orchards—including peaches, apricots, almonds, and sweet cherries. Grapes occupy an even greater acreage, from south of Fresno to Sacramento.

Cotton is grown mainly in the southern one-third of the San Joaquin Valley. The low wetlands of the lower Sacramento Valley are used for rice, those of the Delta for vegetables. Alfalfa and sugar beets are grown throughout the valley.

As the population of California and the entire Pacific Coast has grown, more and more of the huge agricultural production is marketed within the region, and thus dairy, beef, and poultry products have increased relative to crops destined for eastern markets. The San Joaquin Valley has become a major dairy region, supplying milk not only to the Bay Area cities, but also to the cities of southern California.

With the addition of more irrigable land in the San Joaquin Valley over the past 20–30 years, the Valley has developed another function—as a home for irrigation agriculture displaced from other areas by urban growth. As we shall see later in this chapter, farmland (and especially orchard land) has been disappearing in California at a rate of about 50,000 acres (20,000 ha) a year, and much of this has been irrigated land. Two areas particularly overrun have been the orange orchards of the Los Angeles Basin, and the Santa Clara Valley south of San Francisco, which is now much better known as Silicon Valley. Santa Clara grew apricots and other soft fruits. Both

these types of production have found a refuge in the southern San Joaquin Valley, as has the planting of nut crops.

Next in size in terms of irrigated area is the Imperial Valley, and its extension north of Salton Sea, the Coachella Valley. The first of these was cultivated as a private project as early as 1901, but it was swept out of existence again in 1905, when the Colorado River in flood broke its west bank, flowed into the desert, and created the Salton Sea. A new project replaced it, and this, together with the Coachella Project to the northwest, was given new security and added water resources by the opening in 1935 of the government's Hoover Dam, which controls the Colorado, and by the All-American Canal (see Fig. 20-2). Today these lands produce cotton, winter vegetables, alfalfa, and some citrus fruits and dates. The alfalfa is shipped to feed dairy cattle in the Los Angeles Lowland and used, along with such grain crops as sorghum, to fatten beef cattle brought to huge feedlots within the Imperial Valley.

This brings us to the third of California's major areas of irrigation agriculture. The agricultural lands of the Los Angeles Lowland, and to a lesser extent those of the smaller Ventura Lowland to the northwest, have been greatly reduced by urban growth, but are still important. They exhibit, where not covered over by cities, the same vertical progression in land use as that in the Central Valley. Once the major orange-producing area of the continent, the region now ranks second within California (the southeastern Central Valley is first). It is still, especially in the Ventura Lowland, the major producer of lemons. Walnuts and avocados, vegetables, greenhouse and nursery products are important, and there are many vineyards in the area. Large numbers of dairy cattle were kept within the urban areas until 1950—the often huge, feedlot farms have now been displaced to the fringes. Most of the feed must be shipped in from the Imperial and Central valleys.

The Santa Clara Valley, as we have just seen, was once famous for orchards and vegetable farms. Today, it is almost completely urbanized. Other valleys within the Coast Ranges, however, are still important for agriculture. Further south, the Salinas Valley shares with the Imperial Valley the production of much of the enormous quantities of lettuce consumed by North Americans. Large vineyards for wine have recently been planted there. The small valleys north of San Francisco and San Pablo bays, especially the Napa and Sonoma valleys, are producers of quality wines. Their vineyards, the only ones in California not under irrigation, are being encroached upon by urban expansion and second homes—one more case of beautiful, mountain-bordered valleys that invite the destruction of their own attractiveness.

California's agricultural output is one of the marvels of the American economy. As we have seen, it is essentially irrigation-based. Without irrigation, California would be (as it largely is in the nonirrigated hill country, and was in Spanish and later Mexican times) a ranching state, with cattle and sheep moving north to Oregon or east to the Agricultural Interior for fattening, rather than to the adjacent valley lands (Fig. 23-2).

But with thousands of acres withdrawn each year from agricultural use for conversion to other, mainly urban, purposes, how long can these Californian miracles of productivity continue? The loss is not only one of prime farmland, but also of a roughly equal area of grazing land; that is, it affects both ends of the value-scale of farm land.

It was to grapple with this problem that, as we saw in Chapter 5, California introduced its Williamson

Act (officially the California Land Conservation Act) in 1965. Goodenough reports:

Under the Williamson Act landowners and city or county governments are joined in a voluntary contract under which both give up certain benefits for other advantages. Landowners forgo the possibility of development on their land during the time of the contract; in return their property tax assessment is related to the income-producing ability of the land, thus resulting in lower taxes. One survey reveals that the Williamson Act offers farmers an average 83% per acre reduction in property taxes compared with taxes based on the current fair market value. . . . The local government obviously forgoes a proportion of its property tax base and return for retaining the land in agriculture or other open space.[9]

By 1989, the agricultural area contracted under the terms of the act had stabilized at about 15 million acres (6.1 million ha). But there is also a steady trickle of farmers contracting *out* of the scheme—a difficult and, eventually, costly exercise—because of the de-

[9]R. Goodenough, "Room to Grow? Farmland Conservation in California," *Land Use Policy* [London] (Jan. 1992). Quotation from p. 27.

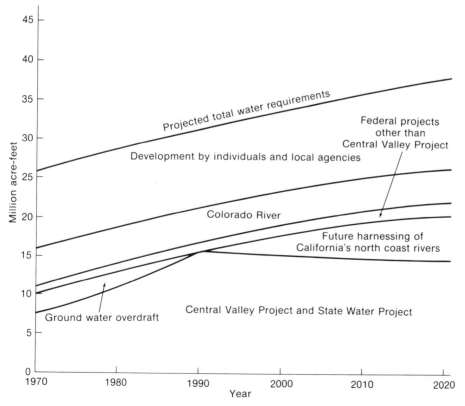

Fig. 23-2. California water supply. The problem of providing for the ever-increasing water needs of California, and especially southern California, has occupied California planners for several decades. Looking ahead to the year 2000 and beyond, the possible water sources for the needs of an increasing population are indicated by the diagram.

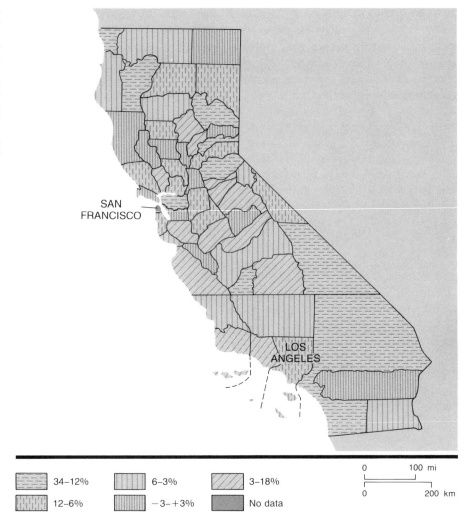

Fig. 23-3. California: Percentage change in the category Land in Farms, by counties, 1982–87. (Reproduced by permission from R. Goodenough, "The Nature and Implications of Recent Population Growth in California," *Geography*, vol. 77 [1992], 123–33.) Factors involved are population growth, urban and service area expansion, shortage of water, and abandonment of marginal holdings.

34–12%	6–3%	3–18%
12–6%	–3–+3%	No data

velopment prospects they can no longer resist (see Fig. 23-3.)

Meanwhile, Californian agriculture continues to be put under pressure by the problem that has plagued it since long before the cities began to encroach on the farms: Where is the water for all this agriculture to come from? And could that water be better used elsewhere? We shall return shortly to this theme.

If, however, we continue for a moment to consider farming as a user not of water but of space, it has three major competitors in California. The first of these is forestry. There are 16.3 million acres (6.6 million ha) of commercial timberland in the state, mainly on the western slopes of the Sierra Nevada, and on the coastal hills north of San Francisco. This timber was in immediate demand, once the gold rush

began in 1849, and the first and most prominent targets for the axman were the magnificent redwoods. It was, in fact, the depredations of the lumbermen in the redwoods that may be said to have brought the modern conservation movement into being in the 1850s, the forerunner of the present Sierra Club. Most of the remaining redwoods are now protected, but in the northern part of the state there is an extensive forest-products industry.

The second alternative to agriculture in the use of space is tourism. In 1986, it is estimated that California earned some $36 billion from travelers' expenditures, double the figure for 1978. It claims to attract more overseas visitors than any other state, but by far the largest part of its tourist receipts come from other U.S. citizens wanting to share the California experience.

Many of these tourists no doubt come to see the cities but, like the other Mountain and Western states, California's attraction is, in the first place, scenic. Its national and state parks cover almost 5 million acres (2 million ha); its national forests virtually all of the Sierra Nevada, and much of northern California apart from the Central Valley. Yosemite National Park, the main tourist site, attracts close to 3 million visitors a year, most of them to a very limited section of its 760,000 acres (308,000 ha). Largest of California's parks is Death Valley, in the desert southeast, which draws its visitors in winter, when snow blocks the Sierras. There is no real clash between tourism and agricultural land use at present, although, with the state's ever-increasing population, such a clash cannot be ruled out in the future.

The third big space-user in California—specifically, in southern California—is the U.S. Defense Department. At least seven large military reservations occupy the desert lands of the southeast, and the same holds true for neighboring southwestern Nevada. These are not lands that are ever likely to be needed for cultivation, although they have probably all been used in the past for livestock grazing. But they eliminate from consideration and planning a sizeable part of a state whose land-use problems are certainly not going to become any easier to solve as its population mounts toward 30 and 40 million in the twenty-first century.

California's Industry

In 1940, before the United States entered World War II, industry employed some 300,000 workers in California. The food-processing industries dominated the state's manufacturing; they gave employment to almost three times as many workers as any other industrial group. But war, and especially war in the Pacific, dramatically changed that situation. To fight the 1941–45 war against Japan, and later the Korean and Vietnam wars, the United States had to build arsenals and bases, shipyards and supply depots along its Pacific coast.

The military effort added to the volume of what was already a long-standing migration to California. Some Americans moved to work in the war industries; others to find employment in the service trades which sprang up as the population grew. Each process fed the other, with the U.S. government as initial paymaster. In World War II, California obtained war supply contracts worth over $17 billion, or nearly 10 percent of the U.S. total. Of this amount, roughly 60 percent was for aircraft built by southern California's six major manufacturers, and 25 percent was for ships, such as the Liberty ships mass-produced by Henry J. Kaiser. After a brief episode of peace, the Korean war broke out in 1950: the nation needed fresh military supplies and, by 1957, the aircraft and missile industries in the state employed over 400,000 workers.

The postwar years actually saw three different developments, which interlocked with each other to produce the industrial California of today with its 2 million-plus employees in manufacturing (Table 23-5).

1. The development of civil aviation after 1945 brought a huge new volume of business to the well-established aircraft manufacturers of southern California.
2. Defense requirements and space exploration ensured a continuation of government spending in the fields of aerospace and electronics.
3. The rapid rise of California's population created consumer markets within the state on an unprecedented scale.

Table 23-5. Western United States: Percentage Shares in Nonagricultural Employment, by States, 1985

	Total Work Force	Construction	Manufacturing	Transport, Public Utilities	Wholesale & Retail Trade	Finance, Insurance, Real Estate	Services	Government
California	10,965	4.4	19.1	5.2	23.8	6.7	24.1	16.4
Arizona	1,278	8.6	14.2	4.9	24.6	6.3	23.3	17.1
Oregon	1,029	3.1	19.4	5.6	25.2	6.5	20.9	19.1
Washington	1,708	4.7	17.2	5.5	24.7	5.8	21.8	20.1
Alaska	231	8.2	5.2	8.2	19.9	5.6	19.5	29.4
Hawaii	423	4.0	5.2	7.8	27.4	7.6	26.0	22.0
U.S., 50 states	97,614	4.8	19.8	5.4	23.7	6.1	22.5	16.8

The outcome of these developments was that, between 1950 and 1963, manufacturing generated 660,000 new jobs within the state, some 60 percent of which were in aerospace. This was, in a sense, no less than the increase in population demanded, if there was to be employment for the newcomers crowding into the state. But California at this stage had a lesson or two to learn from its heavy reliance on government orders and aerospace. In 1967, aerospace gave employment to nearly 600,000 workers—or one in three of all industrial employees at that date. In 1971, the number was down to 440,000. In 1977, it was up again to 560,000. The state was receiving 17–18 percent of U.S. defense supply contracts, but it was also experiencing life on the roller coaster.

Throughout these years, additionally, the nature of the defense demand was changing—away from ships and aircraft, and toward missiles and electronics. At the same time, the share of industrial output for defense purposes was declining, relative to civilian demand: in fact, the proportion of output for government purposes was roughly halved during the 1970s, as civilian demand rose. But by this time California had built up a work force of such technical sophistication that it was in a position to lead the way into all manner of civil applications of equipment originally created for defense purposes.

For California had not secured its defense contracts on any random basis, but because of the acknowledged strengths of its scientists and engineers, backed by a set of fine research institutions. It was this combination of funding and skills that enabled the state to become the pace setter in the new high-tech industries. Los Angeles claims today to have the largest concentration of high-tech firms and workers in North America. The San Francisco Bay Area ranks second, with most of the firms concentrated where once the orchards grew, in Silicon Valley, Santa Clara County, and San Jose. About one-third of the state's high-tech output comes from Santa Clara.

This is the glamorous end of California manufacturing today: it is also the roller coaster end. Defense budgets fluctuate from year to year and, in the 1990s, the roller coaster has run into one of its periodic dips. The ending of the cold war, the need to impose federal budget restraints, and the financial difficulties of the big civil airlines have all produced a downturn in aerospace. For the first time in many years, if ever, California has been losing jobs and workers. So we should not forget the more staid industries that sup-ply 30 million people or more here on the Pacific rim of America. There is a large clothing industry, for California is a leader in many types of leisure wear. The food-processing industries are still active, and giving new meanings year by year to the word "processing." There are the chemical industries supplied by the state's oil fields, and the plants assembling some of the motor vehicles upon which 30 million people depend: at a recent count, there were 18.3 million of them.

California's has been an industrialization for which the materials once considered basic—coal and iron ore—are almost entirely lacking. It is true that the state has a small steel production—to which the best-known contributor is the Fontana Works, built among the orange groves east of Los Angeles. But the state's coal supplies are meager, and steel output is based largely on scrap. California's more obvious natural resource is its oil; oil fields are scattered widely through the southern part of the state, and new strikes have recently taken place, most of them along the continental coastal shelf, under water. Hydroelectricity is also available in quantity, thanks to the construction of numerous flood control and irrigation schemes.

California's population at the 1980 census was 91.3 percent urban, which is the highest urban proportion for any state (compare New Jersey, 89.0; Rhode Island, 87.0). Even more striking, it was over 94 percent metropolitan, and even allowing for the fact that some of California's MSAs contain counties that include huge areas of desert and mountain, this figure does call attention to the drawing power, in the migration to California, of the two great conurbations: the Los Angeles lowland (including the large separate MSA of San Diego) and the San Francisco Bay Area. It is in these areas that newcomers have settled; it is here, too, that industry tends to concentrate, and commercially one area dominates the southern part of the state as the other dominates the central and northern parts.

Central California: Cities and Industries

The gold strikes in the foothills of the Sierras in 1848 focused attention sharply on central California. The San Francisco Bay Area offered the obvious entry point to the goldfields, for most of the early arrivals came by sea and, from the bay, reached the interior by river transport up to Sacramento or Stockton. When, in 1869, the first transcontinental railroad

Table 23-6. Central California: Population of Metropolitan Areas, 1970, 1980, and 1990 (in thousands) and Percentage of Spanish Origin, 1980

	1970	1980	1990	Percentage of Spanish Origin (1980)
San Francisco–Oakland–San Jose CMSA	4754	5368	6253	12.3
Oakland PMSA	1628	1762	2083	10.5
San Francisco PMSA	1482	1489	1604	11.1
San Jose PMSA	1065	1295	1498	17.5
Santa Cruz PMSA	124	188	230	14.7
Santa Rosa–Petaluma PMSA	205	300	388	6.9
Vallejo–Fairfield–Napa PMSA	251	334	451	10.0
Bakersfield MSA	330	403	543	21.6
Fresno MSA	413	515	667	29.3
Modesto MSA	195	266	371	15.0
Sacramento MSA	848	1100	1481	9.6
Salinas–Seaside–Monterey MSA	247	290	356	25.9
Stockton MSA	291	347	481	19.2
Visalia–Tulare–Porterville MSA	188	246	312	29.8

reached California, the importance of the sea route was reduced but the line still terminated on the bay shore, at Oakland and, by that time, there could be no challenge to the importance of the Bay Area cities. Today, as Table 23-6 shows, these cities together contain 6 million people.[10]

The city of San Francisco itself contains a dwindling proportion of these inhabitants (Table 23-2). Its site, on the hilly peninsula between the Pacific and the bay, is too confined to allow for much further growth. Although it quickly overcame the devastating setback of the 1906 earthquake, it lost its lead in population to Los Angeles in the decade that followed. Since then, it is in Oakland, San Jose, and the cities of the bay's north shore that most growth has occurred. In particular San Jose, at the southern end of the Bay Area, has become a sprawling metropolitan area surrounding an urban nucleus which, in 1950, had a population of just 95,000.

These Bay Area cities today occupy an area that extends north and south for 100 miles (160 km), and east to west for over 30 (48 km). They are strung around the shores of the bay and linked—all too tenuously when the traffic builds up—by a series of

bridges. San Francisco suffers from the same problems as New York City, but on a larger scale: the problem of linking together parts of a single conurbation whose various functional zones are separated from one another by water barriers. Even though large areas of the bay shore have been reclaimed and put to use for industry, airport, or harbor works, the problem remains severe.

The functional differences around the bay began early. For all traffic moving to or from the interior—the goldfields or the Central Valley—the eastern bay shore was a better handling area than San Francisco itself: it was low-lying, whereas the peninsula rose steeply from the shoreline, and it soon became the terminal of not one but several rail routes. Oakland therefore developed as the principal cargo handling area and, in due course, became what it is today—one of the nation's largest container ports. Between them, San Francisco and Oakland handle some 60 million tons of freight a year, about one-half of it as foreign trade. Smaller ports specialize in a single commodity, like Crockett with its sugar refinery processing Hawaiian cane, or oil terminals and refineries.

San Francisco itself has seen its port role gradually decline, especially with the virtual disappearance of passenger shipping. Many of its piers have been closed to freight traffic for, like those of Manhattan Island, they do not offer enough space for container handling: in fact, they make more money for the city

[10]Although Santa Cruz PMSA is treated officially as part of the San Francisco CMSA, Santa Cruz itself is not a bay city, but a city of the Pacific coast. When we are dealing with the bay cities, therefore, its population is properly deducted from the CMSA total shown in Table 23-6.

San Francisco: The city, the bay, and the bridges viewed from the southwest. *San Francisco Convention and Visitors Bureau*

as tourist attractions or marinas. Its preeminent function is commercial. Ever since its stock exchange opened in 1862 it has been the region's business center—"the city" to all those who know it. It draws in commuters to its offices much as Oakland draws them into its factories, shipyards, and military bases.

With outlying suburbs and centers expanding, it became clear that the transport links between them needed strengthening. Out of this need grew the Bay Area Rapid Transit system (BART). The first section opened in 1972, and the intention has been to extend the lines stage by stage. BART has become the testbed or model for many other cities faced with similar problems.

At the southern end of the bay is the San Jose metropolitan area, with over 120,000 industrial employ-

ees. Many of them work in factories that stand where, only a few years ago, there were orchards or salt marshes. They include the remarkable collection of high-tech plants grouped in what is known as Silicon Valley, after the critical industrial material they use—the silicon chip. As we have already noted, the San Jose PMSA (or Santa Clara County—the two are the same) contains one-third of California's electronics and computer industry.

Inland from the bay lie the cities of the Central Valley. Of these, the largest is Sacramento, partly because it lay on the earliest route and the first railroad line across the Valley to the coast; partly because it became the capital of the most populous state in the Union. The other cities—Stockton, Modesto, Fresno, Bakersfield—are the urban centers serving one of the world's most productive agricultural areas, and they have grown with the spread of irrigation and the increase in output from the Central Valley farmlands.

Southern California: Cities and Industries

Between San Francisco and Los Angeles, judged as cities, there is little comparison: San Francisco occupies one of the world's finest urban settings; Los Angeles sprawls disjointedly across 50 mi (80 km) of plains and foothills, and it is separated from the coast and from its port, near Long Beach, by a belt of suburban development interspersed with oil fields. Yet it is the southern metropolis that has outgrown the northern.[11] This simply reflects what has happened in the state as a whole: in 1900, the fourteen southern counties of California contained 25 percent of the state's population, and today the proportion is well

over 65 percent. The 1990 census recorded a population for the Los Angeles–Long Beach PMSA of 8.8 million, an increase of 18 percent over the figure for 1980. By southern California standards, however, this increase was modest: the main growth in the conurbation is now so far from the original core—the pueblo of Los Angeles, founded in 1781—that it is taking place in and around secondary centers, which have meanwhile themselves been designated as metropolitan areas (see Table 23-7). It was in this way that the Anaheim–Santa Ana PMSA, southeast of Los Angeles proper, recorded a population increase of 101.8 percent for the decade of the 1960s,[12] and by 1980 had increased to 1,932,000 inhabitants. As we saw in Chapter 3, the old definition of a metropolitan area has become inadequate in southern California; the Bureau of the Census now recognizes a Standard

[11]On the Los Angeles area there is a wealth of literature, formal and informal. A fascinating book with which to start is Reyner Banham, *Los Angeles: The Architecture of Four Ecologies* (London and Baltimore, 1973), which is much more geographical than it sounds; S. Bottles, *Los Angeles and the Automobile: The Making of the Modern City* (Berkeley, 1991); and N. Klein and M. Schiesl, eds., *20th Century Los Angeles: Power, Promotion, and Social Conflict* (Claremont, Calif., 1990), dealing with the various ethnic communities. More formal geographical works are: H. J. Nelson, *The Los Angeles Metropolis* (Dubuque, Iowa, 1983), and R. Steiner, *Los Angeles: The Centrifugal City* (Dubuque, Iowa, 1981).

[12]Most of this growth took place after Disneyland was opened. The German immigrants who, in 1857, founded Anaheim would find today that agriculture had become dry-lot dairying and that their farmlands were occupied by the homes of 2 million people and industries employing 200,000. To the list of such standard locational attractions as coalfields, water supply, or transport we can therefore now add another item: location governed by amusement park. The Disneyland recipe has now been repeated near Orlando, Florida.

On the same theme, see R. Kling, S. Olin, and M. Poster, eds., *The Transformation of Orange County Since World War II* (Berkeley, 1991).

Table 23-7. Southern California: Population of Metropolitan Areas, 1970, 1980, and 1985 (in thousands) and Percentage of Spanish Origin, 1980

	1970	1980	1985 (est.)	Percentage of Spanish Origin (1980)
Los Angeles CMSA	9981	11,498	12,738	24.0
Anaheim–Santa Ana PMSA	1421	1,933	2,123	14.8
Los Angeles–Long Beach PMSA	7042	7,478	8,109	27.6
Oxnard–Ventura PMSA	378	529	600	21.4
Riverside–San Bernardino PMSA	1139	1,558	1,907	18.6
San Diego MSA	1358	1,862	2,133	25.6
Santa Barbara–Santa Maria–Lompoc MSA	264	299	332	18.5

Los Angeles Region: Percentage Increases in Population, by Counties and Decades, 1880–1980

County	1880–90	1890–1900	1900–10	1910–20	1920–30	1930–40	1940–50	1950–60	1960–70	1970–80
Los Angeles	201	69	196	86	136	26	49	46	17	6
Orange	—	46	74	78	93	10	66	226	102	36
Riverside	—	—	94	44	61	30	62	80	49	45
San Bernardino	227	10	102	30	82	20	75	79	36	31
San Diego	306	0	75	81	86	38	92	85	31	37
Santa Barbara	65	20	46	48	59	8	39	72	56	13
Ventura	98	43	28	56	91	27	63	74	90	40

Consolidated Statistical Area, covering four PMSAs totaling 11.5 million inhabitants in 1980, and 14.5 million in 1990. For a number of years in the 1960s, migration into the five counties surrounding Los Angeles approached 200,000 a year. In retrospect, the crest of the wave seems to have occurred in 1963, a year in which 210,000 newcomers arrived.

The Los Angeles conurbation is first a matter of record-breaking statistics; then it is a question of homes and jobs for 14 million people. The statistics are soon dealt with; it seems as if everybody in southern California knows them. On 5 percent of the state's land area is concentrated one-half of its population and its economy—46 percent of its personal income, 56 percent of its manufacturing shipments, and 60 percent of its wholesale trade sales. Between 1975 and 1985, it added over 260,000 new jobs in manufacturing (an increase of 29 percent), while New York City lost 6 percent and Chicago lost 21 percent during the same period. The port of Los Angeles–Long Beach handles foreign trade that amounts to some 40 million tons annually (out of a total of 80 millions of traffic), and that is second only to New York's in value. Its three leading trading partners are Japan, Taiwan and Korea.

But now we must ask what these statistics mean in human terms: What kind of employment does the region offer? For southern California, as a whole, the employment structure is not exceptional, although within the region there are local differences. Manufacturing, for example, is largely concentrated in the old industrial zone southeast of the center of Los Angeles, in the port area, and in the new southeastern suburbs. The Anaheim MSA has the highest proportional employment in industry, wholesaling, and construction, and Oxnard–Ventura in government. Employment in services is everywhere high by U.S. standards—almost 5 percent above the national average—but this is readily explained (1) by a high average standard of living and (2) by the high proportion of elderly or wealthy people requiring such services. The construction industry is prominent everywhere in the newer suburbs. Apart from the 50,000 persons employed in "entertainment and rec-

San Diego, California: San Diego combines the maritime functions of U.S. naval base, fishing port, and leisure cruising center; its huge harbor has ample space for all three. *Port of San Diego*

reation,"[13] Los Angeles has important functions in relation to the oil industry, insurance, and foreign trade and has, in the last two decades, challenged San Francisco's long-standing role as the banking center of the West Coast.

There is an immense variety of industries in southern California, and some of them are of considerable size: in Los Angeles, for example, there are more than 60,000 workers employed in the manufacture of clothing, for the West Coast has developed its own fashions. But in terms of economic activity, nothing within the region compares with the aerospace industry and its offshoots. Natural conditions favored the early growth of an aircraft industry: in 1947, when it employed 150,000 workers throughout the United States, 65,000 of them were in the Los Angeles area. By the late 1960s, employment in the successor industries was over 450,000.

Since that time, as we noted earlier in this chapter, there have been some rapid fluctuations in this employment level. The problem, we can recognize, is twofold: (1) so much of the output of the industry is ordered and paid for by the government, and so depends on budget policy, and (2) progress in technology is so rapid that a good deal of wastage, human or material, along the way is inevitable. In this industry, workers learn to live with uncertainty.

This is a lesson other areas, such as southern New England (Chapter 12) have had to learn. It emerges all the more clearly if we move on from Los Angeles to San Diego. Such has been the growth of Los Angeles that it is easy to overlook San Diego, 125 mi (200 km) to the south. Yet San Diego, founded before Los Angeles, has one of the world's great natural harbors (that of Los Angeles is largely man-made), and in 1990 had a population of 2.5 million, so that in any other company it would stand out as a major urban center. It grew up as a naval base (and still serves as such), as the home port for the big southern California tuna fishing fleet, and as a place to retire to in the sun. But then came an industrial boom based on the

air and spacecraft industry, and its fortunes began to fluctuate. By 1957, it was estimated that more than 80 percent of the city's employment in manufacturing was in "defense-oriented industry." What this meant, in local terms, can be seen in the census of manufactures: in 1958, 46,000 workers were employed in the transportation equipment industries; in 1963, the figure was 13,000; and in 1972 it had risen to 24,000.

In Los Angeles itself, the outgrowths of air and space travel can be seen in a proliferation of electronics firms, which have a much greater range of products and market than the original defense contracts allowed. Once again, as in the case of the San Francisco Bay Area, this development has been related to the existence of a number of the nation's best-known universities—the "Ph.D. producers"—and to the grant of very large sums of government money for research.

Sprawling as it does over the plains that lie below and between the San Gabriel, Santa Monica, and Santa Ana mountains, the Los Angeles conurbation developed what seemed likely to become the typical urban pattern of this motor transport–electricity–amenity–planning second half of the twentieth century, much as "dark, Satanic mills" epitomized the century of the Industrial Revolution. The sequence of development has been an unusual one by earlier standards. The core of the metropolitan area—Los Angeles proper—grew around the old Spanish pueblo of that name and developed much as any other American city was developing in the early twentieth century, if at a faster rate. By the 1920s, however, there had begun the growth of separate cities located from 5–15 mi (8–24 km) from the original center—cities such as Pasadena, Santa Monica, and Long Beach—which offered a wide range of services to the suburban population and which had the effect of drawing off both business and cultural life from the central area of the conurbation. By the end of World War II, central Los Angeles itself boasted administrative offices, old established industries, and slums, but little else. Shortly afterward, the core area suffered a further functional setback when it became one of the world's largest crossroads. Meanwhile, the peripheral cities were developing suburbs and central business districts of their own. The core seemed to have become largely superfluous, and it was fashionable to refer to Los Angeles as a group of suburbs looking for a city.

But since then, there has been a revival in central

---

[13]Very few attempts have been made by geographers to analyze the most famous of all Los Angeles's industries, but there is one at least: M. Storper and C. Christopherson, "Flexible Specialization and Regional Industrial Agglomerations: The Case of the U.S. Motion Picture Industry," *Annals,* Assoc. Amer. Geog., vol. 77 (1987), 104–17. The authors reveal, incidentally, that southern California's dominance of the industry has waned. The rest of the United States now plays an increasing part. Since 1960, Los Angeles's share of world production activity has fallen from one-half to one-third of the total.

The Spacecraft industry in southern California: General Dynamics'
"clean room," where stages of the Centaur space vehicle are assembled. *General Dynamics*

Los Angeles. One of the problems the city confronted
in developing a genuine central business district was
a city ordinance limiting the height of buildings because of earthquake danger. Under these circumstances, it would have been difficult ever to develop
the kind of concentration of business activity found
in other major American city centers. However, with
the development of better building techniques, and a
better understanding of how earthquakes do their
damage, the height limit has been waived, and business has responded by creating a new townscape of
office blocks and dense occupance levels. City interests have also made an effort to introduce some cultural functions at the urban crossroads: the city center has improved greatly in appearance, and
development has been impressive and swift along

the axis that leads west from the old core—the so-
called "Miracle Mile" along Wilshire Boulevard.

But whether the central area is thriving or mori-
bund, an open-plan metropolis of this kind depends
heavily on its road network, and the more so because
urbanization in southern Californian style has almost entirely taken the form of separate single-family dwellings, which occupy a maximum of space and
demand a maximum of service roads. The early system of streetcar lines and suburban railroads built to
serve the then-existing ring of towns around Los Angeles has long since withered away, and has been replaced by a network of freeways. Nowadays, new
suburbs spring up along the line of an advancing
freeway much as they did seventy or eighty years ago
along the rail lines and the underground or subway

routes around London and New York. With an apparently insatiable demand for new freeways before them and the example of San Francisco's BART system behind, Los Angeles planners talk nostalgically about the old suburban railroads and the possibility of their revival.

Los Angeles has been described as "the Ultimate City."[14] To take this title too seriously, however, would be a mistake, for it would be to deny the very lesson this city teaches—that urban patterns change with time, technology, and standards of living, and there is never, therefore, likely to be an ultimate in the process.

Los Angeles is also often referred to, either in hope or in horror, as "the City of the Future." But this has been going on for so long now that it is high time that we asked: Where is this future and its cities, of which Los Angeles is supposed to be the prototype? Although it has been obvious for, say, half a century that Los Angeles (or, rather, urban southern California) was different from other cities in structure and layout, it is today noticeable that we do not have large numbers of imitation southern Californias arising elsewhere. For sheer urban sprawl, Houston, Texas, rivals Los Angeles, but the conditions there (Chapter 5) are quite special—as special in their own way as those of Los Angeles, and quite different. Phoenix, Arizona, with its new satellites, could be taken as a miniature southern California.[15] But it is hard to think of other examples. The truth is that southern California is the product of a particular set of circumstances, that it developed in its own way, and that there is no reason why it should not, nor any reason why other areas must imitate it.

The Future of Southern California

Inevitably, the question arises how long California, and especially southern California, can maintain its breakneck rate of population growth and economic development. Employment openings have so far kept up with population increase, and although, in the immediate neighborhood of the cities, space is becoming a problem, settlement is continually being carried into new areas on the desert edges. The limiting factor is water, and the biggest problem is to decide how best to use it.[16] At present, one water use in the American West far outstrips all others in its demands, and that is irrigation. In some areas, it is estimated that as much as nine out of every ten gallons of water withdrawn from surface or underground sources are used for irrigation, leaving only the remaining gallon for the whole range of urban and industrial uses. To cut their water bills, California's industries have introduced all manner of conservation practices, using water several times over instead of discharging it.[17] Meanwhile, the state's farmers irrigate over 8 million acres (3.2 million ha).

We have seen that in the Central Valley there is water to spare at the northern end of the Valley, and a shortage at the southern end. Exactly the same is true of the state as a whole: the northern part has the water and the southern part has the greater weight of population and demand. Consequently, southern California has had to reach further and further afield for its water supplies (Fig. 23-4). Its two main sources have been the Colorado River to the east and Owens Valley, 250 mi (400 km) away to the north, both of which are linked by aqueduct to Los Angeles. But the amount of water the region can obtain from the Colorado is strictly rationed by the Colorado River Compact (see pp. 429–30). To crowd 10 or 12 million people into a desert or semidesert area 200 mi (320 km) from a major river is to pose a problem of supply and demand that might baffle any engineer.

Since southern California is where all these people wish to live, what solution for this problem can be found? There are, at present, four possibilities:

1. *To obtain usable water from the sea.* Technically this is now possible, and pilot plants exist; the main

[14]Christopher Rand, *Los Angeles: The Ultimate City* (New York, 1967).

[15]When J. F. Hart wrote "Urban Encroachment on Rural Areas," *Geographical Review,* vol. 66 (1976), 1–17, he reported that, between 1945 and 1975, Phoenix grew in area from 17 sq mi (44 sq km) to 248 sq mi (642 sq km). Today, the figure is nearly 400 sq mi/1000 sq km).

[16]On California's water situation the basic reference is the splendid volume, William Kahl, ed., *The California Water Atlas* (Los Angeles, 1979). Among useful books on this topic are J. Humlum, *Water Development and Water Planning in the Southwestern United States* (Aarhus, Denmark, 1969); D. Seckler, ed., *California Water: A Study in Resource Management* (Berkeley, 1975); N. Hurdley, Jr., *Water in the West* (Berkeley, 1975); K. D. Frederick and J. C. Hanson, *Water for Western Agriculture* (Washington, D.C., 1982); and, for historical background, D. J. Pisani, *From the Family Farm to Agribusiness: The Irrigation Crusade in California and the West, 1850–1931* (Berekeley, 1984).

[17]Perhaps the best example of this conservation of water is offered by the Fontana steelworks, which have reduced the amount of water required to make one ton of steel to as little as 1500 gallons. For purposes of comparison, this is about the amount of water required to raise three pounds of rice or a pound of cotton: a pound of beef requires some 4000 gallons.

Fig. 23-4. Southern California: Water supply. The map shows the aqueducts that supply Los Angeles and the canal system that brings Colorado River water to the Coachella and Imperial valleys. The demand for water in the area continues to increase, whereas the supply is limited. Transport costs are increased by the high evaporation rate from open canals and by the many relief obstacles. The highest and lowest points in the Lower 48—Mt. Whitney and Death Valley—are within the area shown.

questions here are those of cost and scale. But it seems clear that at least some of the extra supply needed in the coming decades will be provided by this means.

2. *To reduce the irrigated acreage.* This has already happened in the vicinity of the expanding California cities, as individual farmers sell out and their irrigation water becomes available for other purposes. Yet in the state as a whole, the irrigated area has never been larger than it is today, and we have already taken note of such schemes as the West Side extension in the San Joaquin Valley, which have raised the state's irrigated acreage still higher.

This irrigated land, of course, produces an important part of the United States' food.[18] It is also the source of California's agricultural wealth, so it might

be argued that it is contributing to the gross product of the state and should be left as it is. But this argument is a weak one, as Kelso pointed out: "Per 1000 gallons of water withdrawn per day, manufacturing in California provides 4.33 times as much personal income and produces 68 times as much product value as does farming. On the other hand, farming annually uses 8.6 times as much water per employee as does manufacturing."[19]

In other words, irrigation agriculture is not the most productive use of water in California, nor is it the most remunerative. Much of this irrigation agriculture is based on water rights, which, if they were made over to a municipal or industrial user, would bring in a sizeable income in water rates. Irrigation agriculture depends on cheap water, and in California, water can only be cheap if its use is noncompetitive. It is therefore possible to argue that the immense agricultural output of the state is marketed competitively only because the prime resource—water—is excluded from normal economic calculations of true market value. If irrigation farmers had to bid for water against other users, they would quickly be put out of business.

At the very least, they might use the water to grow different crops. It might be supposed that, when it comes to irrigation, one crop is very much like another in its demands. This is far from being the case. For example, figures for Kern County, in the Central Valley, indicate that the average application of irrigation water, in acre-feet per acre, varies greatly (Table 23-8).

In water-using terms, the "villains" in this list are alfalfa, apples, and irrigated pasture. This is because these crops, unlike rice, can be grown (and, of course, in other parts of the United States *are* grown) without the aid of irrigation. In California, they are grown under irrigation partly because this increases output: alfalfa fields in the Midwest, for example, normally produce three cuttings per year, whereas those in the Imperial Valley produce six, and at higher yields to the acre. But partly, also, the crops are irrigated because the water used is being provided at low cost. If the farmer had to buy water on an open market, he would not be able to afford the luxury of spreading it on crops such as these. Alfalfa—or, for that matter,

---

[18]The U.S. Department of Agriculture estimates that in the United States as a whole some 10 to 11 percent of the nation's cropland is irrigated, and that this irrigated area produces 50 percent of all U.S. vegetables and 56 percent of all potatoes, and that it supports 58 percent of the nation's orchards.

[19]M. M. Kelso, "Theory and Practice in Public Resource Development and Allocation in the Western States," *Proceedings,* Western Farm Econ. Assoc. Annual Meeting (1959), p. 7.

For a comparable argument about the area around Tucson, Arizona, see Chapter 20.

Table 23-8. Water Requirements of California's Irrigated Crops

Acre-feet	Crops
1.5	Barley
2.0	Hay
2.5	Carrots, lettuce, onions, beans
3.0	Olives, potatoes, silage
3.3	Corn
3.5	Citrus, plums, tomatoes, cotton, sugar beet
4.0	Grapes, peaches, almonds
4.5	Alfalfa, apples
5.0	Irrigated pasture
7.0	Rice

Source: Giannini Foundation, *Crop Production and Water Supply Characteristics of Kern County* (Berkeley, 1980) p. 26.

the animal products it goes to create—would be shipped into California from unirrigated lands elsewhere, and the western water could be used to relieve other shortages.

3. *To transfer more water from north to south.* This is the Central Valley Project over again, on a larger scale, and the transfer is already taking place: so much so that water users in northern California are complaining of periodic shortages. The state has built its own water project, in parallel with the Central Valley Project: the first section harnessed the Feather River, which flows into the Central Valley from the east. The Project can be extended, and work has already begun, by damming and reversing the rivers—the Eel, the Trinity, the Klamath—which flow west from the Coast Ranges to the Pacific.

4. *To import water into the region from outside.* Basic to all southern California's water problems is the fact that the Colorado River, nearly 1500 mi (2400 km) long and draining six states, has a mean annual flow of only 14 million acre-ft (17.5 billion m³). It has therefore been proposed that this flow might be doubled by transferring 15 million acre-ft (18.3 billion m³) of water from a river that could lose that amount and never show it—the Columbia in the Pacific Northwest. The Columbia's flow is 175 million acre-ft (214 billion m³) a year, an amount that, given all foreseeable developments within the basin, the northwestern region could never use.

It is proposed (by the Southwest, of course) to build a north–south transfer line, and the necessary studies have already been carried out by the Bureau of Reclamation. It would admittedly be costly, and whether it will ever be built depends, in part, on a comparison of costs with the seawater conversion project already mentioned. But technically it is perfectly feasible; the Feather River scheme already transfers water over a distance half as great as that involved, and there is no reason why the longer transfer from the Columbia should not succeed and become the first leg of a great, continent-wide water "grid." Solution of the remaining problems lies not with the engineers, but rather with the politicians.

The survival and growth of a southern Californian population is a matter of national interest. This is by no means because of universal admiration for the area's life-style: quite the contrary. It is endlessly portrayed on film and television as if it were the American norm, but it certainly is not that. Its future is significant precisely because it is abnormal: it is, in fact, unique. As a great sociological laboratory, it is important to us all: How will it develop? Will it survive the pressures that crime, immigration, and poverty impose on a society which is, at the same time, famous worldwide for wealth and glamour? Is this particular society not being tested, as the engineers say, to destruction?

Today, the Los Angeles area is the goal of the most disparate groups of newcomers to be found anywhere in our world. It holds the largest Hispanic urban population north of Mexico City: it holds 1 million African Americans and yet, at the same time, it also has twice as many Asian inhabitants as New York or any other U.S. city. It has ambitious young Americans looking for stardom, and retired or frail Americans looking for a quiet life in the sun. All these people converge on an area where earthquake is always threatening, where forest fire is a regular hazard, where smog prevents breathing, and large numbers of people do not speak the national language. Why they do this is something of a mystery. Since they have done it, 12 or 14 million of them, we should at least be able, the rest of us, to learn a few lessons about living together.

24
Hawaii

Hawaii: Lumahai Beach on Kauai, whose outstanding scenic quality led to its choice as one of the locations for the movie *South Pacific. Hawaii Visitors Bureau*

More than 2000 mi (3200 km) west of San Francisco lies the capital of the archipelago state of Hawaii, a chain of islands rising from the deep floor of the mid-Pacific in a series of low coral reefs and mountainous volcanoes. There is no physical connection with the rest of North America, but politically, these islands constitute the fiftieth state of the United States; culturally, they have been affected by America for a century and a half, and economically, they play a part in the continental circulation of people and commodities. They therefore merit a place in a geography of North America and, logically enough, that place is after the consideration of California, the part of the continent to which they are linked by air and by sea.

The full extent of the island chain is over 1500 mi (2400 km) from end to end, or 22° of longitude. But the western three-quarters consist of minute islands, many of them uninhabited, and it is within the easternmost 400 mi (640 km) that virtually the whole of the state's 6450 sq mi (16,700 sq km) are to be found, predominantly in five major islands. They lie in a southeast–northwest line in which the most easterly island, the one that is actually called Hawaii, is the largest. Most of the population, however, is concentrated on Oahu, the fourth of the five islands, where the capital of the state, Honolulu, is situated. Out of a 1990 state population of 1,108,000, some 815,000 lived in the metropolitan area of Honolulu, and 375,000 in the city itself.

Small though these islands appear in the vastness of the Pacific, they are full of natural interest. Among other features, Hawaii possesses five volcanoes, (a so-called "Ring of Fire") the highest of which, Mauna Kea, rises to 13,796 ft (4194 m), and a second, Mauna Loa, is estimated to have discharged greater quantities of lava than any other volcano anywhere, with spectacular recent eruptions in 1954 and 1960. But in spite of these periodic outbursts, there is a rich and varied flora, including nearly 1 million acres

(400,000 ha) of commercial forest. The variety is the result of another of Hawaii's geographical features, one it shares with a number of other mountainous islands of the Pacific—remarkably abrupt climatic changes from one side of an island to the other. The Hawaiian chain lies in the belt of influence of the northeasterly trade winds, and each island has a windward and a leeward side. The windward side of the mountains catches the full force of the trades and the rainfall they bring, and in at least four of the islands there are stations that record over 200 in (5000 mm) of rainfall a year, with authenticated annual totals of well over 450 in (11,250 mm). On the leeward side of the islands, by contrast, there are stations with long-term averages of less than 15 in (375 mm). Honolulu itself, sheltered in the lee of the Koolau Range, has a modest 25 in (632 mm) of rainfall; otherwise it could hardly attract tourists as it does. Temperatures vary little throughout the year; in Honolulu, the January and August means are 72 and 81°F (22 and 27°C), respectively. These climatic conditions ensure a year-round tourist traffic and so, too, does the exotic vegetation and even the volcanoes, which are today sufficiently well monitored to serve rather as a kind of outsize fireworks display than as a threat to safety. Only the occasional unpredictable tsunami (tidal wave) does real damage, as in 1946 and 1960, and Hawaiians can boast the longest average life span of any state's inhabitants in the Union: 74.1 years for men and 80.3 for women, against national averages of 70.1 and 77.6 respectively.

The original Hawaiians were Polynesians, or Pacific islanders. Together with other stocks originating in mainland Asia or Japan, these elements formed 62 percent of the population at the 1990 census: the five most "Asian" counties in the United States (with 52–72 percent of their population in this category) are all in Hawaii. The islands have attracted in recent decades two steady streams of immigrants, one from

Hawaii: The volcanoes. In the foreground is Mauna Loa (13,680 ft, 4148 m), and in the distance is Mauna Kea (13,796 ft, 4194 m), the highest point in the islands. *U.S. Geological Survey*

Asia and the other consisting of white Americans, attracted to the islands by retirement or leisure prospects. Economic opportunity and amenity are both on offer.

Hawaii has 2 million acres (810,000 ha) in farms, but as the average farm size on these small islands is 426 acres (172 ha), much of the farm work is plantation agriculture. It was to work on the plantations, especially cutting sugarcane, that many of the Asian immigrants originally came to the islands, and sugar remains the most valuable agricultural product, with pineapples—also plantation grown—second in rank. Dairy farming is carried on but, given the lack of space for cattle pastures, much of it is dry-lot farming, as it is around, say, Los Angeles or Miami. Both stock and feeds had initially to be imported, either from the mainland or from Asia, but good progress has been made in the use of native grasses for feed, and in converting discarded parts of the sugarcane and pineapple crops into forage for cattle.

As on the mainland, the farm population has been steadily falling, thanks to mechanization and the lure of the city, but Hawaiian farming itself is protected by one of the most complete systems of land-use zoning to be found in any of the fifty states. Such are the limits set in these small islands by climate and terrain and such, on the other hand, are the demands for building land for residential and tourist purposes, that, soon after statehood was achieved in 1959, the new state government introduced a system of land-use categories and controls that put Hawaii in the forefront of American land-use planning, and insured most of its farmland against takeover.

Given its climate and its position, however, Hawaii has two sources of income on which it can comfortably rely—defense expenditures and tourism. In the days before World War II, these two were represented by the naval installations at Pearl Harbor, the landlocked bay just west of Honolulu made famous by the Japanese attack of December 1941, and by the

Hawaii: Harvesting pineapples at Wahiawa on central Oahu. The fruit is harvested year round, but during the peak season in summer the machines operate day and night. *Castle & Cooke*, Dole Pineapple Division

cruise ships that brought wealthy tourists from California ports. Today, the defense functions remain, although the exclusively naval role of Hawaii has been modified in an air age that, in a tragic sense, it helped to inaugurate: in 1980, the defense payroll for the state amounted to nearly $1,150 million, which was more than that for New York State or Ohio. As for tourism, it has undergone a process of democratization. Fast air travel has brought the islands within reach of people who have only a week or so of vacation, and cheap fares have provided a tropical paradise for people of limited means. With Bermuda and the Bahamas to the east and Hawaii to the west, few Americans today need be without one.

The American involvement with Hawaii dates back to the 1820s and the arrival of a number of groups of missionaries. From that time, until 1898, the American influence grew, step by step, and it was as much to resolve internal tensions as to resist the pressure of other foreign powers that, in 1898, the islanders asked for annexation by the United States. The islands were created a territory in 1900 and a state in 1959.

25
The Pacific Northwest

The Pacific Northwest: The city of Tacoma, Washington, backed by Mt. Rainier, at 14,410 ft (4381 m) the highest mountain in the north-western United States. *U.S. Department of the Interior. Photo by Lyn Topinka*

The Region and Its Resources

As they run northward, the two great mountain chains of western North America converge, confining the narrow tip of the High Plateau Province between them. Behind this formidable double barricade some 10 million Canadians and Americans live in almost complete physical isolation from the remainder of the continent. Even to the south, the line of the Pacific coast troughs, which elsewhere provides a lowland route, is interrupted by the wild mass of the Klamath Mountains, which mark the Oregon–California boundary and which are dominated by the huge double cone of Mt. Shasta. There is no single line of easy access into this remote corner of the continent, economically vital though it is as a major producer of timber and hydroelectricity.

Most of the population is to be found in the Pacific coast lowlands, and much of it in the line of port cities whose largest members are Vancouver (1,602,000), Seattle–Tacoma (2,559,000), and Portland (1,478,000). Inland, the population is scattered, for the most part, in long streamers up the valley routes through the mountains. But circumstances of topography and history have drawn into the sphere of the Pacific Northwest two larger settled areas that are really parts of the High Plateaus. One is the Columbia Basin of central Washington and southern British Columbia, the "Inland Empire" whose capital is Spokane (356,000); it is linked with the coastal region largely by grace of the Columbia River Gorge behind Portland. Of this Inland Empire, the fertile, south-facing valleys of southern British Columbia form a physical part, since the Columbia and its tributary, the Kootenay, follow zigzag courses that cut across the international boundary and give a geographical unity to the interior lowlands.

The links between the Pacific coast and the second inland area—the Snake River Plains—are even more tenuous. These plains of southern Idaho are physically a part of the Great Basin of Utah and Nevada, and economically they are, in some ways, part of the hinterland of Salt Lake City. The Snake River itself is far from forming a bond between the plains and the coastal centers, for it cuts through the mountain barrier of eastern Oregon in a 5500-ft (1670-m) gorge that is all but impassable. But the plains are linked with the Pacific Northwest in the historical association of the Oregon Trail, now represented by the railroad and the road, which climb through the labyrinthine valleys of the Blue Mountains. As we shall shortly see, both the Snake River Plains and the Inland Empire lie on what in the past has always been the market side of the Pacific Northwest, and their trade follows the same general direction as that of the coast.

Few regions of North America present a greater variety of physical conditions than the Pacific Northwest. Though the general structure conforms to the north–south, fold-and-fault pattern of the major mountain ranges, the pattern has been disturbed by widespread lava flows; the mountains of northern Oregon run from east to west; and the Columbia and Snake, once clear of the Rockies, conform to no drainage pattern but their own. Although most of the terrain of the interior is rugged, the lowlands of the Snake and the central Columbia Basin, leveled by lava, present favorable surfaces for agriculture at elevations between 500 and 3000 ft (150 and 900 m). British Columbia is less favored by its Fraser River; comparable low-level surfaces are scarce in the southern, populated part of the province and only appear in the Nechako Plateau around Prince George, so that agriculture in interior British Columbia is largely restricted to the "trenches" that lie between the ranges of its mountainous southeast.

The variety of climatic conditions within the Pacific Northwest has already been suggested in Chap-

The Columbia River at Crown Point, Oregon, looking eastward. *Oregon State Highway Department*

ter 1. The coastal mountains have the continent's highest precipitation rate (most of it falling in winter, when snows are very heavy), whereas the plateaus are semiarid or arid. Wenatchee and Yakima, in the irrigated orchard region of the western Columbia Basin, have 8.7 and 6.8 in (217 and 170 mm) of rain per annum, respectively, and the Okanagan Valley 10–12 in (250–300 mm). East and south across the basin, precipitation increases to 15 and then to 25 in (375–625 mm), making possible the wheat growing for which this moister section of the basin is famous (Fig. 25-1.)

The agricultural significance of these abrupt climatic differences may be summarized in a few general statements. (1) Agriculture in the Pacific Northwest can be divided into valley agriculture, for which the water must usually be supplied by irrigation; and plateau agriculture, for which rainfall may be adequate, but the frost-free season is short on account of the altitude. (2) Over much of the Northwest, the warm-season rainfall is inadequate for crop growth

(between 2 and 4 in [50 and 100 mm] in the central Columbia Basin), but the heavy snows of winter form a most valuable reservoir. In the interior, if not on the coast, the meltwater is released into the rivers in a manner well suited to the irrigation farmer's needs; peak flow on the Columbia usually occurs in June. (3) Because of the proximity of areas of high and low rainfall, and because of the seasonal character of precipitation in the region, some areas require both winter drainage and summer irrigation for agricultural production.

This combination of physical circumstances also contributes largely to the existence of the Northwest's three prime natural resources: forests, fish, and waterpower. The mountain slopes, with their heavy rainfall, are clothed with the continent's principal reserves of virgin softwoods: the Northwest has been called the "Sawdust Empire." Its salmon fisheries constitute an important element in regional income—an element of sufficient value for the depletion of the fisheries to cause acute concern, and re-

The Cascade Range: Mount St. Helens erupting on May 18, 1980. The spectacular eruption not only brought devastation locally but spread a thick layer of dust for many miles leeward, across the state of Washington. *U. S. Geological Survey*

quire the Canadian and U.S. governments to take in hand restoration of the salmon runs. Finally, the Columbia, Snake, and Fraser together possess a large part of the continent's hydroelectric potential. The harnessing of this potential has been progressing steadily over the past half-century: on the Columbia, it dates from the Great Depression, when the great dams at Grand Coulee and Bonneville were authorized as relief measures under the public works program of 1933. In British Columbia, power sites are widely scattered throughout the province and are gradually being developed as demand increases. The principal river, the lower Fraser, is not used for power generation, however, because of its importance to the salmon fisheries.

Agriculture in the Pacific Northwest

Although fur traders and gold miners occupied the limelight in the first half of the nineteenth century, the first genuine agricultural settlements on the Pacific coast came quietly into being in the fertile Willamette Valley in the 1830s.[1] The area rapidly gained in popularity, so that the 1840s saw a marked increase in the number of settlers trekking west along the Oregon Trail and the beginnings of settlement in the central Columbia Basin. In British Columbia, however, there was little agricultural development before the miners arrived in 1858. From these varied beginnings, local specializations have grown, until today we can recognize four main types of farming:

1. *Dairy farming,* with subsidiary poultry, fruit, and vegetable production. This type of farming occupies the main lowland areas from the Fraser Valley and

[1]See S. M. and E. F. Dicken, *Two Centuries of Oregon Geography,* vol. 1, *The Making of Oregon* (Portland, Oreg., 1979); M. A. Bowen, *The Willamette Valley: Migration and Settlement on the Oregon Frontier* (Seattle, 1979); and J. R. Gibson, *Farming the Frontier: The Agricultural Opening of the Oregon Country, 1786–1846* (Vancouver, B.C., 1985).

508

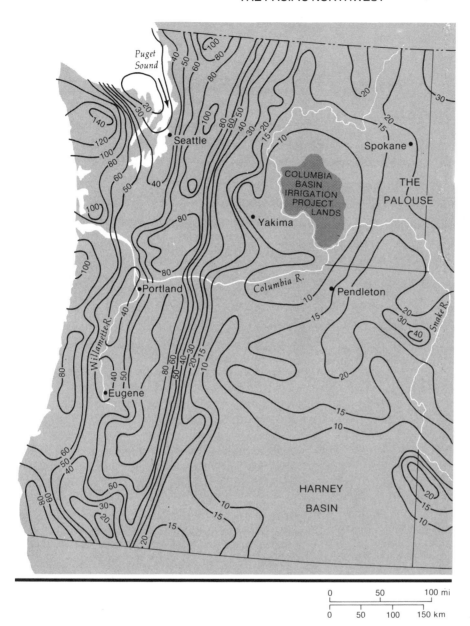

Fig. 25-1. Precipitation map of the northwestern United States. The map emphasizes the marked rainshadow effect of the Olympic Mountains over Puget Sound and of the Cascades over the deserts of the Harney and Columbia basins. Figures represent inches of rainfall.

Puget Sound to the southern Willamette Valley. The great urban centers of the lowlands provide markets for dairy produce, and fodder crops, irrigated and nonirrigated, occupy a large part of the cultivated land. Fruit and vegetable crops account for much of the remainder.

2. *Fruit and vegetable growing*, with subsidiary dairy farming. This second type is a combination of the same elements in different proportions. Away from the principal cities, the market for dairy produce is less favorable, so that in these areas the pro-

duction of fruit and vegetable crops, the Northwest's main agricultural specialty, assumes major importance.

It is this type of farming that is found in the valleys that open into the central Columbia Basin. Sheltered but dry, the river and lakeside terraces are suitable for irrigated orchards. The largest orchard areas are those of the Wenatchee, Yakima, and Okanagan valleys, where apples and pears are the principal fruit crops, but the range of lesser crops is very wide. It includes nuts, hops, holly grown for Christmas deco-

The Wenatchee Valley, Washington. Young apple orchards. The town of Wenatchee claims to be the "Apple Capital of the World." Most of the orchards are situated on river terraces. *Washington State Apple Commission*

rations, and such little-known fruits as youngberries and boysenberries (named for the Northwesterners who developed them), as well as the more familiar blackberries, raspberries, and strawberries. From the pear orchards of the Rogue River Valley in southwestern Oregon to the apple lands of south-central British Columbia, farmers in the sheltered valleys of the Northwest have developed temperate fruits as cash crops in much the same way as the Californians in more southerly latitudes have specialized in subtropical fruits. The areas form two halves of a Pacific Coast Fruit Belt.

3. *Commercial wheat farming.* The roughly circular, mountain-enclosed Columbia Basin is outlined on the west and north by the course of the Columbia River, on the east by the Rocky Mountains in the Idaho Panhandle, and on the south by the Blue Mountains. The Basin is lowest and driest along its western margin at the foot of the Cascades. There the characteristic vegetation is sagebrush and bunch grass. But eastward and southeastward, as elevation and precipitation gradually increase, the natural grasses become thicker and taller. It is this grassland, in the eastern half of the Basin and along its south-

eastern margin at the foot of the Blue Mountains, that has been turned into the third major wheat region of the continent.

The soils, derived from basic lava whose weathered materials were blown eastward to form loess, are fertile. During glacial times, meltwaters rushing southwestward eroded deep channels into the western parts of the Basin. These, together with many outcrops of dark-colored lava rock, have given rise to the name "Channeled Scablands." In the eastern section, between Spokane and the Snake River, the loessial material was formed into great dunes known as the Palouse Hills; they and their area are known simply as The Palouse. It is The Palouse that is the heart of the wheat region. Most of the crop is fall-planted winter wheat. Dry-fallow farming is practiced, and dry peas are an important second crop. Although loessial soils tend to resist erosion, a century of cultivation of the steep slopes has resulted in serious erosion.

4. *Ranching.* On those parts of the region that are too dry for wheat and too rugged for irrigation, such as the Channeled Scablands, stock raising replaces cultivation, as it does everywhere in the Mountain

West. The ranges spread over much of eastern Oregon, the uncultivated parts of the Snake and Columbia lowlands, and the Fraser and Nechako plateaus of British Columbia. Because of the heavy tree cover in the mountains, however, there is a more definite upper limit to the ranching zone than in other parts of the West; the high grazings are limited in extent, and ranching in the Northwest belongs mainly to the middle altitudes.

These, then, have been the four types of regional agriculture found in the Northwest. When the Columbia Basin Project for 1 million acres (405,000 ha) of new irrigated lands was conceived, the question naturally arose as to which of these four types of farming should be practiced on it. It was in 1952 that the first irrigation water from the Columbia flowed into an area whose annual rainfall of 8–10 in (200–250 mm) had seemed to render it useless for agriculture. The scheme makes use of the waters of Lake Roosevelt, impounded behind the largest of all western dams—the Grand Coulee. Close by the dam, a former course of the Columbia River cuts across a bend that the present river makes as it turns from a westerly direction to a southerly one before joining the Snake (see Fig. 25-2). This former course, which was carved out by the waters of a Columbia swollen by ice meltwater, is the Grand Coulee, after which the dam is named.[2]

The irrigation engineers have made use of this feature to carry water to the dry western end of the Columbia Bend. From Lake Roosevelt, water is pumped up into the reservoir above the Grand Coulee dam. From this reservoir, water can be distributed with the gradient to the project land lying south of the Coulee, about one-half of which has been brought into production so far. From an engineer's point of view, it is a remarkably ingenious and successful scheme.

By 1990, the project farmers were raising a wide variety of crops on an irrigable area of 557,500 acres (226,000 ha), of which about 524,000 (212,000 ha) actually received water that year. The list of crops epitomized Northwest specialty-crop cultivation: it included such exotics as spearmint and radish seed, nectarines and asparagus. But the major crops by area were alfalfa, wheat, corn, and potatoes. The

[2]On the historical background to settlement in the Columbia Bend, see D. W. Meinig, *The Great Columbia Plain: A Historical Geography* (Seattle, 1968). On the Columbia Basin Project itself, the Bureau of Reclamation has issued various publications and so, too, have the state Agricultural Extension Service and Washington State University at Pullman, Wash.

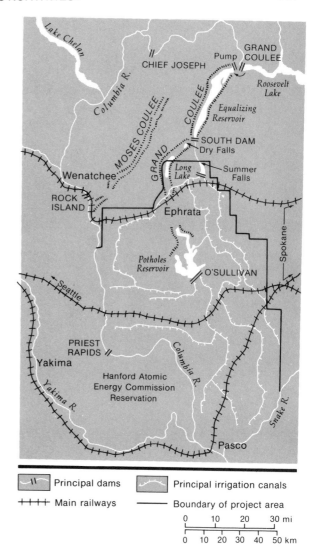

Fig. 25-2. The Columbia Basin Project. Across the surface of the lava plateau, the huge meltwater Columbia and its overflow channels cut deep valleys, often steep sided, called *coulees*. If these are dammed, they can be converted into natural reservoirs.

acreage under tree fruit plantings, mainly apples, has been steadily increasing; in the early post-1952 years, newcomers needed returns from other, annual, crops to pay their way while the tree fruits were reaching bearing age. On the other hand, sugar beets, which figured prominently in the early years, and as lately as 1976 occupied 39,000 acres (15,800 ha) disappeared entirely from later plantings, after the local sugar refinery closed in 1979. These and other fluctuations are recorded in Table 25-1. Beef, rather than dairy cattle, have been favored by project farmers; they are largely concentrated in feed lots within the

The Grand Coulee Dam, Washington State. On the left, the Columbia River is ponded back by the dam to form Lake Roosevelt. At the right, or southern, end of the dam are visible the pipes through which water is pumped to a canal, which in turn will carry it into Banks Lake (background right), the holding basin for the irrigated lands of the Columbia Basin Project. *U.S. Bureau of Reclamation*

Table 25-1. Columbia Basin Project: Crop Areas, Yields, and Value of Crops, 1967, 1981, and 1990

	Irrigated Acreage			Yield per Acre			Percentage of Total Value of All Crops		
	1967	1981	1990	1967	1981	1990	1967	1981	1990
Alfalfa	129,123	121,431	139,399	5.4 tons	5.7 tons	6.4 tons	19.1	16.2	16.3
Wheat	105,054	119,846	101,264	74 bu	94 bu	105 bu	13.4	15.6	19.1
Potatoes	49,191	33,330	51,504	359 cwt	456 cwt	474 cwt	24.7	22.3	25.1
Sugar beets	33,740	—	—	21.6 tons	—	—	14.0	—	—
Corn	6,486	53,238	46,514	110 bu	135 bu	137 bu	1.2	8.3	4.8
Barley	10,343	28,886	2,719	58 bu	91 bu	98 bu	0.8	3.0	0.2
Apples	1,597	5,655	25,793	5.2 tons	11.1 tons	11.6 tons	1.5	4.5	16.9
Cropland irrigated	433,056	517,941	529,685						

Source: Bureau of Reclamation, *Crop Production Report* (Ephrata, Washington, 1967, 1981, and 1990).

Note: 1 acre = 0.405 hectares.

project, for example at Quincy on its northwestern edge.[3]

In the Pacific Northwest, however, and especially on fertile, irrigated lands like those of this project, it is less a question of what farmers can grow than of what they can sell. The market factor dominates the life of the region: specifically, the need to market a large part of their produce outside the region must dominate all planning by farmers throughout the Northwest, and by farmers of the newly irrigated lands in particular. We shall shortly return to this problem.

Forests and Industries

Manufacturing in the Pacific Northwest is dominated by the forest-products industries (Fig. 25-3). Forestry supports thousands of sawmills, from portable one-man plants to giant combinations of mill, furniture factory, and by-products industries, which in one case at least have given rise to a whole town—Longview, Washington—of 31,000 inhabitants. It chokes the smaller harbors on the coast with timber and makes its presence felt even in the heart of so large a city as Portland, where the rafts of lumber are towed down the Willamette beneath the city's bridges. It is a major employer in every state. It provides bulk freight for the railways and part-time employment for farmers in remote valleys of the Cascades.

The Columbia Basin Project: Summer Falls. This picture is included to show the amount of water necessary to irrigate half a million acres (200,000 ha) of farmland, for the falls are simply a natural part of the route of the main supply canal for the project, as shown in Fig. 25.2. *U.S. Bureau of Reclamation*

[3]As can be seen from the figures so far quoted, in area, the Columbia Basin project is little more than half completed. The uncompleted half is to the east, and it is slightly more humid; it may be asked why no progress has been made in bringing this under irrigation in the past thirty years. The answer is certainly not shortage of water. Rather it is that there is no point in pressing forward with applications to Congress for construction funds unless there is a probability of a high take-up rate, by farmers in the "dry half," of the irrigation water offered. But there is no likelihood that any arrangement made to supply irrigation water to the new lands in the 1990s could possibly be so generous as that offered to the original project farmers in the 1950s. We should, therefore, have to visualize two sets of farmers on the same project with one set paying substantially more for water than the other—that, or else a situation in which those with low-cost water would have their contracts redrawn to pay far higher rates. Neither of these alternatives seems even a remote possibility. In addition, so long as acreage limitations on federal water supply exist, no matter how slackly applied, the big farmers of the dry half would have to anticipate a total transformation of both ownership and farming system if they joined the project. The only balancing incentive for this would be a few more bushels of wheat to the acre and a measure of added security in their operations. But they can do without irrigation, and they probably will.

The Pacific Northwest, with its massive firs and pines, is essentially a sawtimber producer, and its leading forest product has until now been basic lumber rather than pulp and paper. The last two or three decades have seen a great improvement in the technique of exploiting the forest resource, principally (and most necessarily) in the U.S. section of the region, where the smallholders of the forest often left a trail of desolation behind them after buying, logging off, and then abandoning the land. The practice of "tree farming" by the large firms has placed the industry on sounder footing; and in British Columbia, the provincial government exercises vigilant control. Nevertheless, the situation is serious in the best and most accessible timber areas; whatever the average cut for the region as a whole, these accessible lands continue to be worked at considerably above replacement rate. There are other hazards, too; near Tillamook, on the Oregon coast, an area of prime Douglas fir 300,000 acres (120,000 ha) in extent was

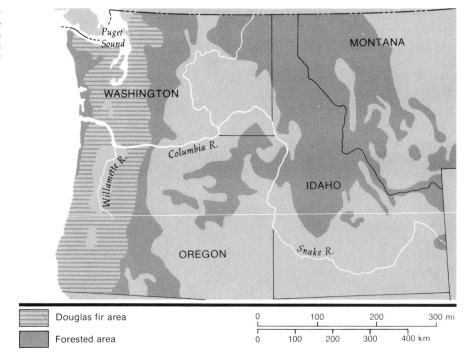

Fig. 25-3. Forests of the North-western United States. Notice the correspondence between this map of forests, the rainfall map (Fig. 25-1), and a relief map of the region.

Douglas fir area

Forested area

0   100   200   300 mi

0   100   200   300   400 km

wiped out by three separate fires between 1933 and 1945.

Over the years, the organization of the industry has altered considerably. There were once hundreds of small mills, individually operated and logging the forests more or less at random. Larger mills were to be found either at tidewater locations (such as Longview and on the eastern shore of Vancouver Island), or at natural assembly points on the river system (such as Lewiston or Coeur d'Alene in Idaho). More recently, however, several changes have taken place:

1. *The forest products industry has seen a decline in the number of small operators,* and a trend toward larger units and more systematic cutting.

2. *A pulp and paper industry has developed.* This development brings a greater degree of sophistication to the forest products industry: in economic terms, it increases the value added by manufacture over what is earned within the province or state by producing raw lumber alone. But its growth depends on demand in distant markets, and on the reliability of supply of the raw material. In the Northwest, there is no shortage of trees; what has to be ensured is their accessibility to the forester and, if capital is to be invested in a paper mill, that accessibility to supplies

must be assured for several decades ahead. With improvement of communications, especially by road, the forest area upon which mills can draw has been enlarged to the point where the investment becomes safe.

3. *There has been an increasing integration within the industry* between saw mills, pulp manufacture, and paper mills. Economically, this has been welcome, because it has led to a fuller utilization of the tree when it is felled. Geographically, it has led to the further concentration of the industry at a few major centers, where lumber, plywood, pulp, paper, chipboard, and other by-products are all produced together.

With these changes in organization, there has been a shift in the industry's location from the coast to the interior. In British Columbia, for example, the hard-worked—because easily accessible—forests of the coast no longer bear the brunt of the felling: in 1986, for example, 66 percent of the province's log production and 72 percent of its lumber output came from interior forest areas.

By the mid-1980s, there were twenty-four pulp and paper mills in British Columbia, 85 percent of whose production was sold outside the province. The forest products industry gave employment to around

Northwestern forestry: Forest products industries in the Pacific Northwest have given rise to entire towns such as Longview, Washington, which now has a population of 30,000 and lies on the north shore of the Columbia River estuary. *Weyerhaeuser Company*

75,000 workers, and accounted for 45 percent of the province's shipments of manufactured goods.

Two other main categories of industry are to be found in the Pacific Northwest. One of these is food processing, which deals with the output of the salmon fisheries and the region's vegetable crops. The other category is more varied and consists of industries attracted by the Northwest's electric power. Although the Northwest is rich in such minerals as silver, lead, and copper, it has had little coal[4] and imports petroleum, so that industry depends heavily on waterpower, the region's main industrial asset. In these circumstances, it is not surprising to find that one of the industries that most firmly established it-

[4]The past tense here is awkward but necessary: the Pacific Northwest was a region without coal until after World War II, but since then British Columbia has become a major producer (25 million tons in 1989).

British Columbia: The Alcan aluminum works at Kitimat. Kitimat is 136 mi (218 km) by road southeast of Prince Rupert, and is located on a coastal inlet, Douglas Channel. Alumina (aluminum oxide refined from bauxite, or aluminum ore) arrives by sea from Australia or Japan; power is supplied by the hydroelectric plant at Kemano, 40 mi (70 km) farther south, which is backed by a still-extending system of dams and reservoirs among the coastal mountains. *Alcan Smelters and Chemicals, Ltd.*

self here during World War II was light-metal smelting. Aluminum plants were established at numerous points, such as Troutdale, near Portland, and Spokane. Although the demand for these metals depended to a rather dangerous degree on defense expenditures, the industry weathered various recessions and, with the construction of the giant Kitimat smelter on the coast of British Columbia, the region reaffirmed its faith in it.

Apart from these regional specializations, the coastal cities possess a range of industries roughly appropriate to their size (see Table 25-2). Vancouver, the third city of Canada and the only large city on Canada's west coast, has grown impressively in the postwar years. Apart from its importance as a market and as a manufacturing center, it is steadily adding to its significance as Canada's "back door"; traffic to and from the interior, including the western Prairies, is routed via the west coast rather than the east. With the rapid rise of Canadian (not to speak of United

States) trade with Japan in the past ten years, Vancouver, in common with U.S. northwestern ports, has also now a much-enlarged role in trans-Pacific trade.[5] In a typical recent year, its port handled nearly 60 million tons of freight, of which more than 50 million tons were in foreign trade, particularly coal and wheat.

Seattle, on Puget Sound, dominates a coastal lowland that contains a number of smaller cities, and it has outstripped Tacoma, further up the Sound, which was the original Pacific coast terminal of the first transcontinental railroad to the Northwest. Seat-

[5]There are two aspects to the British Columbia–Asia connection—migration and trade. On the former, see Audrey Kobayashi, ed., "Focus: Asian Migration to Canada," *The Canadian Geographer*, vol. 32 (1988), 351–62; and D. Chuen Yan Lai, *Chinatowns: Towns Within Cities in Canada* (Vancouver, B.C., 1988). On the latter, see R. Blain and G. Norcliffe, "Japanese Investment in Canada and Canadian Exports to Japan, 1965–84," *The Canadian Geographer*, vol. 32 (1988), 141–50.

Table 25-2. Province of British Columbia: Leading Industrial Groups, by Employment, 1987

Industrial Group	Employment
1. Wood industries	42,425
2. Paper & allied products	17,662
3. Food industries	15,896
4. Fabricated metal products	9,688
5. Printing & publishing	9,303
6. Primary metal industries	6,446
7. Transportation equipment	5,888
8. Machinery industries	5,417
9. Electrical & electronic products	5,006
10. Nonmetallic mineral products	3,991
Total, provincial industrial workers	142,512

tle's industry is dominated by the Boeing aircraft company; even the lumber industries take second place here. Not that this is to Seattle's advantage, however: no major manufacturing center, not even San Diego, has experienced such ups and down of fortune and employment as has Seattle, tied to the fluctuating output levels of the aircraft industry. Over a few years in the 1960s, Boeing's employment in the Seattle area ran as high as 101,000 and as low as 53,000. In several years in the 1970s, Seattle had the highest unemployment rate of any metropolitan area in the nation—more than double the rate for the country as a whole. Fortunately it has its service, administrative, and port functions to fall back upon, and thanks to its location and to the strategic importance of the Northlands today, it has thrived on its connection with Alaska.

The advantages enjoyed by the port of Seattle on the seaward side, however, are balanced by the limitations of its hinterland. In its rivalry with Portland for the industry and commerce of the Northwest, it suffers a severe handicap: it possesses no Columbia Gorge. Portland stands at the seaward end of the single water-level route from the interior, whereas east of Seattle, winter snow may pile to rooftop height in the passes. The Columbia Gorge has been of inestimable value to Portland, and even those railroads that make for Puget Sound prudently provide themselves with an alternative route down the Columbia. Portland, with the Inland Empire at its back and the fertile Willamette Valley on its flank, has great commercial importance within the region. Its industries, although concentrated on lumber, pulp, and paper, also include shipbuilding, machinery, and textiles.

Basic Problems of the Pacific Northwest

Such is the geography of this formerly isolated region beyond the mountains, a region in which frontiers of settlement are still advancing. But isolation poses economic problems, and without a consideration of those we can have little understanding of how this region fits into the geography of the continent as a whole.

There are two main economic problems. The first is that, at least until World War II, the Pacific Northwest possessed only "colonial" status in relation to the East. It provided a vivid example of the way in which—to use again the terms we have introduced—the Periphery stood or stands in a dependent relationship to the economic core. The marks of this status were (1) that the region was a primary producer whose resources were largely exploited by outside financial interests and (2) that the "balance of trade" of the region showed a large outflow of bulky raw materials and an inflow of manufactured goods of higher value, but smaller bulk. Further, the value of the region's total "exports" was greater than its material "imports," the balance being made up by invisibles—services rendered by the East.

This "colonial" position, analogous to that of the former African colonies of the European powers, was well indicated by the balance of trade drawn up in 1942 by the regional planning commission covering the states of Washington, Oregon, Idaho, and Montana (Table 25-3). This survey was a one-off wartime effort, and it seems a great pity that no comparable studies have been made since then, for they would certainly tell an interesting story.

The figures tell their own tale: a huge volume of timber and agricultural produce left the region, while

Table 25-3. Balance of Trade of the Pacific Northwest States Annual Average, 1934–39

Commodity	Quantity Balance (1000s of Tons)	Value Balance (1000s of Dollars)
Products of Agriculture	+3600	+143,085
Animals and Products	+456	+93,366
Products of Mines	−6771	−69,630
Products of Forests	+8285	+178,318
Manufactures and Miscellaneous	−262	−221,798
	+5308	+123,341

Source: The External Trade of the Pacific Northwest, Pacific Northwest Regional Planning Commission, Portland, Oregon, 1942, pp. 23 and 25.

Grain harvest in the rolling Palouse country of eastern Washington State. *Washington State University*

a small quantity of manufactured goods entering it more than balanced, in value, 8 million tons of forest products. At the same time, had the prewar Northwest not been deficient in power and fuel (the "Products of Mines" received were largely petroleum), the "export surplus" of the region would have been much larger.

The war and the postwar increase in the Pacific Northwest's population altered this situation, to a certain extent, by stimulating local industries and enlarging the local market. When a region exports its raw materials to have them processed elsewhere, then the increase in value that manufacture brings is lost to that region. It is therefore possible to judge the economic progress of a "colonial" area, whether in Africa or North America, by the extent to which it is processing its own raw materials. By this criterion, the Pacific Northwest has made real progress; its manufacturing is constantly increasing in range and size. Perhaps the best single indication of this, as we have already noted in the case of British Columbia, is the growth of a northwestern pulp and paper industry alongside the old forest-products industry, which produced sawtimber and plywood, commodities with only a low value added by manufacture. The greater the degree of processing, the larger the increment of value obtained by the region from its resources.

However, the Northwest has not been so favored

in this respect as some other regions. The basic need of any area of the periphery is likely to be capital, and capital is much more forthcoming in an area with a fast-yielding resource to exploit—oil is the obvious case—than with the much slower-yielding northwestern resources of forests and fisheries.

The second problem of the Pacific Northwest is simply that of distance from markets. It is a problem that hangs over the region's agriculture and forestry. We have already noted that a very high proportion of the output of some agricultural items must be sold outside the region: this is inevitable because they are specialty crops produced for a national market. But to a considerable degree, the same is true of such a basic product as wheat, because of the limited market afforded by the 10 million people who live within the region. And what is true of agriculture is much more true of forest products.

Like California, the Northwest must sell in distant markets. But unlike California, the region has additional handicaps. California's produce at least has the advantage that much of it consists of crops for which the state is the only, or the major, producer in North America, so that it has a certain rarity value. But the Pacific Northwest, less favored climatically, is seeking to market produce—wheat, timber, fruit—in regions that produce, or could produce, these commodities themselves. To market their products, northwestern farmers must carry them past the doors of their competitors to reach the East. As an Oregonian once put it, "It's always bad business to carry wheat through wheat country."

Two examples may be briefly mentioned: apples and timber. British Columbia produces 40 percent of the Canadian apple crop and Washington some 20 percent of that of the United States. Growers have secured markets for these apples, in spite of distance, by attention to appearance, packaging, and flavor. But such advantages are hard to maintain and are ill suited to hold off the challenge of growers in Ontario, Michigan, or New York, 2000 mi (3200 km) nearer their markets. The result for Washington growers has been declining sales in the East,[6] and to add to their difficulties, changes in eating habits have over the years reduced the U.S. per capita consumption of apples. And as if those were not enough in the way of problems, Table 25-2 reveals that the irrigated orchards on the Columbia Basin Project lands are only now coming into full apple production, thus adding to the marketing difficulty.

In the case of forest products, the problem is the same; only the competitor is different. Washington and Oregon have to compete in the main U.S. markets with the softwood producers of the southeastern states. To the main selling areas, the southern wood has to travel only one-half or one-third as far as that from the Pacific Northwest, some of which, in fact, moves via the Panama Canal to the East Coast.

In this situation, it is not distance alone that tells against the Pacific Northwest, and it is certainly not lack of transport facilities, for, besides coastal shipping lines, no less than six transcontinental railroads were built through the area—more indeed than the traffic of the region made strictly necessary. In describing this region earlier as "formerly isolated," we focus attention on the extent to which its improving communications—by air and by express highway—have reduced that isolation, and brought the rest of the continent effectively closer.

There is also the problem of freight rates. Since, as we saw in Chapter 6, these are not based on distance alone, there is always the possibility that an adjustment of a cent or two will open or close a market to the Northwest. (This happened several times with the market for Washington grain in the southeastern states of the United States.) And the general level of

freight rates is also affected by the fact that a region that ships out very large quantities of lumber and wheat, while importing manufactures and petroleum, cannot use the same trucks or pipelines in both directions. It is a region with a large number of expensive "returned empties."

The problem for British Columbia is of the same kind, but with an extra dimension. As Table 25-4 shows, more than half of the province's various forest products, in most years, are sold in the United States; that is, in a foreign market that has its own dynamics and its own regulations. It is widely recognized that the market for British Columbia building materials depends on the annual number of housing starts in the United States—a number over which, needless to say, no Canadian has any control whatever. In years when the market is depressed,

Table 25-4. British Columbia: Major Export Markets for Forest Products, 1982–85 (in millions of Canadian dollars)

Export Item	1982	1983	1984	1985
Lumber				
U.S.	1059	1847	1881	2115
Japan	385	342	351	398
U.K.	135	140	133	94
Other E.E.C.	110	99	86	76
Rest of World	157	204	217	166
Subtotal	1846	2632	2668	2848
Pulp				
U.S.	404	376	468	458
Japan	290	274	328	317
U.K.	77	57	79	64
Other E.E.C.	518	435	533	422
Rest of World	230	285	354	334
Subtotal	1520	1427	1762	1595
Newsprint				
U.S.	457	483	520	670
Japan	2	4	5	4
U.K.	6	6	11	16
Other E.E.C.	1	3	3	5
Rest of World	100	82	102	137
Subtotal	567	578	641	832
All Other Forest-Based Products				
U.S.	322	470	526	596
Japan	126	159	224	195
U.K.	76	91	84	80
Other E.E.C.	83	87	77	74
Rest of World	62	78	128	116
Subtotal	669	885	1039	1060
All Forest-Based Products				
U.S.	2242	3177	3395	3839
Japan	803	779	908	913
U.K.	294	294	307	254
Other E.E.C.	713	623	699	577
Rest of World	552	649	801	752
	4603	5522	6110	6334

[6]The importance of holding on to these eastern markets is indicated by the fact that the northwestern states count on selling nearly 40 percent of their fruit in the New England and Middle Atlantic states.

American producers are sure to call for controls on the import of foreign building materials. This is one problem, at least, that Washington and Oregon are spared.

What can be done to improve this situation? There are three possibilities:

1. *Achieve a more favorable level of freight rates.* The Pacific Northwest is a region whose producers, businessmen, and newspapers are all very "freight-rate conscious." But in this respect, the region has been well served by its transport companies, for they realize that there will be no freight for them to carry unless freight rates can be held down, and they have cooperated with the shippers as fully as they could:

The Pacific Northwest, as a region, is fortunate in that its environmental surroundings and its economic base are such that its economic interests coincide with those of its serving railroads. Thus the region's geographical "disadvantage" of a location remote from many of the nation's principal consuming centers is minimized.... Tapering and blanket structures and holddowns keep long-haul rates on basic raw materials and some of their processed derivatives at levels which permit the region's goods to be sold in distant markets despite varying elasticities and inelasticities of demand, the availability of substitute products, or the existence of competing sources of supply.[7]

2. *Market new products.* When the Columbia Basin Project was created, it seemed all too likely that the cultivation of a million new acres (400,000 ha) of irrigated farmland would merely add to the region's market problems by increasing its surpluses. But whatever criticism may reasonably be leveled at the project (as, for example, that the federal subsidy to its farmers is too large or that its farm units are too small), it cannot be said to have failed on the marketing side, and the reason for its success is the point to notice. We have already seen that two of the project's main crops are alfalfa and potatoes. Both are bulky products with a low value-weight ratio; both have experienced marketing difficulties in the past; in addition, potato consumption per capita in North America has been static or declining for years. Yet both crops have been successfully marketed by a simple conversion process: the alfalfa is converted into pellets (and travels in this form as far as Japan), and the potatoes are transformed into frozen french fries. The result is, in effect, a new product from an old crop.

Indeed, it could almost be argued—although it would be unfair to press the point—that the officials who planned this great project in the 1940s have been saved from the embarrassment of unsaleable crops by the invention of products that did not even exist when their original calculations were made. The markets for these new products are highly expansible, and they typify the kind of sensible adaptation to local conditions a remote region needs to make in the competitive agriculture of today.

3. *Find new markets.* In spite of its diminishing isolation within the continent, the Pacific Northwest will always need a long reach if it is to sell its products within the domestic market. Between this region and the main centers of consumption in eastern North America lie the great spaces of the Plains and the Prairies.

It had always been hoped that this disadvantage would one day be balanced by the opening of markets overseas, to which access would be cheap and easy. The Northern Pacific, the first railroad to enter the region from the east, adopted as its badge a Chinese symbol, as a kind of declaration that the Far East was its true goal.

Yet for three-quarters of a century, that symbol mocked its users: the Asian markets proved illusory or, at best, unreliable. World War II, which left Japan prostrate and was shortly followed by the closure of China behind Communist doors, only served to set back to an even more remote time the day when the expectation of markets in Asia could realistically be entertained.

Now, however, the dream is a reality: North America's trade with the Far East is booming. Japanese imports—cars and electronic equipment—must be off-loaded somewhere, and the ports of the Northwest are as good a place as any. More to the point, the Asian nations have developed a voracious appetite for a whole range of primary products—coal from western Canada, alfalfa and wheat from Washington and Oregon. It was during the 1960s that this change made itself felt; wheat exports from Washington and Oregon doubled, and virtually every bushel of them was destined for Asian markets. Now at last the Pacific Northwest is selling products for which the market direction is reversed: instead of having to "ship wheat through wheat country" to the east, it is at the front of the line for shippers moving goods west.

The case of British Columbia is, in some respects, special. The main agricultural areas of western Can-

[7]R. J. Sampson, *Railroad Shipments and Rates from the Pacific Northwest* (Eugene, Oreg., 1961), pp. 59, 60.

ada are east of the Rockies, and farmland in the province is limited by relief and aridity. It is not so much domestic agricultural output, therefore, that British Columbia is shipping, but rather Prairie produce that has a genuine choice of route, to the west coast or to the east. It is a case in which the forces acting within the gravity model depend neither on population nor on output, but on freight rates and ease of access to the east and west coast ports.

Consequently, we can recognize an economic continental divide, from which goods, like water, flow east and west. This divide was for a long time close to the Pacific coast. The hinterland of British Columbia's ports was limited by physical barriers to movement and the small number of transport routes through the mountains. British Columbia's own northeast, the Peace River Country, had easier access to Edmonton than to Vancouver, and the valleys of the south to Spokane.

This situation has now changed, thanks largely to the improvement of transport routes and some vigorous campaigning about freight rates. The Trans-Canada Highway has narrowed the mountain belt; the Hart Highway links Vancouver with the Peace, and the Pacific Great Eastern Railway, which for years was a music-hall joke of a line, starting nowhere and ending in the wilderness, has been renamed (British Columbia Railroads) and completed from Vancouver to Dawson Creek. The Prairie oil fields and gas supplies are linked by pipeline not only to the Canadian east, but also to the Pacific Coast.

With these changes, the "divide" now lies well east of the mountains, and it is to the British Columbia ports that coal and wheat from Alberta move, as well as from the province's own mines and fields.

It seems likely that, if British Columbia continues to press its advantage as Canada's western door on the world, this process may go further: in the twenty-first century the "divide" may lie still further east. At present, British Columbia handles about 20 percent of Canada's exports, and 10 percent of its imports. About two-thirds of this traffic is moving to or from the United States and Japan. With demand increasing in Asia, and population increasing on the West Coast of the United States, those figures of 20 and 10 percent look low: the future can surely only see them increase.

Let us now return, at the end of this chapter, to consider once more the theme of core and periphery that has underlain our treatment of the regions of this continent. In the historic pattern of those regions, the Pacific Coast has belonged clearly to the periphery; the core region has always been in the East. But in Chapter 23, we considered the change of balance between East and West and, in particular, the focusing of power and the concentration of control functions in the West, specifically, in central and southern California.

What has developed there is, for the present, a secondary core. It is a compound of natural resources, resultant wealth, and population growth, and it remains a question whether, in the twenty-first century, the secondary core may become the primary and, in ways not yet evident to us, displace the historic core.

That remains to be seen. What concerns us here is the role of the Northwest in this growth of influence of the Pacific shores. We have already noted the gradual growth in maturity of the region's economy, although we must recognize that, in the U.S. section, at least, there is for the present no question of a challenge to the core, either eastern or western.

In Canada, the situation has developed rather further. What we have to consider there is a combination of forces: Alberta, with its mineral wealth, and British Columbia, with its increasing population and westward connections, as well as its minerals and power. Neither of these provinces individually poses a serious present threat to the primacy of the old Québec–Ontario core, but together they are formidable, and if they choose to act together under the single banner of "Western" interests, they are a force to be reckoned with, either at home in the West or in Ottawa.

South of the international boundary, the states of Washington and Oregon have still some distance to go before they are free from their old colonial status, Boeing or no Boeing. As they develop away from their subservience to the East, we cannot fail to notice that one area where they can find markets for their produce and, if they need to, raise capital is California. The 30 million people in that state, 90 percent of them urban, certainly provide market opportunities for what is still predominantly a region of primary production.

This leads us to a final comment. There are two alternative futures for the U.S. Northwest that we can, at present, envision. One of them is that it becomes a dependency, not now of the old core but of the new; that it is regarded by and, if they are given the opportunity, treated by Californians as an extension of their own state, well stocked with resources and providing plenty of overflow room for a population that is now crowding itself out of its living space. Certainly California covets the Northwest's water, flowing extravagantly to the sea through the Columbia–Snake system. It needs the Northwest's grain, being itself a deficit area, and it can put to good use Northwestern forest products. In this scenario, the Northwest region serves the Pacific Coast as storehouse and spare room, in continuing dependency. Was it to this future that its author was objecting when he or she created the immortal bumper-sticker: *Don't Californicate Oregon?*

The second alternative emerges from the remarkable efforts made in the 1970s by the environmentalist interests in these states. In a wave of legislation, they sought to create an "Ecotopia" in the Northwest. In other words, they asserted their freedom to be different; to surrender some of the short-term benefits of quick development in favor of slower, more controlled growth. In one sense, they were deliberately choosing to place themselves on the periphery. In another, they were postponing their own economic coming of age, in a nation where to be first and fastest has always been accepted as the highest good. Whether such eccentricity can survive we have yet to see.

In a region that offers few of the major incentives to modern American migrants—great mineral wealth, expanding job markets, unlimited sunshine—they have decided to be different: to remain much as they are. Free from major regional problems—except when a Mount St. Helens blows its top off—they would prefer not to import any from outside. This may be an unusual form of wisdom, but let no one call it folly.

INDEX

acid rain, 110, 157
"actual cultivators," 97, 118, 134, 148
Adirondack Mts., 9, 16, 160, 219, 236, 237, 242
aerospace/aircraft industries, 120, 155, 174, 198, 260, 306, 329, 365, 431, 489, 495, 517
African-Americans, 49–54; distribution, 50–53, 76–77, 226–27, 229, 232, 248, 298, 302–3, 317, 319, 320, 325, 326–27, 348, 360, 432; economic status of, 50–51, 54, 77, 170, 232, 313, 321–22, 325, 327–28, 334, social characteristics of, 48, 49, 52–53, 76–77, 197, 226–27, 298, 319, 322, 326. *See also* ghetto; Harlem; slavery
agribusiness, 141, 356
Agricultural and Rural Development Act (Canada), 201
airlines, 116, 179, 190–93, 490
Akron, Ohio, 288, 289
Alabama: agriculture, 314, 318–19, 323, 324; forests, 317; mining in, 156; transport in, 180. *See also* South
Alaska, 469–73; earthquake, 462, 469, 471, 472; fisheries, 469–70; forests, 34, 469; geographical position, 84, 454, 469, 472; history and settlement, 101, 461, 465, 469; petroleum in, 91, 122, 162, 165, 364, 399, 462, 471; physical features, 12, 15, 21, 469; population, 57, 458, 469; transport in, 193, 462. *See also* North
Alaska Highway, 14, 465
Albany, N.Y., 221, 227
Alberta: farming and ranching in, 397, 459; forests, 34; geographical position, 210; history and settlement, 49, 460, 461; industry, 155, 170; mining in, 158, 408; petroleum in, 110, 124, 163, 165, 166, 399; population, 395; water resources, 141. *See also* Prairie Provinces
Albuquerque, N.M., 67, 408, 416, 430, 431
alfalfa, 390, 410, 486, 498, 511, 520
Allegheny Front, Plateau, 9, 228, 336
aluminum, 260–61, 296, 329, 464, 516
Amtrak, 177, 186
Anaheim, Calif., 493, 494
Appalachia, 123, 171, 202, 204, 212, 215, 335–51, 452, 468

Appalachian coalfield, 9, 156, 157, 158, 159, 160, 167, 219, 265, 288, 291, 292, 317, 338, 343, 346, 348, 350
Appalachian Mts., 6, 9, 11, 25, 35, 134, 213, 214, 215, 219, 273, 283, 316, 336, 439
Appalachian Regional Development Act, Commission, 201, 336, 340, 341, 344–45, 348–49, 350
Arctic, 5, 22, 25, 34, 122, 165, 372, 455, 456, 458, 463
Arizona: climate, 133, 414, 431; farming and ranching in, 428; forests, 412; history and settlement, 426, 427; Indian reservations in, 422, 435; industry, 155, 287, 431; mining in, 406–7, 428; physical features, 14, 15, 424; population, 57, 60, 428; soils and vegetation, 38, 40; water resources, 108, 428–30, 484
Arkansas, 25. *See also* South
Arkansas River, 180
Aroostook, 241–42, 440, 442, 451
asbestos, 237, 258
Athabasca tar and oil, sands, 165, 166, 167, 401, 462
Atlanta, Ga.: industry, 329, 334; population, 68, 329; role and services, 210, 330–31; transport role, 187, 191, 192, 331
Atlantic City, N.J., 232, 233
automobile industry, 155, 161, 171, 172–73, 198, 260, 265, 288, 296, 297–98, 306, 329, 490
azonal soils, 30, 31

Baltimore, Md., 229–31; industry, 153, 154, 160, 219, 230; population, 221; port, 164, 184, 207, 222, 229–31; role and services, 221, 223, 229, 230; site and settlement, 229, 231; transport in, 184, 222, 229–30
Banff, Alta., 105, 414, 415
barbed wire, 381
Baton Rouge, La., 184, 361, 365
bauxite, 261, 317, 329, 363
Beaufort Sea, 167, 463
Birmingham, Ala., 156, 160, 329–30
Black Americans. *See* African-Americans
Black Hills, 12
Bluegrass Region (Ky.), 11, 271, 284, 317, 337

Blue Ridge, 8, 212, 213, 337
Boston, Mass.: industry, 173, 228, 244, 246, 248; population, 242; port, 184, 248–49; role and services, 70, 80, 222 n.8, 248–49; site and settlement, 70, 71, 222, 248; transport in, 72, 186, 248
Bridge subsidy (Canada), 116
British Columbia: climate, 21, 25, 404; farming and ranching in, 459, 506, 508–13; fisheries, 507–8; forests, 34, 110, 412, 413, 513–15; geographical position, 196, 521; industry, 413, 515, 516, 518; markets, search for, 131, 399, 518–21; mining in, 158, 399, 405, 406, 407, 408, 515; physical features, 15, 22, 506; population, 56, 197; transport in, 186, 521; water power, 167, 507, 515
Buffalo, N.Y., 68, 69, 155, 160, 223, 229, 264, 296
Butte, Mont., 405, 407, 416

Calgary, Alta., 68, 155, 208, 393, 418
California, 475–99; climate, 25, 27, 60, 133, 414, 477; farming and ranching in, 131, 479, 480–87; field labor force, 479, 482–83; forests, 34, 488, 489; freight rates, 115, 131; history and settlement, 44, 60, 94, 426, 476–77, 478–79; industry, 73, 119, 120, 123, 151, 154, 165, 173, 198, 246, 287, 480, 489–90; military reservations, 489; mining in, 379, 405, 407, 426, 476, 479; petroleum in, 161, 162, 364, 490; physical features, 14, 506; planning in, 92; population, 57, 60, 68–69, 200, 210, 479, 480, 489, 490; population, Hispanic, 54, 426, 480, 483; position, as core area, 210; soils and vegetation, 477, 478; tourism, 488; water projects, 484, 499; water resources, 106, 108, 476, 478, 487, 488, 497–99. *See also* Central Valley, Calif.
Canadian Transportation Commission, 114–16
Canadian Wheat Board, 118, 119
Cape Cod, 6, 236
Carolina, North *and* South: agriculture, 317, 323, 325, 332; climate, 60; economic fortunes of, 210, 331; forests, 317; history and settlement, 317; in-

523

Carolina, North *and* South (*continued*)
dustry, 153, 173, 330, 332; physical
features, 316; population, 327. *See*
also South
Cascade Range, 15, 510
Catskill Mts., 9, 219
"cattle kingdom," 380, 428
center of population (U.S.), 60 n.11
Central Arizona Project, 430, 484
central business district (CBD), 73, 74,
81, 82, 225, 227, 296, 298, 299, 300
central place theory, 65, 67
Central Valley, Calif., 16, 25, 39, 106,
129, 130, 476, 477, 483–86, 493
Champlain, Lake, Lowland, 9, 236, 240,
253, 278
Channeled Scablands, 510
Charlotte, N.C. *See* Metrolina
Chesapeake Bay, 6, 165, 212, 213, 229,
232, 317
Chicago, 298–303; industry, 155, 160,
172, 173, 274, 287, 288, 299, 300;
population, 67–68, 299; population,
African-American, 51, 302–3; population, Hispanic, 303; port, 223, 261,
300; role and services, 67, 181, 187,
192, 196, 207, 265, 299, 300; site and
settlement, 65, 72, 287, 291, 299, 300;
transport in, 180, 187, 191, 299, 300,
301
Chicano, 433
Chinook, 375
Christaller, W., 303, 304
Cincinnati, Ohio, 64, 65, 172, 273, 287,
305
Civil War (1861–65), 46, 50, 84, 88, 154,
199, 208, 229, 312, 313, 362
Clay Belts, 56, 138, 458, 461
Clean Air Act (1970), 112, 157
Cleveland, Ohio, 65, 69, 155, 160, 264,
296–97
coal, 43, 91, 121, 123, 124, 125, 155,
156–59, 160, 165, 179, 181, 184, 219,
229, 244, 258, 265, 285, 288, 292,
295, 296, 313, 329, 338–39, 343, 346,
393, 399, 408, 428, 446, 462, 515
Coal Mine Health and Safety Act
(1969), 157
Coast Ranges, 16, 477, 486, 499
Colorado: climate, 197; farming and
ranching in, 140, 390; geographical
position, 200; history and settlement,
42; industry, 160; mining in, 122, 405,
406, 407, 408; petroleum in, 122, 165;
physical features, 200; population,
197, 416; water resources, 390
Colorado Plateau, 14, 404, 410, 425
Colorado River, Basin, 14, 106, 107, 233,
390, 412, 429, 430, 484, 486, 497, 499
Colorado River Compact, 429, 497
Columbia Basin Project, 106, 168, 410,
511–13, 519, 520
Columbia River Basin, 12, 14, 15, 21,
106, 107, 108, 167, 184, 408, 464,
499, 506, 507, 508, 510
Commodity Credit Corporation, 281–82
Connecticut. *See* New England

Connecticut River, Valley, 9, 233, 236,
247
conservation, definition of, 108
container traffic, 184–85, 225, 226, 247
copper, 282, 405, 406, 428, 446, 447,
462
Core, economic: Canada, 209, 210, 252,
253, 263, 266, 354, 438, 439, 441,
446, 454; U.S., 207–8, 209, 210, 212,
219, 221, 236, 270, 288, 319, 336,
354, 368, 439, 476, 521
core-fringe, core-periphery, 207–10,
270, 369, 448, 454, 476, 517, 521
corn, 128, 132, 136, 139, 214, 240, 272,
273, 274, 275, 276, 278, 281, 312,
314, 511
Corn Belt, 25, 128, 131, 132, 139, 148,
269, 272–77, 279, 280, 281, 283, 284,
291, 298, 308, 378, 390, 392
cotton, 50, 117, 131, 132, 136, 172, 312,
314, 318, 319, 320, 322, 323, 332,
341, 358, 362, 390, 428–29, 476, 479,
481, 486
Cotton Belt, 128, 131, 132, 133, 362
counter-urbanization, 57
Cropland Reserve Program, 118
Crow's Nest Pass Agreement, 115
Cumberland Plateau, 9, 336

dairy farming, dairy regions, 131, 139,
213, 215, 216, 240, 258–64, 274, 278,
279, 281, 283, 298, 324, 443, 486,
503, 508
Dakota, North *and* South: climate, 372;
farming and ranching in, 129, 130;
mining in, 407; population, 57. *See*
also Great Plains
Dallas, Tex.: industry, 155, 173; population, 328; role and services, 209, 306,
368; site and settlement, 366; transport in, 191, 368
Death Valley, 14, 15, 25, 68, 403, 404,
414, 476
Delaware, 214, 220
Delaware Bay, River, 107, 160, 212, 213,
221, 228, 232, 233
Delmarva Peninsula, 214, 325
Denver, Colo.: industry, 393, 417; population, 77, 393; role and services, 67,
199, 210, 390, 393, 408, 416, 417; site
and settlement, 397; transport role,
417
Department of Regional Economic Expansion (DREE), 200, 449
deregulation, of oil and industry, 124; of
transport, 116, 186
Des Moines, Iowa, 175, 308
Detroit, Mich.: industry, 160, 171, 173,
297–98; population, 69, 298; population, African-American, 298; site and
settlement, 65, 271, 297
District of Columbia. *See* Washington,
D.C.
drift, glacial, 17–18, 20, 236, 273, 282,
283
Driftless Area, 20–21

dry farming, 381, 384, 405, 410, 428,
486
Duluth, Minn., 261, 292
Dustbowl. *See* Great Plains
dust mulch, 381, 382

Eastern Townships, 56, 257, 258
Economic Development Administration,
201
"Ecotopia," 451
Edmonton, Alta., 68, 163, 164, 167, 393,
399, 418
Ellis Island, 81, 224, 480
El Paso, Tex., 408, 416, 428, 430, 431,
432
energy policy, 121–23, 124, 165, 166
Environmental Control Act (Vermont),
111
Environmental Impact Statement, 90,
111, 112
Environmental Land and Water Man-
agement Act (Fla.), 111
Environmental Protection Ordinance
(1974), 111
Environmental Quality Improvement
Act (1970), 111
Erie Canal, 129, 134, 178, 184, 198, 221,
222, 227, 228, 239, 271, 295
Erie, Lake, 110, 111, 216. *See also* Great
Lakes
Eskimos. *See* Inuit
Everglades, 6, 133, 355, 356, 357, 359

Fall Line, 6, 212, 213, 214, 232 n.17,
258, 334
family farms, 118, 144, 148–49, 272,
276, 277, 278, 319, 321, 358, 449,
481, 482
farm inputs and outputs, 133–35, 142,
145
farm population, 49, 56, 57, 144, 309,
328, 385, 503
farm regions, 145–46
farm size, 145–46
Feed Freight Assistance (Canada), 115–
16
fertilizers, 110, 135, 137–38, 145, 214,
291, 319, 322, 329, 342, 364, 408
Florida: agriculture, 115, 147, 355, 356;
climate, 24, 25, 354, 356, 357, 359,
414, 431; forests, 354, 363; geographical position, 67, 354, 365; history and
settlement, 359, 423; industry in, 154,
173; mining in, 364; physical features,
6; population, 54, 57, 76, 197, 327,
354, 359; tourism in, 358–59; water,
conflicts of interest, 357, 359. *See also*
Gulf Coast
Foreign Investment Review Agency
(Canada), 174
forests, 33–37, 218, 239, 338, 354, 363,
404, 412, 444, 464–65, 470, 488, 502,
507, 513. *See also* National Forest
Service
forest product industries, 36–37, 155,
170, 174, 237, 260, 282, 332, 338,

363, 405, 412, 444, 464–65, 488, 513–15

Fort Worth, Tex., 155, 209, 366, 368

Fraser River, 12, 14, 15, 508, 509

freight rates, 114–16, 141, 186, 222, 229, 248, 296, 314, 344, 381, 481, 519, 520, 521

French Canadians, 45, 48–49, 76, 86–88, 252, 253–58, 261–63, 267, 427, 440, 458, 460

frontier hypothesis, 208–9

functional clustering, functional exclusion, 73

Fundy, Bay of, 9, 48, 439

fur trade, 254, 256, 270, 306, 378, 405, 508

Gaspe, 9, 87, 253, 438, 439, 447, 456

gentrification, 81

Georgia: agriculture, 326; economic fortunes of, 173, 334, 345; forests, 317; industry, 329, 330; population, 347. *See also* South

German immigrants, 44, 46, 212, 213, 280, 283, 440, 493 n.12

ghetto, 76–78, 227, 432. *See also* African-Americans; Harlem; Hispanic Americans

glaciation, glaciers, 9, 10, 11, 16–22, 33, 273, 278

gold, 106, 178, 236, 378, 379, 405, 407, 426, 427, 446, 462, 467, 476, 479, 490

Golden Crescent, 183, 365

Golden Sands, Wis., 140

Grand Canyon, 14, 105, 414, 416, 424, 430. *See also* Colorado River

Grand Coulee, 21, 167, 508, 511

Grapes. *See* vines, vineyards

grasslands, 33, 34, 37–39, 45, 129, 278, 354, 372, 378, 380, 381, 392, 404, 426, 510

gravity model, 304, 309, 399

grazing permits, 411, 412, 418

Great Basin, 14, 21, 22, 42, 404, 408, 410, 506

Great Lakes: cities of, 293–303; climate, 22, 23, 25, 30, 109, 280–81, 440; physical features, 19–20, 169, 291; pollution of, 108, 110–11, 291; ports, 168, 228, 288, 295, 296, 297; as routeway, 159, 178, 181, 223, 264, 265, 271, 290, 296; traffic on, 165, 180, 181, 183, 265, 285, 291–93, 294–95

Great Lakes Compact (1985), 108

"Great Lakes of the South," 344

Great Plains, 371–93; central places in, 384, 386–87, 392–93; climate, 25, 27, 132, 372, 373–78, 381 n.5; cultivation techniques on, 140, 381, 382, 384, 385, 395; dustbowls, 110, 111, 117, 382, 479; farming and ranching on, 148, 276, 380; geographical position, 397; history and settlement, 42, 372, 378–82, 384; physical features, 11–12; population, 56, 197, 385–86; soils and vegetation, 33, 38–39, 372, 378; water resources, 108, 379–80, 381, 385,

387–88, 390–91. *See also* Prairie Provinces

Great Salt Lake, 21

"green belt," 103

growth poles, 198, 201, 330, 331, 332, 345, 349

g-scale, 205 n.8

Gulf Coast: agriculture, 139, 356–58; climate, 22, 23, 24, 354; defense installations, 365; forests, 34, 354; geographical position, 67, 184, 361; history and settlement, 423; industry, 155, 160, 170, 183, 198; petroleum on, 155, 220, 354, 364; physical features, 6, 10, 11; population, 354, 359, 360, 361; soils and vegetation, 354; tourism, 358–9, transport on, 164, 183. *See also* Florida, Louisiana; Texas

Halifax, N.S., 56, 439, 440, 443, 448

Hamilton, Ont., 56, 160, 264, 265, 281, 295

Hampton Roads, 158, 159, 219, 222

Harlem, 52, 72, 226–27, 326

Hartford, Conn., 63, 175, 247

Hawaii, 92, 357, 501–4

high-tech industries, 73, 172, 173, 228, 231, 246, 248, 289, 300, 398, 490, 492

"hinge" cities, 209, 270, 306, 368, 392

Hispanic Americans, 48, 54–55, 70, 170, 232, 303, 360, 368, 431–32, 433. *See also* Spanish settlement in America

Homestead Act (Canada), 99, 148, 395

Homestead Act (U.S.), 97, 98, 148, 384

Honolulu, 502

Houston, Tex.: industry, 155, 183, 366, 367; population, 354, 366; port, 184, 361, 363, 366; role and services, 102–3, 365; site and settlement, 70, 366, 367; transport role, 183, 361

Hudson Bay, 9, 16, 19, 22, 33, 34, 258, 394, 397

Hudson River, Valley, 6, 9, 20, 129, 212, 213, 218, 221, 222, 223, 233

Hudson's Bay Co., 95, 395

Huron, Lake, 180, 279. *See also* Great Lakes

hurricanes, 23, 359, 360

hydroelectricity, 167–69, 201, 258, 265, 343, 350, 387, 490, 508

Idaho: agriculture, 410, 442; forests, 412; mining in, 407, 408; physical features, 12, 14, 506

Illinois: agriculture, 110, 129, 224, 275–76, 284; forests, 34; mining in, 156; physical features, 11. *See also* Corn Belt; Midwest

immigration: from Asia, 44, 48, 81, 480; to Canada, 44, 45, 49, 56, 57, 81, 265, 438, 440, 458; from Europe, 43–45, 60, 75, 106, 155, 184, 212, 226, 238, 243, 247, 271. *See also* German immigrants; Hispanic Americans; Irish immigrants; Mexico, immigration from; Scandanavian immigrants; Spanish

settlement in America; Ukrainian immigrants

Imperial County, Valley, 410, 482, 483, 486, 498

indenture, 50

Indiana: agriculture, 275–76; history and settlement, 97, 271, 272; mining in, 156, 157. *See also* Corn Belt; Midwest

Indianapolis, 67, 71, 308

Indians, American; Native Americans: continental, 42–43, 128, 197, 271; in Canada, 42, 43, 167, 455, 458, 467–69; in U.S., 14, 43, 95, 132, 170, 239, 380, 422–23, 425, 428, 431, 434, 472, 478

"Inland Empire," 506, 517

Interstate Commerce Commission (ICC), 114–16, 186, 187

Intracoastal Waterway, 180, 183, 363

intrazonal soils, 30

Inuit, 42, 455, 458, 467–69, 472

Iowa: agriculture, 110, 129, 130, 274–76; history and settlement, 45; population, 57. *See also* Corn Belt; Midwest

Irish immigrants, 81, 248, 440

iron ore, 156, 159–60, 161, 170, 180, 181, 193, 219, 228, 230, 259–60, 291, 292, 295, 296, 313, 329, 446, 462, 464

irrigation, 106, 127, 130, 132–33, 139–40, 142, 214, 274, 277, 322, 358, 373, 384, 387, 390, 391, 397, 410, 422, 425, 427, 428, 481, 483–86, 497–98, 507, 509, 511–13

Jasper, Alta., 105, 414, 415

Kanawha River, Valley, 155, 288, 289, 296

Kansas: agriculture, 129, 130, 274, 390; climate, 274, 372; industry, 155, 172; water resources, 140. *See also* Great Plains

Kansas City, Kans.–Mo.: industry, 287, 306–7; role and services, 209, 306, 307, 308, 368; site and settlement, 306, 307; transport role, 186, 305, 306, 307

Kentucky: agriculture, 284; coal in, 91, 156, 157, 158, 345, 348; geographical position, 312; history and settlement, 271, 339; physical features, 336; soils and vegetation, 33. *See also* Appalachia; Appalachian coalfield

Kentucky River, 184

Kitimat, B.C., 167, 464, 516

kudzu, 324 n.10

Labrador, 9, 16, 25, 33, 160, 161, 193, 198, 258–59, 260, 438, 446, 454, 462, 463–64

La Grande, River, P.Q., 167, 259, 464

Lancaster, Pa., 214, 216, 229

Land Management, U.S. Bureau of, 101, 117, 411

Laramide Revolution, 12

Las Vegas, Nev., 68, 414, 416, 480

Laurentide ice sheet, 16–21
lead, 405, 408, 446, 447, 462
Local Development District (LDD), 347–48
London, Ont., 56, 186, 265
Long Island, 19, 213, 214, 218, 223, 225, 236, 241
Los Angeles, Calif.: agriculture in lowlands, 356, 486; industry, 160, 162–63, 173, 490, 494; population, 44, 67–68, 480, 490, 493, 499; population, African-American, 499; population, Hispanic, 54, 423, 499; port, 184, 494; role and services, 67, 70, 78, 199, 494–95, 497; site and settlement, 14, 16, 70, 209, 479, 493, 495; transport role, 186, 189, 191
Louisiana: agriculture, 323, 355, 356; forests, 363; history and settlement, 84, 94; industry, 364; mining in, 364; petroleum in, 162, 364. See also Gulf Coast
Louisiana Purchase (1803), 94, 97, 222
Louisville, Ky., 305–6
Lowell, Mass., 80, 152 n.1, 246, 247

Mackenzie River, Valley, 11, 34, 399, 451, 458, 462, 464, 465
Maine: agriculture, 241, 242; forests, 239, 242, 444; industry, 246; physical features, 236, 439; population, 246. See also New England
Manitoba: agriculture, 397; history and settlement, 378; mining in, 462; soils and vegetation, 30. See also Prairie Provinces
Manufacturing Belt, 120, 153, 154, 155, 172, 173, 221, 230, 284, 285, 307, 337
maquiladora plants, 175, 436
Maritime Provinces, Canada, 437–52; agriculture, 441–44; climate, 25, 440; dual economy of, 449; fisheries, 439, 441; forests, 444–45; freight rates, 115, 196; geographical position, 55, 87, 253, 438–39, 447, 449, 452; history and settlement, 84, 257, 439, 440; industry, 160, 446, 448; mining in, 446–47; petroleum in, 124, 447; physical features, 6, 9, 236, 439; population, 48, 56, 200, 438, 447; soils and vegetation, 441; tourism in, 447, 451
Maryland: agriculture, 214; history and settlement, 317; physical features, 33; population, 232
Massachusetts: agriculture, 241; history and settlement, 50; industry, 244, 245, 246; industry, textile, 73. See also New England
Mead, Lake, 105
meat-packing industry, 172, 287, 300, 398
mechanization, farm, 135–37, 144–45, 319–20, 325, 381, 481
Megalopolis, 212, 221, 232–34, 236, 238, 242, 334, 359
Memphis, Tenn., 331
Merrimack River, Valley, 243, 244, 246

Mesabi Range, 159, 291, 295
Metis, 43, 467, 468
Metrolina, 210, 331, 332, 333
Mexico, frontier with U.S., 4, 54, 426, 427, 428, 435–36; immigration from, 54, 433, 434, 479, 480; relations with U.S., 4, 426, 435; U.S. industry in, 436; water disputes, 107, 428. See also Hispanic Americans; Spanish settlement in America
Miami, Fla.: population, 54, 358, 360; role and services, 359; site and settlement, 70, 358, 359; tourism in, 358–59; transport role, 191, 361
Michigan: agriculture, 242, 279, 280; industry, 155, 171, 172–73, 297; mining in, 160, 282, 406; physical features, 19, 33, 282. See also Great Lakes; Midwest
Michigan, Lake, 155, 279, 280, 291, 299, 300. See also Great Lakes
Midwest: agriculture, 132, 272–84; climate, 22–23, 272; definition of, 270; history and settlement, 45, 137, 147, 240, 263, 271, 303, 304; industry, 173, 284–90; physical features, 18–19, 272; soils and vegetation, 33, 273; urban hierarchy in, 276, 303–5, 306, 308
Milwaukee, Wisc., 290, 298
Minneapolis–St. Paul, Minn., 155, 180, 186, 209, 274, 306, 308
Minnesota: agriculture, 278; history and settlement, 45, physical features, 9. See also Great Lakes; Midwest
missions, Spanish, 64, 426, 427, 477, 478, 479, 483
Mississippi: agriculture, 314, 323, 324; forests, 317; population, 327. See also South
Mississippi River, 6, 20, 106, 159, 165, 178, 180, 254, 273, 299, 305, 316, 341, 361
Mississippi Valley, 6, 11, 31, 33, 137, 147, 148, 275, 316–17, 321, 323, 356
Missouri: agriculture, 274, 284; soils and vegetation, 33. See also Midwest
Missouri River, Valley, 16, 106, 180, 184, 273, 274, 305, 387–88
Mohawk Gap, River, Valley, 215, 218, 221, 227, 228, 265
molybdenum, 406, 407
Montana: farming and ranching in, 380; forests, 412; mining in, 90, 406; physical features, 12. See also Great Plains
Montreal, P.Q.: industry, 155, 164, 260; population, 48, 68; port, 164, 184, 207, 223, 261; rivalry with Toronto, 68, 252, 263, 266–67; role and services, 60, 222 n.8, 253, 261, 266, 278, 398; site and settlement, 71; transport role, 186, 187, 261

Nantucket, 19, 20, 236
Nashville Basin, 11, 317, 324
National Environmental Policy Act (1969), 111

National Forest Service, 101, 105, 110, 117, 410, 411, 412–13, 489
national parks, 101, 104, 110, 416, 418, 489
natural gas. See petroleum and natural gas
Nebraska: climate, 372; farming and ranching in, 140, 272, 274, 277, 278; physical features, 12, 31, 278; population, 309; water resources, 140, 391. See also Great Plains
"neoplantations," 320, 321, 330
Nevada: economy, 90, 92; history and settlement, 101; mining in, 405, 406, 407; physical features, 14; population, 57, 60
Newark, N.J., 225, 227
New Bedford, Mass., 237, 244, 247
New Brunswick. See Maritime Provinces, Canada
New England, 236–50; agriculture, 236, 239, 240–42, 325; climate, 240; fisheries, 236, 237; forests, 35, 236, 239, 240; geographical position, 196, 236, 238, 244; history and settlement, 42–43, 64, 65, 67, 95, 129, 152, 238, 239, 242, 247, 249–50, 257–58, 438; industry and research, 120, 152, 154, 238, 243–48, 285, 495; industry, textile, 73, 153, 154, 172, 219, 238, 244–45, 495; physical features, 6, 9, 16, 18, 236, 237, 239; population, 48, 239, 242, 246, 247, 257, 451; soils and vegetation, 33, water power, 243, 244
"new ethnicity," 48, 54
Newfoundland: agriculture, 443, 444; climate, 25; fisheries, 441, 449; forests, 444–45; geographical position, 55, 439, 447, 448; history and settlement, 440, 449; mining in, 446, 447; petroleum in, 124, 447; physical features, 9, 439. See also Maritime Provinces, Canada
New Hampshire: agriculture, 241; geographical position, 212; industry, 244, physical features, 9; population, 246. See also New England
New Haven, Conn., 247
New Jersey: agriculture, 214, 242; industry, 165, 220, 225; physical features, 213, 225, 234; ports, 225
New Mexico: agriculture and ranching in, 428; Indian reservations in, 422, 428; industry, 431; mining in, 406, 428; petroleum in, 428; physical features, 21
New Orleans, La., 361–62; port, 180, 184, 361, 365; role and services, 60; site and settlement, 271; transport role, 134, 362
New York City, 223–27; CBD, 225, 227; industry, 155, 173; population, 44, 67–68, 82; population, African-American, 226–27; population, Hispanic, 76, 227; port, 164, 184, 222, 223, 224–25; role and services, 60, 78, 221, 225, 228; site and settlement, 70, 71, 223–

24, 226; transport role, 187, 189, 191, 222, 223, 225

New York State: agriculture, 129, 213–14, 216, 218, 278; history and settlement, 218, 272; industry, 219, 220; physical features, 6, 9, 19, 219, 345; water resources, 234

Niagara Cuesta, Falls, 16, 265, 271, 296

Niagara fruit belt, 280, 281, 282

nickel, 462

nordicity, 85 n.3, 454

North Carolina. *See* Carolina, North *and* South

North, 453–73; agricultural possibilities, 257 n.4, 458–61; definition of, 454; forestry in, 464; future of, 85, 458–69; hydroelectricity in, 464; mining in, 462–64; physical conditions in, 455–58; tourism, 465–66; transport in, 193, 465–66

Northwest Territories (NWT), 56, 454, 467, 469

Nova Scotia. *See* Maritime Provinces, Canada

nuclear energy, 122, 169–70

Oakland, Calif. *See* San Francisco

"oak openings," 33

"oasis farming," 409–10, 428

Ohio: agriculture, 134, 283; history and settlement, 98, 129, 272; industry, 155, 159, 283, 288–90; physical features, 11, 95 n.1, 283, 336; urban growth in, 308. *See also* Great Lakes

Ohio River, Valley, 16, 155, 160, 178, 180, 183, 271, 273, 285, 289

oil. *See* petroleum and natural gas

Okanagan Valley, B.C., 281, 404, 507

Oklahoma: climate, 25, 372, 374; history and settlement, 386 n.12, 425; petroleum in, 162; population, 386, 480; soils and vegetation, 30. *See also* Great Plains

Olympic Mts., 16

Ontario: agriculture, 104, 134, 263, 278, 279, 280, 281; climate, 280; energy in, 167, 169; history and settlement, 129, 263, 280; industry in, 155, 264–65; mining in, 291, 408, 462; petroleum in, 124; physical features, 19, 279; population, 56–57, 200; soils and vegetation, 263, 279; transport in, 186. *See also* Great Lakes; Toronto

Ontario, Lake, 216, 265, 279

oranges, 115, 131, 356, 357, 358, 479, 486

Oregon: agriculture, 508–13; climate, 25, 509; fisheries, 507; forests, 36, 110, 513–15; history and settlement, 138, 197, 379, 508; industry, 37, 513–18; markets, search for, 518–21; physical features, 14, 15, 506; planning in, 92; population, 197; relations with California, 521–22; soils and vegetation, 39; water resources, 108, 507, 515. *See also* forest product industries

Oregon Trail, 14, 178, 305, 379, 506, 508

Ottawa, 68, 83, 88, 261, 400

Ouachita Mts., 11, 316, 317

Ozark Mts., 11, 284, 316, 317

Pacific Coast fruit belt, 510

Palliser's Triangle, 394, 395

Palouse, the, 129, 130, 510

paper. *See* forest product industries

Park Belt, 45, 394, 395, 397

Paterson, N.J., 227, 228

"Payment in Kind" (PIK), 118, 147

Peace River, 56, 199, 408, 454, 458, 459–61, 464, 521

pedalfer, pedocal (soils), 30

Pennsylvania: agriculture, 213–14, 216, 218, 278; history and settlement, 45, 213, 218; industry, 155, 159, 219, 283, 288–90; mining in, 45, 156, 229, 349; petroleum in, 161, 162; physical features, 8, 336; population, 218; water needs, 234

Pennsylvania Dutch, 214

permafrost, 456, 457

petroleum and natural gas: technology of extraction, 157, 158, 160, 161–67, 170, 220, 364–65, 406, 447, 462, 472; in Canada, 121, 124–25, 161–67, 174, 198, 258, 265, 399–400, 408, 447, 462; in U.S., 43, 116, 121, 124, 161–67, 184, 185, 198, 228, 229, 285, 317, 364, 367, 408, 462, 490

Philadelphia, 228–29; industry, 154, 219, 228–29; population, 221, 222 n.8, 229; port, 164, 184, 222, 225; role and services, 229; site and settlement, 70, 228, 229; transport role, 184, 222, 228

Phoenix, Ariz., 61, 199, 408, 429, 430, 431, 497

Piedmont, Appalachian, 6, 8, 129, 155, 172, 210, 213, 214, 215, 218, 229, 236, 316, 317, 323, 329, 334

Piedmont, Colo., 210, 287, 390, 397, 417

pine barrens, 133, 212, 232

pipelines, 116, 160, 164, 165, 167, 180, 220, 260, 408, 456, 462, 470, 472, 521

Pittsburgh, Pa.: industry, 153, 155, 160, 161, 219, 228, 286, 288–89, 338; population, 68; role and services, 64, 65, 215, 289; site and settlement, 271, 288, 289; transport role, 180, 288

"Pittsburgh Plus," 160–61, 286, 289, 290

placer mining, 106

"polaricity," 454

pole of growth. *See* growth poles

police power, 91, 102

pollution, 90, 108, 110, 123, 157, 175, 232, 234, 289, 295

Portland, Ore., 184, 506, 517

"post-industrial city," 65, 154 n.2

potatoes, 214, 241, 390, 410, 442, 451, 511, 520

Potomac River, 8, 206, 212, 213, 231, 233

poultry, broiler fowls, 148, 214, 240, 241, 279, 280, 324–25, 443

Prairie Farms Rehabilitation Act, Administration, 104, 119, 385, 396–97

Prairie Provinces, 393–401; climate, 372, 393–94; cultivation techniques for, 139, 385; exports from, 118; farming and ranching in, 118–19, 129, 395, 397, 459–60; geographical position, 397, 398–99, 400–1; history and settlement, 45, 96, 394–96, 397; industry, 398; petroleum in, 164, 521; population, 56, 397; water resources, 397. *See also* Alberta; Great Plains; Manitoba; Saskatchewan

praesidio, presidio, 64, 426

Preemption Act (1841), 99, 148

Prince Edward Island. *See* Maritime Provinces, Canada

propulsive industry, 198, 261

Provo, Utah, 160

Providence, R.I., 247

public domain, public lands, 71, 89, 94–102, 110, 116–17, 365, 412, 418, 471

Puget Sound, 16, 131, 210, 509

Québec: agriculture, 253, 256, 257, 258, 456; climate, 253; forests, 260; geographical position, 87, 454; history and settlement, 138, 252–56, 469; hydroelectricity in, 124, 167, 258–59, 461; industry, 260; mining in, 160, 259–60, 408, 462; physical features, 9, 253, 439; population, 48–49, 56, 257, 467; separatism in, 49, 86–87, 197, 261–63; soils and vegetation, 253. *See also* French Canadians

Québec City, P.Q., 186, 252, 253, 254, 261

quotas: agricultural, 117, 145, 322, 330; immigration, 44, 434; petroleum, 162

railroads: Class I, 185; and land grants, 99–100, 141, 178, 396; and settlement, 71, 141, 178–79, 219, 265, 272, 283, 289, 295, 300, 301, 304, 306, 387, 395, 416, 479, 496; and traffic, 114–16, 129, 141, 158, 159, 179, 185–86, 222, 239, 296, 299, 361, 380, 461, 465, 480, 521

Range Livestock Belt, 133

rapeseed, 390, 460

Reclamation, U.S.Bureau of, 106, 117, 141, 387, 430, 484

recreational land, 102, 104–6, 140, 218–19, 405

rectangular (gridiron) survey, 71–72, 263, 272

Red River, Minn.–Man., 95, 390, 394, 397, 442

redwoods, 34–35, 488

refineries, petroleum refining, 164, 165, 220, 260, 265, 296, 300, 363, 365, 367, 399, 448, 449, 491

region, definition, 202–3

regionalism, definition of, 198–200, 201
regional diversity, 196–98
regional planning, 88–89
Rhode Island, 90; agriculture, 241; industry, 244, 245, 246. *See also* New England
rice, 50, 131, 135, 326, 356, 486
Richmond, Va., 233, 332
Rio Grande, River, Valley, 106, 107, 356, 358, 359, 423, 425, 426, 427, 428, 436
roads, 72, 114, 141, 144, 179, 187–90, 214, 222, 223, 228, 271, 297, 337, 344–45, 346, 414, 415, 416, 465, 496
Rochester, N.Y., 228
Rocky Mts., 10, 11, 12, 22, 34, 373, 404, 408
Rocky Mt. Trench, 12, 13

Sacramento River. *See* Central Valley, Calif.
Saguenay River, P.Q., 258, 260, 261
St. John, N.B., 439, 448
St. John's, Newfoundland, 439, 449
St. Lawrence River, 106, 155, 178, 252, 253, 261
St. Lawrence Seaway, 160, 180–81, 182, 184, 222–23, 229, 253, 258, 287, 295, 296
St. Lawrence Valley, 9, 23, 30, 56, 153, 155, 167, 252–63, 264, 278, 464
St. Louis, Mo.: industry, 155, 165, 274, 306; population, 67–68, 69; role and services, 64, 196, 206, 209, 306; site and settlement, 271, 305; transport role, 306
St. Maurice River, P.Q., 258, 260, 261, 265
Salt Lake City, Utah, 160, 199, 408, 410, 417–18, 506
Salton Sea, 404, 486
San Antonio, Tex., 365, 366, 368
San Diego, Calif., 25, 186, 423, 426, 479, 490, 495, 517
San Francisco, Calif.: industry, 173; population, 44, 68, 76, 490, 491; population, Hispanic, 423; port, 184, 479, 490, 491; role and services, 78, 80, 479, 491; 492; site and settlement, 16, 71, 491; transport role, 187, 492
San Joaquin River. *See* Central Valley, Calif.
San José, Calif., 173, 490, 491, 492
Santa Clara County, Valley, 486, 490, 492
Santa Fé, N.M., 305, 423, 426
Santa Fé Trail, 178, 427
Sarnia, Ont., 161 n.6, 164, 265, 290
Saskatchewan: farming and ranching in, 390, 397; mining in, 158, 399, 412; physical features, 99; population, 395, 467. *See also* Prairie Provinces
Saskatchewan Rivers, North *and* South, 106, 397
Scandinavian immigrants, 44, 45, 46, 212, 213
Scranton, Pa., 215, 219, 229, 296

Seattle, Wash.: industry, 516; population, 506; port, 184, 517; role and services, 516–17; site and settlement, 516
second homes, 61, 218, 246, 418, 487
section, quarter-section (survey units), 95–96, 98, 101, 148, 256, 272, 277, 303, 384, 396, 461
sectionalism, 199, 270
severance taxes, 90–91, 348, 399, 470
sharecroppers, 313, 319, 320, 321, 325
Shield, Canadian *or* Laurentian, 9–10, 45, 55, 56, 133, 167, 196, 252, 258, 282, 378, 397, 398, 455, 463
Sierra Nevada, 14, 15, 133, 417, 477, 478, 479, 484
Silicon Valley, 151, 171, 173, 490, 492
silver, 405, 407, 446
slavery, slaves, 49, 50, 98 n.4, 101, 131, 132, 148, 208, 247, 312, 313, 314, 317, 318
small-arms manufacture, 285
Snake River, 14, 184, 410, 442, 506
Soil Bank, 117, 118, 147
Soil Conservation Service (SCS), 110, 117, 147–48, 344
soil erosion, 110, 117, 119, 140, 148, 313, 341, 376, 378
Soo (Sault Ste. Marie), 106, 159, 160, 180, 181, 182, 254, 264, 292, 295, 397. *See also* Great Lakes
sorghums, 148, 272, 274, 358, 390, 486
South, 311–34; agriculture in, 109, 131, 148, 275, 322–26; climate, 25, 27, 30, 313, 317, 325; definition of, 312, 313; economic fortunes of, 171, 208, 313, 314, 315, 319, 329, 330, 334; forests, 317; geographical position, 178, 184; history and settlement, 45, 50, 67, 95, 109, 198; industry, 155, 325, 326–29; industry, textile, 153, 172, 245, 330; physical features, 6, 313, 316–17; population, 326, 330; population, African-America, 50–52, 313, 318–20, 325, 326, 327–28, 334; soils and vegetation, 316; urban growth in, 319, 322, 328
South Carolina. *See* Carolina, North *and* South
soybeans, 132, 148, 272, 274, 275, 276, 279, 280, 323
Spanish Americans. *See* Hispanic Americans; Mexico; Spanish settlement in America
Spanish settlement in America, 4, 33, 67, 94, 354–55, 360, 368, 423–24, 426–27, 477, 478
Spokane, Wash., 408, 416, 506
standard of living, 4, 87, 171, 172, 197, 337, 359, 494
steel: technology involved, 159, 160–61, 170, 172, 183, 289; in Canada, 156, 160–61, 264, 265, 291, 295, 296, 446, 451; in U.S., 153, 155, 156, 160–61, 219, 220, 229, 230, 285–86, 287, 288–90, 291, 292, 295, 296, 300, 329, 330, 367, 408, 431, 451, 490
Strip Mining Control and Reclamation Act (1977), 157, 348

stubble mulch, 381
sugar: beet, 357, 390, 410, 486, 511; cane, 128, 131, 356, 357, 491, 503
sulfur, 123, 157, 364
"Sun Belt," 60–61, 171, 288, 359
Superior, Lake, 159, 160, 161, 180, 219, 292, 466. *See also* Great Lakes
Superior Upland, 9, 16, 20
support prices, farm, 117, 145, 281
Susquehanna River, 20, 213, 221, 233
Syracuse, N.Y., 221, 228

taconite, 159, 161
Tampa, Fla., 359, 361
tar sands. *See* Athabasca tar sands
tenancy, farm, 145, 323. *See also* sharecroppers
Tennessee: hydroelectricity in, 154, 167–69; industry, 341–42; mining in, 317; physical features, 317. *See also* Appalachia; South
Tennessee River, 8, 88–89, 107, 155, 167, 169, 170, 180, 317, 341, 342
Tennessee Valley Authority (TVA), 122, 155, 168, 200, 316, 322, 331, 341–44, 346, 349–50, 387
Texas, 365–68; climate, 354, 372, 374; defense installations in, 368; farming and ranching, 318, 322, 323, 355, 366, 380, 382 n.6, 390; geographical position, 365; history and settlement, 210, 314, 355, 365, 426; industry, 165, 173, 198, 287, 330, 398; mining in, 158, 364; petroleum in, 162, 364, 368; physical features, 6, 372; population, 54, 68, 327, 359; ports, 362–63; soils and vegetation, 30, 33, 38; urban growth in, 199, 366–68; water resources, 106, 140, 390. *See also* South; Spanish settlement in America
textile industry, textiles, 73, 153, 155, 172, 219, 229, 244–45, 247, 260, 261, 330, 332, 517
Three Mile Island, Pa., 170
Thunder Bay, Ont., 292
tobacco, 50, 117, 131, 136, 213, 214, 280, 281, 312, 314, 317, 318, 319, 325, 332
Toledo, Ohio, 159, 292
Toronto, Ont.: industry, 155, 164, 265; population, 56, 68, 265; port, 223, 265, 296; rivalry with Montreal, 68, 252, 263, 266–67; role and services, 79, 81, 196, 207, 265, 266; site and settlement, 70, 251, 296; transport role, 72, 186, 187, 295, 296
"transactional city," 64
Trans–Canada Highway, 188, 397, 521
transhumance, 412
Trenton, N.J., 160, 213, 219, 221, 228
tsunami, 502
truck farming, 135, 213, 214, 216, 228, 283
tundra, 34, 472

Ukrainian immigrants, 44, 45
uranium, 43, 406, 462

Utah: history and settlement, 42, 60; industry, 155; mining in, 406; physical features, 424

Vancouver, B.C.: population, 44, 56, 68, 76, 81, 506; port, 184, 516; role and services, 210, 516; site and settlement, 70; transport role, 399, 408
vegetables, 129–30, 213, 214, 239, 240, 279, 280, 283, 356, 357, 358, 390, 410, 429, 442, 443, 476, 481, 486, 509. *See also* truck farming
Vermont: agriculture, 129, 240, 278; population, 246; quarrying in, 237. *See also* New England
VIA RAIL, 186
vines, vineyards, 216, 280, 281, 481, 486–87
Virginia: agriculture, 129, 214, 325, 332; climate, 25; geographical position, 212; history and settlement, 50, 213, 312, 317, 438; physical features, 8; population, 232, 327. *See also* South
volcanoes, 14, 15, 502
Von Thünen, 303

Washington (state): climate, 25, 507, 509; farming and ranching in, 131, 242, 508–13; fisheries, 507; forests, 36, 110, 513–15; markets, search for, 131, 518–21; geographical position, 196; markets, search for, 131, 518–21; physical features, 15, 22, 506; population, 60; relations with California, 521; soils and vegetation, 39, 510; water resources, 108, 167, 507, 515. *See also* forest product industries
Washington, D.C., 57, 64, 89, 186, 187, 221, 231–32
Welland Canal, 180, 181, 182, 265, 295
West Virginia: coal mining in, 156, 158, 339, 345, 348; geographical position, 312; industry, 155, 288, 289; physical features, 9, 336; population, 57, 340, 348; prospects, 348. *See also* Appalachia; coal
Western Grain Transportation Act (1983), 115
wheat, 117, 118, 119, 128, 129, 130, 138, 184, 214, 272, 274, 276, 381, 389–90, 395–96, 410, 442, 461, 480, 483, 507, 511, 518
Wheat Belts, 128, 129, 131, 132, 277, 359, 366, 388, 392, 510
Wheeling, W.Va., 160, 289
Wilderness Act (U.S.), 418
Wilderness Road, 178
Willamette River, Valley, 16, 131, 508, 509, 517
Williamson Act (1965), 104, 487
Windsor, Ont., 173, 186, 252, 265, 297

Winnipeg, Man.: climate, 24; industry, 155, 398; population, 49, 398, 399; role and services, 196, 209, 398; transport role, 397, 398
Wisconsin: agriculture, 129, 139, 140, 278, 442; history and settlement, 45; physical features, 9, 17, 19, 20, 140; planning in, 92; soils and vegetation, 33, 282. *See also* dairy farming; Midwest
women in the labor force, 170–71, 172, 246
"Workshop of the continent," 153
Wyoming: farming and ranching in, 392; mining in, 158, 393, 408, 409; petroleum in, 162; physical features, 14; population, 57, 200

Yellowstone National Park, 12, 104, 105, 110, 414, 416. *See also* national parks
Yosemite National Park, 15, 104, 105, 110, 475. *See also* national parks
Youngstown, Ohio, 68, 288, 289
Yukon, 56, 105, 407, 454, 456, 469
Yukon River, 14, 464

zinc, 405, 408, 446, 447, 462
zonal soils, 30
zoning, 73, 79, 102–3, 258, 503